Amino Acids
Biosynthesis and Genetic Regulation

BIOTECHNOLOGY

BIOTECHNOLOGY SERIES

1. R. Saliwanchik	*Legal Protection for Microbiological and Genetic Engineering Inventions*
2. L. Vining (editor)	*Biochemistry and Genetic Regulation of Commercially Important Antibiotics*
3. K. Herrmann and R. Somerville (editors)	*Amino Acids: Biosynthesis and Genetic Regulation*

Other volumes in preparation

D. Wise (editor)	*Organic Chemicals from Biomass*
A. Demain (editor)	*Biology of Industrial Microorganisms*
A. Laskin (editor)	*Applications of Isolated Enzymes and Immobilized Cells to Biotechnology*
S. Dagley (editor)	*Biodegradation*

AMINO ACIDS
Biosynthesis and Genetic Regulation

Edited by

Klaus M. Herrmann
Purdue University
West Lafayette, Indiana

Ronald L. Somerville
Purdue University
West Lafayette, Indiana

1983
ADDISON-WESLEY PUBLISHING COMPANY
Advanced Book Program
Reading, Massachusetts
LONDON · AMSTERDAM · DON MILLS, ONTARIO · SYDNEY · TOKYO

Library of Congress Cataloging in Publication Data
Main entry under title:

Amino acid biosynthesis and genetic regulation.

(Biotechnology series ; 3)
Includes bibliographical references and index.
1. Amino acid synthesis. 2. Microbiological synthesis. 3. Genetic regulation. I. Herrmann, Klaus M., 1937- . II. Somerville, Ronald L., 1935- . III. Series: Biotechnology ; 3.
[DNLM: 1. Amino acids—Biosynthesis. 2. Amino acids—Metabolism. 3. Genetics, Biochemical. 4. Gene expression regulation. W1 B1918W no. 3 / QU 60 A51435]
QP561,A464 1983 574,19′296 83-8829
ISBN 0-201-10520-9

MANUFACTURED IN THE UNITED STATES OF AMERICA

ABCDEFGHIJ-HA-89876543

To

Charles Yanofsky

who contributed to the education of both of us

CONTRIBUTORS

Stanley W. Artz, Department of Bacteriology, University of California, Davis, California

Annette Baich, Department of Biological Sciences, Southern Illinois University, Edwardsville, Illinois

Jnanendra K. Bhattacharjee, Department of Microbiology, Miami University, Oxford, Ohio

Jean E. Brenchley, Genex Corporation, Gaithersburg, Maryland

Joseph M. Calvo, Section of Biochemistry, Molecular and Cell Biology, Cornell University, Ithaca, New York

Helen Camakaris, Department of Microbiology, University of Melbourne, Victoria, Australia

Georges N. Cohen, Département de Biochimie et Génétique Moleculaire, Institut Pasteur, Paris, France

Laszlo N. Csonka, Department of Biological Sciences, Purdue University, West Lafayette, Indiana

Raymond E. Cunin, Laboratorium voor Erfelijkheidsleer en Microbiologie, Vrije Universiteit Brussel, and Research Institute CERIA, Brussels, Belgium

Rowland H. Davis, Department of Molecular Biology and Biochemistry, University of California, Irvine, California

Jeffrey F. Gardner, Department of Microbiology, University of Illinois, Urbana, Illinois

Craig C. Garner, Department of Biochemistry, Purdue University, West Lafayette, Indiana

Klaus M. Herrmann, Department of Biochemistry, Purdue University, West Lafayette, Indiana

Donald Holzschu, Department of Bacteriology, University of California, Davis, California

Gunter B. Kohlhaw, Department of Biochemistry, Purdue University, West Lafayette, Indiana

Nicholas M. Kredich, Duke University Medical Center, Durham, North Carolina

Steven P. Lynn, Department of Microbiology, University of Illinois, Urbana, Illinois

Shigeru Nakamori, Central Research Laboratories of Ajinomoto Company, Inc., Kawasaki, Japan

Jean-Claude Patte, Institut de Microbiologie, Centre d'Orsay, Université Paris, SUD, Orsay, France

James Pittard, Department of Microbiology, University of Melbourne, Victoria, Australia

Lawrence J. Reitzer, Department of Biology, Massachusetts Institute of Technology, Cambridge, Massachusetts

Stuart A. Rosenfeld, Laboratory of Renewable Resources Engineering, Purdue University, West Lafayette, Indiana

Robin J. Rowbury, Department of Botany and Microbiology, University College, University of London, London, U.K.

Ronald L. Somerville, Department of Biochemistry, Purdue University, West Lafayette, Indiana

George V. Stauffer, Department of Microbiology, University of Iowa, Iowa City, Iowa

Anna L. Tan-Wilson, Department of Biological Sciences, State University of New York at Binghamton, Binghamton, New York

H. Edwin Umbarger, Department of Biological Sciences, Purdue University, West Lafayette, Indiana

Fumihiro Yoshinaga, Ajinomoto Company, Inc., Tokyo, Japan

CONTENTS

FOREWORD

The systematic investigation of how the twenty common amino acids are biosynthesized was well under way long before the field of biotechnology began to take shape. From its earliest beginnings, however, biotechnology has drawn heavily on the conceptual and technical advances that originated in the laboratories of investigators in the amino acid field. There is every reason to suppose that the techniques of biochemistry and molecular genetics that have served to spark progress in this area of scientific endeavor will be widely exploited as biotechnology continues to expand its horizons. The reader of this book will acquire a solid understanding of amino acid biosynthesis, how intermediates are diverted from central metabolism to form the amino acids, how this metabolic flow is regulated, and what experimental strategies are used to identify and bypass control points in biosynthetic pathways.

In one of the best recent surveys of amino acid biosynthesis and regulation, [*Ann. Rev. Biochem. 47,* 533–606 (1978)], our colleague Ed Umbarger expressed the opinion that a coherent overview by a single individual of this vast field was last feasible in 1973. More than one contributor to this volume might be prepared to testify that the exponential increase in new material has stretched to the limit the ability of practicing scientists to remain current even with respect to a few of the amino acids. It is our hope that this collection of reviews will serve to orient the reader to the current literature by providing a suitable base of background information as well as a series of bibliographical references to key papers.

The pathways to all of the amino acids are known. The structures of most of the intermediates have been verified by chemical synthesis. Sometimes there are two pathways to one amino acid (e.g., *lys, leu, phe, tyr*). Some of these species-related routes have been fairly recently discovered, and it is possible that careful studies of organisms that have not been investigated in detail may reveal additional variations of known biochemical pathways. The nonuniversality of amino acid biosynthetic pathways is somewhat surprising in view of the basic function of the amino acids.

The mechanisms for regulating carbon flow to the amino acids are manifold. Feedback inhibition was the first to be discovered and remains the major quantitative regulatory mechanism for the biosynthesis of some amino acids. Transcriptional control involves either repression, which directly affects transcription initiation, or attenuation, a translationally-coupled control of transcription termination.

There are several general control mechanisms related to guanosine tetraphosphate and/or growth rate. These mechanisms are not discussed in detail in this volume. This may reflect the difficulty of the experimental approach, and, in particular, the difficulty of approximating physiological reality in the laboratory. It should be noted that the growth rate of *E. coli* in the digestive tract is about 1/50th of the rate in a shake flask with Luria broth as growth medium.

In order to include recent developments in a field that is moving at the forefront of basic science, we tried to enlist as authors persons who have been leading contributors in their specialty. This resulted in manuscripts from four continents and led in turn to some minor communication problems. It also appeared that contemporary intercontinental postal systems are sometimes not much faster than mail at the time of the *Mayflower*.

We thank every contributor. Although we tried to include as much material as possible from each original contribution, in some cases editing led to major rewriting. We are thus responsible for all errors that may have crept into the text. We thank Patricia Carmony for help in eliminating errors from the reference sections, and Dr. Ingrid Krohn of Addison-Wesley for sustained, untiring encouragement. Eunice Carlson retyped the entire book and Michael Poling did most of the art work.

West Lafayette, Indiana
February 1983

KLAUS M. HERRMANN
RONALD L. SOMERVILLE

Amino Acids
Biosynthesis and Genetic Regulation

CHAPTER 1

Regulation of Glutamate and Glutamine Biosynthesis

Stuart A. Rosenfeld
Jean E. Brenchley

I. INTRODUCTION

The synthesis of glutamate and glutamine involves three primary enzymes: glutamate dehydrogenase, glutamine synthetase, and glutamate synthase (Fig. 1.1). Not only do glutamate and glutamine feed directly into protein synthesis, they also serve as the amino and amido donors for a wide range of other nitrogen-containing cellular components. The use of α-ketoglutarate and ammonia as substrates for glutamate and glutamine synthesis places these amino acids at a metabolic junction between carbon utilization and ammonia assimilation. Because of this central position in metabolism, the glutamate and glutamine biosynthetic enzymes may have unique control features in addition to those found for other biosynthetic pathways.

This review emphasizes the regulation of enzyme activity and synthesis in the enteric microorganisms, where both biochemical and genetic studies have yielded the most detailed information about glutamate dehydrogenase, glutamine synthetase, and glutamate synthase. The interconversion of glutamate and glutamine makes a separate discussion of their biosynthesis seem artificial; however, they have been investigated individually, and for

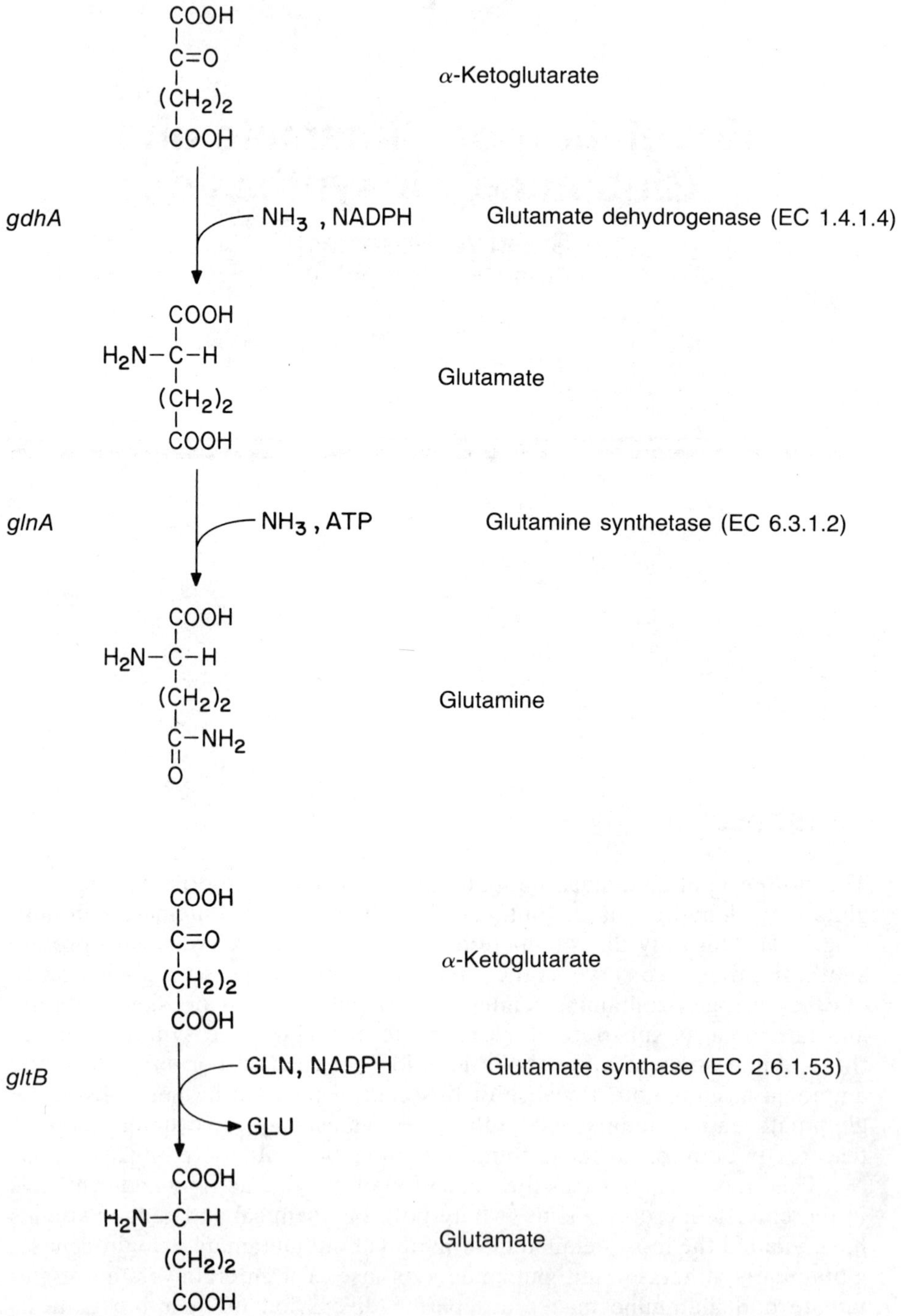

FIGURE 1.1 Pathways for glutamate and glutamine biosynthesis.

clarity, we have divided this review into two sections; the first examines the control of glutamate biosynthesis and the second reviews the regulation of glutamine synthetase.

II. CONTROL OF GLUTAMATE FORMATION

A. Enzyme Structure: Glutamate Dehydrogenase

Glutamate dehydrogenase catalyzes the reductive amination of α-ketoglutarate with ammonia (Fig. 1.1). The enzyme from enteric bacteria requires NADPH, suggesting a biosynthetic function. Other organisms, such as *Saccharomyces* (Holzer and Schneider, 1957) and *Neurospora* (Sanwal and Lata, 1961), have two distinct glutamate dehydrogenases, the first responsible for glutamate catabolism, and the second solely biosynthetic.

The molecular weights of the enzymes isolated from *Escherichia coli* (Sakamoto et al., 1975) and *Salmonella typhimurium* (Coulton and Kapoor, 1973a) are 300,000 and 280,000, respectively. The native enzyme probably exists as an aggregate of six identical monomers (Sakamoto et al., 1975; Coulton and Kapoor, 1973b). The K_m values for both ammonia and α-ketoglutarate are about 1 mM (Sakamoto et al., 1975; Coulton and Kapoor, 1973a, b; Miller and Stadtman, 1972). The relatively high K_m for ammonia suggests that this enzyme is not catalytically active during cell growth on low concentrations of ammonia. Physiological and genetic results summarized below support this hypothesis.

B. Enzyme Structure: Glutamate Synthase

Glutamate synthase catalyzes the formation of glutamate from glutamine and α-ketoglutarate (Fig. 1.1). The native enzyme from *E. coli* (Mantsala and Zalkin, 1976a, b; Miller and Stadtman, 1972; Geary and Meister, 1977) has a molecular weight of 800,000 (Miller and Stadtman, 1972) and probably exists as an aggregate of four catalytically active dimers. Each dimer consists of two nonidentical subunits having molecular weights of 53,000 and 135,000 for the *E. coli* (Mantsala and Zalkin, 1976c) and 51,500 and 175,000 for the *Klebsiella aerogenes* (Trotta et al., 1974) enzyme.

Two features distinguish glutamate synthase from other well-characterized amidotransferases. First, glutamate synthase is an iron–sulfur flavoprotein. The iron–sulfur moiety resides within the large subunit (Rendina and Orme-Johnson, 1978). Perhaps the complex structure of glutamate synthase reflects some hitherto unknown function in addition to its role in glutamate synthesis, since iron–sulfur structures are usually characteristic of enzymes involved in electron transport and oxidation–reduction.

A second distinction between this enzyme and other amidotransferases is the manner in which the catalytic sites are partitioned between subunits.

Glutamine binds to the large subunit of glutamate synthase and transfers the amide group to α-ketoglutarate bound to the smaller subunit; other amidotransferases show the opposite arrangement (Trotta et al., 1974; Mantsala and Zalkin, 1976c). In vitro, the small subunit of glutamate synthase can catalyze an ammonia-dependent amidation of α-ketoglutarate (Mantsala and Zalkin, 1976c). The ammonia-dependent activity of other amidotransferases usually resides within the large subunit (Buchanan, 1973).

C. Physiology of Glutamate Synthesis

1. Effect of Ammonia Availability. Because some microorganisms lacking glutamate dehydrogenase activity did not require glutamate for growth, Meers et al. (1970) sought another route for glutamate formation. Their investigations led to the discovery of glutamate synthase. Enteric organisms grown in media having more than 1 mM ammonia synthesize both glutamate dehydrogenase and glutamate synthase. However, glutamate synthase mutants (Brenchley and Magasanik, 1974; Pahel et al., 1978; Dendinger et al., 1980) cannot grow on low levels of ammonia or on growth rate–limiting nitrogen sources. Thus glutamate synthase is essential during these growth conditions, while glutamate dehydrogenase functions only during growth in excess ammonia. Mutants having neither glutamate dehydrogenase nor glutamate synthase are strict glutamate auxotrophs, the implication being that there is no other major pathway for glutamate biosynthesis.

Since the functions of both glutamate dehydrogenase and glutamate synthase depend upon the ammonia availability, it becomes important to understand how the synthesis of these enzymes is regulated. In *K. aerogenes,* glutamate dehydrogenase is repressed during growth on low ammonia or on a variety of alternative nitrogen sources (Brenchley et al., 1975). *S. typhimurium* and *E. coli* do not show this repression during ammonia limitation (Brenchley et al., 1975; Pahel et al., 1978). Genetic experiments suggest that differences in glutamate dehydrogenase repression between *K. aerogenes* and *E. coli* are related to differences in their respective promoter–operator sequences (Pahel et al., 1978).

It is unclear whether glutamate synthase production responds to changes in ammonia availability. Ammonia limitation did not significantly affect glutamate synthase levels in either *K. aerogenes* or *S. typhimurium* (Meers et al., 1970; Brenchley et al., 1975), whereas *E. coli* W had repressed levels of glutamate synthase in growth rate–limiting concentrations of ammonia (Senior, 1975).

2. Effect of Glutamate and Glutamine on Enzyme Levels. Addition of glutamate or compounds metabolized to glutamate (for example, aspartate, proline) represses glutamate dehydrogenase in *K. aerogenes* (Meers et al.,

1970), *E. coli* (Varricchio, 1969), and *S. typhimurium* (Brenchley et al., 1975). Since glutamate lowers glutamate dehydrogenase synthesis, glutamate limitation might lead to overproduction of this enzyme. Previous difficulties in measuring the effect of glutamate limitation have been circumvented by selecting a *lacZ* fusion to the glutamate dehydrogenase promoter in a mutant with a thermolabile glutamate synthase (Rosenfeld et al., 1982). Glutamate limitation, elicited by growth at an elevated temperature, led to elevated β-galactosidase production. These results suggest that glutamate directly or indirectly (for example, via glutamyl-tRNA) influences glutamate dehydrogenase synthesis via an unknown molecular mechanism.

Glutamate, when added to ammonia-grown cells, does not significantly lower glutamate synthase production; however, compounds metabolized to glutamate are repressive (Brenchley et al., 1975). Starvation for glutamine or growth of cells with glutamate as a sole nitrogen source leads to reduction in glutamate synthase levels (Brenchely et al., 1975). These results suggest that the intracellular pools of glutamine directly or indirectly influence this enzyme's synthesis. Considering the importance of this enzyme, it is clear that further characterizations are needed to understand its regulation.

D. Transfer RNA and Regulation

Current models involving tRNA in the control of gene expression postulate that the cognate amino acylated tRNA is necessary for normal translation of leader regions preceding attenuator sites (Keller and Calvo, 1979; Chapters 10, 14, 18, 20, and 21). Two studies indicate a role for tRNA in regulating glutamate synthesis. The *hisT* mutants of *S. typhimurium* have an altered pseudouridine synthetase I enzyme, resulting in under-pseudouridylation of numerous tRNA species (Cortese et al., 1974; Turnbough et al., 1979). The *hisT* mutation influences the expression of several operons (for example, histidine and leucine), altering their normal attenuation mechanisms (Brenchley and Williams, 1975; Keller and Calvo, 1979). It is unknown whether gutamate biosynthesis is controlled by attenuation. However, *hisT* mutants have threefold higher levels of glutamate dehydrogenase and twofold lower levels of glutamate synthase under certain growth conditions (Rosenfeld and Brenchley, 1980a). Rho mutations, which alter attenuation of various amino acid biosynthetic pathways (Korn and Yanofsky, 1976), lead to a twofold elevation in glutamate synthase in *E. coli* (M. J. Madonna and J. E. Brenchley, unpublished data). Lapointe et al. (1975) reported that a thermolabile glutamyl-tRNA synthetase mutant had tenfold derepressed glutamate synthase and 50-fold elevated glutamine synthetase activities when the strain was grown at a partially restrictive temperature. These results suggest that tRNA affects glutamate biosynthesis. Future analyses of the nucleotide sequence preceding *gdhA* and *gltB*, the structural genes for glutamate dehydrogenase and glutamate synthase, may elucidate the mechanism of tRNA action.

E. Genetics of Glutamate Synthesis

Mutants lacking glutamate dehydrogenase (*gdhA*) have been isolated in *E. coli* (Berberich, 1972; Pahel et al., 1978). *S. typhimurium* (Dendinger et al., 1980; Dendinger and Brenchley, 1980), and *K. aerogenes* (Brenchley and Magasanik, 1974; Bender et al., 1976). Although the mutations are located in homologous sections of the chromosome (20–30 units), fine structure mapping reveals differences. *S. typhimurium* and *E. coli* chromosomes differ in this region owing to a large inversion (Sanderson and Hartman, 1978) that may include the glutamate dehydrogenase structural genes. Such an inversion could influence linkages between *gdhA* and other genes in these organisms.

A detailed study of the *gdhA* region of *S. typhimurium* (Dendinger et al., 1981; Rosenfeld et al., 1982) emloying the transposon Tn*10* as a genetic marker showed that *gdhA* lies between the *pncA* and *nit* genes of *S. typhimurium*. Conjugation experiments provided evidence for the following gene order: *pyrD pncA gdhA nit trp*. Mu *c*ts dl (Ap^r, *lac*) provided a method for selecting numerous strains containing deletions within and outside of *gdhA* (Rosenfeld and Brenchley, 1980b). The deletion mapping of a mutation resulting in a thermolabile glutamate dehydrogenase provided proof that this chromosomal region included the glutamate dehydrogenase structural gene. Additional genetic analyses of *gdhA*::mu *c*ts dl (Ap^r, *lac*) fusion strains established the counterclockwise direction of *gdhA* transcription (Rosenfeld et al., 1982) (Fig. 1.2).

Another class of *S. typhimurium gdh* mutants obtained by Mu mutagenesis reverted to gdh^+ at high frequency with retention of the Mu insertion. Preliminary genetic analyses suggest that the Mu insertion may either be within the *gdhA* promoter or within an activator gene necessary for *gdhA* expression.

Glutamate synthase mutants of *E. coli* (Berberich, 1972; Pahel et al., 1978), *K. aerogenes* (Brenchley et al., 1973; Bender et al., 1976), and *S. typhimurium* (Dendinger et al., 1980) fail to grow on limiting ammonia or on a variety of alternative nitrogen sources. Originally, these mutants were designated Asm^- for "ammonia assimilatory deficient" (Nagatani et al.,

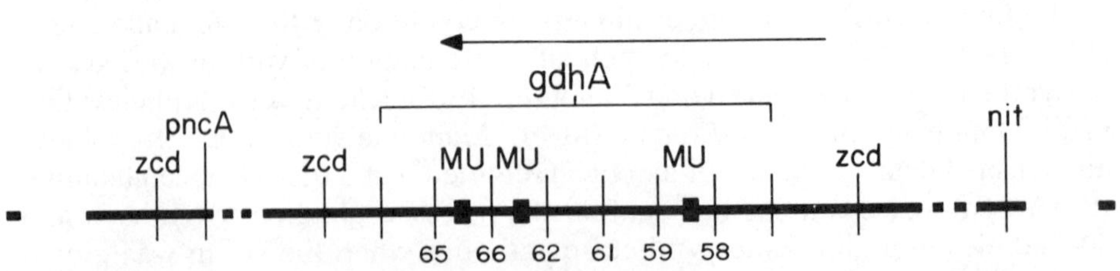

FIGURE 1.2 Genetic map of the *gdhA* region of *S. typhimurium*. The locations of the Tn*10* insertions (designated *zcd*) and the Mu insertions are depicted. Mutations within *gdhA* are represented by the allele numbers below the map, and the direction of *gdhA* transcription is indicated by the arrow above the map.

1971). The mutations causing loss of glutamate synthase have been mapped at a unique site within the chromosome, designated *gltB*. These results imply that both subunits of the enzyme are encoded by closely linked genes. Fuchs et al. (1982) demonstrated that *gltB* was the structural gene for one or both subunits of glutamate synthase in *S. typhimurium,* since mutations causing heat-labile glutamate synthase activity mapped in this region. The genes surrounding *gltB* in *S. typhimurium* were determined by cotransduction experiments utilizing linked Tn*10* elements (Fig. 1.3.). In addition, Mu *cts* dl (Ap^R, *lac*) insertions in *gltB* have been used to construct a deletion map of this gene and to determine the direction of *gltB* transcription (Fig. 1.3; M. J. Madonna et al., manuscript in preparation).

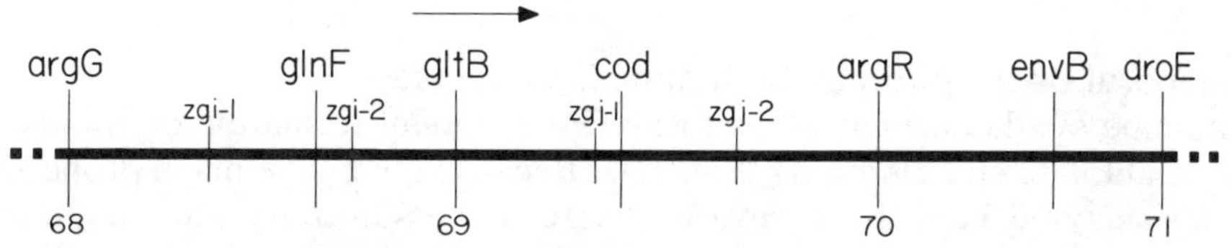

FIGURE 1.3 Genetic map of the *gltB* region (68–71 U) of *S. typhimurium*. The Tn*10* insertions are denoted by the *zgi* designation. The direction of *gltB* transcription is indicated by the arrow above the map.

F. Cloning of the Genes for Glutamate Biosynthesis

Covarrubias et al. (1980a) have identified a plasmid from the Clarke-Carbon *E. coli* gene bank (Clarke and Carbon, 1975) that complements *gdhA* and another that complements *gltB*. The *colE-gltB*$^+$ hybrid plasmid probably contains both subunits for glutamate synthase, since minicells containing the plasmid synthesized both subunits (Lozoya et al., 1980). Whether the genes for the glutamate synthase subunits are contiguous or controlled by the same promoter is still unknown.

Although both the *gdhA*$^+$ and *gltB*$^+$ plasmids have copy numbers of 15–30 (Hershfield et al., 1974), both glutamate dehydrogenase and glutamate synthase are elevated only twofold to threefold in plasmid-containing strains. Perhaps these plasmids are lacking additional control elements necessary for both *gdhA* and *gltB* expression. Plasmids containing the *S. typhimurium gdhA* gene have tenfold to 20-fold elevated glutamate dehydrogenase levels (E. Miller, unpublished data). This could be due to a difference between *gdhA* regulation in *E. coli* and *S. typhimurium* or to the presence of additional genes on the larger *S. typhimurium* plasmid.

G. Future Directions

This review has summarized what we consider to be the most relevant information on glutamate synthesis in enterics. Much additional work is needed to understand how the metabolism of this amino acid is regulated.

One promising approach would be to further exploit the *gdhA–lac* and *gltB–lac* fusions. The techniques that expanded our understanding of *lac* gene expression can now be used to study *gdhA* and *gltB* regulation. Selections for increased or decreased β-galactosidase levels could yield promoter mutations and other regulatory mutations unlinked to *gdhA* or *gltB*. Mutational events leading to elevated glutamate dehydrogenase or glutamate synthase were previously unattainable, since no phenotype could be selected. Repressors, inducers, or novel modulators of the expression of these genes can now be identified. Another powerful approach will involve sequencing of these genes.

III. CONTROL OF GLUTAMINE FORMATION

A. General Description of Glutamine Synthetase

Glutamine synthetase catalyzes the ATP-dependent formation of glutamine from glutamate and ammonia (Fig. 1.1). Because the biochemical properties of this enzyme have been reviewed elsewhere (Ginsburg and Stadtman, 1973; Meister, 1980; Stadtman et al., 1970; Stadtman and Ginsburg, 1974), only a summary will be presented here.

For most microorganisms, only one form of glutamine synthetase exists, and its loss causes a complete requirement for glutamine. Some exceptions are an extreme thermophile, *Bacillus caldolyticus,* and the plant-associated organisms *Rhizobia* and *Agrobacterium*, in which there are two isozymes (Wedler et al., 1978; Darrow, 1980; Fuchs and Keister, 1980). It has been suggested that the isozymes have separate roles during ammonia assimilation in *Rhizobia* (Ludwig, 1980). Although there are biochemical differences in the glutamine synthetases from the related enteric bacteria (Bender and Streicher, 1979), the overall properties are similar. Antibodies prepared against glutamine synthetase from *E. coli* cross-reacted with proteins from other *Enterobacteriaceae, Pseudomonas,* and *Alcaligenes,* a result suggesting that antigenic determinants in this enzyme's structure are common to these organisms (Baumann and Baumann, 1978; Tronick et al., 1973).

The glutamine synthetase from *E. coli* is composed of 12 identical subunits and has a molecular weight of about 600,000. The subunits are arranged in two hexagonal layers (Stadtman and Ginsburg, 1974; Valentine et al., 1968). Glutamine synthetase is assayed in several different ways (Stadtman et al., 1970). One common assay determines its biosynthetic activity directly (Fig. 1.1). The second assay measures the nonphysiological catalysis of a γ-glutamyl transfer in which, in the presence of ADP, divalent metal ions, and arsenate, the γ-glutamyl moiety of glutamine is transferred to hydroxylamine, forming γ-glutamyl hydroxamate and ammonia.

The γ-glutamyl transfer assay facilitates the determination of the adenylylation state of glutamine synthetase. Each subunit of the enzyme can be modified by the covalent attachment of an AMP moiety to a tyrosine residue, and the amount of adenylylation can vary from 0 to 12 for each enzyme molecule. The average extent of modification for an enzyme population can be determined by separately measuring the Mg^{2+}- and Mn^{2+}-dependent γ-glutamyl transferase activities of glutamine synthetase (Stadtman et al., 1970). Both the adenylylation and deadenylylation of glutamine synthetase are catalyzed by the enzyme adenylyltransferase. The direction of this reaction is controlled by the interaction of adenylyltransferase with a small regulatory protein, PII. An unmodified form of PII (PII_A) is required for the adenylylation reaction, whereas a uridylylated form (PII_D) stimulates the deadenylylation activity. This bicyclic cascade has been reviewed by Stadtman et al. (1980).

B. Regulation of Enzyme Activity

The activity of glutamine synthetase from *E. coli* can be influenced by the availability of Mn^{2+} or Mg^{2+}, the concentration of feedback inhibitors, the adenylylation state (Stadtman et al., 1970, 1980), and possibly oxidative inactivation (Levine et al., 1981). The deadenylylated form of the enzyme is dependent on Mg^{2+} for its biosynthetic activity; the adenylylated form is dependent on Mn^{2+} for its activity, and it is more sensitive to feedback inhibition (Stadtman et al., 1970). Feedback inhibition is cumulative in nature. The inhibitors alanine, glycine, histidine, tryptophan, cytidine triphosphate, adenosine monophosphate, carbamyl phosphate, and glucosamine 6-phosphate produce only partial inhibition when tested individually but produce additive inhibition in combination (Woolfolk and Stadtman, 1967). Because glutamine synthetase activity is dependent on metal ions, feedback inhibitors, and adenylylation state, it is difficult to state precisely how glutamine production is controlled. It appears that a fully adenylylated glutamine synthetase in *K. aerogenes* (Foor et al., 1978) and *S. typhimurium* (Bancroft et al., 1978) retains some activity, since mutants unable to deadenylylate the enzyme grow slowly without glutamine. Similar mutants (*glnD*) of *E. coli*, however, have an absolute requirement for glutamine (Bloom et al., 1978). Whether this reflects differences in the activity of the adenylylated enzymes or changes in the metal requirements and sensitivity to feedback inhibition is unknown, but these mutants could be useful for examining the significance of these controls on glutamine production.

The adenylylation of glutamine synthetase appears to be limited to gram-negative bacteria, with the exception of *Streptomyces* (Streicher and Tyler, 1981). A gram-positive organism, *Bacillus subtilis,* controls glutamine formation primarily by feedback inhibition with glutamine as the inhibitor (Deuel and Prusiner, 1974). This shows that the adenylylation of glutamine synthetase is not a physiologically essential process. What advantage adenylylation confers on the gram-negative bacteria is unknown. However, the cascade system is extremely sensitive to small physiological changes, and it may be important for modulating carbon and nitrogen control.

Adenylylation is the result of the reversible covalent modification of tyrosine residues on the glutamine synthetase protein (Heinrikson and Kingdon, 1971). The limited digestion of purified glutamine synthetase into two fragments demonstrated that the AMP binding site resides within a 35,000-dalton carboxy-terminal fragment (Dautry-Varsat et al., 1979; Lei et al., 1979). The recent nucleotide sequence determination of a portion of the glutamine synthetase gene from *S. typhimurium* (Koduri et al., 1980) shows that this tyrosine is about 70 amino acids from the carboxy terminus (R. Hanau et al., manuscript in preparation). Electron micrographs of glutamine synthetase treated with AMP-specific antibodies (Hohman and Stadtman, 1978) showed that the modification is on the outer edge of the hexagonal disc of the protein (Frink et al., 1978).

C. Regulation of Enzyme Synthesis

Because glutamine is used not only for protein synthesis but also as an amino and amido donor for other metabolites, the regulation of glutamine synthetase is coupled with its role in ammonia assimilation. Therefore many studies of glutamine synthetase also involve an examination of the nitrogen utilization process. It is difficult to review the numerous studies on glutamine synthetase control during the last ten years, so we will briefly summarize work already reviewed by Tyler (1978) and will concentrate on the current work.

The initial studies that focused attention on glutamine synthetase and the nitrogen utilization process were carried out with *K. aerogenes*. Physiological experiments demonstrated that the induced levels of histidase, an enzyme involved in histidine degradation, increased in cells grown with limiting nitrogen, even when a good carbon source, such as glucose, was present (Neidhardt and Magasanik, 1957; Prival and Magasanik, 1971). Histidase and glutamine synthetase levels in cells grown in a variety of media showed a correlation between these two activities. Studies of mutants with altered glutamine synthetase activities strengthened this notion (Prival et al.,1973; Brenchley et al., 1973). In addition, an inverse correlation was found between glutamine synthetase and glutamate dehydrogenase levels (Brenchley et al., 1973).

Subsequent genetic studies placed the mutations affecting glutamine synthetase and the regulation of histidase and glutamate dehydrogenase within the structural gene for glutamine synthetase, *glnA* (DeLeo and Magasanik, 1975; Bender and Magasanik, 1977a). Since some of these mutations simultaneously altered glutamine synthetase levels, glutamine synthetase was proposed to regulate autogenously its own synthesis (Bender and Magasanik, 1977a).

Two biochemical studies supported this regulatory function. First, an in vitro transcription assay was used to examine the ability of glutamine synthetase to stimulate synthesis of mRNA from the histidine utilization (*hut*) operon (Tyler et al., 1974). These experiments used a lambda phage carrying the *hut* operon from *S. typhimurium* and RNA polymerase from *E. coli* to compare the quantity of *hut*-specific mRNA synthesized with and without the addition of purified glutamine synthetase from *K. aerogenes*. The addition of deadenylylated, but not adenylylated, glutamine synthetase increased the quantity of *hut* mRNA, a result supporting the genetic evidence that deadenylylated glutamine synthetase was an activator for the expression of the *hut* genes. Second, glutamine synthetase formed a complex with DNA, which could be purified, even though specific binding to control sites was not established (Streicher and Tyler, 1980).

Numerous other experiments suggested that glutamine synthetase functions as a general regulator of both nitrogen utilization (see Tyler, 1978, for review) and its own synthesis. However, it is clear that other factors are also involved. In *K. aerogenes,* mutations unlinked to the *glnA* gene influenced its expression (Prival et al., 1973; Brenchley et al., 1973). Mutations in *glnB* caused glutamine auxotrophy; however, revertants could be isolated with second site mutations linked to *glnA,* causing elevated glutamine synthetase activities (Brenchley et al., 1973). These results suggest that the *glnB* product is needed for *glnA* expression and that mutations near *glnA* can suppress this requirement. Mutations in *glnB,* causing glutamine auxotrophy, apparently alter the PII regulatory protein so that it stimulates adenylylation rather than deadenylylation (Foor et al., 1975, 1980; Reuveny et al.,1981; Leonardo and Goldberg, 1980). It is not clear how this change in PII causes glutamine auxotrophy, particularly since insertions of Tn*5* into *glnB* cause the overproduction of glutamine synthetase rather than a decrease (Foor et al., 1980; Reuveny et al., 1981).

Another class of mutants displaying decreased glutamine synthetase levels are those with low glutamate synthase activities (*Asm*$^-$ or *gltB*) (Brenchley and Magasanik, 1974). As was described earlier, the mutational loss of glutamate synthase prevents growth on limiting nitrogen sources. Many such mutants also have about half the normal level of glutamine synthetase, and revertants have mutations linked to *glnA,* causing overproduction of glutamine synthetase (Brenchley and Magasanik, 1974; Funanage and Brenchley, 1977; Funanage et al., 1978). Whether the loss of glutamate synthase activity directly affects *glnA* expression is not known.

The identification of glutamine auxotrophs with mutations in the *glnF* gene of *S. typhimurium* (Garcia et al., 1977), unlinked to *glnA* and not involved in the adenylylation process, led to a modification of the model that postulated that glutamine synthetase was the sole regulator of its own synthesis. The *glnF* product is indispensible for full *glnA* expression. Revertants of *glnF* mutants were isolated with lesions near the *glnA* gene. These suppressor mutations, originally designated *glnR* in *S. typhimurium* (Kustu et al., 1979) and *glnG* in *E. coli* (Pahel and Tyler, 1979), are in a gene separate from *glnA*.

The identification of a gene near *glnA* involved in glutamine synthetase control prompted a reevaluation of the earlier reports placing the mutations that altered glutamine synthetase and nitrogen utilization control within *glnA*. Until a reexamination of these genetic experiments is complete, it is hard to evaluate the possible function of glutamine synthetase protein in regulation. Measurements of *lacZ* expression from the *glnA–lacZ* fusions show that normal regulation can occur in the absence of the glutamine synthetase protein under some growth conditions (Rothstein et al., 1980). Many of the previous results could be explained by assuming that the mutations studied were actually within regulatory genes. Other mutations do affect both glutamine synthetase protein and its regulation (Bender and Magasanik, 1977b; Bedwell and Brenchley, 1980, and unpublished results). The mechanisms by which these effects are mediated are unknown.

The discovery of *glnA*-linked regulatory genes provided impetus for further analysis of this DNA region. Recently, two regulatory genes closely linked to *glnA* were found (McFarland et al., 1981; Roberts et al., 1981; MacNeil et al., 1981). These genes, designated either *ntrB* and *ntrC* (McFarland et al., 1981) or *glnL* and *glnG* (MacNeil et al., 1981) specify proteins thought to function as a complex in regulating the expression of *glnA* and other nitrogen utilization genes. Mutations in the *glnL* (*ntrB*) gene apparently result in the overproduction of glutamine synthetase, whereas mutations in *glnG* (*ntrC*) diminish the production of glutamine synthetase. The mechanism by which these proteins act is not known, but the *glnF* gene product is involved, since mutations in these newly identified genes suppress lesions in *glnF*. With information from other operons as a model, it is plausible that the *glnF* product is needed to convert a complex of the *glnL* (*ntrB*) *glnG* (*ntrC*) proteins from a repressing form to an activating form (McFarland et al., 1981).

Several models could be proposed on the basis of these observations; however, considering the current complexity of this process, it is clear that thorough genetic and biochemical studies are still needed. The *glnF* product could interact directly with DNA or other proteins, or it could catalyze the synthesis of a small molecular effector analogous to cyclic AMP. Because complementation tests have not been done with *glnF* alleles, it remains possible that more than one gene exists in the *glnF* region. Even when this information is available, significant questions regarding the influence of the *glnB, gltB*, and *glnD* mutations on glutamine synthetase levels will remain.

D. Cloning the *glnA* Gene

To understand *glnA* expression properly requires the nucleotide sequence of the control region. The *glnA* genes from *E. coli* (Covarrubias et al., 1980a, b, c; Backman et al., 1981) and *S. typhimurium* (Koduri et al., 1980) have been cloned. The nucleotide sequence preceding the region encoding the amino terminus has been determined for the *E. coli* (A. Covarrubias, personal communication) and *S. typhimurium* (R. Hannau et al., manuscript in preparation) *glnA* genes. The sequence from *S. typhimurium* shows some dyad symmetry within an A-T-rich region. Of particular interest is the sequence TGGTGC(A), which is repeated three times between nucleotides -133 and -88 prior to a possible transcription initiation site. Since the current models include positive, negative, and autogenous control of *glnA*, an analysis of changes in these repeated sequences could help identify where these potential regulators act.

The DNA sequence downstream from *glnA* also has some interesting features (R. K. Koduri, et al., manuscript in preparation). This region contains an ideal Pribnow heptamer (TATAATG) and binds RNA polymerase in vitro, a condition suggesting the existence of a promoter for another gene, possibly *glnL*. Near the Pribnow heptamer is the sequence TGGTGCA, which is identical to the one repeated three times prior to *glnA* (R. K. Koduri, et al., manuscript in preparation). Between the coding sequence for the carboxy terminus of glutamine synthetase and the Pribnow heptamer are three contiguous glutamine codons. It is not known whether their presence is relevant to regulation. There is some in vivo evidence indicating that transcription from *glnA* can continue into a downstream regulatory gene (Magasanik and Rothstein, 1980; Tyler et al., 1980), and this region could be involved in modulating this process.

IV. PROSPECTUS

Despite the central positions of glutamate and glutamine in metabolism, the mechanisms controlling their production still await clarification. Previous genetic and physiological studies combined with the detailed information from nucleotide sequencing may change this situation quickly. The information provided by understanding the control of glutamate and glutamine formation in the enterics can be applied to other microorganisms. The *gdhA* gene from *E. coli* has been cloned and transferred to *Methylophilus methylotrophus*, which normally lacks glutamate dehydrogenase. An improved cell yield resulted when this strain was grown on methanol in the pesence of high levels of ammonia, presumably because less energy is required in the glutamate dehydrogenase pathway compared to the glutamine synthetase–glutamate synthase pathway (Windass et al., 1980). Other ways to improve single cell protein or to produce glutamate may be anticipated as our basic knowledge of the regulatory mechanisms develops.

The control of glutamine production is interconnected with the overall use of nitrogenous compounds. The regulation of nitrogen fixation may have controls similar to those discussed for other nitrogen utilization pathways (Ausubel et al., 1979; Streicher et al., 1974; Tubb, 1974). Information on the mechanisms by which these controls act will be important for future attempts to transfer the nitrogen fixation capability to new microorganisms or plants.

ACKNOWLEDGMENTS

This review was supported by Public Health Service Grant GM25251 from the National Institute of General Medical Sciences and grant PCM20882 from the National Science Foundation. S.A.R. is a recipient of a David Ross Fellowship from the Purdue Research Foundation, and J.E.B. is a recipient of a Public Health Service Research Career Development Award (GM00449) from the National Institute of General Medical Sciences.

REFERENCES

Ausubel, F. M., Bird, S. C., Durbin, K. J., Janssen, K. A., Margolskee, R. F., and Peskin, A. P. (1979) *J. Bacteriol. 140,* 597–606.
Backman, K., Chen, Y-M., and Magasanik, B. (1981) *Proc. Nat. Acad. Sci. U.S.A. 78,* 3743–3747.
Bancroft, S., Rhee, S. G., Neumann, C., and Kustu, S. (1978) *J. Bacteriol. 134,* 1046–1055.
Baumann, L., and Baumann, P. (1978) *Arch. Microbiol. 119,* 25–30.
Bedwell, D. M., and Brenchley, J. E. (1980) *Abstracts of the Annual Meeting of the American Society for Microbiology,* p. 143.
Bender, R. A., and Magasanik, B. (1977a) *J. Bacteriol. 132,* 100–105.
Bender, R. A., and Magasanik, B. (1977b) *J. Bacteriol. 132,* 106–112.
Bender, R. A., and Streicher, S. (1979) *J. Bacteriol. 137,* 1000–1007.
Bender, R. A., Macaluso, A., and Magasanik, B. (1976) *J. Bacteriol. 127,* 141–148.
Berberich, M. A. (1972) *Biochem. Biophys. Res. Comm. 47,* 1498–1503
Bloom, F. R., Levin, M. S., Foor, F., and Tyler, B. (1978) *J. Bacteriol. 134,* 569–577.
Brenchley, J. E., and Magasanik, B. (1974) *J. Bacteriol. 117,* 544–550.
Brenchley, J. E., and Williams, L. S. (1975) *Ann. Rev. Microbiol. 29,* 251–274.
Brenchley, J. E., Prival, M. J., and Magasanik, B. (1973) *J. Biol. Chem. 248,* 6122–6128.
Brenchley, J. E., Baker, C. A., and Patil, L. G. (1975) *J. Bacteriol. 124,* 182–189.
Buchanan, J. M. (1973) *Adv. Enzymol. 39,* 91–183.
Clarke, L., and Carbon, J. (1975) *Proc. Nat. Acad. Sci. U.S.A. 72,* 4361–4365.
Cortese, R., Kammen, H. O., Spengler, S. J., and Ames, B. N. (1974) *J. Biol. Chem. 249,* 1103–1108.
Coulton, J. W., and Kapoor, M. (1973a) *Can. J. Microbiol. 19,* 427–438.
Coulton, J. W., and Kapoor, M. (1973b) *Can. J. Microbiol. 19,* 439–450.
Covarrubias, A. A., Sanchez-Pescador, R., Osorio, A., Bolivar, F., and Bastarrachea, F. (1980a) *Plasmid 3,* 150–164.
Covarrubias, A. A., Rocha, M., Bolivar, F., and Bastarrachea, F. (1980b) *Gene 11,* 239–251.
Covarrubias, A., Sanchez-Pescador, R., Lozoya, E., Bastarrachea, F., and Bolivar, F. (1980c) in *Glutamine: Metabolism, Enzymology and Regulation* (Mora, J., and Palacios, R., eds.) pp. 123–137, Academic Press, New York.
Darrow, R. A. (1980) in *Glutamine: Metabolism, Enzymology and Regulation* (Mora, J., and Palacios, R., eds.) pp. 139–166, Academic Press, New York.

Dautry-Varsat, A., Cohen, G. N., and Stadtman, E. R. (1979) *J. Biol. Chem. 254,* 3124–3128.
DeLeo, A. B., and Magasanik, B. (1975) *J. Bacteriol. 121,* 313–319.
Dendinger, S. M., and Brenchley, J. E. (1980) *J. Bacteriol. 144,* 1043–1047.
Dendinger, S. M., Patil, L. G., and Brenchley, J. E. (1980) *J. Bacteriol. 141,* 190–198.
Dendinger, S. M., Rosenfeld, S. A., Murphy, C. H., and Brenchley, J. E. (1981) Abstracts of the Annual Meeting of the American Society for Microbiology, p. 168.
Deuel, T. F., and Prusiner, S. (1974) *J. Biol. Chem. 249,* 257–264.
Foor, F., Janssen, K. A., and Magasanik, B. (1975) *Proc. Nat. Acad. Sci. U.S.A. 72,* 4844–4848.
Foor, F., Cedergren, R. J., Streicher, S. L., Rhee, S. G., and Magasanik, B. (1978) *J. Bacteriol. 134,* 562–568.
Foor, F., Reuveny, Z., and Magasanik, B. (1980) *Proc. Nat. Acad. Sci. U.S.A. 77,* 2636–2640.
Frink, R. J., Eisenberg, D., and Glitz, D. G. (1978) *Proc. Nat. Acad. Sci. U.S.A. 75,* 5778–5782.
Fuchs, R. L., and Keister, D. L. (1980) *J. Bacteriol. 141,* 996–998.
Fuchs, R. L., Madonna, M. J., and Brenchley, J. E. (1982) *J. Bacteriol. 149,* 906–915.
Funanage, V. L., and Brenchley, J. E. (1977) *Genetics 86,* 513–526.
Funanage, V. L., Ayling, P. D., Dendinger, S. M., and Brenchley, J. E. (1978) *J. Bacteriol. 136,* 588–596.
Garcia, E., Bancroft, S., Rhee, S. G., and Kustu, S. (1977) *Proc. Nat. Acad. Sci. U.S.A. 74,* 1662–1666.
Geary, L. E., and Meister, A. (1977) *J. Biol. Chem. 252,* 3501–3508.
Ginsburg, A., and Stadtman, E. (1973) in *The Enzymes of Glutamine Matabolism* (Prusiner, S., and Stadtman, E. R., eds.) pp. 9–43, Academic Press, New York.
Heinrikson, R. L., and Kingdon, H. S. (1971) *J. Biol. Chem. 246,* 1099–1106.
Hershfield, V., Boyer, H. W., Yanofsky, C., Lovett, M. A., and Helinski, D. R. (1974) *Proc. Nat. Acad. Sci. U.S.A. 71,* 3455–3459.
Hohman, R. J., and Stadtman, E. R. (1978) *Biochem. Biophys. Res. Comm. 82,* 865–870.
Holzer, H., and Schneider, S. (1957) *Biochem. Z. 329,* 361–369.
Keller, E. B., and Calvo, J. M. (1979) *Proc. Nat. Acad. Sci. U.S.A. 76,* 6186–6190.
Koduri, R. K., Bedwell, D. M., and Brenchley, J. E. (1980) *Gene 11,* 227–237.
Korn, L. J., and Yanofsky, C. (1976) *J. Mol. Biol. 106,* 231–241.
Kustu, S., Burton, D., Garcia, E., McCarter, L., and McFarland, N. (1979) *Proc. Nat. Acad. Sci. U.S.A. 76,* 4576–4580.
Lapointe, J., Delcuve, G., and Duplain, L. (1975) *J. Bacteriol. 123,* 843–850.
Lei, M., Aebi, U., Heidner, E. G., and Eisenberg, D. (1979) *J. Biol. Chem. 254,* 3129–3134.
Leonardo, J. M., and Goldberg, R. B. (1980) *J. Bacteriol. 142,* 99–110.
Levine, R. L., Oliver, C. N., Fulks, R. M., and Stadtman, E. R. (1981) *Proc. Nat. Acad. Sci. U.S.A. 78,* 2120–2124.
Lozoya, E., Pescador, R. S., Covarrubias, A., Vichido, I., and Bolivar, F. (1980) *J. Bacteriol. 144,* 616–621.

Ludwig, R. A. (1980) *J. Bacteriol. 141,* 1209–1216.
MacNeil, T., MacNeil, D., and Tyler, B. (1981) Abstracts of the Annual Meeting of the American Society for Microbiology, p. 151.
Magasanik, B., and Rothstein, D. M. (1980) in *Glutamine: Metabolism, Enzymology and Regulation* (Mora, J., and Palacios, R., eds.) pp. 61–68, Academic Press, New York.
Mantsala, P., and Zalkin, H. (1976a) *J. Biol. Chem. 251,* 3294–3299.
Mantsala, P., and Zalkin, H. (1976b) *J. Biol. Chem. 251,* 3300–3305.
Mantsala, P., and Zalkin, H. (1976c) *J. Bacteriol. 126,* 539–541.
McFarland, N., McCarter, L., Artz, S., and Kustu, S. (1981) *Proc. Nat. Acad. Sci. U.S.A. 78,* 2135–2139.
Meers, J. L., Tempest, D. W., and Brown, C. M. (1970) *J. Gen. Microbiol. 64,* 187–194.
Meister, A. (1980) in *Glutamine: Metabolism, Enzymology and Regulation* (Mora, J., and Palacios, R., eds.) pp. 1–40, Academic Press, New York.
Miller, R. E., and Stadtman, E. R. (1972) *J. Biol. Chem. 247,* 7407–7419.
Nagatani, H., Shimizu, M., and Valentine, R. C. (1971) *Arch. Mikrobiol. 79,* 164–175.
Neidhardt, F. C., and Magasanik, B. (1957) *J. Bacteriol. 73,* 253–259.
Pahel, G., and Tyler, B. (1979) *Proc. Nat. Acad. Sci. U.S.A. 76,* 4544–4548.
Pahel, G., Zelenetz, A. D., and Tyler, B. M. (1978) *J. Bacteriol. 133,* 139–148.
Prival, M. J., and Magasanik, B. (1971) *J. Biol. Chem. 246,* 6288–6296.
Prival, M. J., Brenchley, J. E., and Magasanik, B. (1973) *J. Biol. Chem. 248,* 4334–4344.
Rendina, A. R., and Orme-Johnson, W. H. (1978) *Biochemistry 17,* 5388–5393.
Reuveny, Z., Foor, F., and Magasanik, B. (1981) *J. Bacteriol. 146,* 740–745.
Roberts, G. P., MacNeil, T., Snyder, L., and Tyler, B. (1981) Abstracts of the Annual Meeting of the American Society for Microbiology, p. 151.
Rosenfeld, S. A., and Brenchley, J. E. (1980a) *J. Bacteriol. 143,* 801–808.
Rosenfeld, S. A., and Brenchley, J. E. (1980b) *J. Bacteriol. 144,* 848–851.
Rosenfeld, S. A., Dendinger, S. M., Murphy, C. H., and Brenchley, J. E. (1982) *J. Bacteriol. 150,* 795–803.
Rothstein, D. M., Pahel, G., Tyler, B., and Magasanik, B. (1980) *Proc. Nat. Acad. Sci. U.S.A. 77,* 7372–7376.
Sakamoto, N., Kotre, A. M., and Savageau, M. A. (1975) *J. Bacteriol. 124,* 775–783.
Sanderson, K. E., and Hartman, P. E. (1978) *Microbiol. Rev. 42,* 471–519.
Sanwal, B. D., and Lata, M. (1961) *Can. J. Microbiol. 7,* 319–328.
Senior, P. J. (1975) *J. Bacteriol. 123,* 407–418.
Stadtman, E. R., and Ginsburg, A. (1974) in *The Enzymes* (Boyer, P. D., ed.) 3rd Ed., Vol. X, pp. 755–807, Academic Press, New York.
Stadtman, E. R., Ginsburg, A., Ciardi, J. E., Yeh, J., Hennig, S. B., and Shapiro, B. M. (1970) *Adv. Enzyme Regulation 8,* 99–118.
Stadtman, E. R., Mura, U., Chock, P. B., and Rhee, S. G. (1980) in *Glutamine: Metabolism, Enzymology and Regulation* (Mora, J., and Palacios, R., eds.) pp. 41–59, Academic Press, New York.
Streicher, S. L., and Tyler, B. (1980) *J. Bacteriol. 142,* 69–78.
Streicher, S. L., and Tyler, B. (1981) *Proc. Nat. Acad. Sci. U.S.A. 78,* 229–233.

Streicher, S. L., Shanmugam, K. T., Ausubel, F., Morandi, C., and Goldberg, R. B. (1974) *J. Bacteriol. 120,* 815–821.

Tronick, S. R., Ciardi, J. E., and Stadtman, E. R. (1973) *J. Bacteriol. 115,* 858–868.

Trotta, P. P., Platzer, K. E. B., Haschemeyer, R. H., and Meister, A. (1974) *Proc. Nat. Acad. Sci. U.S.A. 71,* 4607–4611.

Tubb, R. S. (1974) *Nature 251,* 481–485.

Turnbough, C. L., Jr., Neill, R. J., Landsberg, R., and Ames, B. N. (1979) *J. Biol. Chem. 254,* 5111–5119.

Tyler, B. (1978) *Ann. Rev. Biochem. 47,* 1127–1162.

Tyler, B., DeLeo, A. B., and Magasanik, B. (1974) *Proc. Nat. Acad. Sci. U.S.A. 71,* 225–229.

Tyler, B., Bloom, F., and Pahel, G. (1980) in *Glutamine: Metabolism, Enzymology and Regulation* (Mora, J., and Palacios, R., eds.) pp. 69–78, Academic Press, New York.

Valentine, R. C., Shapiro, B. M., and Stadtman, E. R. (1968) *Biochemistry 7,* 2143–2152.

Varricchio, F. (1969) *Biochem. Biophys. Acta 177,* 560–564.

Wedler, F. C., Kenney, R. M., Ashour, A. E., and Carfi, J. (1978) *Biochem. Biophys. Res. Comm. 81,* 122–126.

Windass, J. D., Worsey, M. J., Pioli, E. M., Pioli, D., Barth, P. T., Atherton, K. T., Dart, E. C., Byrom, D., Powell, K., and Senior, P. J. (1980) *Nature 287,* 396–401.

Woolfolk, C. A., and Stadtman, E. R. (1967) *Arch. Biochem. Biophys. 118,* 736–755.

CHAPTER

2

Alanine Biosynthesis and the General Transaminases

Anna L. Tan-Wilson

I. INTRODUCTION

Perhaps because of its apparent simplicity, the biosynthesis of alanine in bacteria has not been studied with the intensity accorded to the biosynthesis of other amino acids. Whereas the regulation of far more complex pathways has been elucidated at several levels, a total description of alanine biosynthesis and its regulation has not been accomplished.

The carbon atoms of alanine are considered to originate from pyruvate in a single reaction (Fig. 2.1) catalyzed by one or more aminotransferases. Although the enzyme(s) responsible for in vivo alanine biosynthesis in bacteria have not been identified conclusively, a group of enzymes exhibit catalytic activities that make them likely candidates for participation in the synthesis of alanine. The best-studied of these enzymes are transaminases A, B, and C and the aromatic transaminase of *Escherichia coli*. The overlapping functions and cross-pathway regulation of these enzymes require that we consider the question of alanine biosynthesis within the context of a larger network, including reactions involving the branched-chain, aromatic, and acidic amino acids.

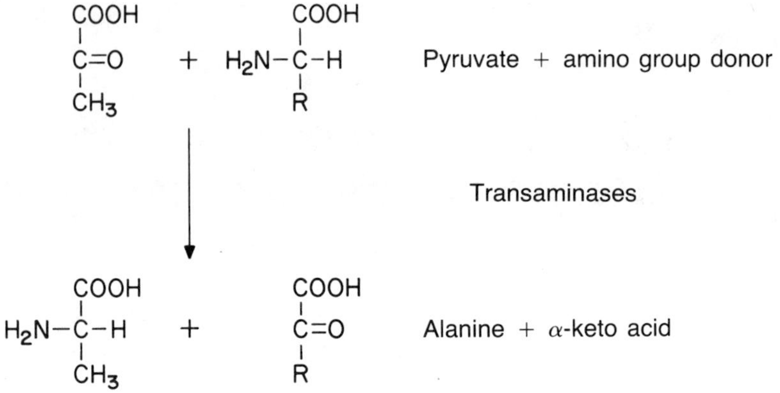

FIGURE 2.1 The biosynthesis of alanine from pyruvate.

II. ALANINE SYNTHESIS IN *ESCHERICHIA COLI*

Extracts of *E. coli* contain two enzyme activities producing alanine in vitro. One catalyzes transamination between pyruvate and glutamate, producing alanine and α-ketoglutarate (Falkinham, 1977); the second catalyzes transamination between pyruvate and valine, producing alanine and α-ketoisovalerate (Rudman and Meister, 1953). These reactions are shown in Figure 2.2.

A. Pyruvate–Glutamate Transamination

An enzyme catalyzing this reaction exclusively has not been identified. Several *E. coli* enzymes undoubtedly contribute to this activity. These are aspartate aminotransferase, sometimes referred to as transaminase A, encoded by the gene designated *aspC*; aromatic amino acid aminotransferase, encoded by the gene designated *tyrB*; and branched-chain amino acid aminotransferase, sometimes referred to as transaminase B, encoded by the gene designated *ilvE*. Prior to the recognition that they were separate proteins (Collier and Kohlhaw, 1972; Powell and Morrison, 1978a; Chesne et al., 1978), the aspartate and aromatic amino acid aminotransferases were considered to be one enzyme, collectively called transaminase A (Rudman and Meister, 1953). The contention that the branched-chain aminotransferase is not one but two enzymes, specific for either valine or isoleucine (Guardiola, 1977), has been disproved (Lee-Peng et al., 1979; Adams et al., 1979).

The initial confusion in assigning catalytic specificity to aminotransferase proteins reflects the overlapping substrate specificities of these enzymes. The aspartate, aromatic, and branched-chain amino acid aminotransferase enzymes have now been purified to homogeneity. Purification procedures, amino acid compositions, and partial amino acid sequences for the aspartate

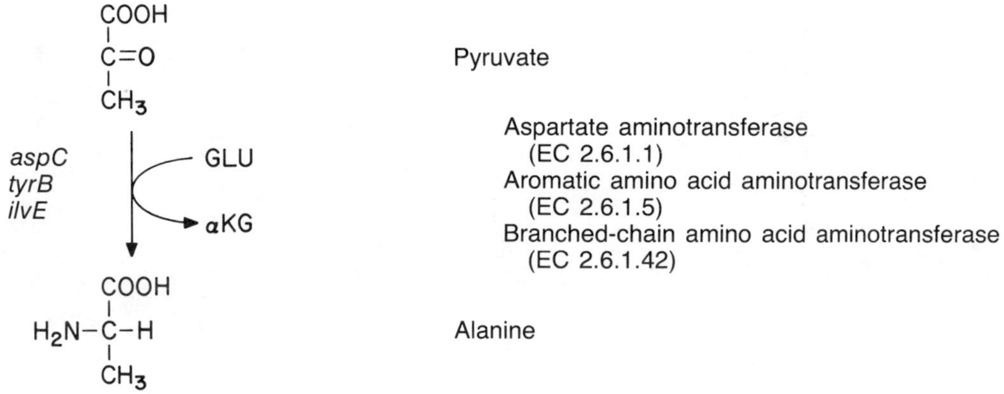

FIGURE 2.2(a) Pyruvate–glutamate transamination in *Escherichia coli.*

and aromatic aminotransferases (Powell and Morrison, 1978a; Mavrides and Orr, 1974; Chesne and Pelmont, 1974; Yagi et al., 1979; Kagamiyama and Toshiharu, 1979) and for the branched-chain amino acid aminotransferase (Lee-Peng et al., 1979; Adams et al., 1979; Monnier et al., 1976) have been published.

Enzyme specificities were investigated by comparing K_m and V_{max} values of purified enzymes. The substrate specificities are summarized in Table 2.1. Listed are the amino acids that react with α-ketoglutarate and the α-keto acids that react with glutamate. The amino acids and α-keto acids in boldface are the preferred substrates. Thus in reactions with α-ketoglutarate, asparatate aminotransferase has preferred specificity for the acidic amino acids, the aromatic aminotransferase has preferred specificity for the aromatic amino acids, and the branched-chain amino acid aminotransferase has preferred specificity for the branched-chain amino acids. However, even when purified to apparent homogeneity, each enzyme catalyzes transaminations between α-ketoglutarate and amino acids not regarded as the preferred substrates. These enzymes also catalyze transamination reactions between -glutamate and α-keto acids other than the preferred ones.

Table 2.1 includes neither the substrate alanine in reactions with α-ketoglutarate nor pyruvate in reactions with glutamate. This is partly due to the lack of reported data. However, the purified aspartate aminotransferase was found not to catalyze the reaction between α-ketoglutarate and alanine (Chesne and Pelmont, 1973).

B. Pyruvate–Valine Transamination—Transaminase C

Transamination between pyruvate and valine or the reverse reaction between α-ketoisovalerate and alanine (or α-aminobutyrate) has been attributed to a fourth aminotransferase in *E. coli,* termed transaminase C (Rudman and Meister, 1953). Unlike the branched-chain amino acid transferase, neither α-ketoglutarate, glutamate, leucine, nor isoleucine serves as a substrate.

TABLE 1. Substrate Specificities of *Escherichia coli* Aminotransferases

	Aspartate Aminotransferase	*Aromatic Amino Acid Aminotransferase*	*Branched-Chain Amino Acid Aminotransferase*
Amino acids reacting with α-ketoglutarate	**Aspartate**	Aspartate	
	Glutamate	Glutamate	Glutamate
	Phenylalanine	**Phenylalanine**	Phenylalanine
	Tyrosine	**Tyrosine**	Tyrosine[1]
	Tryptophan[1]	**Tryptophan**	
	Methionine[1]	Methionine	
	Leucine[1]	Leucine	**Leucine**
			Isoleucine
			Valine
α-Keto acids reacting with glutamate	**Oxaloacetate**	Oxaloacetate	
	α-Ketoglutarate	α-Ketoglutarate	α-Ketoglutarate[2]
	Phenylpyruvate	**Phenylpyruvate**	Phenylpyruvate
	4-Hydroxyphenylpyruvate	**4-Hydroxyphenylpyruvate**	
		α-Ketoisocaproate	**α-Ketoisocaproate**
			α-Keto-β-methyl-*n*-valerate
			α-Ketoisovalerate

1. very weak.
2. reacting with valine.

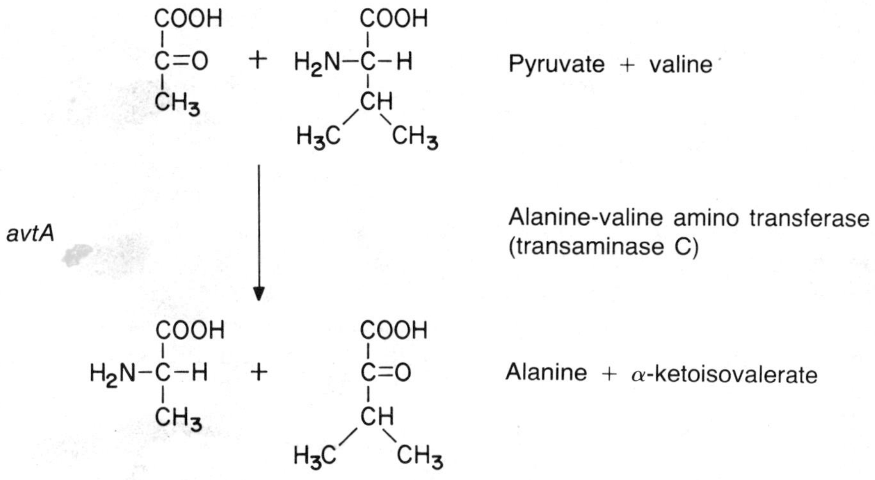

FIGURE 2.2(b) Pyruvate–valine transamination in *Escherichia coli*.

Because transaminase C has so far been detected only in crude extracts, the reaction may either reflect a single enzyme protein or be attributable to one or more enzymes.

Purified branched-chain amino acid aminotransferase has no detectable transaminase C activity (Lee-Peng et al., 1979). Also, *E. coli* strain 11A16, which lacks the branched-chain amino acid aminotransferase, still exhibits transaminase C activity. Strain 11A161 derived from 11A16 even showed an increase in transaminase C activity (Adelberg and Umbarger, 1953). Another *ilvE* mutant, JFM17, lacked alanine-α-ketoisovalerate transaminase (Powell and Morrison, 1978b). If transaminase C activity is the result of several enzymes working in concert, the involvement of the branched-chain amino acid aminotransferase, specific for valine, is unclear. Nor is transaminase C attributable to the aspartate aminotransferase. Extracts of *E. coli* DG27, having aspartate aminotransferase and lacking both the aromatic and branched-chain amino acid aminotransferases (Gelfand and Steinberg, 1977), failed to catalyze transamination between alanine and α-ketoisovalerate (Powell and Morrison, 1978b).

An *E. coli* mutation resulting in the loss of 90% of transaminase C activity does not result in alanine auxotrophy (Falkinham, 1979). Therefore either transaminase C is not essential for alanine biosynthesis in vivo, or other aminotransferase enzymes can function to produce enough alanine for growth.

An enzyme somewhat analogous to the putative *E. coli* transaminase C is the leucine–pyruvate aminotransferase in *Acetobacter suboxydans* (Tachiki and Tochikura, 1973). This enzyme has been purified to homogeneity (Tachiki et al., 1975). With pyruvate as the amino group acceptor, this enzyme shows a marked preference for leucine over valine and isoleucine. However, this enzyme has a much wider specificity than the proposed transaminase C because many other amino acids are substrates to various extents. With leucine as the amino group donor, pyruvate is the preferred acceptor, although α-ketoglutarate and oxaloacetate are also active (Tachiki et al., 1975). Thus this enzyme is similar to the proposed transaminase C in its preference for pyruvate and only one of the three branched-chain amino acids, in this case leucine.

C. Pyruvate Transamination with Other Amino Group Donors

Alanine can be formed from pyruvate in reactions other than the single transamination step between pyruvate and one of the 20 common amino acids. The amino group can come from α-dialkyl-amino acids, ω-amino acids, monoamines, and diamines (Fig. 2.3).

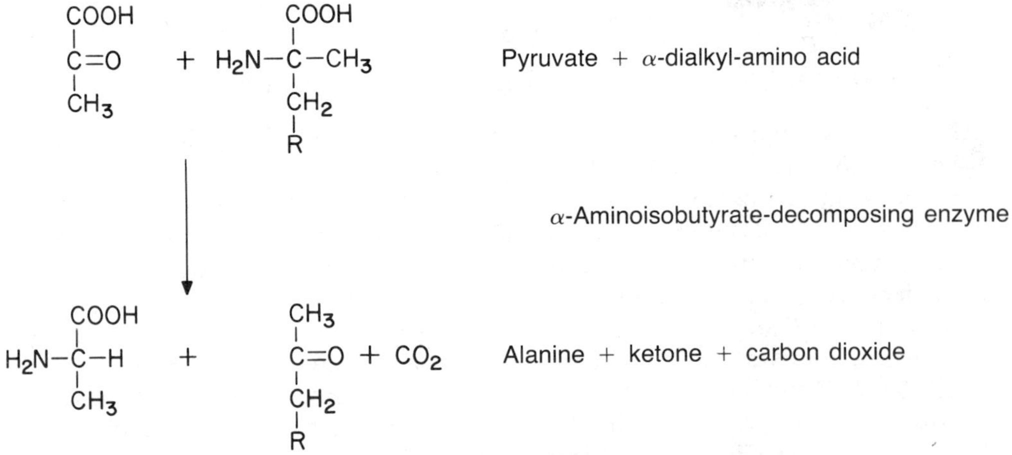

FIGURE 2.3(a) Pyruvate–α-dialkyl-amino acid transamination.

1. α-Dialkyl-Amino Acid-Decomposing Enzyme. An α-aminoisobutyrate-decomposing enzyme, purified from *Pseudomonas* sp. (Bailey and Dempsey, 1967) decarboxylates α-dialkyl-amino acids with transfer of the amino group to enzyme-bound pyridoxal phosphate. The resulting pyridoxamine phosphate then donates the amino group to an acceptor α-keto acid to produce the corresponding amino acid. In this reaction, pyruvate is superior to α-ketoglutarate. This enzyme also catalyzes conventional transaminations between alanine and α-keto acids such as α-ketobutyrate (Sato et al., 1978; Honma and Shimomura, 1974).

2. ω-Amino Acid Aminotransferases. An ω-amino acid–pyruvate aminotransferase, also called β-alanine–pyruvate aminotransferase, has been purified from *Pseudomonas* sp. F-126 (Yonaha et al., 1977). Monocarboxylic amino acids with the amino group at the β, γ, or ε position, but not at the α position, as well as ω-amino acid sulfonates and sulfinates, act as substrates.

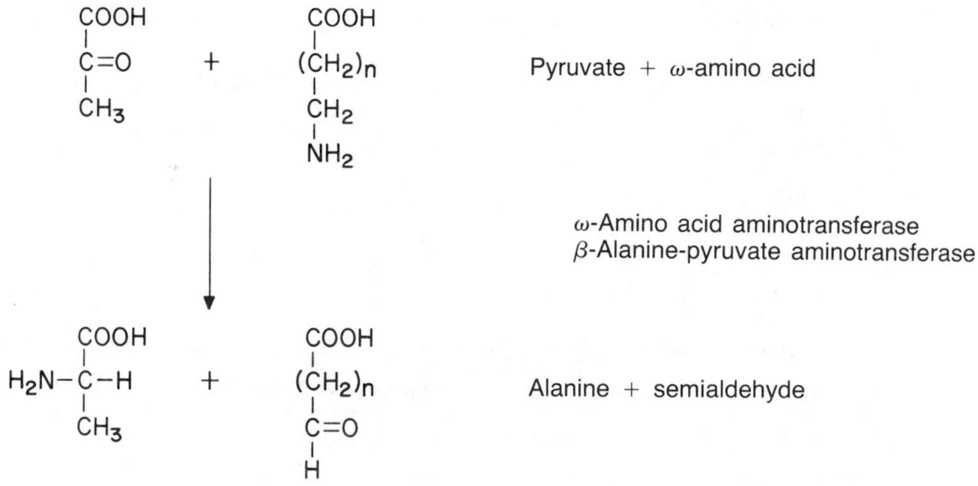

FIGURE 2.3(b) Pyruvate–ω-amino acid transamination.

α,ω-diaminocarboxylic acids, such as lysine, are inert as amino group donors (Hayaishi et al., 1961). This enzyme catalyzes the transamination of monoamines and of diamines. In the reaction with diamines, only one amino group is transferred. Pyruvate is the sole amino group acceptor for all three types of reactions. Thus a wide variety of substrates can transfer amino groups to pyruvate to form alanine (Yonaha and Toyama, 1978, 1979).

Although alanine is a recognized product of the ω-amino acid–pyruvate aminotransferase, it is doubtful that this enzyme has alanine biosynthesis as its main function. The enzyme is induced by growth on β-alanine, indicating its catabolic function. A similar enzyme, present in *Bacillus cereus,* is not pyruvate-specific (Nakano et al., 1977). Both *Pseudomonas* sp. F-126 and *B. cereus* have another type of ω-amino acid aminotransferase, γ-aminobutyrate aminotransferase, with distinct acceptor preference for α-ketoglutarate over pyruvate. Because other species, *Achromobacter superficialis* and *A. polymorph,* have an α-ketoglutarate-specific ω-amino acid aminotransferase (Toyama and Soda, 1972), amino groups from ω-amino acids can be transferred to α-ketoglutarate to form glutamate.

D. Alanine Biosynthesis Not Involving Transamination of Pyruvate

Results of experiments tracing ^{3}H and ^{14}C from various radioactively labeled forms of glucose into alanine in *E. coli* proteins suggest that transamination of pyruvate might not be the only route to alanine (Csonka, 1977). No alternative routes were proposed, although experiments with *Pseudomonas aeruginosa* grown on labeled valine were consistent with a pathway through pyruvate to alanine (Sokatch, 1966).

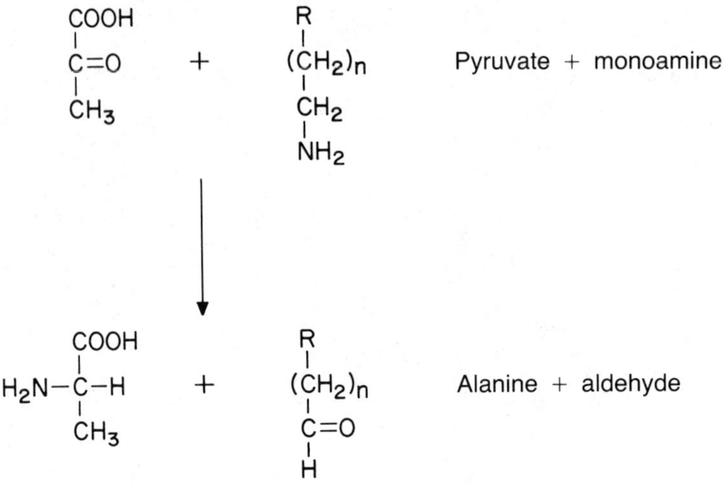

FIGURE 2.3(c) Pyruvate–monoamine transamination.

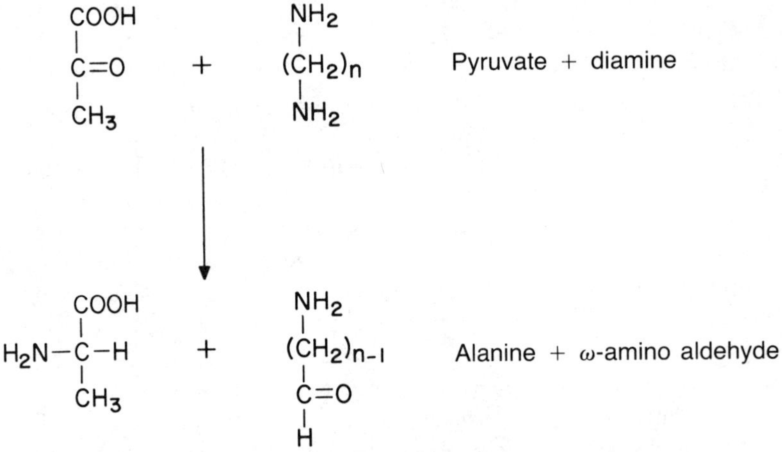

FIGURE 2.3(d) Pyruvate–diamine transamination.

III. COMPARING ALANINE BIOSYNTHES OF EUKARYOTIC AND PROKARYOTIC CELLS

The apparent absence of bacterial enzymes specialized for the biosynthesis of alanine stands in sharp contrast to the situation in mammalian systems. In rat muscle, alanine is formed by transamination between glutamate and pyruvate, catalyzed by a specialized alanine aminotransferase. Other amino acids contribute amino groups to alanine by prior transamination with α-ketoglutarate to produce glutamate. Alanine biosynthesis serves a special purpose in rat muscle. Not only is this amino acid a precursor for protein biosynthesis, alanine is also synthesized in muscle to divert amino acid nitrogen released during exercise so as to prevent ammonia toxicity. Alanine also serves as an alternative to lactate as a glycolytic end product. Alanine is transported to the liver, where its nitrogen is converted to urea. The carbon skeleton is used in gluconeogenesis. The glucose returns to the muscle to replenish the glycogen stores (Snell, 1980).

If a specific bacterial alanine aminotransferase does not exist, one might speculate on the origins of this striking difference between the enzymology of alanine biosynthesis in bacteria versus mammals. Recently, Jensen and Calhoun suggested that primitive aminotransferase(s) possessed a very broad substrate specificity. Gene amplification, followed by divergence, led to the subsequent evolution of enzymes with narrower substrate specificities (Jensen and Calhoun, 1981). Ultimate specialization to the extent that one enzyme has one function and that a given reaction is catalyzed only by one enzyme has distinct advantages for enzyme regulation. However, mutations that severely impair or destroy the catalytic activity of the enzyme would seriously jeopardize the viability of the cell. The general aminotransferases described in this chapter stand somewhere in between the proposed primitive aminotransferase and the completely specialized mammalian alanine aminotransferase. Perhaps the needs of the bacterial cell are easily met by the alanine produced in a variety of reactions catalyzed by several general aminotransferases. There can be no great need for the regulation of a simple, readily reversible, one-reaction "pathway" for alanine biosynthesis. Thus one can speculate that bacteria have retained flexibility in alanine biosynthesis by not evolving a single specialized enzyme.

IV. REGULATION OF ENZYME ACTIVITY

In general, aminotransferases catalyze freely reversible reactions between an α-amino and an α-keto acid, using a ping-pong mechanism (Jensen and Calhoun, 1981). Some enzymes, such as the *E. coli* aspartate and aromatic amino acid aminotransferases, have roughly the same values of K_m for an amino acid and its α-keto acid analog. Specific activities in both directions are also similar (Powell and Morrison, 1978a).

Other enzymes show a distinct preference for either the forward or the backward reaction. Two transaminases from *Pseudomonas aeruginosa,* enzymes DE I and HA I, catalyze reactions involving phenylpyruvate (or 4-hydroxyphenylpyruvate) and α-ketoglutarate to produce phenylalanine (or tyrosine) and glutamate. The K_m of enzyme DE I for phenylpyruvate is 1.7 mM; for phenylalanine, it is 100 mM. The K_m of enzyme DE I for glutamate is 0.7 mM; for α-ketoglutarate, it is 25 mM. On the other hand, the K_m of enzyme HA I for glutamate is 16 mM, while the K_m for α-ketoglutarate is 99 mM. These numbers show that the enzyme DE I functions most efficiently in aromatic amino acid synthesis, while the enzyme HA I functions most efficiently in aromatic amino acid catabolism (Patel et al., 1978).

V. REGULATION OF ENZYME SYNTHESIS

Regulation of the expression of the genes coding for the aminotransferases seems to have evolved in a manner consistent with the characteristic overlapping substrate specificities and functions of these enzymes. Because of the overlapping functions, great care should be taken in interpreting data. In the absence of one enzyme, there are other enzymes that take over the function of the missing one. For instance, *ilvE tyrB* strains do not require added tyrosine or phenylalanine for growth, since the aspartate aminotransferase can take over the function of aromatic amino acid transamination (Gelfand and Steinberg, 1977). Rigorous evidence for enzyme induction or repression should include some measurement of enzyme protein in addition to enzyme-specific activity.

A. Regulation of *Escherichia coli* Aminotransferase Synthesis

1. Aspartate and Aromatic Amino Acid Aminotransferases. The synthesis of the aromatic amino acid aminotransferase is repressed by growth on tyrosine (see Chapter 19), while the synthesis of the aspartate aminotransferase is constitutive. This is seen in the *E. coli ilvE* mutant DG31 and in the *ilvE* and *aspC* double mutant DG34. Both strains depend on the aromatic amino acid aminotransferase for leucine biosynthesis and hence growth. Both strains have reduced plating efficiencies when grown in the presence of tyrosine (Gelfand and Steinberg, 1977).

In a more direct experiment, *E. coli* CU2, a strain that also lacks the branched-chain amino acid aminotransferase, grown in 1 mM tyrosine had aromatic aminotransferase levels that were only 18–35% of those for the same cells grown without tyrosine (Collier and Kohlhaw, 1972). This repression was confirmed by the direct measurement of enzyme protein. The aspartate and aromatic aminotransferases have distinct heat stabilities and can be separated on DEAE-Sephadex. After growth of *E. coli* K-12 in the presence of 1 mM tyrosine, the aromatic aminotransferase levels decreased sharply while the aspartate aminotransferase remained unaffected (Chesne et al., 1978).

Reports, published before the distinction between the aspartate and aromatic aminotransferases became apparent, support the idea of the constitutive nature of aspartate aminotransferase synthesis and the repressibility of the aromatic aminotransferase by tyrosine (Silbert et al., 1963; Brown and Somerville, 1971). In *tyrR* strains of *E. coli,* the synthesis of the aromatic amino acid aminotransferase, as well as of other aromatic amino acid biosynthetic enzymes, is constitutive (Gibson and Pittard, 1968; Wallace and Pittard, 1969; Im et al., 1971; chapter 19 of this volume). Aromatic aminotransferase specific activity is also affected by growth of *E. coli* on leucine (Powell and Morrison, 1978b).

2. Branched-Chain Amino Acid Aminotransferase. The gene encoding the branched-chain amino acid aminotransferase in *E. coli* lies within the *ilv* operon. The major control site of this gene cluster is an attenuator region proximal to the *ilvG* gene (Chapter 13). High concentrations of charged leucyl-, isoleucyl-, and valyl-transfer RNA would act as a multivalent signal for repression by transcription termination (Nargang et al., 1980; Lawther and Hatfield, 1980). Another control over *ilvEDA* expression has been traced to the *ilvG* gene. This locus specifies an isoenzyme of acetohydroxy acid synthase insensitive to feedback inhibition by valine. The *ilvG* gene is silent in the *E. coli* wild type. A frameshift mutation resulting in the expression of the *ilvG* gene product also results in a fivefold to tenfold increase in the expression of the *ilvEDA* genes, presumably by reversal of translational polarity (DeFelice et al., 1977). In *Salmonella typhimurium,* transcription from an internal constitutive promoter results in an mRNA–*ilvEDA* transcript (Blazey and Burns, 1979, 1981). These authors suggest that the low-level expression of *ilvE* under repressing conditions originates from this internal promoter.

3. Transaminase C. Transaminase C activity drops when *E. coli* K-12 is grown with excess alanine or leucine. The effect of the two amino acids is additive. Neither valine nor isoleucine affects transaminase C activity in the wild type (Falkinham, 1979).

This pattern of regulation is different in *E. coli* strain CU2002, which lacks the branched-chain amino acid aminotransferase. Because of the lack of the *ilvE* gene product, isoleucine always had to be added to the medium. Further addition of valine to the medium resulted in inhibition of growth and in an observed decrease in transaminase C activity. The growth inhibition could be reversed by the still further addition of leucine or of α-ketoisovalerate, an earlier intermediate in leucine biosynthesis. The specific activity of transaminase C was not restored to the same levels as were obtained in a medium containing isoleucine alone but came close to the levels noted when this strain was grown on a medium containing isoleucine and leucine (McGilvray and Umbarger, 1974).

The growth inhibition of strain CU2002 by valine has been explained by feedback inhibition of the acetohydroxy acid synthase. Valine also represses transaminase C. Both effects result in limitation of the supply of α-ketoisovalerate and eventually of α-ketoisocaproate (McGilvray and Umbarger, 1974). In cells that depend on the less efficient aromatic aminotransferase, higher concentrations of α-ketoisocaproate would be needed for the last step in leucine biosynthesis. Growth would be inhibited, and the inhibition would be reversed by L-leucine.

Why valine represses transaminase C activity in an *E. coli ilvE* mutant and not in the wild type is not readily apparent. Possibly, in the wild type, valine can be converted to α-ketoisovalerate by both the branched-chain amino acid aminotransferase and transaminase C. Only the latter reaction leads to alanine accumulation. In an *ilvE* mutant, all transamination reactions involving valine would also produce alanine, whose buildup would repress transaminase C.

A plausible role for transaminase C is to mediate alanine and α-ketoisovalerate synthesis, the latter required for leucine synthesis. Both alanine and leucine cause transaminase C repression. The first step in alanine catabolism in *E. coli* is its conversion to D-alanine, not transamination (Franklin and Venables, 1976). The fact that a 90% loss of transaminase C activity does not lead to alanine auxotrophy does not weaken this proposal, since other enzymes could serve in alanine synthesis. Insofar as valine metabolism is concerned, transaminase C would favor the conversion of valine to α-ketoisovalerate and thence to leucine. The branched-chain amino acid aminotransferase would favor the use of α-ketoisovalerate in the biosynthesis of valine.

B. Regulation of Aminotransferases with Both Biosynthetic and Catabolic Functions

One special characteristic of aminotransferases is that even when they catalyze only one reaction, the same reaction can be biosynthetic in one direction and degradative in the reverse direction. This applies to the metabolism of the branched-chain amino acids in pseudomonads (Massey et al., 1976). This metabolic arrangement contributes to cellular economy by avoiding duplication of protein biosynthesis. In this case, the type of regulation that works in an organism that uses the enzyme unidirectionally would not be suitable. In *E. coli* K-12, the branched-chain amino acid aminotransferase is produced at very low levels by growth in high concentrations of leucine, isoleucine, and valine (Kline et al., 1974). Clearly, repression by an amino acid of a biosynthetic enzyme, also meant for catabolism of the same amino acid, would be incompatible with a catabolic role. Aminotransferases that function only in the first step of a catabolic pathway are often found to be inducible by substrate.

For example, the α-aminoisobutyrate-decomposing enzyme and the ω-amino acid–pyruvate aminotransferase described earlier are induced by the presence of α-aminoisobutyrate and of β-alanine, respectively (Bailey and Dempsey, 1967; Nakano et al., 1977; Toyama and Soda, 1972). However, induction by an amino acid of a catabolic enzyme, also functioning in the biosynthesis of the same amino acid, would prevent sufficient buildup of the amino acid pools unless there were separate compartments for each pool.

In *Pseudomonas putida,* branched-chain amino acid aminotransferase synthesis is constitutive. Branched-chain α-keto acid dehydrogenase, the enzyme catalyzing the second step in valine catabolism, is induced by branched-chain α-keto acids. The product of that step in turn induces enzymes catalyzing two later reactions in the pathway (DePaul-Marshall and Sokatch, 1972). The same branched-chain amino acid aminotransferase catalyzes the first step in leucine catabolism. D-leucine in the medium induces the synthesis of the three enzymes in the D-leucine catabolic pathway (Massey et al., 1974). The latter two enzymes are needed to degrade the α-ketoisocaproate produced from leucine in the transaminase reaction. Thus this organism keeps the aminotransferase always available for either biosynthetic or catabolic purposes. Enzymes leading to the production of the α-keto acid substrate for biosynthesis or away from the α-keto acid product during degradation can be regulated. The direction of reaction of the aminotransferase, determined by substrate availability, is being controlled. The synthesis of the enzyme is not.

Different species have evolved different patterns of regulation for identical situations. In *P. fluorescens,* one aminotransferase is responsible for synthesis and catabolism of leucine, isoleucine, and valine. The enzyme is induced by leucine, isoleucine, valine, or their corresponding α-keto acids (Puuka et al., 1972). Thus a particular enzyme is induced by substrates for both the biosynthetic and the degradative reactions it catalyzes. Again, "neutrality" is retained, but the cell need not synthesize the enzyme when neither reaction is required.

In some cases, aminotransferases catalyze the first step in catabolism but not the last step in the biosynthesis of the same amino acid. This applies to tryptophan metabolism in *Clostridium sporogenes* (O'Neill and De Moss, 1968), *Pseudomonas* sp. (Yuji, 1980), and *K. aerogenes* (Paris and Magasanik, 1981).

In other cases, aminotransferases catalyze the last step in biosynthesis but not the first step in catabolism. In *E. coli* K-12, alanine is presumably synthesized by an aminotransferase. Yet, catabolism of alanine occurs by first racemization to D-alanine followed by oxidation. Both alanine racemase and D-alanine dehydrogenase are regulated by catabolite repression (Franklin and Venables, 1976). A different pattern applies in *Rhodopseudomonas acidophila,* which utilizes an alanine-glyoxylate aminotransferase as the first step in alanine catabolism (Herbert et al., 1980).

VI. CLONING AND SEQUENCING OF AMINOTRANSFERASE GENES

The genes coding for the general aminotransferases in *E. coli* have been mapped. The *tyrB* gene, which encodes the aromatic aminotransferase, is cotransducible with *metA* and *malB* and maps in the order *metA-malB-tyrB* at 91 min (Gelfand and Rudo, 1977) on the linkage map of *E. coli* K-12 (Bachmann and Low, 1980). Gene *aspC*, which encodes the aspartate aminotransferase, is cotransducible with *serC* and *pyrD*. The gene order is *serC aspC pyrD*. Gene *aspC* maps at 20 min (Gelfand and Rudo, 1977). Gene *ilvE* maps at 84 min. The gene is in the *ilv* operon, the gene order being *ilvGEDA* (McCorkle et al., 1978). A mutation affecting transaminase C–specific activity is located near 83 min on the *asn* side of the *ilv* gene cluster, away from the *metE* side (Falkinham, 1979).

A number of *E. coli* strains bearing plasmids that carry the *ilvE* gene have been constructed (McCorkle et al., 1978; Subrahmanyam et al., 1980). Approximately 10-fold to 40-fold elevation of the branched-chain amino acid aminotransferase is obtained in extracts of these strains (Chapter 13).

The sequence of a potential internal promoter preceding *ilvE* and the first 94 nucleotides of the *ilvE* gene have been presented (Lawther et al., 1981). The nucleotide sequence of the *E. coli ilv* attenuator region has been determined (Nargang et al., 1980; Lawther and Hatfield, 1980). The sequence includes a region upstream of *ilvG* that, upon transcription, would yield a leader transcript specifying a peptide of 32 residues, of which 14 are multiple isoleucine, valine, and leucine residues. A transcription termination site precedes *ilvG*.

Curiously enough, this putative peptide would also have four alanine residues, two in tandem. There are no reports of *ilv* operon control by alanine. Lawther and Hatfield's model, based on the nucleotide sequence in the attenuator region, precludes a regulatory role for alanine. However, Falkinham (1979) reported that transaminase C mutants were inhibited by alanine and that this inhibition could be reversed by isoleucine. α-Ketobutyrate instead of isoleucine was not effective in overcoming the growth inhibition, a finding suggesting that alanine, in the absence of transaminase C activity, prevents the flow of intermediates from α-ketobutyrate to isoleucine. The transaminase C–deficient strains grew as well as the wild type in a medium without alanine. If transaminase C at high alanine concentrations were to favor the transamination of alanine with α-ketoisovalerate, a decrease in alanine levels would accompany its conversion to pyruvate. In transaminase C–deficient strains, high alanine (and thus high-charged alanyl transfer RNA) concentrations would persist and possibly favor transcription termination.

VII. CONCLUSION

In this review, I have presented seemingly unrelated observations scattered throughout the literature that, taken together, could imply that branched-chain amino acid metabolism is regulated in part by alanine and that alanine biosynthesis could be regulated by branched-chain amino acids. The metabolism of the branched-chain amino acids shares common features with that of the acidic and the aromatic amino acids by virtue of the aspartate, aromatic, and branched-chain amino acid aminotransferases, enzymes with overlapping functions. When studied in this context, alanine biosynthesis in bacteria will assume much greater significance.

REFERENCES

Adams, C. W., Lawther, R. P., and Hatfield, G. W. (1979) *Biochem. Biophys. Res. Comm. 89,* 650–658.

Adelberg, E. A., and Umbarger, H. E. (1953) *J. Biol. Chem. 205,* 475–482.

Bachmann, B. J., and Low, K. B. (1980) *Microbiol. Rev. 44,* 1–56.

Bailey, G. B., and Dempsey, W. B. (1967) *Biochemistry 6,* 1526–1533.

Blazey, D. L., and Burns, R. O. (1979) *Mol. Gen. Genet. 177,* 1–11.

Blazey, D. L., and Burns, R. O. (1981) *Fed. Proc. 40,* 1649.

Brown, K. D., and Somerville, R. L. (1971) *J. Bacteriol. 108,* 386–399.

Chesne, S., and Pelmont, J. (1973) *Biochimie 55,* 237–244.

Chesne, S., and Pelmont, J. (1974) *Biochimie 56,* 631–639.

Chesne, S., Monnier, N., and Pelmont, J. (1978) *Biochimie 60,* 403–407.

Collier, R. H., and Kohlhaw, G. (1972) *J. Bacteriol. 112,* 365–371.

Csonka, L. N. (1977) *J. Biol. Chem. 252,* 3392–3398.

DeFelice, M., Squires, C., Levinthal, M., Guardiola, J., Lamberti, A., and Iaccarino, M. (1977) *Mol. Gen. Genet. 156,* 1–7.

DePaul-Marshall, V., and Sokatch, J. R. (1972) *J. Bacteriol. 110,* 1073–1081.

Falkinham, J. O., III (1977) *J. Bacteriol. 130,* 569–570.

Falkinham, J. O., III (1979) *Mol. Gen. Genet. 176,* 147–149.

Franklin, F. C. H., and Venables, W. A. (1976) *Mol. Gen. Genet. 149,* 229–237.

Gelfand, D. H., and Rudo, N. (1977) *J. Bacteriol. 130,* 441–444.

Gelfand, D. H., and Steinberg, R. A. (1977) *J. Bacteriol. 130,* 429–440.

Gibson, F., and Pittard, J. (1968) *Bacteriol. Rev. 32,* 465–492.

Guardiola, J. (1977) *Mol. Gen. Genet. 158,* 157–164.

Hayaishi, O., Nishizuka, Y., Tatibana, M., Takeshita, M., and Kuno, S. (1961) *J. Biol. Chem. 236,* 781–790.

Herbert, R. A., Siefert, E., and Pfennig, N. (1980) *Arch. Microbiol. 125,* 105–109.

Honma, M., and Shimomura, T. (1974) *Agr. Biol. Chem. 38,* 953–958.

Im, S. W. K., Davidson, H., and Pittard, J. (1971) *J. Bacteriol. 108,* 400–409.

Jensen, R. A., and Calhoun, D. H. (1981) *CRC Critical Rev. Microbiol. 8,* 229–266.

Kagamiyama, H., and Toshiharu, Y. (1979) *Biochem. Biophys. Res. Comm. 89,* 1347–1353.

Kline, E. L., Brown, C. S., Coleman, W. G., Jr., and Umbarger, H. E. (1974) *Biochem. Biophys. Res. Comm. 57,* 1144–1151.

Lawther, R. P., and Hatfield, G. W. (1980) *Proc. Nat. Acad. Sci. U.S.A. 77,* 1862–1866.

Lawther, R. P., Calhoun, D. H., Adams, C. N., Hauser, C. A., Gray, J., and Hatfield, G. W. (1981) *Proc. Nat. Acad. Sci. U.S.A. 78,* 922–925.

Lee-Peng, F.-C., Hermodson, M. A., and Kohlhaw, G. B. (1979) *J. Bacteriol. 139,* 339–345.

Massey, L. K., Conrad, R. S., and Sokatch, J. R. (1974) *J. Bacteriol. 118,* 112–120.
Massey, L. K., Sokatch, J. R., and Conrad, R. S. (1976) *Bacteriol. Rev. 40,* 42–54.
Mavrides, C., and Orr, W. (1974) *Biochim. Biophys. Acta 336,* 70–78.
McCorkle, G. M., Leathers, T. D., and Umbarger, H. E. (1978) *Proc. Nat. Acad. Sci. U.S.A. 75,* 89–93.
McGilvray, D., and Umbarger, H. E. (1974) *J. Bacteriol. 120,* 715–723.
Monnier, N., Montmitonnet, A., Chesne, S., and Pelmont, J. (1976) *Biochimie 58,* 663–675.
Nakano, Y., Tokunaga, H., and Kitaoka, S. (1977) *J. Biochem. 81,* 1375–1381.
Nargang, F. E., Subrahmanyam, C. S., and Umbarger, H. E. (1980) *Proc. Nat. Acad. Sci. U.S.A. 77,* 1823–1827.
O'Neill, S. R., and DeMoss, R. D. (1968) *Arch. Biochem. Biophys. 127,* 361–369.
Paris, C. G., and Magasanik, B. (1981) *J. Bacteriol. 145,* 257–265.
Patel, N., Stenmark-Cox, S. L., and Jensen, R. A. (1978) *J. Biol. Chem. 253,* 2972–2978.
Powell, J. T., and Morrison, J. F. (1978a) *Eur. J. Biochem. 87,* 391–400.
Powell, J. T., and Morrison, J. F. (1978b) *J. Bacteriol. 136,* 1–4.
Puukka, M., Lönnberg, H., and Nurmikko, V. (1972) *Acta Chem. Scand. 26,* 1271–1273.
Rudman, D., and Meister, A. (1953) *J. Biol. Chem. 200,* 591–604.
Sato, S., Honma, M., and Shimomura, T. (1978) *Agr. Biol. Chem. 42,* 2341–2346.
Silbert, D. F., Jorgensen, S. E., and Lin, E. C. C. (1963) *Biochim. Biophys. Acta 73,* 232–240.
Snell, K. (1980) *Biochem. Soc. Trans. 8,* 205–213.
Sokatch, J. R. (1966) *J. Bacteriol. 92,* 72–75.
Subrahmanyam, C. S., Noti, J. D., and Umbarger, H. E. (1980) *J. Bacteriol. 144,* 279–290.
Tachiki, T., and Tochikura, T. (1973) *Agr. Biol. Chem. 37,* 1439–1448.
Tachiki, T., Moriguchi, M., and Tochikura, T. (1975) *Agr. Biol. Chem. 39,* 43–50.
Toyama, S., and Soda, K. (1972) *J. Bacteriol. 109,* 533–538.
Wallace, B. J., and Pittard, J. (1969) *J. Bacteriol. 97,* 1234–1241.
Yagi, T., Kagamiyama, H., Motosugi, K., Nozaki, M., and Soda, K. (1979) *FEBS Lett. 100,* 81–84.
Yonaha, K., and Toyama, S. (1978) *Agr. Biol. Chem. 42,* 2363–2367.
Yonaha, K., and Toyama, S. (1979) *Agr. Biol. Chem. 43,* 1043–1048.
Yonaha, K., Toyama, S., Yasuda, M., and Soda, K. (1977) *Agr. Biol. Chem. 41,* 1701–1706.
Yuji, K. (1980) *Agr. Biol. Chem. 44,* 2013–2019.

CHAPTER 3

Proline Biosynthesis

Laszlo N. Csonka
Annette Baich

I. INTRODUCTION

Proline is unique among the building blocks of proteins in that it is a secondary amine and therefore properly designated as an imino acid. 1-Pyrroline 5-carboxylate, excreted by two proline auxotrophs of *Escherichia coli,* was the first intermediate identified in proline biosynthesis (Vogel and Davis, 1952). Isotope studies showed glutamate to be a precursor of proline in both microorganisms and animal tissues (Stetten, 1955). γ-Glutamyl phosphate was first proposed as an intermediate in the conversion of glutamate to 1-pyrroline 5-carboxylate by Baich (1969).

Although the proline biosynthetic pathway in the Enterobacteriaceae (Fig. 3.1) comprises only four reactions, its enzymology and genetics are poorly understood. The slow progress in understanding proline biosynthesis can be attributed to the lability of γ-glutamyl phosphate (Katchalsky and Paecht, 1954), which was unambiguously identified as the first intermediate of the pathway only very recently (Hayzer and Leisinger, 1981). Because of the lability of γ-glutamyl phosphate, it has been difficult to reliably assay the first two enzymes of the pathway or probe the mechanisms which regulate their synthesis.

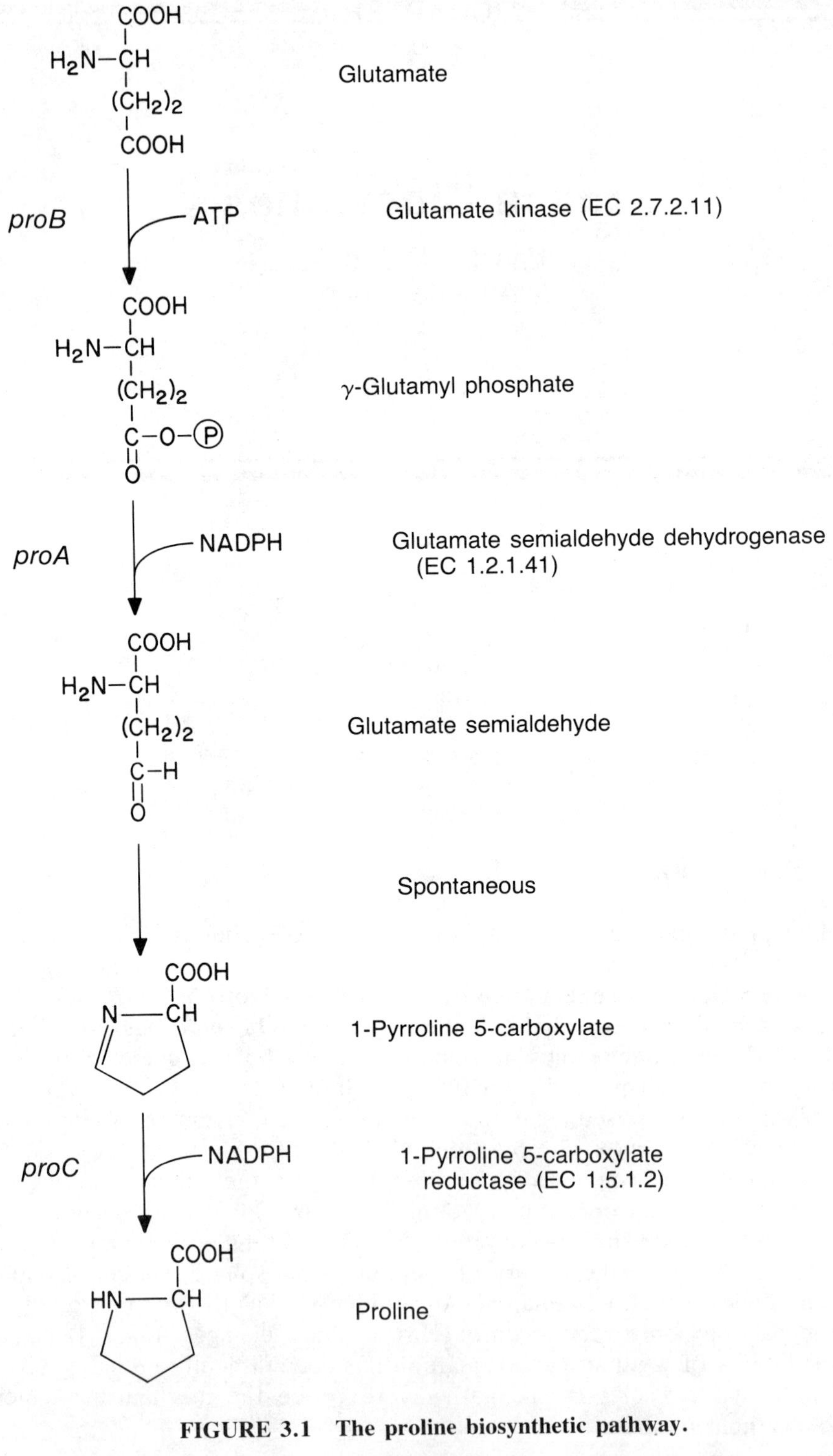

FIGURE 3.1 The proline biosynthetic pathway.

Only three of the four reactions of proline biosynthesis from glutamate are enzymatically catalyzed. The third reaction in the sequence, the cyclization of glutamate semialdehyde, proceeds spontaneously (Wu and Seifter, 1975). Proline biosynthesis is analogous to that of homoserine (Chapter 9). In both cases a dicarboxylate—glutamate or aspartate—is first phosphorylated. The phosphorylation product then undergoes two successive NADPH-linked reductions. However, in the proline pathway, glutamate semialdehyde first cyclizes to form a Schiff's base before the second reduction step. The first two reactions in proline biosynthesis also resemble the initial steps of arginine biosynthesis, in which N-acetylglutamate is converted via N-acetyl γ-glutamyl phosphate to N-acetylglutamate semialdehyde (Chapter 4). The function of the acetyl group in N-acetyl glutamate semialdehyde might be to block the cyclization of glutamate semialdehyde, and thus allow arginine biosynthesis.

Proline synthesis from glutamate has also been demonstrated in a variety of animal tissues and cell lines, including HeLa and KB cells (Eagle et al., 1965), guinea pig (Ishibashi et al., 1968), chicken liver (Shen et al., 1973), human lung fibroblasts (Shen and Strecker, 1975), Chinese hamster ovary cells (Baich, 1977; Smith et al., 1980; Lodato et al., 1981), cartilage (Smith and Phang, 1978), and blowfly mitochondria (Wadano, 1980). Proline can also be synthesized from arginine through ornithine. In many cells, both glutamate and ornithine are precursors of proline. Ornithine appears to be the preferred precursor in chicken (Clemmons, 1973), kidney cells (Baich, 1977), lactating mammary glands (Mezl and Knox, 1977), and liver fluke (Goldberg et al., 1980).

In this chapter, we summarize the properties of the enzymes of proline biosynthesis, the organization of the proline biosynthetic genes, the regulation of proline biosynthesis, and mutations that result in proline overproduction. We emphasize what is known about the pathway in *E. coli* and *Salmonella typhimurium,* although some of the available information from other bacteria and higher organisms is included.

II. THE ENZYMES OF PROLINE BIOSYNTHESIS

A. Glutamate Kinase

The first unique reaction of proline biosynthesis is the ATP-dependent conversion of glutamate to γ-glutamyl phosphate. In the assay of the enzyme that catalyzes this reaction, the product, γ-glutamyl phosphate, is converted to γ-glutamyl hydroxamate (Baich, 1969), which is quantitated spectrophotometrically as the $FeCl_3$ complex (Elliot, 1955; Krishnaswamy et al., 1962). The assay is not specific for the first enzyme in proline biosynthesis, because a glutamate kinase reaction is also involved in glutamine synthesis (Krishnaswamy et al., 1962). In *Brevibacterium flavum,* two glutamate kinases have been identified and purified (Yoshinaga et al., 1966; 1967): one is inhibited by glutamine and the other is inhibited neither by glutamine nor by proline (Yoshinaga et al., 1975). *E. coli* also contains two glutamate kinase activities; one is an activity of glutamine synthetase, and the other is the proline-sensitive activity of the first enzyme of proline biosynthesis (Hayzer and Moses, 1978b). The specific assay for the proline biosynthetic glutamate kinase in *E. coli* cell extracts exploits its proline sensitivity (Baich, 1969; Hayzer and Leisinger, 1980).

Like N-acetylglutamyl phosphate of arginine biosynthesis (Chapter 4), γ-glutamyl phosphate appears to be always enzyme bound (Baich, 1969) to a complex containing the first two enzymes of proline biosynthesis (Gamper and Moses, 1974; Hayzer and Moses, 1978a). The molecular weight of this complex is 350,000 (Hayzer and Moses, 1978b).

A partially purified preparation of glutamate kinase from *E. coli* is 80% inhibited by 1 mM proline (Baich, 1969). Fructose-6-phosphate, glucose-6-phosphate and AMP also inhibit the enzyme (Baich, 1970). The significance of this inhibition remains unclear. A more highly purified preparation of glutamate kinase from *Pseudomonas aeruginosa* has an apparent molecular weight of 84,000 and is 50% inhibited by 5 mM proline (Krishna and Leisinger, 1979).

B. Glutamate Semialdehyde Dehydrogenase

The second enzyme of proline biosynthesis, glutamate semialdehyde dehydrogenase (Baich, 1971; Hayzer and Moses, 1978b), is assayed spectrophotometrically at 340 nm by measuring the phosphate-dependent reduction of NADP by glutamate semialdehyde. The D,L-isomer of glutamate semialdehyde is prepared from D,L-allo-δ-hydroxylysine by $NaIO_4$ oxidation (Williams and Frank, 1975; Wu and Seifter, 1975).

Two problems limit the utility of the assay. First, cell extracts of *E. coli* contain a phosphate-independent, NADP specific glutamate semialdehyde dehydrogenase whose in vivo function is obscure (Hayzer and Leisinger, 1982). Second, glutamate semialdehyde equilibrates rapidly and spontaneously with 1-pyrroline 5-carboxylate, the substrate of the terminal reaction in proline biosynthesis. Therefore, in assays of cell extracts the NADPH formed in the oxidation of glutamate semialdehyde is reoxidized in the reduction of 1-pyrroline 5-carboxylate to proline and no net NADPH is produced (Hayzer and Leisinger, 1980). This complication has been circumvented by separating the two competing activities by DEAE cellulose chromatography (Hayzer and Leisinger, 1980).

Glutamate semialdehyde dehydrogenase has been purified to near homogeneity from *E. coli* and *P. aeruginosa*. The *E. coli* enzyme, purified over 1200-fold, consists of four identical subunits of molecular weight 47,000 each (Hayzer and Leisinger, 1982). The substrate specificity of the pure *E. coli* enzyme is not reported, but in cell extracts, the apparent K_m values for glutamate semialdehyde/1-pyrroline 5-carboxylate, NADP and phosphate are 2.2, 0.18 and 11 mM, respectively, and the *E. coli* enzyme has an absolute specificity for NADP (Baich, 1971). The enzyme from *P. aeruginosa*, purified 150-fold, has a molecular weight of about 100,000 (Krishna et al., 1979). The *P. aeruginosa* enzyme utilizes either NADP or NAD as cofactors, with K_m values of 0.31 and 0.36 mM, respectively. The Km for glutamate semialdehyde/1-pyrroline 5-carboxylate is 4–8 mM, and the Km for phosphate is 0.28 mM.

C. 1-Pyrroline 5-Carboxylate Reductase

The last step in proline biosynthesis is the NADPH-linked reduction of 1-pyrroline 5-carboxylate to proline, catalyzed by 1-pyrroline 5-carboxylate reductase. The enzyme is assayed by following the oxidation of NADPH. The *E. coli* enzyme, purified 200-fold, has a molecular weight of 320,000 (Rossi et al., 1977b) and a K_m of 0.15 mM for 1-pyrroline 5-carboxylate. Although the enzyme can use NADH in vitro (K_m = 0.23 mM), in vivo it probably uses NADPH (K_m = 0.03 mM) exclusively (Csonka and Fraenkel, 1977). The enzyme is inhibited by proline concentrations in excess of 15 mM (Rossi et al., 1977b). The in vivo importance of this phenomenon is questionable.

The enzyme from *P. aeruginosa,* purified 56-fold, has a molecular weight of 95,000, and K_m values for 1-pyrroline 5-carboxylate, NADPH and NADH of 0.01, 0.2 and 0.05 mM, respectively (Krishna et al., 1979). Thus, in contrast to the *E. coli* enzyme, that from *P. aeruginosa* shows less preference for NADPH than for NADH. A 60-fold purified 1-pyrroline 5-carboxylate reductase from *Clostridium sporogenes* has a specificity for NADH (Costilow and Cooper, 1978). It is not clear whether this complete specificity for NADH is a unique feature of biosynthesis in this obligate anaerobe or whether the enzyme normally functions as part of proline catabolism only. A partially purified 1-pyrroline 5-carboxylate reductase from *Neurospora crassa* is about 16 times more active with NADPH than with NADH (Yura and Vogel, 1955; 1959; Meister et al., 1957).

D. Prolyl tRNA Synthetase

A final reaction associated with the ultimate function of proline as a component of proteins is its charging to prolyl tRNA. Prolyl tRNA synthetase of *E. coli* B has been purified 200-fold (Lee and Muench, 1969), and its substrate specificity (Papas and Mehler, 1970) and kinetic parameters (Papas and Mehler, 1971) examined in detail. The enzyme is a dimer of subunit molecular weight 47,000 (Lee and Muench, 1969). Only the dimeric form is active. The enzyme is cold-labile due to a dissociation into the subunits at low temperatures.

III. CONTROL OF PROLINE BIOSYNTHESIS

A. Genes

In *E. coli,* the genes encoding glutamate kinase, glutamate semialdehyde dehydrogenase and 1-pyrroline 5-carboxylate reductase are *proB, proA* and *proC,* respectively (Strecker, 1957; Baich and Pierson, 1965; Rossi et al., 1977b; Hayzer and Leisinger, 1980). The *proB* and *proA* genes are located at minute 6 and *proC* at minute 9 of the *E. coli* linkage map (Bachmann and Low, 1980). A positive selection for *E. coli proB* or *proA* mutants is based on the increased resistance of such mutants to 4-nitropyridine-1-oxide (Inuzuka et al., 1976a,b). The biochemical basis for the enhanced resistance is unknown. The *proBA* cluster of *E. coli* has been cloned into the multicopy plasmid pBR322. Restriction endonuclease mapping indicates that the two genes are present on a 4.5 Kb DNA fragment (Mahan et al., 1982), which also carries *phoE,* an unrelated gene, encoding an outer membrane pore protein (Tommassen et al., 1982; M. J. Mahan and L. N. Csonka, in preparation).

In *S. typhimurium,* an analogous set of genes required for proline biosynthesis has been identified (Miyake and Demerec, 1960; Itikawa and Demerec, 1968). The structural gene for 1-pyrroline 5-carboxylate reductase is *proC,* but it is yet unknown which enzymes are encoded by the other two genes (Sanderson and Hartman, 1978). In *S. typhimurium,* Tn*10* insertion mutations in *proB* are polar for the expression of *proA,* indicating that the two genes constitute an operon whose direction of transcription is from *proB* to *proA* (Mahan and Csonka, in preparation). Some proline auxotrophs of *Klebsiella pneumoniae* (S. G. Kustu, in preparation) can be restored to proline prototrophy by the introduction of an *E. coli* plasmid carrying the *proBA* cluster. This result implies a deficiency in glutamate kinase or glutamate semialdehyde dehydrogenase in these mutants (LeRudulier et al., 1982; S. S. Yang and L. N. Csonka, in preparation). Mutants lacking all three enzymes of the proline pathway have been obtained in *P. aeruginosa* (Krishna and Leisinger, 1979; Krishna et al., 1979); the genes *proA, proB* and *proC,* named in the order of the three reactions, are unlinked (Holloway et al., 1979). Proline auxotrophs of yeast fall into three complementation groups corresponding to the three biosynthetic enzymes (Brandriss, 1979).

In *E. coli,* a temperature-sensitive mutant in *proS,* the structural gene for prolyl tRNA synthetase, has been isolated and the mutation localized at minute 5 of the bacterial chromosome (Bohmann and Isaakson, 1980).

B. Genetics of Proline Regulation

Our understanding of the control of expression of the *pro* genes is much less complete than with most of the other amino acids. One report suggests that the synthesis of glutamate kinase in *E. coli* is negatively regulated (Tristram and Thurston, 1966), and another that 1-pyrroline 5-carboxylate dehydrogenase is subject to positive control (Gaudie and Tristram, 1970). However, neither conclusion seems to be substantiated by other studies which indicate invariant levels of all three enzymes, regardless of the presence or absence of proline in the growth medium (Baich and Pierson, 1965; Baich, 1971; Condamine, 1971; Rossi et al., 1977b; Hayzer and Leisinger, 1980).

The specific activity of glutamate semialdehyde dehydrogenase in strains carrying the *proBA* cluster on a multicopy plasmid is 20-fold higher than that of wild type strains. The specific activity of glutamate kinase is increased only three-fold (Hayzer and Leisinger, 1980).

C. Feedback Inhibition of Glutamate Kinase

The major regulatory point of proline biosynthesis in *E. coli* appears to be at the level of the catalytic activity of the first enzyme in the pathway. The proline-sensitivity of the glutamate kinase was first noted in *proC* mutants which excrete 1-pyrroline 5-carboxylate; the excretion is inhibited by addition of proline (Strecker, 1957; Baich and Pierson, 1965) or azetidine 2-carboxylate (Baich and Smith, 1968). Subsequently, it has been demonstrated in vitro that glutamate kinase is inhibited by proline (Baich, 1969; Hayzer and Leisinger, 1980).

IV. PROLINE OVERPRODUCING MUTANTS

Proline overproducing strains occur among mutants resistant to toxic analogs of proline (Tristram and Thurston, 1966; Baich, 1969; Condamine, 1971; Ratzkin et al., 1978; Csonka, 1981). Mutants with higher than wild type levels of proline are analog resistant because proline reverses toxicity. The structures of some proline analogs used for such selections are shown in Figure 3.2. Azetidine 2-carboxylate not only inhibits glutamate kinase (Baich and Smith, 1968), but it is also a substrate for prolyl tRNA synthetase (Papas and Mehler, 1970); its incorporation into proteins causes aberrant structures (Fowden and Richmond, 1963; Grant et al., 1975). 3,4-Dehydro-

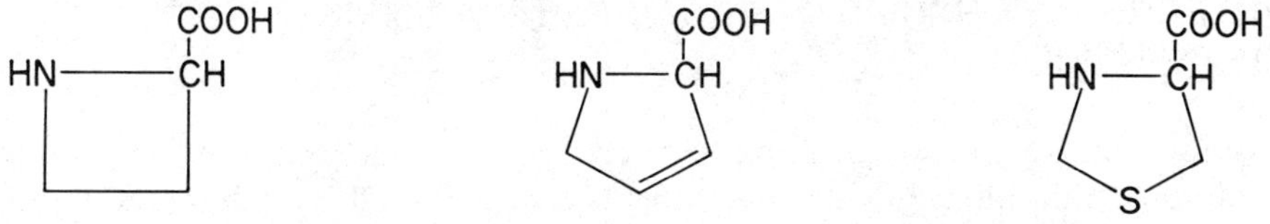

I II III

FIGURE 3.2 Proline analogs. I, azetidine 2-carboxylate; II, 3,4-dehydroproline; III, 4-thiazolidine carboxylate.

proline is also an inhibitor of glutamate kinase (Tristram and Thurston, 1966).

One problem in obtaining proline overproducing strains by selecting analog resistant mutants raises from the fact that these antimetabolites are taken up by various transport systems which normally function in the uptake of proline (Ratzkin et al., 1978; Anderson et al., 1980; Csonka, 1982). Most *S. typhimurium* and *E. coli* analog resistant mutants result from an inactivation of the major proline transport system, which is required for proline uptake when the amino acid serves as a sole carbon or nitrogen source (Ratzkin et al., 1978; Wood, 1981; Csonka, 1981). Mutations which result simultaneously in analog resistance and proline overproduction are closely linked to the *proBA* cluster (Condamine, 1971), a characteristic that distinguishes them from mutations in the transport system. This observation is the basis for a special procedure to select such proline overproducing mutants (Csonka, 1981). The starting strain for the selection is a mutant of *S. typhimurium* deleted for the chromosomal *proBA* cluster, but harboring *proBA* on an *E. coli* plasmid. Mutants resistant to azetidine 2-carboxylate or 3,4-dehydroproline are selected, then mated with another *S. typhimurium proBA* deletion. By simultaneously selecting for proline prototrophy and analog resistance, mutations linked to *proBA* genes carried on the plasmid are obtained exclusively.

Since glutamate kinase is inhibited by proline and its analogs, a likely mutation leading to proline overproduction would be one which desensitizes the enzyme to feedback inhibition. This assumption is supported by experiments with *proC* mutants that also carry mutations that render glutamate kinase insensitive to feedback inhibition. The excretion of 1-pyrroline 5-carboxylate from such cells is unaffected by azetidine 2-carboxylate, 3,4-dehydroproline or proline in the medium (Baich and Pierson, 1965; Tristram and Thurston, 1966; Condamine, 1971). Glutamate kinase of such analog resistant mutants is insensitive to proline inhibition (Baich, 1969). In strains heterozygous for the *proBA* region, mutations to proline overproduction and analog resistance are dominant; complementation tests with such diploid strains indicate that the mutations are in *proB* (Condamine, 1971).

V. INTERRELATIONSHIPS OF PROLINE AND ARGININE BIOSYNTHESIS

In some organisms proline can be derived from exogenous ornithine, which is an intermediate in arginine biosynthesis and degradation (Chapters 4 and 5). The conversion of ornithine to proline can occur by two metabolic routes: in one the δ-amino group of ornithine is lost, and in the other the α-amino group is lost (Fig. 3.3).

A. Ornithine δ-Aminotransferase

The conversion of ornithine to glutamate semialdehyde by transamination removing the δ-amino group is catalyzed by ornithine δ-aminotransferase, which is found in a number of Gram positive bacteria (Scher and Vogel, 1957), fungi (Vogel, 1953; Vogel and Bonner, 1954; Vogel and Kopac, 1960;

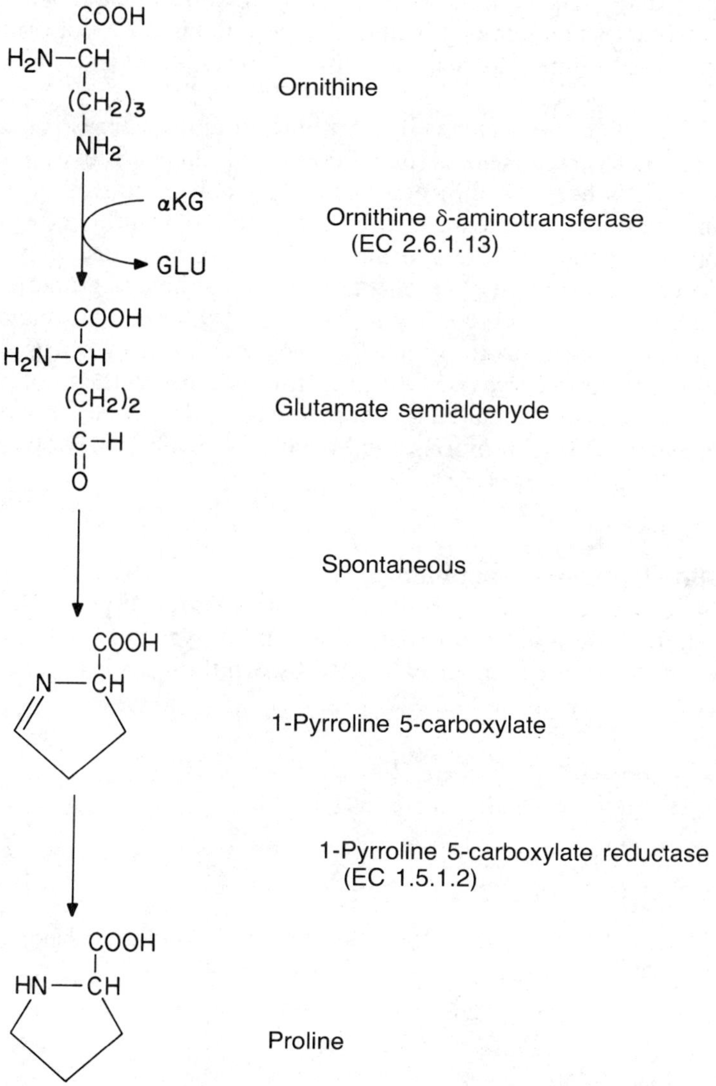

FIGURE 3.3(a) Conversion of ornithine to proline in fungi.

Middelhoven, 1968; Piotrowska et al., 1969) and animal tissues (Quastel and Witty, 1951). The purified enzyme from *N. crassa* has a K_m of 2.0 mM for ornithine and 1.7 mM for α-ketoglutarate (Vogel and Kopac, 1960). Ornithine δ-aminotransferase purified from *Klebsiella aerogenes* (Friedrich and Magasanik, 1978) has a higher affinity for N-acetylornithine than for ornithine (Friedrich et al., 1978). Canaline, a breakdown product of canavanine, is an irreversible inhibitor of ornithine δ-aminotransferase (Rahiala et al., 1971; Rosenthal, 1978; Kito et al., 1978). Gabaculine and 4-aminohex-5-yonate also act as enzyme-activated irreversible inhibitors of ornithine δ-aminotransferase (Jung and Seiler, 1978).

Although exogenously provided ornithine can be readily converted to proline in a number of organisms by ornithine aminotransferase or ornithine cyclase (Section V.B.), it is not certain if ornithine is an obligatory intermediate in proline biosynthesis of any organism. The conversion of ornithine to proline may be the consequence of the degradation of ornithine to glutamate semialdehyde (Chapter 5). Thus, ornithine δ-aminotransferase may be more properly regarded as an enzyme of arginine or ornithine degradation rather than proline biosynthesis. Accordingly, mutants of *N. crassa* lacking ornithine δ-aminotransferase are prototrophs, but they cannot grow on ornithine as the sole nitrogen source (Davis, 1968a,b). In yeast, the synthesis of ornithine δ-aminotransferase is stimulated by exogenous arginine (Middelhoven, 1967; Deschamps et al., 1979; Brandriss and Magasanik, 1980; 1981) or by nitrogen starvation (Middelhoven, 1968).

B. Ornithine Cyclase (Deaminating)

Clostridia convert ornithine to proline in a one-step reaction which involves the loss of the α-amino group (Costilow and Laycock, 1969; Muth and Costilow, 1974). The reaction is catalyzed by ornithine cyclase. The enzyme

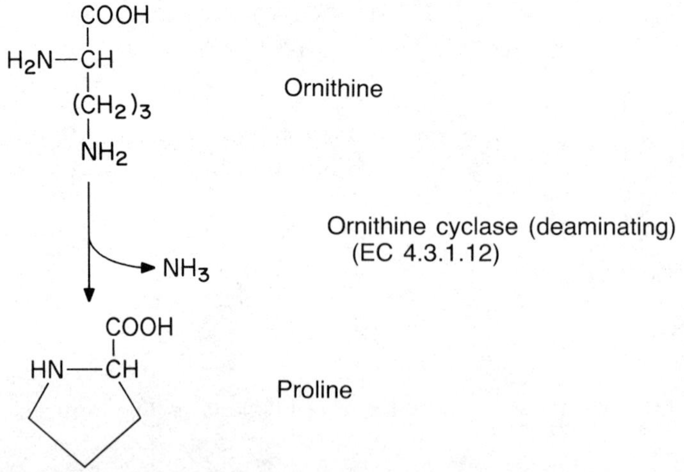

FIGURE 3.3(b) Conversion of ornithine to proline in *Clostridia*.

uses NAD to oxidize ornithine during the removal of the α-amino group and the NADH formed is used to reduce the enzyme-bound intermediate to proline (Muth and Costilow, 1974). The three higher plants *Nicotiana tabacum, Datura stramonium,* and *Lupinus angustifolius* convert ornithine to proline by loss of the α-hydrogen, hence by inference, loss of the α-amino group. The reaction presumably proceeds in three distinct steps: transamination to α-keto-δ-aminovalerate, spontaneous cyclization analogous to glutamate semialdehyde cyclization, and reduction of 1-pyrroline 2-carboxylate to proline (Fig. 3.3c). However, the plant data are also consistent with the single step mechanism proposed for *Clostridia*.

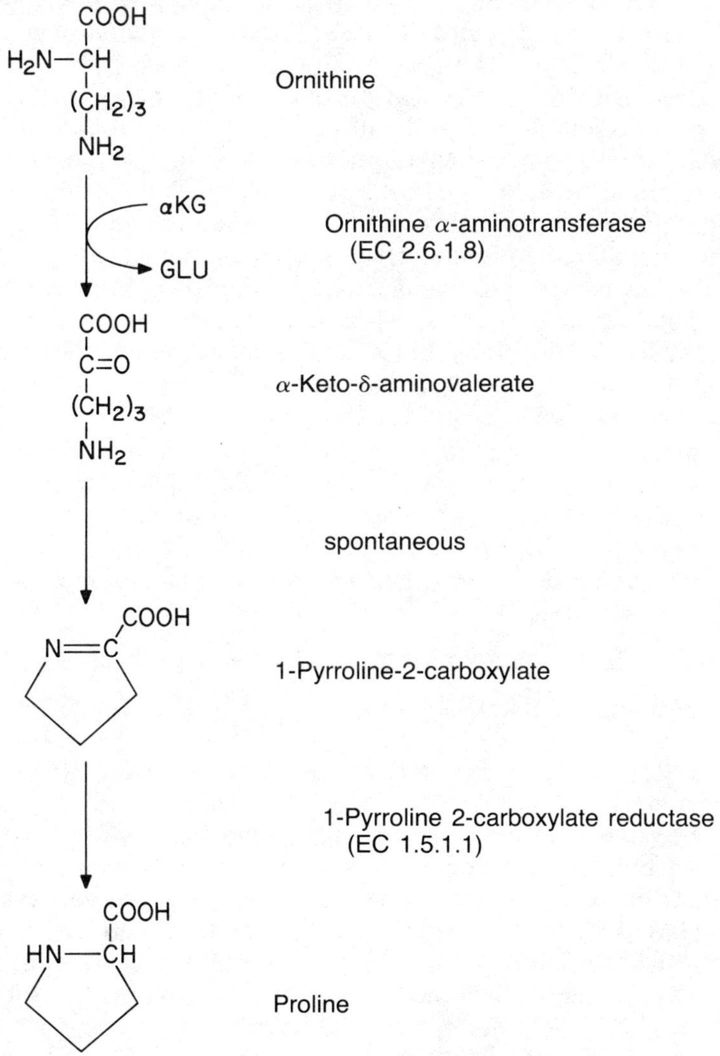

FIGURE 3.3(c) Conversion of ornithine to proline in higher plants.

C. Suppression of Mutations in the Proline Pathway by Mutations in the Arginine Pathway

The parallels between the initial steps of proline and arginine biosynthesis provide the basis for an interesting class of proline overproducing mutants which, at the same time, suppress mutations in the first two steps of proline biosynthesis in *E. coli* and *S. typhimurium* (Itikawa et al., 1968; Kuo and Stocker, 1969; Berg and Rossi, 1974; Rossi et al., 1977a). The fourth step of arginine biosynthesis is the conversion of N-acetylglutamate semialdehyde to N-acetylornithine by N-acetylornithine aminotransferase encoded in *argD* (Chapter 4). The fifth step in the pathway is the conversion of N-acetylornithine to ornithine by N-acetylornithinase (ornithine acetyltransferase). In *argD* mutants, N-acetylglutamate semialdehyde accumulates sufficiently to act as a substrate for N-acetylornithinase, thereby undergoing conversion to glutamate semialdehyde. Thus *argD* mutations suppress *proB* or *proA* mutations. Although *argD* mutants lack one arginine biosynthetic enzyme, they grow at a reduced rate without arginine, because a number of other non-specific transaminases can offset the defect so as to sustain cellular arginine requirements. In *argD proB/A* double mutants, proline synthesis proceeds via the first three enzymes of arginine synthesis, which are not subject to regulation by proline. Therefore, *argD* mutants are proline overproducers: they excrete proline and are resistant to azetidine 2-carboxylate and 3,4-dehydroproline (Berg and Rossi, 1974; Rossi et al., 1977a; Wood, 1981). However, the double mutants cannot grow in the presence of arginine, since arginine inhibits the synthesis of N-acetylglutamate semialdehyde (Baich and Vogel, 1962; Cybis and Davis, 1974).

The main requirement for suppression of *proB/A* mutations through mutations in the arginine pathway is that N-acetylglutamate semialdehyde be converted to glutamate semialdehyde. Therefore, perhaps other classes of mutants may be obtained, wherein suppression is due to an increase in the specific activity of N-acetylornithinase or an increase in its affinity for N-acetylglutamate semialdehyde (J. L. Ingraham, personal communication).

VI. PROLINE DEGRADATION

Proline is degraded by a two-step oxidation via 1-pyrroline 5-carboxylate to glutamate (Frank and Ranhand, 1964). The first reaction is coupled to electron transport and uses oxygen as the ultimate electron acceptor; the second uses NAD as cofactor. In *E. coli* and *S. typhimurium* both reactions are catalyzed by the bifunctional enzyme proline dehydrogenase-pyrroline 5-carboxylate dehydrogenase. The enzyme has been purified from *E. coli* (Scarpulla and Soffer, 1978) and *S. typhimurium* (Menzel and Roth, 1981a,b). The *Salmonella* enzyme is a membrane-bound flavoprotein of subunit molecular weight 130,000 with a K_m of 50-80 mM for proline, 11 mM for 1-pyrroline 5-carboxylate and 0.1 mM for NAD.

The structural gene for the bifunctional enzyme is *putA* (Ratzkin and Roth, 1978; Wood, 1981). Most *putA* mutants defective in proline dehydrogenase have also lost pyrroline 5-carboxylate dehydrogenase activity. Proline catabolism requires in addition to a functional *putA* product a specific proline permease encoded in *putP* (Ratzkin et al., 1978; Ratzkin and Roth, 1978). Mutations in *putA* or *putP* result in inability to use proline as the sole source of nitrogen or carbon.

Proline coordinately induces *putA* and *putP* in *S. typhimurium* (Dendinger and Brill, 1970; Ratzkin et al., 1978), *E. coli* (Condamine, 1971; Wood, 1981) and *K. aerogenes* (Prival and Magasanik, 1971). Presumably, *putA* is autogenously regulated, and the *putA* product is also assumed to be the repressor for *putP*, since some mutations conferring constitutivity on *putA* and *putP* are located in *putA* (Menzel and Roth, 1981c). In the three enterics, expression of *putA* and *putP* is also subject to catabolite repression. In *K. aerogenes,* but not in the other two species, the proline degradative enzymes are also subject to nitrogen control; the enzymes are induced in ammonia deficient media even under conditions of catabolite repression (Prival and Magasanik, 1971). Although the proline degradative enzymes are repressed by glucose in media devoid of proline, there seems to be some turnover of the intracellular proline, because proline is approximately five-fold increased in exponentially growing *putA* as compared to wild type cells (L. N. Csonka, in preparation).

The regulation of the enzymes of proline degradation is more intricate in those organisms which catabolize arginine via ornithine and glutamate semialdehyde/pyrroline 5-carboxylate. *Bacillus subtilis* has two distinct pyrroline 5-carboxylate dehydrogenases, one induced by proline and another induced by arginine (DeHauwer et al., 1964). In yeast, where proline is an intermediate in arginine degradation, pyrroline 5-carboxylate dehydrogenase is induced by proline but not by arginine (Brandriss and Magasanik, 1980; 1981).

VII. PROLINE OVERPRODUCTION AND OSMOTIC TOLERANCE IN BACTERIA

In a wide variety of organisms, from bacteria (Tempest et al., 1970; Brown and Stanley, 1972; Measures, 1975) to plants (Storey and Wyn-Jones, 1977; Tully et al., 1979; Dix and Pierce, 1981) and some marine animals (Fyhn, 1976) the cytoplasmic proline concentrations increase in response to environmental osmotic stress. That the increased proline levels confer enhanced osmotic stress tolerance has been suggested, because proline specifically stimulates growth of *Enterobacteriaceae* in media of inhibitory osmolarity (Christian, 1955a,b).

This observation was used as the basis for the isolation of osmotolerant mutants of *S. typhimurium* among proline overproducing mutants constructed as described in Section IV (Csonka, 1981). In a number of these mutants, increases in osmotolerance were correlated with the intracellular proline levels. It is not clear why elevated intracellular proline levels confer increased osmotolerance. Perhaps proline acts as an inert osmotic balancer which prevents dehydration of cells (Flowers et al., 1977). Alternatively, proline may interact with proteins and membranes and promote maintenance of their conformation in solution of high osmolarity (Schobert, 1977).

An unexpected property of proline overproducing, osmotolerant mutants is that the intracellular proline levels respond to the osmotic strength of the medium; mutants produce increasing amounts of proline as the osmotic strength of the medium increases (Csonka, 1981; LeRudulier et al., 1982). In a mutant of very pronounced osmotolerant phenotype, free proline production rose from 43 μmoles to 1.3 mmoles per g of cellular protein, when 0.6 M NaCl was added to cells growing in a minimal medium (L. N. Csonka, in preparation).

Wild type *S. typhimurium,* unlike its proline overproducing derivatives, does not increase the intracellular proline levels upon exposure to osmotic stress (Csonka, 1981). However, like other species of bacteria (Tempest et al., 1970; Brown and Stanley, 1972; Munro et al., 1972; Makemson and Hastings, 1979), wild-type as well as the proline overproducing mutants of *S. typhimurium* increase the levels of glutamate and glutamine in response to osmotic stress (Csonka, 1981). Normal regulation of proline synthesis is abolished in proline overproducing mutants, probably due to loss of feedback inhibition of glutamate kinase. The mechanism responsible for the increased production of proline as a consequence of growth in media of elevated osmotic strength is not clear.

ACKNOWLEDGMENTS

We thank S. G. Kustu and C. Lindquist for valuable discussions and comments on the manuscript.

REFERENCES

Anderson, R. R., Menzel, R. R., and Wood, J. M. (1980) *J. Bacteriol. 141,* 1071–1076.
Bachmann, B. J., and Low, K. B. (1980) *Microbiol. Rev. 44,* 1–56.
Baich, A. (1969) *Biochim. Biophys. Acta 192,* 462–467.
Baich, A. (1970) *Biochem. Biophys. Res. Comm. 39,* 544–550.
Baich, A. (1971) *Biochim. Biophys. Acta 244,* 129–134.
Baich, A. (1977) *Som. Cell Gen. 3,* 529–538.
Baich, A., and Pierson, D. J. (1965) *Biochim. Biophys. Acta 104,* 397–404.
Baich, A., and Smith, F. I. (1968) *Experientia 24,* 1107.
Baich, A., and Vogel, H. J. (1962) *Biochem. Biophys. Res. Comm. 7,* 491–496.

Berg, C. M., and Rossi, J. J. (1974) *J. Bacteriol. 118,* 928–939.
Bohmann, K., and Isaakson, L. A. (1980) *Mol. Gen. Genet. 177,* 603–605.
Brandriss, M. C. (1979) *J. Bacteriol. 138,* 816–822.
Brandriss, M. C., and Magasanik, B. (1980) *J. Bacteriol. 143,* 1403–1410.
Brandriss, M. C., and Magasanik, B. (1981) *J. Bacteriol. 145,* 1359–1364.
Brown, C. M., and Stanley, S. O. (1972) *J. Appl. Chem. Biotechnol. 22,* 363–389.
Christian, J. H. B. (1955a) *Aust. J. Biol. Sci. 8,* 75–82.
Christian, J. H. B. (1955b) *Aust. J. Biol. Sci. 8,* 490–497.
Clemmons, J. J. (1973) *Comp. Biochem. Physiol. 44B,* 259–267.
Condamine, H. (1971) *Ann. Inst. Pasteur, Paris 120,* 126–143.
Costilow, R. N., and Cooper, D. (1978) *J. Bacteriol. 134,* 139–146.
Costilow, R. N., and Laycock, L. (1969) *J. Bacteriol. 100,* 662–667.
Csonka, L. N. (1981) *Molec. Gen. Genetics 182,* 82–86.
Csonka, L. N. (1982) *J. Bacteriol. 151,* 1433–1443.
Csonka, L. N., and Fraenkel, D. G. (1977) *J. Biol. Chem. 252,* 3382–3391.
Cybis, J. J., and Davis, R. H. (1974) *Biochem. Biophys, Res. Comm. 60,* 629–634.
Davis, R. H. (1968a) *J. Bacteriol. 96,* 383–388.
Davis, R. H. (1968b) *J. Bacteriol. 96,* 389–395.
de Hauwer, G., Lavalle, R., and Wiame, J. M. (1964) *Biochim. Biophys. Acta 81,* 257–269.
Dendinger, S. and Brill, W. J. (1970) *J. Bacteriol. 103,* 144–152.
Deschamps, J., Dubois, E., and Wiame, J. M. (1979) *Molec. Gen. Genet. 174,* 225–232.
Dix, P. T., and Pierce, R. S. (1981) *Z. Pflanzenphysiol. 102,* 243–248.
Eagle, H., Washington, C. L., and Levy, M. (1965) *J. Biol. Chem. 240,* 3944–3950.
Elliott, W. H. (1955) in *Methods in Enzymology,* Vol. II (Colowick, S. P., and Kaplan, N. O., eds.) pp. 337–342, Academic Press, New York.
Flowers, T. J., Troke, P. F., and Yeo, A. R. (1977) *Ann. Rev. Plant Phys. 28,* 89–121.
Fowden, L. and Richmond, M. H. (1963) *Biochim. Biophys. Acta 71,* 459–461.
Frank, L., and Ranhand, B. (1964) *Arch. Biochem. Biophys. 107,* 325–331.
Friedrich, B., and Magasanik, B. (1978) *J. Bacteriol. 133,* 680–685.
Friedrich, B., Friedrich, C. G., and Magasanik, B. (1978) *J. Bacteriol. 133,* 686–691.
Fyhn, H. J. (1976) *Comp. Bioch. Physiol. 53A,* 19–30.
Gamper, H., and Moses, V. (1974) *Biochim. Biophys. Acta 354,* 75–87.
Gaudie, D., and Tristram, H. (1970) *J. Gen. Micro. 63,* XIII–XIV.
Goldberg, M., Flescher, E., Gold, D., and Lengy, J. (1980) *Comp. Biochem. Physiol. 65B,* 605–613.
Grant, M. M., Brown, A. S., Corwin, L. M., Troxler, R. F., and Franzblau, C. (1975) *Biochim. Biophys. Acta 404,* 180–187.
Hayzer, D. J., and Leisinger, T. (1980) *J. Gen. Microbiol. 118,* 287–293.
Hayzer, D. J., and Leisinger, T. (1981) *Biochem. J. 197,* 269–274.
Hayzer, D. J., and Leisinger, T. (1982) *Eur. J. Biochem. 121,* 561–565.
Hayzer, D. J., and Moses, V. (1978a) *Biochem. J. 173,* 207–217.
Hayzer, D. J., and Moses, V. (1978b) *Biochem. J. 173,* 219–228.
Holloway, B. W., Krishnapillai, V., and Morgan, A. F. (1979) *Microbiol. Rev. 43,* 73–102.
Inuzuka, M., Miyano, H., and Tomeda, M. (1976a) *Antimicrob. Ag. Chemotherapy 10,* 325–332.

Inuzuka, M., Toyama, H., Miyano, H., and Tomeda, M. (1976b) *Antimicrob. Ag. Chemotherapy 10,* 333–343.
Ishibashi, S., Ide, T., and Tsurufuji, S. (1968) *Biochim. Biophys. Acta 165,* 296–299.
Itikawa, H., and Demerec, M. (1968) *J. Bacteriol. 95,* 1189–1190.
Itikawa, H., Baumberg, S., and Vogel, H. J. (1968) *Biochim. Biophys. Acta 159,* 547–550.
Jung, M. J., and Seiler, N. (1978) *J. Biol. Chem. 253,* 7431–7439.
Katchalsky, A., and Paecht, M. (1954) *J. Am. Chem. Soc. 76,* 6042–6044.
Kito, K., Sanada, Y., and Katunuma, N. (1978) *J. Biochem. 83,* 201–206.
Krishna, R. V., Leisinger, T. (1979) *Biochem. J. 181,* 215–222.
Krishna, R. V., Beilstein, P., and Leisinger, T. (1979) *Biochem. J. 181,* 223–230.
Krishnaswamy, P. R., Pamiljans, V., and Meister, A (1962) *J. Biol. Chem. 237,* 2932–2940.
Kuo, T. T., and Stocker, B. A. D. (1969) *J. Bacteriol. 98,* 593–598.
Lee, M.-L., and Muench, K. H. (1969) *J. Biol. Chem. 244,* 223–230.
LeRudulier, D., Yang, S. S., and Csonka, L. N. (1982) *Biochim. Biophys. Acta,* (in press).
Lodato, R. F., Smith, R. J., Vallee, D., Phang, J. M., and Aoki, T. T. (1981) *Metabolism 30,* 908–913.
Mahan, M. J., Csonka, L. N., Valentine, R. C. (1982) *Abstracts of the Annual Meeting, American Society of Microbiology,* p. 150. K85.
Makemson, J. C., and Hastings, J. W. (1979) *Appl. Env. Microbiol. 38,* 178–180.
Measures, J. C. (1975) *Nature 257,* 398–400.
Meister, A., Radhakrishnan, A. N., and Buckly, S. D. (1957) *J. Biol. Chem. 229,* 789–800.
Menzel, R., and Roth, J. (1981a) *J. Biol. Chem. 256,* 9755–9761.
Menzel, R., and Roth, J. (1981b) *J. Biol. Chem. 256,* 9762–9766.
Menzel, R., and Roth, J. (1981c) *J. Mol. Biol. 148,* 21–44.
Mestichelli, L. J. J., Gupta, R. N., and Spenser, I. D. (1979) *J. Biol. Chem. 254,* 640–647.
Mezl, V. A., and Knox, W. E. (1977) *Biochem. J. 166,* 105–113.
Middlehoven, W. J. (1964) *Biochim. Biophys. Acta 93,* 650–652.
Middlehoven, W. J. (1967) *Biochem. J. 106,* 32P.
Middlehoven, W. J. (1968) *Biochim. Biophys. Acta 156,* 440–443.
Miyake, T., and Demerec, M. (1960) *Genetics 45,* 755–762.
Munro, G. F., Hercules, K., Morgan, J., and Sauerbier, W. (1972) *J. Biol. Chem. 247,* 1272–1280.
Muth, W. L., and Costilow, R. N. (1974) *J. Biol. Chem. 249,* 7463–7467.
Papas, T. S., and Mehler, A. H. (1970) *J. Biol. Chem. 245,* 1588–1595.
Papas, T. S., and Mehler, A. H. (1971) *J. Biol. Chem. 246,* 5924–5928.
Piotrowska, M., Sawicki, M., and Weglenski, P. (1969) *J. Gen. Microbiol. 55,* 301–305.
Prival, M. J., and Magasanik, B. (1971) *J. Biol. Chem. 246,* 6288–6296.
Quastel, J. H., and Witty, R. (1951) *Nature 167,* 556.
Rahiala, E. L., Kekomaki, M., Janne, J., Raina, A., and Raiha, N. C. R. (1971) *Biochim. Biophys. Acta 227,* 337–343.
Ratzkin, B., Garbnar, M., and Roth, J. (1978) *J. Bacteriol. 133,* 737–743.
Rosenthal, G. A. (1978) *Life Sciences 23,* 93–98.

Rossi, J. J., Vender, J., and Berg, C. M. (1977a) *Biochem. Genetics 15,* 287–296.
Rossi, J. J., Vender, J., Berg, C. M., and Coleman, W. H. (1977b) *J. Bacteriol. 129,* 108–114.
Sanderson, K. E., and Hartman, P. E. (1978) *Microbiol. Rev. 42,* 471–519.
Scarpulla, R. C., and Soffer, R. L. (1978) *J. Biol. Chem. 253,* 5997–6001.
Scher, W. I., Jr., and Vogel, H. J. (1957) *Proc. Natl. Acad. Sci. U.S.A. 43,* 796–803.
Schobert, B. (1977) *J. Theor. Biol. 68,* 17–26.
Shen, T. F., and Strecker, H. J. (1975) *Biochem. J. 150,* 453–461.
Shen, T. F., Bird, H. R., and Sunde, M. L. (1973) *Poultry Sci. 52,* 676–682.
Smith, R. J., and Phang, J. M. (1978) *Metabol. Clin. Exp. 27,* 658–694.
Smith, R. J., Downing, S. J., and Phang, J. M. (1980) *Proc. Natl. Acad. Sci. U.S.A. 77,* 5221–5225.
Stetten, M. R., in McElroy, W. D., and Glass, B. (1955) *Amino Acid Metabolism,* 277–290. Johns Hopkins Press, Baltimore, Md.
Strecker, H. J. (1957) *J. Biol. Chem. 225,* 825–834.
Storey, R., and Wyn-Jones, R. G. (1977) *Phytochem. 16,* 447–453.
Tempest, D. W., Meers, J. L., and Brown, C. M. (1970) *J. Gen. Microbiol. 64,* 171–185.
Tommassen, J., Overduin, P., Lugtenberg, B., and Bergmans, H. (1982) *J. Bacteriol. 149,* 668–672.
Tristram, H., and Thurston, C. F. (1966) *Nature 212,* 74–75.
Tully, R. E., Hanson, A. D., and Nelson, C. E. (1979) *Plant Physiol. 63,* 518–523.
Vogel, H. J. (1953) *Proc. Nat. Acad. Sci. U.S.A. 39,* 578–583.
Vogel, H. J., and Davis, B. D. (1952) *J. Am. Chem. Soc. 74,* 109–112.
Vogel, H. J., and Bonner, D. M. (1954) *Proc. Nat. Acad. Sci. U.S.A. 40,* 688–694.
Vogel, R. H., and Kopac, M. J. (1960) *Biochim. Biophys. Acta 37,* 539–540.
Wadano, A. (1980) *Experientia 36,* 1028–1029.
Williams, I., and Frank, L. (1975) *Analy. Biochem. 64,* 85–97.
Wood, J. M. (1981) *J. Bacteriol. 146,* 895–901.
Wu, G. Y., and Seifter, S. (1975) *Analyt. Bioch. 67,* 413–421.
Yoshinaga, F., Konishi, S., Okumura, S., and Katsuya, N. (1966) *J. Gen. Appl. Microbiol. 12,* 219–228.
Yoshinaga, F., Yoshihára, Y., Okumura, S., and Katsuya, N. (1967) *J. Gen. Appl. Microbiol. 13,* 25–37.
Yoshinaga, F., Tsuchida, T., and Okumura, S. (1975) *Agr. Biol. Chem. 39,* 1269–1273.
Yura, T., and Vogel, H. J. (1955) *Biochim. Biophys. Acta 17,* 582.
Yura, T., and Vogel, H. J. (1959) *J. Biol. Chem. 234,* 335–338.

CHAPTER 4

Regulation of Arginine Biosynthesis in Prokaryotes

Raymond Cunin

I. INTRODUCTION

Arginine is synthesized in eight steps, starting from glutamate (Fig. 4.1). The first reaction of the pathway is the acetylation of the amino group of glutamate. Acetylation prevents the cyclization of glutamate semialdehyde, a proline precursor. Prokaryotes exploit two enzymatic strategies for the acetylation and the subsequent deacetylation of acetylornithine. Enterobacteriaceae and bacillaceae follow a linear pathway involving N-acetylglutamate synthase and acetylornithinase (Fig. 4.1). In pseudomonads, blue-green algae, photosynthetic bacteria, and other organisms such as *Thermus aquaticus,* the acetyl group is recycled by ornithine acetyltransferase (Fig. 4.1; Hoare and Hoare, 1966; Udaka, 1966; Degryse et al., 1976). This energetically more efficient process is also observed in fungi such as *Saccharomyces cerevisiae, Neurospora crassa,* and *Chlamydomonas reinhardii* (see Chapter 5). Thus acetylglutamate synthase performs an anaplerotic function by replenishing the acetylglutamate pool. This difference in the mechanism of glutamate acetylation is reflected in the targets for feedback inhibition by arginine. Only acetylglutamate synthase is feedback-inhibited

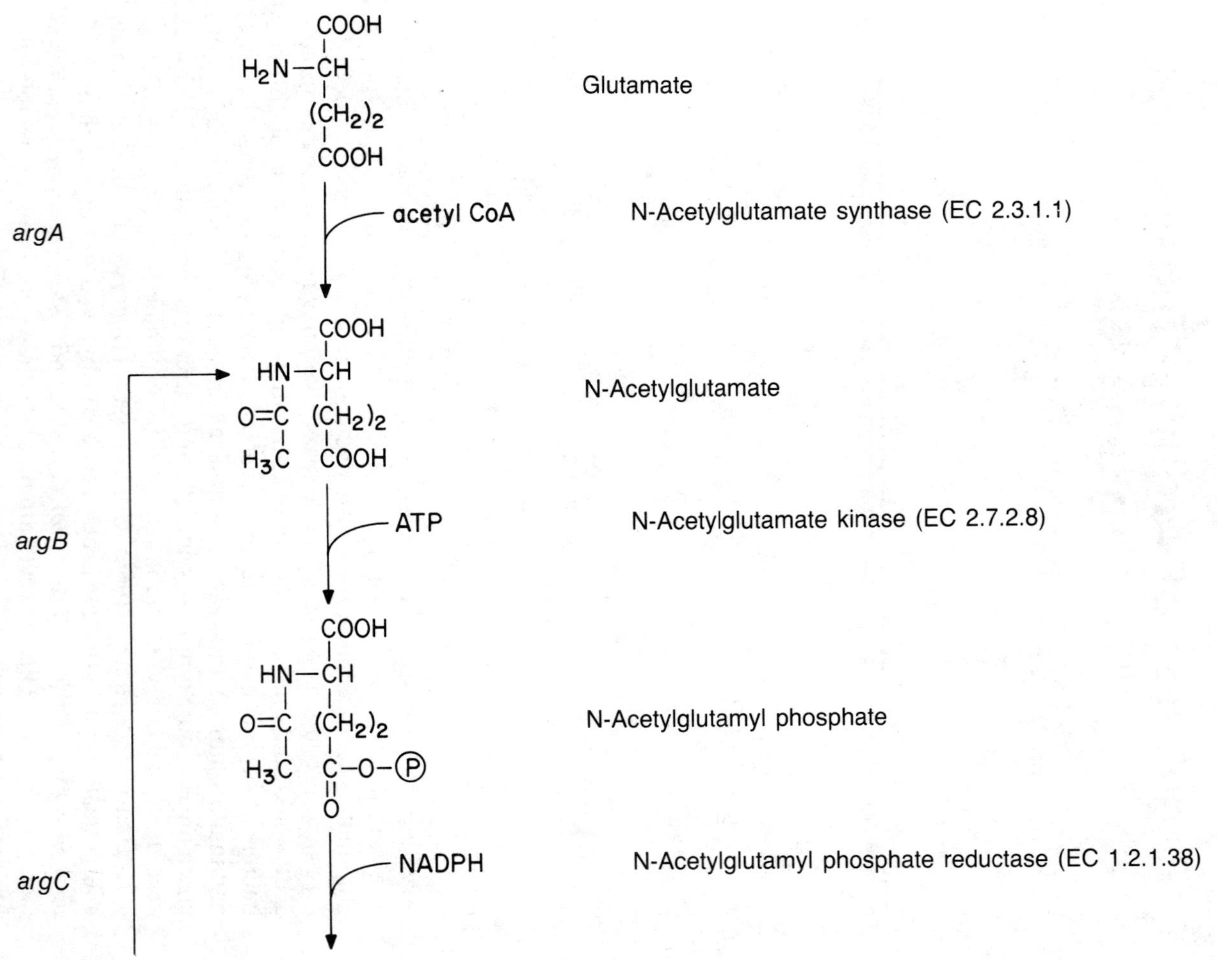
argA
argB
argC
COOH
H2N—CH
(CH2)2
COOH
acetyl CoA
HN—CH
O=C
H3C
ATP
C—O—P
O
NADPH
Glutamate
N-Acetylglutamate synthase (EC 2.3.1.1)
N-Acetylglutamate
N-Acetylglutamate kinase (EC 2.7.2.8)
N-Acetylglutamyl phosphate
N-Acetylglutamyl phosphate reductase (EC 1.2.1.38)

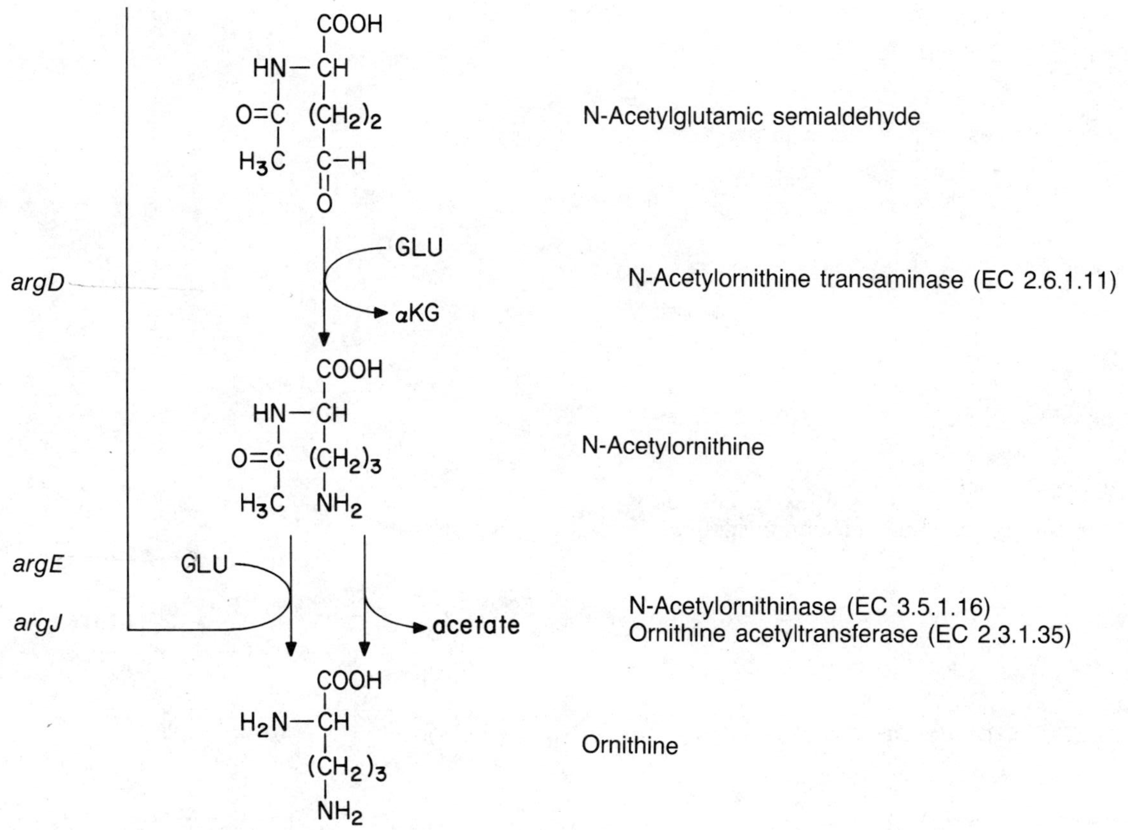

FIGURE 4.1(a) **The biosynthesis of ornithine from glutamate.**

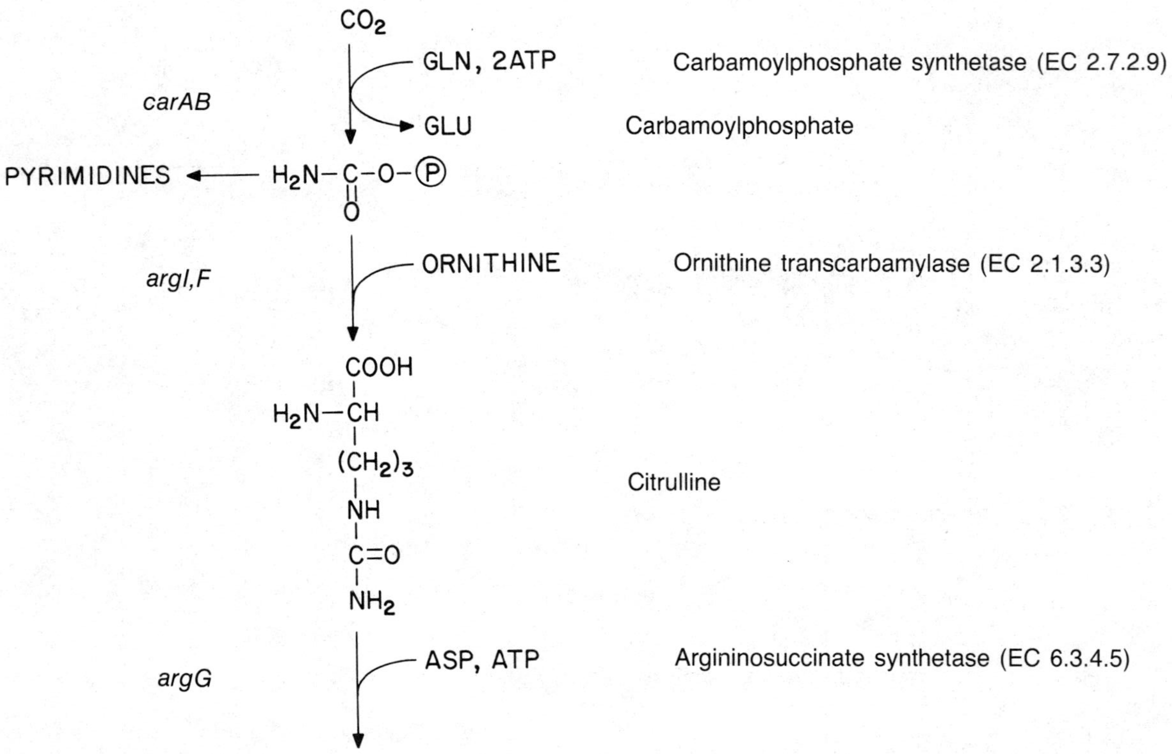
CO2
GLN, 2ATP
Carbamoylphosphate synthetase (EC 2.7.2.9)
carAB
GLU
Carbamoylphosphate
PYRIMIDINES
H2N-C-O-P
O
ORNITHINE
Ornithine transcarbamylase (EC 2.1.3.3)
argI,F
COOH
H2N-CH
(CH2)3
NH
C=O
NH2
Citrulline
ASP, ATP
Argininosuccinate synthetase (EC 6.3.4.5)
argG

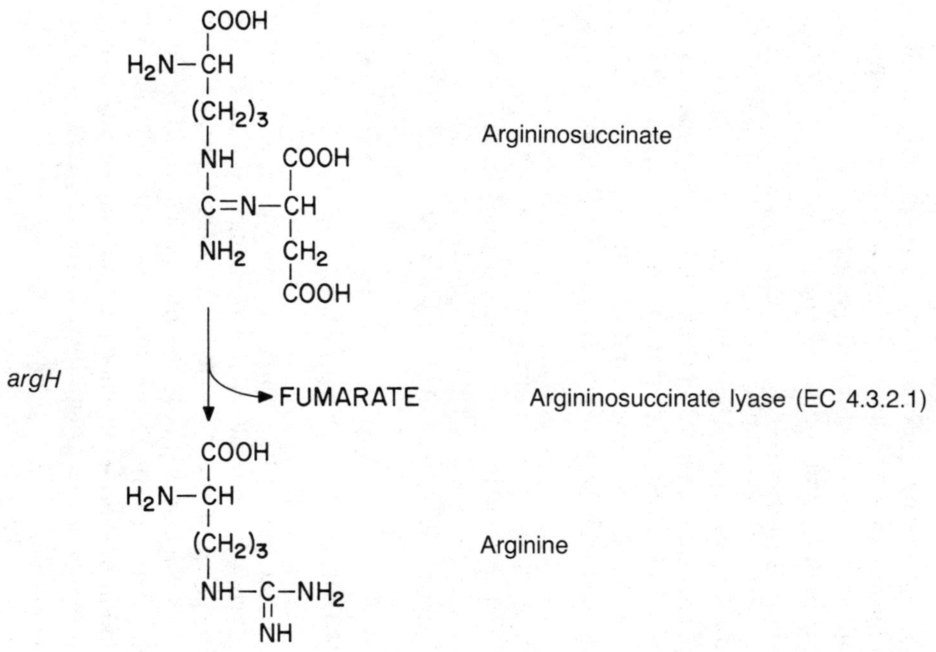

FIGURE 4.1(b) **The biosynthesis of arginine from ornithine.**

by arginine in enteric bacteria and bacilli, whereas the second enzyme of the pathway and sometimes both the first and second enzymes are feedback-inhibited by arginine in prokaryotes that recycle the acetyl group. In the latter organisms, it is not clear whether the enzymatic activity corresponding to reaction 5 is due to the presence of a vestigial biosynthetic acetylornithinase or to a carboxypeptidase, as seems to be the case in *S. cerevisiae* (Degryse, 1974).

Arginine biosynthesis shares one intermediate, carbamoylphosphate, with pyrimidine biosynthesis. In enterobacteriaceae, a single enzyme—carbamoylphosphate synthetase, provides for both pathways (Fig. 4.1). This situation has implications both for the control of enzyme activity (Section III.B) and for the regulation of enzyme synthesis (Section IV.B.3). *Bacillus subtilis* has two carbamoylphosphate synthetases, one for each pathway, but there is no strict channeling of the carbamoylphosphate that each synthesizes: strains deficient in one enzyme utilize carbamoylphosphate from the other pathway. This situation resembles that found in *S. cerevisiae* but not in other fungi such as *N. crassa,* where the two pools of carbamoylphosphate are effectively channeled (Lacroute et al., 1965; Bernhardt and Davis, 1972).

Ornithine and, to a lesser extent, arginine are also precursors for polyamine biosynthesis. In pseudomonads, spermine, spermidine, putrescine, and cadaverine participate in the feedback inhibition of N-acetylglutamate synthase (Haas and Leisinger, 1974) (see Section III.A).

II. GENETIC ORGANIZATION

The genes coding for arginine biosynthetic enzymes are scattered over the chromosome. This scattering is extreme in pseudomonads (Haas et al., 1977) (Fig. 4.2), where the genes of most biosynthetic pathways are unlinked (Holloway et al., 1979). In *Escherichia coli,* the four genes *argECBH* are clustered (Fig. 4.2) (Glansdorff, 1965). This cluster seems characteristic of enteric bacteria; it is found in *Klebsiella pneumoniae* (Matsumoto and Tazaki, 1970) and in *Salmonella typhimurium,* where the equivalent genes are designated *argACHF* (Sanderson, 1972). *Proteus* species and *Serratia marcescens* harbor an *argECBGH* gene cluster (Prozesky et al., 1973; Matsumoto et al., 1975). Clustering is also observed in *B. subtilis,* but it involves *arg* genes different from those in enteric bacteria (Baumberg, 1976).

The K-12 strain of *E. coli* has two genes coding for ornithine carbamoyltransferase, *argF* and *argI* (Glansdorff et al., 1967), which may have originated by duplication of the ancestor of *argI,* the only allele present in other enteric bacteria. The recent discovery that *argF* is flanked by IS-1 elements suggests that *argF* could well have been acquired horizontally and inserted as a transposon into the *E. coli* K-12 chromosome (Hu and

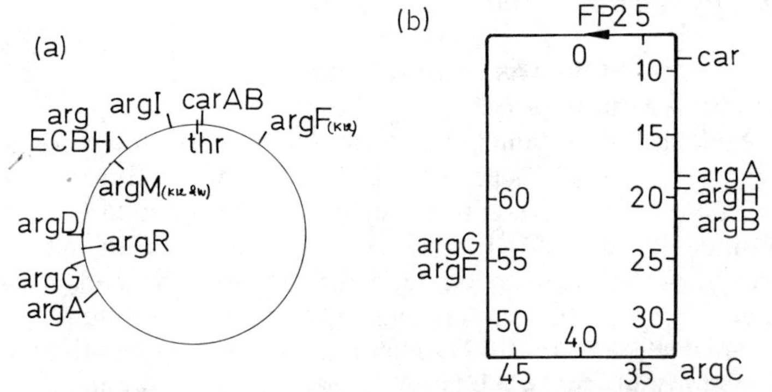

FIGURE 4.2 Chromosomal localization of the genes of arginine biosynthesis (a) in *E. coli* and (b) in *P. aeruginosa* (Haas et al., 1977; Holloway et al., 1979). Transfer times are based on conjugation mediated by the sex factor FP2.

Deonier, 1981). The common ancestry of *argF* and *argI,* which could be predicted from the interchangeability of their protein products in the trimeric active enzymes (Legrain et al., 1972), has been confirmed by DNA heteroduplex analysis (Kikuchi and Gorini, 1975) and by a comparison of their amino-terminal amino acid and nucleotide sequences (Gigot et al., 1977; Piette et al., 1981b; Moore, 1981). There is also a remarkable similarity to the catalytic subunit of aspartate carbamoyltransferase encoded by *pyrB*, a gene located very close to *argI*. This enzyme of the pyrimidine pathway catalyzes a reaction equivalent to that of ornithine carbamoyltransferase. A common ancestry for *pyrB* and *argI* seems likely.

In *E. coli* W (and probably in strain K-12), a cryptic, arginine-inducible acetylornithine-δ-transaminase can be activated in *argD* strains by mutations at a locus called *argM* (Bacon and Vogel, 1963). The amino acid composition and peptide maps of the *argD* and *argM* products are strikingly similar, a fact suggesting that both *argD* and *argM* arose from a single gene (Billheimer et al., 1979).

Among the relatively limited number of prokaryotic organisms investigated to date, the arginine metabolic pathway exhibits a wide variety of patterns of gene arrangements. This contrasts with histidine or tryptophan biosynthesis, where operons have been conserved (Bonner et al., 1965; Crawford, 1975). This raises the question of what selective forces shaped the evolution of metabolic pathways and which evolutionary advantages might accrue from one or the other type of genetic organization. The relative metabolic flexibility inherent in the arginine system with its separate genes may reflect a superiority over clustering, when one considers potential interactions with other biosynthetic pathways and arginine catabolism (Abdelal, 1979).

III. REGULATION OF ENZYME ACTIVITY

A. N-acetylglutamate Synthase (*argA*) and Acetylglutamate Kinase (*argB*)

Arginine inhibits acetylglutamate synthase, the first enzyme of the pathway, in all prokaryotes investigated so far (Vyas and Maas, 1963; Haas et al., 1972). In *E. coli,* 0.02 mM L-arginine causes 50% inhibition of the enzyme (Leisinger and Haas, 1975). In *Pseudomonas aeruginosa* and in several other prokaryotes that form acetylglutamate via ornithine acetyltransferase, the first two enzymes of arginine biosynthesis are inhibited by arginine (Udaka, 1966; Haas et al., 1972). This allows a quite sensitive control of the pathway through the modulation of the N-acetylglutamate pool size (Haas and Leisinger, 1975). At moderate arginine concentrations, N-acetylglutamate synergistically enhances the inhibition of the first enzyme by arginine while counteracting the inhibition of the second enzyme. When the N-acetylglutamate pool is depleted, feedback inhibition of the first enzyme is relieved while the second enzyme is inhibited. The other product of the first reaction, coenzyme A, also enhances inhibition of the first enzyme by arginine (Haas and Leisinger, 1974). In *P. aeruginosa,* spermine and, to a lesser extent, spermidine, cadaverine, and putrescine inhibit N-acetylglutamate synthase synergistically with N-acetylglutamate. In this way, polyamines, end products of the arginine pathway, drain a sizable proportion of ornithine and arginine and modulate carbon flow without complete shutoff, since their inhibitory effect depends on a sufficient N-acetylglutamate pool. Feedback inhibition by arginine of the second anabolic enzyme seems to be the only regulatory mechanism controlling arginine biosynthesis in blue-green algae and photosynthetic bacteria (Carr, 1973; Hoare and Hoare, 1966). No data are available on feedback inhibition of the first enzyme in these organisms. The first two enzymes of the pathway are targets in vivo of several arginine analogs, including arginine hydroxamate (Leisinger et al., 1974). This compound has been used successfully to enrich for arginine-excreting mutants of *B. subtilis* (Kisumi et al., 1971; Kato et al., 1977).

B. Carbamoylphosphate Synthetase (*carAB*)

In most prokaryotes, a single carbamoylphosphate synthetase provides this intermediate for both the arginine pathway and the pyrimidine pathway (Piérard et al., 1965; Prozesky and Coetzee, 1966; Abdelal and Ingraham, 1969). A notable exception is *B. subtilis* (Potvin and Gooder, 1975). The activity of the enzyme is modulated by a set of antagonistic effectors that ensure adequate supplies of carbamoylphosphate for both pathways (Piérard, 1966; Anderson and Meister, 1966).

In *E. coli,* the enzyme is inhibited by UMP and less effectively by other nucleotides. This inhibition is antagonized by ornithine (Piérard, 1966) and by inosine-5′-monophosphate and other purines (Anderson and Meister, 1966).

The subunit encoded by *carA* (50,000 daltons) binds glutamine (Trotta et al., 1971), while the subunit encoded by *carB* (140,000 daltons) has ammonium-dependent activity and is the target of the allosteric effectors. These effectors influence the distribution of the enzyme between at least three different conformational states; one of these, facilitating substrate binding, is stabilized by the positive effectors, whereas another one is stabilized by the inhibitor UMP (Anderson, 1977). Carbamoylphosphate synthetase from *S. typhimurium* and *E. coli* are very similar enzymes (Abdelal and Ingraham, 1975).

IV. REGULATION OF ENZYME SYNTHESIS

In enterobacteriaceae, the synthesis of all arginine biosynthetic enzymes is under arginine control (Vogel, 1961; Gorini et al., 1961; Maas, 1961; Kelln and O'Donovan, 1976; Prozesky, 1969; Kelln and Zak, 1978). In *Bacillus,* the synthesis of at least five of the anabolic enzymes is subject to arginine regulation (Lehner and Jones, 1962; R. H. Vogel and Vogel, 1963; Harwood and Baumberg, 1977). In pseudomonads, arginine represses the synthesis of only one biosynthetic enzyme, ornithine carbamoyltransferase, while inducing N-acetylornithine aminotransferase (Voellmy and Leisinger, 1978). In blue-green algae, arginine biosynthesis seems to be regulated exclusively through feedback inhibition by arginine of the second enzyme of the pathway (Hoare and Hoare, 1966; Carr, 1973).

So far, *E. coli* is the only prokaryote in which the expression of the *arg* genes is even partially understood at a molecular level. The following sections will therefore deal mainly with experiments performed with *E. coli.*

A. Control of Enzyme Synthesis in the *arg* Regulon of *E. coli*

All the *arg* genes are repressible by arginine (Gorini et al., 1961; Maas, 1961; Vogel, 1961). The term "repression," as opposed to induction, of enzyme synthesis was first coined to describe a regulatory mechanism in arginine biosynthesis (Vogel, 1958). Repression is mediated by a protein encoded by the *argR* gene (see below); *argR* expression itself is also subject to repression (T. Eckhardt, personal communication).

Expression of the *arg* genes scattered around the chromosome is not strictly coordinated in response to changes in intracellular arginine concentration. Repression–derepression ratios for the various enzymes are given in Table 4.1. Such organizational units in which one regulatory gene governs the expression of unlinked but functionally related structural genes are termed regulons (Maas and Clark, 1964).

TABLE 4.1 Repression–Derepression Ratios[1] in the Arginine Regulon of *E. coli*

Transcription Units	*Strain*	
	K-12	*B*
argA	250	—
argCBH	60 (50–70)	6–10
argD	20	5
argE	17	3
argF	300	Not present
argI	600	18
argG	—	—
carAB	3.8	
	26[2]	
argR	10	

From Elseviers et al. (1972), Jacoby (1972), and Piérard et al. (1980).
1. Specific enzyme activities in extracts of isogenic *argR*/*argR*$^+$ strains grown in excess arginine.
2. Bacteria grown in excess arginine and uracil.

1. The Arginine Repressor (*argR*). Mutations in *argR* cause constitutive synthesis of all arginine biosynthetic enzymes (Maas, 1961; Gorini et al., 1961; Vogel, 1961). Mutations in *argR* are *trans*-recessive (Maas et al., 1964; Maas and Clark, 1964; Kadner and Maas, 1971). *E. coli* K-12 *argR* mutants have been isolated by two different selection techniques, resistance to canavanine (Novick and Maas, 1961) or utilization of acetylhistidine as a source of histidine in the presence of arginine (Baumberg, 1970). Such mutants exhibit high enzyme levels close to those observed under conditions of physiological derepression, that is, growth of a bradytroph without arginine, growth of a mutant on limiting arginine (Crabeel et al., 1975), or growth of a wild-type strain on an arginine-free medium (Novick and Maas, 1961). In this respect, *argR* mutants differ from *trpR* mutants, which can still be derepressed significantly by tryptophan starvation. This different response of genetically derepressed mutants toward end product limitation reflects a fundamental difference between the types of regulatory loops involved in the control of these two pathways. Mutants with atypical *argR* products have been isolated, and their study has shed some light on the mechanism of repression. In *E. coli* W, such an *argR** mutant with a diminished repression–derepression range has been isolated (Vogel and Bacon, 1966). Both the degree of repression and the degree of derepression are affected, a result suggesting that *argR** codes for a repressor with an altered affinity for arginine. Fully derepressed *argR* mutants can be obtained from the *argR** strain by exposure to canavanine (Vogel et al., 1978).

The *argR* genes of *E. coli* B and K-12, $argR_B$ and $argR_K$, code for repressors with different properties. Under conditions of arginine limitation, the arginine biosynthetic enzyme levels are slightly lower in *E. coli* B than in K-12. Increasing arginine concentrations in *E. coli* B cause at first a slight repression, then, above a threshold value, a moderate induction of the synthesis of the enzymes, a condition suggesting that the $argR_B$ product becomes inhibited by excess arginine (Karlström and Gorini, 1969). An elevation of the temperature or a single mutation can cause the $argR_B$ allele to behave like $argR_K$ or give rise to genetic derepression with enzyme levels equivalent to these observed in $argR_K$ mutants (Jacoby and Gorini, 1969; Karlström and Gorini, 1969). Merodiploids combining dissimilar but active *argR* show intermediate control patterns: in $argR_B/argR_K$ diploids, the $argR_B$ allele is dominant under conditions of arginine limitation and recessive when arginine is in excess, a result that is consistent with subunit mixing in an oligomeric *argR* product (Jacoby and Gorini, 1969; Kadner and Maas, 1971).

The *argR* product is also involved in the induction by arginine of *argM* encoding acetylornithine transaminase (Bacon and Vogel, 1963). The inducible character of *argM* is attributable to the nature of the regulatory sequence preceding *argM*.

The $argR_K$ product has been partially purified (Udaka, 1970; Kelker et al., 1976; Kelker and Eckhardt, 1977). Estimates of its cellular concentration range from 200 molecules per cell (Kelker et al., 1976) to 67 molecules per cell (Cleary et al., 1977) or 40 molecules per cell (Piérard et al., 1980). On the basis of the nucleotide sequence of *argR* (T. Eckhardt, personal communication), the *argR* product should be a short protein (12,000 daltons) similar to the *trpR* product (Singleton et al., 1980; Gunsalus and Yanofsky, 1980). It has been suggested that the active repressor is associated with an arginine-binding protein that could also participate in the arginine uptake system (S. Udaka, personal communication).

2. Multiplicity of Operator Sites. The multiplicity of operator sites has for some time constituted an obstacle to the isolation of mutants causing *cis*-dominant constitutive expression of *arg* genes. The first mutants of this type were isolated as phenotypic revertants of polar mutants in the *argECBH* cluster, and all of them involve sizable chromosomal rearrangements: duplications (Glansdorff and Sand, 1968; Elseviers et al., 1969), deletions (Cunin et al., 1969), and insertions (Charlier et al., 1978). Operator mutants of *argI* and *argC* have been obtained by an elegant procedure based on relief from $argR_B$ repression of streptomycin-induced suppression of *arg* nonsense mutations (Jacoby, 1971, 1972). The *argCo* mutants show constitutive expression of all four genes of the *argECBH* cluster (Jacoby, 1972), a result supporting previous evidence that *argCBH* constitutes an operon (Cunin et al., 1969). These results also show that the *argCBH* cluster shares regulatory sites with *argE,* as inferred from the analysis of deletion mutants within the cluster (Elseviers et al., 1972). The *argCo* mutants provided evidence for the existence of operons transcribed divergently from promoters that face each other. Another selection procedure for operator constitutive *arg* mutants combines localized mutagenesis (Hong and Ames, 1971) with the ability of N-acetylornithinase (*argE*) to deacetylate N-acetyl amino acids such as acetylhistidine with enough efficiency to support the growth of *his* auxotrophs when the arginine system is derepressed (Bretscher and Baumberg, 1976). This method has yielded *cis*-dominant regulatory mutants affecting either *argECBH* or *argE* expression (Crabeel et al., 1974; Bretscher and Baumberg, 1976; Boyen et al., 1978), confirming that the regulatory sites of *argE* and *argCBH* are closely interwoven.

In a more general perspective, the demonstration of the long-inferred existence of operator loci for several genes of the arginine regulon shows that it is regulated as a set of individual operons responding to varying extents to a common regulatory signal.

B. Mechanism of Repression of Enzyme Synthesis

1. Repression of Gene Transcription. That repression of enzyme synthesis occurs at the level of transcription was shown for the arginine system in vivo by comparing mRNA levels to the corresponding enzyme levels in cultures grown in the presence or absence of exogenous arginine (Rogers et al., 1971; Cunin and Glansdorff, 1971). The involvement of *argR* was demonstrated (Cunin and Glansdorff, 1971; Krzyzek and Rogers, 1972). Hybridization experiments, measuring independently *argE* and *argCBH* mRNA levels, confirmed that repression is essentially transcriptional (Cunin et al., 1975). The basis for differential repressor effectiveness for *argE* and *argCBH* lies at the level of transcription. Transcriptional control in the *arg* regulon has also been shown for *argA, argF,* and *argI* (Sens et al., 1977a, b).

The exact mechanism by which the arginine repressor blocks transcription is unknown. From the nucleotide sequence of *arg* control regions and from RNA polymerase binding experiments, one can infer that in some cases the arginine repressor excludes RNA polymerase from *arg* promoters (Piette et al., 1981a) (see Section V).

2. The Nature of the Corepressor. The characterization of the molecular complex actually involved in repression required the partial purification of the repressor (Kelker et al., 1976) and the development of a purified in vitro transcription system for some of the *arg* genes (Pannekoek et al., 1975). The partially purified repressor is free of arginyl-tRNA synthetase. Addition of arginine and repressor to an in vitro system represses transcription of the *argCBH* operon (Cunin et al., 1976) and of the *carAB* operon (Lissens et al., 1980), the indication being that free arginine is the corepressor. These experiments also rule out autogenous regulation in which each enzyme participates in controlling its own biosynthesis.

Neither arginyl-tRNA synthetase nor $tRNA^{Arg}$ seem to be involved in the regulation of transcription. Arginyl-tRNA synthetase (*argS*) mutants, selected for resistance to canavanine on an arginine-free medium, are normally repressible (Hirshfield et al., 1968). Moreover, the levels of charging of the five isoacceptor $tRNA^{Arg}$ species are not significantly different under repressive and derepressive conditions (Celis and Maas, 1971). Experiments monitoring the levels of arginyl-tRNA and *argECBH* mRNA during transitions from repression to derepression and vice versa tend to exclude a role for $tRNA^{Arg}$ in regulation (Krzyzek and Rogers, 1976a).

It has been claimed, however, that *argS* mutants selected on a minimal medium are derepressed (Williams and Williams, 1973) and that inhibition of arginyl tRNA synthetase by certain biosynthetic precursors was correlated with derepression of arginine biosynthesis (Yem and Williams, 1971). Other studies do not confirm these results and suggest possible artifacts in the determination in vitro of $tRNA^{Arg}$ charging (Charlier and Gerlo, 1976). The apparent repression of enzyme synthesis by L-canavanine when $tRNA^{Arg}$ is predominantly charged with this analog (Faanes and Rogers, 1972) can be ascribed to posttranscriptional regulation (see Section IV.C), since canavanine does not repress transcription (Rogers et al., 1971).

3. Cumulative Repression of *carAB* Transcription. The synthesis of carbamoylphosphate synthetase is under control of both arginine and pyrimidines. The effect of arginine and uracil together corresponds approximately to the sum of their separate effects (Piérard et al., 1965). The involvement of the arginine repressor, already inferred from in vivo experiments, has been confirmed in vitro (Lissens et al., 1980). The levels of *carAB* mRNA and enzyme activity show an almost perfect correlation over the whole repression–derepression range, a result strongly suggesting that the control of enzyme synthesis is exclusively transcriptional (Piérard et al., 1980).

C. The Question of a Second Control System

Although repression of transcription by arginine accounts in large measure for the shutoff of gene expression, the possibility remains that a second control mechanism modulates the rate of enzyme synthesis in the low expression range (Cunin and Glansdorff, 1971; Rogers et al., 1971; Krzyzek and Rogers, 1972). Indeed, the accurate determination of *argE* and *argCBH* mRNA levels with DNA probes having negligible background hybridization is compatible with the notion that a portion of the mRNA produced under repressive conditions is not translated: a threefold to fourfold discrepancy is observed between the ratios of RNA levels and those of enzyme specific activities (Cunin et al., 1975). This discrepancy contrasts markedly with the parallelism observed between *carAB* mRNA and carbamoylphosphate synthetase levels (Piérard et al., 1980). Equivalent data are not available at the moment for other genes of the *arg* regulon. The mechanism responsible for the comparatively small adjustment of the rate of expression of *argE* and *argCBH* under repressive conditions remains undefined, as does its physiological significance.

1. The Apparent Absence of Attenuation. Attenuation occurs in several biosynthetic systems, either as the sole mode of regulation (histidine, threonine, isoleucine–valine, leucine, phenylalanine) or superimposed upon repression (tryptophan). In the *argECBH* cluster, a small RNA species produced predominantly under repression conditions has been considered to represent attenuated transcripts (Krzyzek and Rogers, 1976b). However, it could not be shown that the smaller transcript corresponds to the operator-proximal part of the *arg* cluster. Rather, hybridization experiments allowing independent measurements of in vivo *argCBH* mRNA detected no relative increase of the amount of transcripts corresponding to the *argC* operator-proximal region in the presence of excess arginine. In a purified transcription system, the distal part of the *argCBH* operon appears to be transcribed with the same frequency as the proximal region (Beny et al., 1982). Under similar conditions, a majority of attenuated transcripts of the tryptophan and histidine operons are found (Lee et al., 1976; Artz and Broach, 1975).

Starvation for arginine of *argR* mutants does not elicit higher levels of *arg* enzymes, in contrast to systems such as the *trp* operon endowed with both repressor and attenuation controls. On the basis of available data, no attenuation mechanism seems to be involved in the regulation of the arginine pathway, at least in the case of *argECBH*. This is confirmed by the nucleotide sequences of *arg* control regions (Section V.B): the leader sequences preceding the coding parts of *argF*, *argCBH*, and maybe *argE* do not exhibit structural features usually associated with attenuation control.

2. Translational Control. At an early stage in the study of the regulation of arginine biosynthesis, translational control was envisioned as an alternative to control at the level of transcription (Vogel et al., 1971). More recently, some of the early observations have been reinterpreted to allow for a posttranscriptional modulation of gene expression, superimposed on repression. Most of the experiments devised to assess the translational yield of *arg* mRNA under different steady state conditions are in fact indirect measurements of *arg* mRNA stability and do not distinguish between arginine-specific effects and general effects on bulk mRNA stability, related to the overall rate of protein synthesis (Pato et al., 1973). Indeed, it has been shown that *argECBH* mRNA stability can be enhanced by leucine starvation and that addition of arginine does not enhance the decay rate of that mRNA (Krzyzek and Rogers, 1976a). Similarly, mutations in a ribosomal protein thought to influence *arg* gene expression through translation initiation (Vogel et al., 1978) generally affect mRNA stability. Experimental data suggesting a translational control by arginine of the functional half-life of *argF* mRNA (McLellan and Vogel, 1970) have been contradicted (Hall and Gallant, 1973).

Another phenomenon that might have been ascribed to a translational regulation is the transient lack of correspondence between the apparent rate of enzyme synthesis and the level of mRNA during a transition from nonrepressing to repressing conditions. This phenomenon has been observed with acetylornithinase, ornithine carbamoyltransferase, and argininosuccinase (Lavallé, 1970; Krzyzek and Rogers, 1976a; Boyen, 1978) as well as in other systems (tryptophan synthase, Lavallé and de Hauwer, 1970; aspartate kinase I–homoserine dehydrogenase, Boy et al., 1973). When this phenomenon was discovered (Lavallé, 1970), it led to the notion that, upon arginine addition, while *arg* mRNA drops to its repressed steady state level in a matter of minutes (Krzyzek and Rogers, 1976a), enzyme synthesis is transiently reduced to almost zero by a specific translation inhibitor. The inhibitor is probably an arginine-repressible molecule that becomes progressively diluted after the onset of repression. Quantitatively, however, this phenomenon can be accounted for by enzyme inactivation (turnover). Upon arginine addition, turnover becomes important in the balance of enzyme content and can mimic a transient translational control, as was shown for *carAB* gene expression (Piérard et al., 1980).

Today, no satisfactory explanation accounts for the discrepancy between *argECBH* mRNA and corresponding enzyme levels in the low expression range. The mechanism of second site control remains a matter of conjecture.

V. CLONING AND NUCLEOTIDE SEQUENCING

A. Gene Enrichment and Cloning into Multicopy Plasmid Vectors

Following the methodologies developed by Gottesman and Beckwith (1969), Press et al. (1971), Schrenk and Weisberg (1975), and Casadaban (1975), most of the arginine genes of *E. coli* or their control regions fused to *lacZ* have been incorporated into specialized transducing bacteriophages. These vectors have facilitated measurements of in vivo transcription, allowed in vitro regulation studies, and opened the way to physical maps of several *arg* genes and of their regulatory regions through electron microscopy of heteroduplex DNA molecules (Kikuchi and Gorini, 1975; Mazaitis et al., 1976; Charlier et al., 1978).

Several genes or regions of the *arg* regulon of *E. coli* have been cloned into multicopy plasmid vectors (Table 4.2). The *P. aeruginosa argE* and *argF* (maybe together with *argG*) genes have been cloned into an R′ plasmid

TABLE 4.2 Genes or Regions of the *arg* Regulon of *E. coli* K-12 Cloned into Multicopy Plasmids

Gene or Region	*Type Fragment*	*Size (kb)*	*Vector*	*Plasmid Name*	*Reference*
argECBH	EcoRI	17	colE1[1]	colE1*argECBH*	Crabeel et al. (1977)
argECBH	EcoRI	9.5	pMB9[2]	pMC7	Crabeel et al. (1977)
argECBH regulatory region	HindIII	1.8	pBR322[2]	pMC31	Crabeel et al. (1979)
Regulatory region with sundry control mutants	HindIII	1.8	pBR322	pMC31-1 to 13	M. Crabeel (unpublished)
argECBH regulatory region + IS2:II	HindIII	1.12	pBR322	pA07	M. Crabeel (unpublished)
argECBH regulatory region + IS2:1	HindIII	0.8	pBR322	pA010	R. Cunin (unpublished)
argECBH regulatory region with ▽102 deletion	HindIII EcoRI	4.0	pBR322	pSU102	R. Cunin and L. Strackx (unpublished)
argF	BamHI	1.6	pBR322	pPJ2	Moore et al. (1978)
				pMC23, pMC20	Crabeel et al. (1979)
argI	EcoRI	4.0	pBR322	pYO-1	Y. Ota (unpublished)
carAB	HindIII	5.6	pBR322	pMC50	M. Crabeel (unpublished)
				pMC40	
carA	EcoRI HindIII	3.2	pMB9	pGW1	M. Crabeel and G. Weyens (unpublished)
				pGW2	

1. Hershfield et al. (1974).
2. Rodriguez et al. (1977).

by complementing corresponding auxotrophic lesions of *E. coli* (Hedges and Jacob, 1977). A 9.3-kb EcoRI fragment containing the *argC* gene of *B. subtilis* has been cloned into pBR322. The resulting plasmid, which complements an *argC* mutant of *E. coli,* is unstable and segregates shorter plasmids containing 1.4 kb of *B. subtilis* DNA (N. H. Mann and S. Baumberg, personal communication).

B. Nucleotide Sequences of *arg* Regulatory Regions

1. Regulation of Divergent Transcription in the *argECBH* Bipolar Operon. The *argECBH* gene cluster is transcribed divergently from the promoters situated between *argE* and *argC* (Pouwels et al., 1974; Panchal et al., 1974). *Cis*-dominant regulatory mutations affecting the expression of both wings of the cluster occur in the same location (Jacoby, 1972; Elseviers et al., 1972). In two strains with *argCB* deletions (Fig. 4.3) extending into the control region, *argE* expression is abolished, although the deletions are external to *argE*. The repressibility of *argH* is reduced, but its potential expression is virtually unchanged (Elseviers et al., 1972).

Transcription of *argCBH* is more repressible (60-fold to 70-fold) than that of *argE* (17-fold); this could reflect the existence of two operators (Cunin et al., 1975).

Among a large set of *cis*-dominant operator-constitutive mutations, some exhibit constitutive expression of both *argE* and *argCBH*, while a few display constitutive expression of *argE* only. Another category of *cis*-dominant mutants exhibit a promoter-down effect on *argCBH* and a concomitant increase of *argE* transcription (Bretscher and Baumberg, 1976; Boyen et al., 1978). The control region thus appears to contain two overlapping operators.

The control region for the *argECBH* cluster has been cloned into plasmid pBR322 on a 1.8-kb HindIII fragment (pMC31 in Table 4.2). The sequence of an 850-bp stretch starting from one of the HindIII sites has been established. The portion of the sequence relevant to this discussion is shown in Fig. 4.3. A 60-bp stretch located 315 bp from the HindIII site in *argE* displays two overlapping imperfect palindromes and has been assigned to the control region by the localization within its limits of three IS-2 insertions (Piette et al., 1980), which affect the regulation of both wings of the cluster (Charlier et al., 1978). Deletion Δsup102 destroys part of the palindromes (Fig. 4.3); this destruction confirms that they are part of the regulatory region of the cluster.

The *argCBH* promoter has been localized on the *argE* side of the dyad-symmetry region. Its −35 region (Siebenlist et al., 1980) contains a HincII restriction site, which is effectively protected by RNA polymerase against endonucleolytic cleavage (Piette et al., 1981a). Two promoter-down mutants in which the rate of *argCBH* transcription has been reduced twofold to

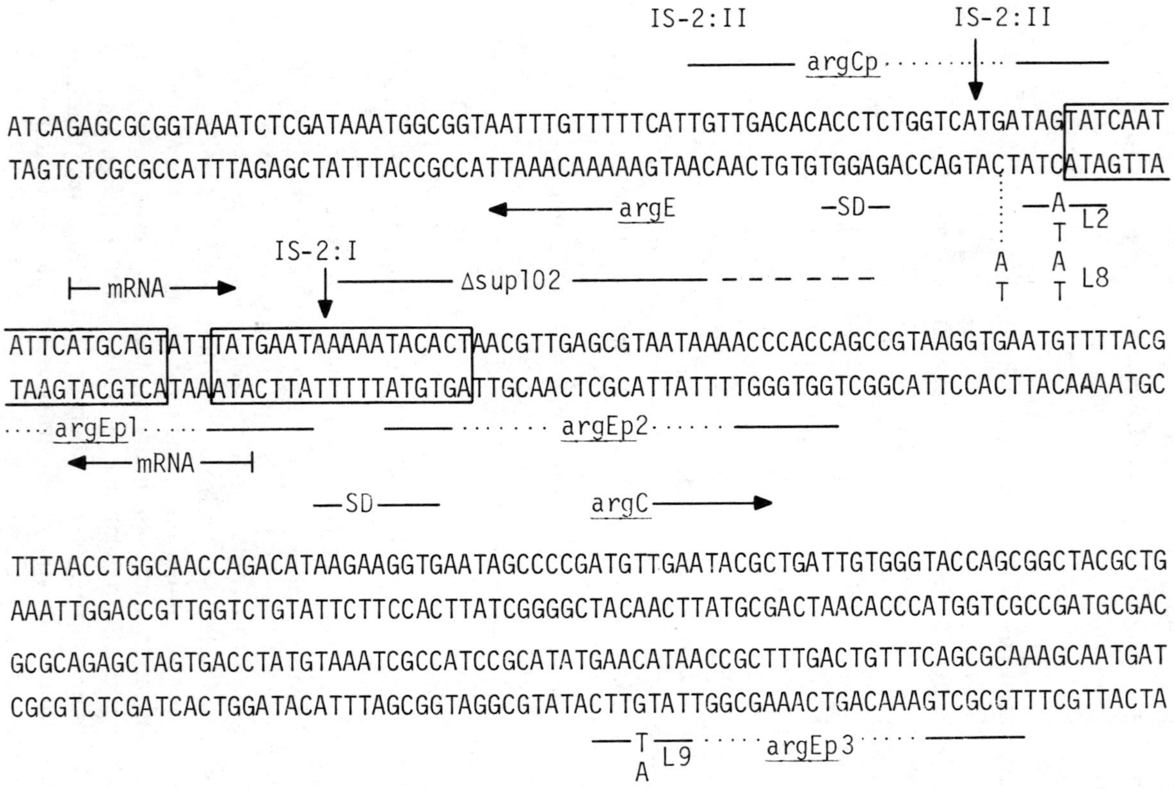

FIGURE 4.3 Nucleotide sequence of the control region of the *argECBH* cluster. *argCp* is the promoter for *argCBH; argEp1* to *argEp3* are the putative promoter sites for *argE*. The solid lines emphasize −35 and −10 sequences. The boxed sequences are ARG boxes. The symmetrical stretches are indicated. SD are the putative Shine-Dalgarno sequences. IS2:I and IS2:II are the location of IS2 insertions, with Roman numerals referring to the orientation relative to *argE*. Transcription starts are indicated by |—mRNA→.

threefold (Bretscher and Baumberg, 1976; Boyen et al., 1978) exhibit a GC → AT transition in the Pribnow box (Pribnow, 1975) matching the −35 region defined by the protection experiment. One of the mutants (L8) harbors a second GC → AT transition four nucleotides upstream of the first one (Fig. 4.3). Because the Pribnow box of the *argCBH* promoter overlaps the dyad-symmetry region, a condition suggesting that repression may occur through a competition between the repressor and the RNA polymerase for binding to the DNA.

The NH_2-terminal amino acid sequence of the *argC* enzyme is not known. The first translation initiation codon (ATG) allowing an open reading frame for at least 144 codons is located 116 bp downstream from the promoter. No "typical" attenuation is suggested by this sequence.

The *argE* promoter was first localized within the limits of the Δsup102 deletion. The deletion removes two out of three putative sites for *argE* promoters, which can be read from the sequence on the basis of their resemblance to the ideal promoter sequence (Siebenlist et al., 1980). S-1 nuclease mapping experiments have confirmed that *argEp2* (Fig. 4.3) is the *argE* promoter. The localization of the translation initiation codon for *argE* is tentative and relies only on the existence of an open reading frame up to the end of the cloned fragment (about 100 codons).

The two promoters face each other. Thus a certain degree of interference between *argE* and *argCBH* transcription is anticipated. Transcription initiation at the two promoters should be mutually exclusive on steric grounds, so initiation of divergent transcription should take place in alternating fashion. Antagonistic effects between transcription of the two wings of the cluster would depend on the frequency of initiation at each promoter. Such a face-to-face disposition of promoters was originally proposed for the *bioABFCD* region (Guha et al., 1971) and was later observed in this cluster (Otsuka and Abelson, 1978).

Mutant L9, previously characterized as *argEo* (Fig. 4.3), is a single base change in the Pribnow box of a promoter-like sequence (*argEp3*) 200 bp upstream from the palindromes (Piette et al., 1981a). This base substitution is analogous to one of the promoter-up mutations in the *lacUV5* mutant (Gilbert, 1976) and might be expected to create or to activate a promoter for *argE*. Transcription initiated at that promoter might displace an operator-bound repressor and largely escape control (Reznikoff et al., 1969). It is possible that in the wild type a low constitutive rate of initiation at this promoter is responsible for the lower repressibility of *argE* transcription compared to *argCBH*.

2. The Regulatory Region of *argF*. Gene *argF* has been cloned into pBR322 on a 1.65-kb BamH1 fragment (Moore et al., 1978; Crabeel et al., 1979) (Table 4.2). The nucleotide sequence of a 530-bp stretch of DNA, 230 nucleotides preceding *argF* and 306 within the gene itself, has been established (Piette et al., 1981b) (Fig. 4.4). The unambiguous localization of the translation

```
                 ——————······argFp1····——————————

argF  TCCAATGTCTTTCTGCTTCTGCAGAGAATCGGAGGCAGATTACGATTATTTTCACACAACGGACGGGTTTGCCTCCACCTTTGTAAGAAA
argI  GGG-T TG CA G-   CA C T TT   GTCCT  C  -    TC     G        - CACT ACA                 G C

               ——————————······argFp2····————————

argF  GAATTGTGAAATGGGGTTGCAAATGAATAATTACACATATAAAGTGAATTTTAATTCAATAAGTGGCGTTCGCCATGCGAGGATAAAATG
argI     T  A  CTT A AC               C TC        T                 T-- A       A     CAG     GATCT

                     -35                    -10         |—mRNA—>
                                                         |—mRNA—>
```

FIGURE 4.4 DNA sequences preceding the *argF* and *argI* genes. For the *argI* regulatory region, only those bases are shown that differ from the *argF* sequence; — indicates a base to be deleted in *argF* for optimal alignment of the two sequences. The boxed sequences are the ARG boxes *argFo1*/*argIo1* and *argFo2*/*argIo2* (see Fig. 4.5); *argFp1* and *argFp2* are putative promoters for *argF*; —mRNA→ indicates a transcription start; |→ indicates the translation start for both *argF* and *argI*.

initiation codon is based on amino-terminal amino acid sequence of the corresponding protein (Gigot et al., 1977), which was in turn confirmed and extended by the nucleotide sequence.

Two putative promoter loci can be identified on the basis of their similarity to the prototype sequence (Siebenlist et al., 1980). One of them, *argFp2*, has its −10 region centered 30 bp upstream of the protein startpoint and is closely followed by a ribosome entry site (Shine and Dalgarno, 1974). Promoter *argF2* overlaps completely with a 40-bp sequence containing two successive 18-bp palindromes showing some analogies with sequences in the control region of *argECBH* (Fig. 4.3, boxed sequences) and of *argR* (T. Eckhardt, personal communication).

If these sequences constitute part or all of an operator, their total overlap with *argFp2* is compatible with the high repressibility of *argF* in vivo. S-1 nuclease mapping confirms that *argFp2* initiates *argF* transcription in vivo; *argFp2* is located 130 bp upstream of the protein start; it coincides with a PstI restriction site protected by RNA polymerase (Piette et al., 1981b). This is in agreement with observations that PstI inhibits *argF* transcription in vitro (Moore et al., 1978, 1982). The function of the leader RNA initiated at *argFp2* remains obscure. Its sequence does not exhibit any of the features usually associated with control by attenuation (Yanofsky, 1981).

3. The ARG Box, the Basis of the Differential Repressibility of the Arginine Genes. A comparison between the regulatory regions of the *arg* genes sequenced so far allowed us to identify, upstream of and close to the coding part of the genes, one or two copies of a conserved 18-bp sequence showing dyad symmetry: the ARG box (Cunin et al., 1981) (Figs. 4.3, 4.4, 4.5). Two ARG boxes are centered 72 and 52 bp upstream of the protein start of *argF*, overlapping completely the *argF* promoter (Fig. 4.4). The nucleotide sequence of the *argI* control region (Moore, 1981; Piette et al., 1982) shows a very high homology with that of *argF*. Again, two ARG boxes precede the coding region and overlap completely with the promoter. The *argCBH* wing of the *argECBH* cluster is also preceded by two ARG boxes, one of them overlapping with the Pribnow box (Fig. 4.3). The *argE* promoter also overlaps with these ARG boxes. A single ARG box is found 23 bp upstream of the putative *argR* promoter (T. Eckhardt, personal communication).

The presence of these hyphenated ARG boxes suggests that they are involved in gene regulation as parts of operator loci. Sequence analysis of several operator mutants has revealed nucleotide changes within the ARG boxes of the *argECBH* cluster (Piette et al., 1982). Thus base compositions of ARG boxes and their degree of homology with the prototype ARG box sequence (Fig. 4.5) may be the basis of the differential repressor effectiveness in the arginine regulon. Other factors, such as the number of ARG box

argFo1	A A T G A A T A A • T T A C A C A T A	
	T T A C T T A T T • A A T G T G T A T	
argFo2	A G T G A A T T T • T A A T T C A A T	
	T C A C T T A A A • A T T A A G T T A	
argCo1	T A T C A A T A T • T C A T G C A G T	
	A T A G T T A T A • A G T A C G T C A	
argCo2	T A T G A A T A A • A A A T A C A C T	
	A T A C T T A T T • T T T A T G T G A	
argRo	T T T G A C A A T • A T T A A T C A G	
	A A A C T G T T A • T A A T T A G T C	
argIo1	A A T G A A T A A • T C A T C C A T A	
	T T A C T T A T T • A G T A G G T A T	
argIo2	A T T G A A T T T • T A A T T C A T T	
	T A A C T T A A A • A T T A A G T A A	
	A A T G A A T A T • T □ A T □ C A □ T	

FIGURE 4.5 The ARG box; symmetries are underlined. In the consensus sequence (last line), nucleotides in large type are those showing at least 75% conservation. Nucleotides in small type exhibit 50% conservation.

copies in a given operator or their position relative to a given promoter, may also influence the efficiency of repression.

VI. CONCLUDING REMARKS

Prokaryotes may be divided into two categories according to the acetylation–deacetylation strategy they follow in the arginine biosynthetic pathway. In the organisms that recycle the acetyl group, biosynthesis is regulated predominantly through feedback inhibition of the first enzyme, whereas regulation of enzyme synthesis prevails in those prokaryotes in which the acetyl group is not recycled.

With the exception of pseudomonads, most studies on the regulation of arginine biosynthesis have been performed on organisms of the second group: enterobacteriaceae and bacillaceae. In enterobacteriaceae, the genes of arginine biosynthesis are dispersed on the chromosome and are subject to repression via a common regulatory molecule. The amplitude of their response to this control appears to be related to variations in the operator nucleotide sequences. Posttranscriptional control may also operate, but its relatively small amplitude makes its characterization difficult.

No growth rate effect on arginine biosynthesis has been detected in enterobacteriaceae, but there is such a control in the prokaryotes able to use arginine as a source of carbon or nitrogen. The in vitro expression of *argH* and *argA* of *E. coli* is stimulated by guanosine tetraphosphate (Kelker et al., 1976; Kelker and Eckhardt, 1977), and *argE* is repressed (Yang et al., 1974). The mode of action of the tetraphosphate is still obscure, but the opposite effects on *argE* and *argH* might be related to the genetic organization of the control region of *argECBH*.

Studies on the arginine regulon have emphasized the dramatic changes in regulatory patterns that can be brought about by single mutations: a single revertible base pair change in the regulatory gene can make the whole regulon inducible rather than repressible by arginine (*E. coli* B versus *E. coli* K-12); a small difference between the regulatory region of *argM* and that of the other *arg* genes causes the former to be induced while the other genes are repressed by arginine. Larger chromosomal rearrangements of *arg* genes have been used to study the origin of divergently transcribed clusters (Charlier et al., 1979).

ACKNOWLEDGMENTS

The contributions of T. Eckhardt and J. Piette were determinant in the discovery of the ARG box. I am very much indebted to N. Glansdorff for his constructive reading of the manuscript, to A. Jacobs for her help with the figures, to the whole arginine group in Brussels for fruitful discussions, and to S. Baumberg, R. Celis, D. Haas, R. Kelln, H. Matsumoto, O. Prozesky, and S. Udaka for communicating new results or reminding me of older ones.

Work on the sequences was supported by Grant S2/5-AME21 of the Belgian NFWO.

REFERENCES

Abdelal, A. T. (1979) *Ann. Rev. Microbiol. 33,* 139–168.
Abdelal, A. T., and Ingraham, T. L. (1969) *J. Biol. Chem. 244,* 4033–4038.
Abdelal, A. T., and Ingraham, T. L. (1975) *J. Biol. Chem. 250,* 4410–4417.
Anderson, P. M. (1977) *Biochemistry 16,* 587–592.
Anderson, P. M., and Meister, A. (1966) *Biochemistry 5,* 3164–3169.
Artz, S. W., and Broach, J. R. (1975) *Proc. Nat. Acad. Sci. U.S.A. 72,* 3453–3457.
Bacon, D. F., and Vogel, H. J. (1963) *Cold Spring Harbor Symp. Quant. Biol. 28,* 437–438.
Baumberg, S. (1976) in *Second Symposium on the Genetics of Industrial Microorganisms* (MacDonald, K. D., ed.) pp. 369–389, Academic Press, New York.
Baumberg, S. (1970) *Mol. Gen. Genet. 106,* 162–173.

Beny, G., Cunin, R., Glansdorff, N., Boyen, A., Charlier, D., and Kelker, N. (1982) *J. Bacteriol. 151,* 58–61.

Bernhardt, S. A., and Davis, R. H. (1972) *Proc. Nat. Acad. Sci. U.S.A. 69,* 1868–1872.

Billheimer, J. R., Shen, M. Y., Carnevale, H. N., Horton, H. R., and Jones, E. E. (1979) *Arch. Biochem. Biophys. 195,* 401–413.

Bonner, D. M., DeMoss, J. A., and Mills, S. E. (1965) in *Evolving Genes and Proteins* (Bryson, V., and Vogel, H. J., eds.) pp. 305–318, Academic Press, New York.

Boy, E., Theze, T., and Patte, J. C. (1973) *Mol. Gen. Genet. 121,* 77–81.

Boyen, A. (1978) in "Quantitative Studies *in vivo* and *in vitro* of *argECBH* Expression in *E. coli* K-12", Ph.D. thesis, University of Brussels, Brussels, Belgium.

Boyen, A., Charlier, D., Crabeel, M., Cunin, R., Palchaudhuri, S., and Glansdorff, N. (1978) *Mol. Gen. Genet. 161,* 185–196.

Bretscher, A. P., and Baumberg, S. (1976) *J. Mol. Biol. 102,* 205–220.

Carr, N. G. (1973) in *The Biology of Blue-Green Algae* (Carr, N. G., and Whitton, B. A., eds.) p. 39, Blackwell Scientific Publications, Oxford, England.

Casadaban, M. J. (1975) *Proc. Nat. Acad. Sci. U.S.A. 72,* 809–813.

Celis, T. F. R., and Maas, W. K. (1971) *J. Mol. Biol. 62,* 179–188.

Charlier, D., and Gerlo, E. (1976) *Eur. J. Biochem. 70,* 137–145.

Charlier, D., Crabeel, M., Palchaudhuri, S., Cunin, R., Boyen, A., and Glansdorff, N. (1978) *Mol. Gen. Genet. 161,* 175–184.

Charlier, D., Crabeel, M., Cunin, R., and Glansdorff, N. (1979) *Mol. Gen. Genet. 174,* 75–88.

Cleary, M. L., Garvin, R. T., and James E. (1977) *Mol. Gen. Genet. 157,* 155–165.

Crawford, I. P. (1975) *Bacteriol. Rev. 39,* 87–120.

Crabeel, M., Charlier, D., Boyen, A., Cunin, R., and Glansdorff, N. (1974) *Arch. Int. Physiol. Biochim. 82,* 973–974.

Crabeel, M., Charlier, D., Cunin, R., Boyen, A., Glansdorff, N., and Pierard, A. (1975) *J. Bacteriol. 123,* 898–904.

Crabeel, M., Charlier, D., Glansdorff, N., Palchaudhuri, S., and Maas, W. K. (1977) *Mol. Gen. Genet. 151,* 161–168.

Crabeel, M., Charlier, D., Cunin, R., and Glansdorff, N. (1979) *Gene 5,* 207–231.

Cunin, R., and Glansdorff, N. (1971) *FEBS Lett. 18,* 135–137.

Cunin, R., Elseviers, D., Sand, G., Freundlich, G., and Glansdorff, N. (1969) *Mol. Gen. Genet. 106,* 32–47.

Cunin, R., Boyen, A., Pouwels, P., Glansdorff, N., and Crabeel, M. (1975) *Mol. Gen. Genet. 140,* 51–60.

Cunin, R., Kelker, N., Boyen, A., Yang, H.-L., Zubay, G., Glansdorff, N., and Maas, W. K. (1976) *Biochem. Biophys. Res. Comm. 69,* 377–382.

Cunin, R., Eckhardt, T., Piette, J., Boyen, A., and Glansdorff, N. (1981) *Arch. Int. Physiol. Biochim. 89,* B94.

Degryse, E. (1974) *FEBS Lett. 43,* 285–288.

Degryse, E., Glansdorff, N., and Piérard, A. (1976) *Arch. Int. Physiol. Biochim. 84,* 599–601.

Elseviers, D., Cunin, R., and Glansdorff, N. (1969) *FEBS Lett. 3,* 18–20.

Elseviers, D., Cunin, R., Glansdorff, N., Baumberg, S., and Ashcroft, E. (1972) *Mol. Gen. Genet. 117,* 349–366.

Faanes, R., and Rogers, P. (1972) *J. Bacteriol. 112,* 102–113.

Gigot, D., Glansdorff, N., Legrain, C., Piérard, A., Stalon, V., Konigsberg, W., Caplier, I., Strosberg, A., and Herve, G. (1977) *FEBS Lett. 81,* 28–32.
Gilbert, W. (1976) in *RNA Polymerase* (Losick, R., and Chamberlain, M., eds.) pp. 193–205, Cold Spring Harbor Laboratory, New York.
Glansdorff, N. (1965) *Genetics 51,* 167–179.
Glansdorff, N., and Sand, G. (1968) *Genetics 60,* 257–268.
Glansdorff, N., Sand, G., and Verhoef, C. (1967) *Mut. Res. 4,* 743–751.
Gorini, L., Gundersen, W., and Burger, M. (1961) *Cold Spring Harbor Symp. Quant. Biol. 26,* 173–182.
Gottesman, S., and Beckwith, J. R. (1969) *J. Mol. Biol. 44,* 117–127.
Guha, A., Saturen, Y., and Szybalski, W. (1971) *J. Mol. Biol. 56,* 53–62.
Gunsalus, R. P., and Yanofsky, C. (1980) *Proc. Nat. Acad. Sci. U.S.A. 77,* 7117–7121.
Haas, D., and Leisinger, T. (1974) *Biochem. Biophys. Res. Comm. 60,* 42–47.
Haas, D., and Leisinger, T. (1975) *Eur. J. Biochem. 52,* 377–383.
Haas, D., Kurer, V., and Leisinger, T. (1972) *Eur. J. Biochem. 31,* 290–295.
Haas, D., Holloway, B. W., Schamböck, A., and Leisinger, T. (1977) *Mol. Gen. Genet. 154,* 7–22.
Hall, B. G., and Gallant, J. A. (1973) *J. Mol. Biol. 73,* 121–124.
Harwood, C. R., and Baumberg, S. (1977) *J. Gen. Microbiol. 100,* 177–188.
Hedges, R. W., and Jacob, A. E. (1977) *FEMS Microbiol. Lett. 2,* 15–19.
Hershfield, V., Boyer, H. W., Yanofsky, C., Lovett, M. A., and Helinski, D. R. (1974) *Proc. Nat. Acad. Sci. U.S.A. 71,* 3455–3459.
Hirschfield, I. N., DeDeken, R., Horn, P. C., Hopwood, D. A., and Maas, W. K. (1968) *J. Mol. Biol. 35,* 83–93.
Hoare, D. S., and Hoare, S. L. (1966) *J. Bacteriol. 92,* 375–379.
Holloway, B. W., Krishnapillai, V., and Morgan, A. F. (1979) *Microbiol. Rev. 43,* 73–102.
Hong, J., and Ames, B. N. (1971) *Proc. Nat. Acad. Sci. U.S.A. 68,* 3158–3162.
Hu, M., and Deonier, R. C. (1981) *Mol. Gen. Genet. 181,* 222–229.
Jacoby, G. A. (1971) *J. Bacteriol. 108,* 645–651.
Jacoby, G. A. (1972) *Mol. Gen. Genet. 117,* 337–348.
Jacoby, G. A., and Gorini, L. (1969) *J. Mol. Biol. 39,* 73–87.
Kadner, R. J., and Maas, W. K. (1971) *Mol. Gen. Genet. 111,* 1–14.
Karlström, O., and Gorini, L. (1969) *J. Mol. Biol. 39,* 89–94.
Kato, J., Kisumi, M., Takagi, T., and Chibata, I. (1977) *Appl. Env. Microbiol. 34,* 689–694.
Kelln, R. A., and O'Donovan, G. A. (1976) *J. Bacteriol. 128,* 528–535.
Kelln, R. A., and Zak, V. L. (1978) *Mol. Gen. Genet. 161,* 333–335.
Kelker, N., and Eckhardt, T. (1977) *J. Bacteriol. 132,* 67–72.
Kelker, N. E., Maas, W. K., Yang, H.-L., and Zubay, G. (1976) *Mol. Gen. Genet. 144,* 17–20.
Kikuchi, A., and Gorini, L. (1975) *Nature 256,* 621–624.
Kisumi, M., Kato, J., Sugiura, M., and Chibata, I. (1971) *Appl. Micr. 22,* 987–991.
Krzyzek, R. A., and Rogers, P. (1972) *J. Bacteriol. 110,* 945–954.
Krzyzek, R. A., and Rogers, P. (1976a) *J. Bacteriol. 126,* 348–364.
Krzyzek, R. A., and Rogers, P. (1976b) *J. Bacteriol. 126,* 365–376.
Lacroute, F., Piérard, A., Grenson, M., and Wiame, J. M. (1965) *J. Gen. Microbiol. 40,* 127–142.

Lavallé, R. (1970) *J. Mol. Biol. 51,* 449–451.
Lavallé, R., and de Hauwer, G. (1970) *J. Mol. Biol. 51,* 435–447.
Lee, F., Squires, C. L., Squires, C., and Yanofsky, C. (1976) *J. Mol. Biol. 103,* 383–393.
Legrain, C., Halleux, P., Stalon, V., and Glansdorff, N. (1972). *Eur. J. Biochem. 27,* 93–102.
Lehner, H. I., and Jones, M. E. (1962) *Biochim. Biophys. Acta 65,* 360–362.
Leisinger, T., and Haas, D. (1975) *J. Biol. Chem. 250,* 1690–1693.
Leisinger, T., O'Sullivan, C., and Haas, D. (1974) *J. Gen. Microbiol. 84,* 253–260.
Lissens, W., Cunin, R., Kelker, N., Glansdorff, N., and Piérard, A. (1980) *J. Bacteriol. 141,* 58–66.
Maas, W. K. (1961) *Cold Spring Harbor Symp. Quant. Biol. 26,* 183–191.
Maas, W. K., and Clark, A. J. (1964) *J. Mol. Biol. 8,* 365–370.
Maas, W. K., Maas, R., Wiame, J. M., and Glansdorff, N. (1964) *J. Mol. Biol. 8,* 359–364.
Matsumoto, H., and Tazaki, T. (1970) *Jap. J. Microbiol. 14,* 129–141.
Matsumoto, H., Hosogaya, S., Suzuki, K., and Tazaki, T. (1975) *Jap. J. Microbiol. 19,* 35–44.
Mazaitis, A. J., Palchaudhuri, S., Glansdorff, N., and Maas, W. K. (1976) *Mol. Gen. Genet. 143,* 185–196.
McLellan, W. L., and Vogel, H. J. (1970) *Proc. Nat. Acad. Sci. U.S.A. 67,* 1703–1709.
Moore, S. K. (1981) in "Nucleotide Sequence of the *argF* and *argI* Regulatory Region of *E. coli* K-12", Ph.D. thesis, University of South Carolina, Columbia, S.C.
Moore, S. K., James, E., James, P. M., and Fareed, G. (1978). *Gene 4,* 261–278.
Moore, S. K., Garvin, R. T., and James, E. (1982) *Gene 16,* 119–132.
Novick, R. P., and Maas, W. K. (1961) *J. Bacteriol. 81,* 236–240.
Otsuka, A., and Abelson, J. (1978) *Nature 276,* 689–694.
Panchal, C. J., Bagchee, S. N., and Guha, A. (1974) *J. Bacteriol. 117,* 675–680.
Pannekoek, H., Cunin, R., Boyen, A., and Glansdorff, N. (1975) *FEBS Lett. 51,* 143–145.
Pato, M. L., Bennett, P. M., and von Meyenburg, K. (1973) *J. Bacteriol. 116,* 710–718.
Piérard, A. (1966) *Science 154,* 1572–1573.
Piérard, A., Glansdorff, N., Mergeay, M., and Wiame, J. M. (1965) *J. Mol. Biol. 14,* 23–36.
Piérard, A., Lissens, W., Halleux, P., Cunin, R., and Glansdorff, N. (1980) *J. Bacteriol. 141,* 382–385.
Piette, J., Cunin, R., Crabeel, M., Boyen, A., Glansdorff, N., Squires, C., and Squires, C. L. (1980) *Arch. Int. Physiol. Biochim. 88,* B243.
Piette, J., Boyen, A., Crabeel, M., Cunin, R., and Glansdorff, N. (1981a) *Arch. Int. Physiol. Biochim. 89,* B126–127.
Piette, J., Cunin, R., Crabeel, M., and Glansdorff, N. (1981b) *Arch. Int. Physiol. Biochim. 89,* B127–128.
Piette, J., Cunin, R., Van Vliet, F., Charlier, D., Crabeel, M., Ota, Y., and Glansdorff, N. (1982) EMBO J. *1,* 853–857.
Potvin, B., and Gooder, H. (1975) *Biochem. Genet. 13,* 125–143.
Pouwels, P. H., Cunin, R., and Glansdorff, N. (1974) *J. Mol. Biol. 83,* 421–424.

Press, R., Glansdorff, N., Miner, P., DeVries, J., Kadner, R., and Maas, W. K. (1971) *Proc. Nat. Acad. Sci. U.S.A. 68,* 795–798.
Pribnow, D. (1975) *Proc. Nat. Acad. Sci. U.S.A. 72,* 784–788.
Prozesky, O. W. (1969) *J. Gen. Microbiol. 55,* 89–102.
Prozesky, O., and Coetzee, J. (1966) *Nature 209,* 1262–1263.
Prozesky, O. W., Grabow, W., van der Merwe, S., and Coetzee, J. N. (1973) *J. Gen. Microbiol. 77,* 237–240.
Reznikoff, W. S., Miller, J. H., Scaife, J. G., and Beckwith, J. R. (1969) *J. Mol. Biol. 43,* 201–213.
Rodriguez, R. L., Tait, R., Shine, J., Bolivar, F., Heyneker, H., Betlach, M., and Boyer, H. W. (1977) in *Miami Winter Symposia,* (Scott, W. A., and Werner, R., eds.) Vol. XIII, pp. 73–84, Academic Press, New York.
Rogers, P., Krzyzek, R., Kaden, T. M., and Arfman, E. (1971) *Biochem. Biophys. Res. Comm. 44,* 1220–1226.
Sanderson, K. E. (1972) *Bacteriol. Rev. 36,* 558–586.
Schrenk, W. J., and Weisberg, R. A. (1975) *Mol. Gen. Genet. 137,* 101–107.
Sens, D., Natter, W., and James, E. (1977a) *J. Bacteriol. 130,* 642–655.
Sens, D., Natter, W., Garvin, R. T., and James, E. (1977b) *Mol. Gen. Genet. 155,* 7–18.
Shine, J., and Dalgarno, L. (1974) *Proc. Nat. Acad. Sci. U.S.A. 71,* 1342–1346.
Siebenlist, U., Simpson, R. B., and Gilbert, W. (1980) *Cell 20,* 269–281.
Singleton, C. K., Roeder, W. D., Bogosian, G., Somerville, R. L., and Weith, H. L. (1980) *Nucleic Acid Res. 8,* 1551–1560.
Trotta, P. P., Burt, M. E., Haschemeyer, R. H., and Meister, A. (1971) *Proc. Nat. Acad. Sci. U.S.A. 68,* 2599–2603.
Udaka, S. (1966) *J. Bacteriol. 91,* 617–621.
Udaka, S. (1970) *Nature 228,* 336–338.
Voellmy, R., and Leisinger, T. (1978) *J. Gen. Microbiol. 109,* 25–35.
Vogel, H. J. (1958) in *The Chemical Basis of Heredity* (McElroy, W. D., and Glass, B., eds.) p. 276, The Johns Hopkins Press, Baltimore, Md.
Vogel, H. J. (1961) in *Control Mechanisms and Cellular Processes* (Bonner, D., ed.) p. 23, Ronald Press, New York.
Vogel, H. J., and Bacon, D. F. (1966) *Proc. Nat. Acad. Sci. U.S.A. 55,* 1456–1459.
Vogel, R. H., and Vogel, H. J. (1963) *Biochim. Biophys. Acta 69,* 174–176.
Vogel, R. H., McLellan, W. L., Hirvonen, A. P., and Vogel, H. J. (1971) in *Metabolic Pathways* (Vogel, H. J., ed.) Vol. V, pp. 463–488, Academic Press, New York.
Vogel, R. H., Devine, E. A., and Vogel, H. J. (1978) *Mol. Gen. Genet. 162,* 157–162.
Vyas, S., and Maas, W. K. (1963) *Arch. Biochem. Biophys. 100,* 542–546.
Williams, A. L., and Williams, L. S. (1973) *J. Bacteriol. 113,* 1433–1441.
Yang, H. L., Zubay, G., Urm, E., Reiness, G., and Cashel, M. (1974) *Proc. Nat. Acad. Sci. U.S.A. 71,* 63–67.
Yanofsky, C. (1981) *Nature 289,* 751–758.
Yem, D. W., and Williams, L. S. (1971) *J. Bacteriol. 107,* 589–591.

CHAPTER
5

Arginine Synthesis in Eukaryotes

Rowland H. Davis

I. ENZYMES AND THEIR LOCALIZATION IN *NEUROSPORA CRASSA* AND *SACCHAROMYCES CEREVISIAE*

This chapter will explore the fully developed arginine pathways of the fungi *Neurospora crassa* and *Saccharomyces cerevisiae,* which are called *Neurospora* and yeast, respectively, in what follows. The steps of arginine synthesis and degradation and their relation to proline and polyamine synthesis in *Neurospora* and yeast are shown in Figs. 5.1, 5.2, and 5.3. Certain differences between these fungi in enzyme localization, urea degradation, and control mechanisms will be pointed out.

A. Ornithine Synthesis

The carbon skeleton of ornithine is derived from glutamate via the acetylated intermediates seen in prokaryotes. However, the fungi have a cyclic pathway in which a transacetylase transfers the acetyl group from acetylornithine to glutamate. This "acetylglutamate cycle" (Fig. 5.1) conserves bound acetyl groups as ornithine is formed. A quantitatively minor acetylglutamate synthase reaction maintains the bound acetyl pool by forming acetylglutamate from glutamate and acetyl-coenzyme A.

N. crassa	*S. cerevisiae*	
		Glutamate
arg14 or *arg13*	*arg2*	N-Acetylglutamate synthase (EC 2.3.1.1)
		N-Acetylglutamate
arg6	*arg6*	N-Acetylglutamate kinase (EC 2.7.2.8)
		N-Acetylglutamyl phosphate
arg6	*arg5*	N-Acetylglutamyl phosphate reductase (EC 1.2.1.38)

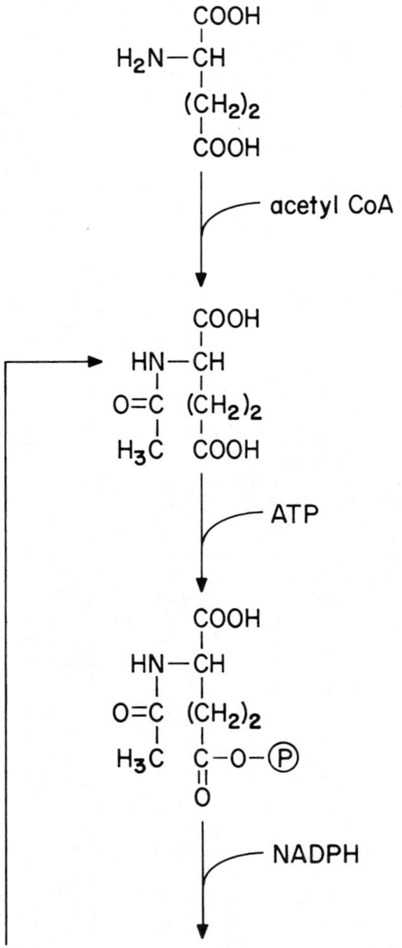

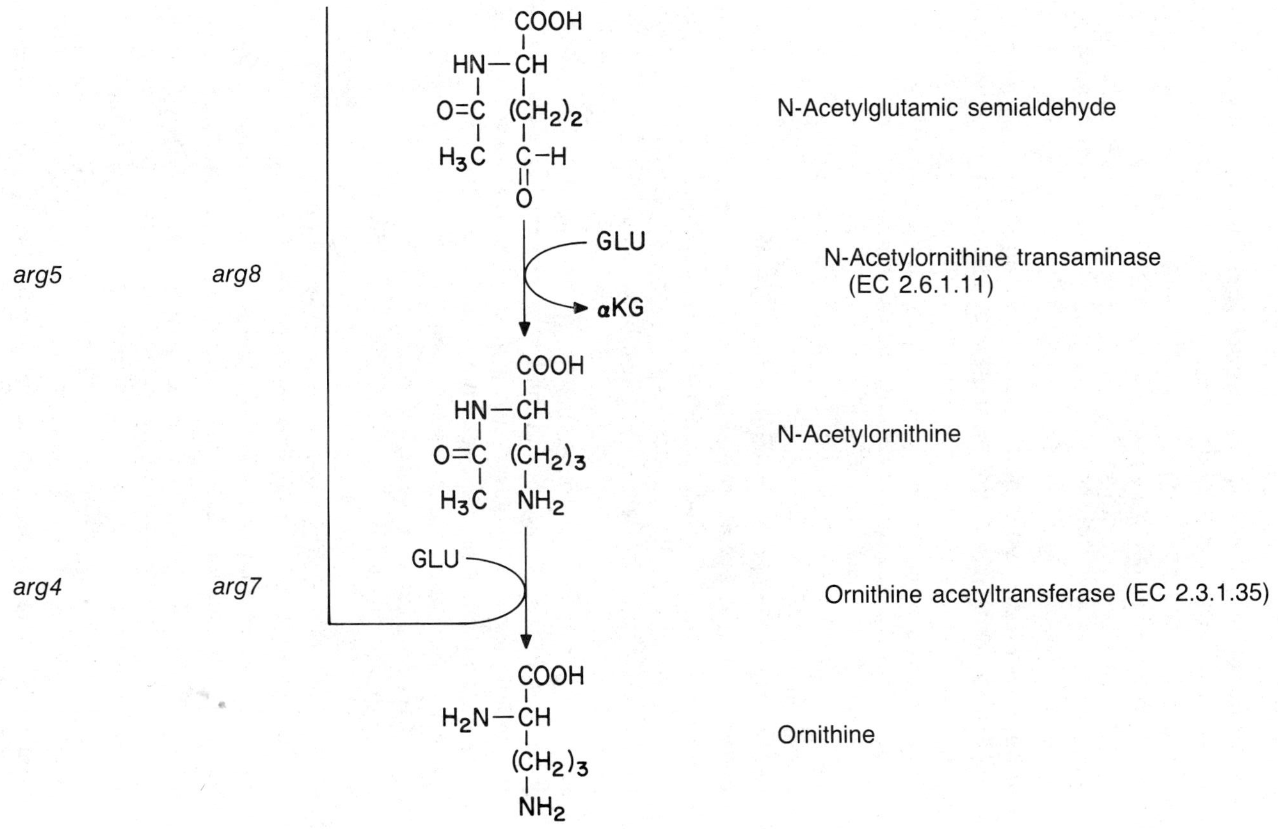

FIGURE 5.1 Biosynthesis of ornithine from glutamate.

Acetylglutamate synthase has been found in mitochondria of yeast (Wipf and Leisinger, 1977, 1979; Jauniaux et al., 1978) and *Neurospora* (R. L. Weiss, personal communication). Little information on affinity constants is available, but the enzyme of yeast is inhibitable and repressible by arginine. A yeast mutant, *arg2*, is presumed to lack the enzyme. It is a strict auxotroph, as would be expected of cells lacking a vital function.

A two-step sequence transforms acetylglutamate to acetylglutamic semialdehyde (Hilger et al., 1973; Cybis and Davis, 1974, 1975; Minet et al., 1979; Wolf and Weiss, 1980). This requires first the phosphorylation of the γ-carboxyl group of glutamate in an ATP-Mg-requiring reaction, usually assayed by hydroxamate formation in the presence of hydroxylamine. A subsequent NADPH-dependent reduction yields the aldehyde, with the liberation of inorganic phosphate. In both fungi, the genes encoding these two enzyme activities are adjacent. Single nonsense mutations in both fungi can eliminate both activities (Davis, 1979; Minet et al., 1979); a *cis*-dominant, operator-like mutation that renders both enzymes constitutive is known in yeast (Jacobs et al., 1980).

It is possible that the mature enzymes, which are physically separable, are translated initially as a single polypeptide from a long mRNA (Jacobs et al., 1980). It is also possible that, despite their separability, the enzymes form a specific aggregate within mitochondria where they are found and can thus handle the labile acetylglutamyl phosphate strictly as an enzyme-bound intermediate. In vitro, the kinase reaction requires high levels (>2 mM) of acetylglutamate, leading one to suspect that chemical and/or conformational requirements may not be met in the usual assay mixture.

Acetylglutamate semialdehyde is transformed to N-acetylornithine by acetylornithine transaminase (Cybis and Davis, 1975; Jauniaux et al., 1978). The amino donor is glutamate. The reaction requires pyridoxal phosphate and is often assayed in the reverse direction.

The acetylglutamate cycle is completed by acetylornithine–glutamate acetyltransferase (DeDeken, 1963). This reversible reaction liberates ornithine and regenerates acetylglutamate, and most of the glutamate destined for arginine synthesis enters the pathway here. The evidence for this is that acetyltransferase-deficient mutants in both fungi grow very poorly without arginine (Davis, 1979), despite the occurrence of acetylglutamate synthase and an "acetylornithinase" (see p. 85) and less specific hydrolases in crude extracts. The mutant phenotypes further suggest that low activity of acetylglutamate synthase is sufficient in the wild type to meet the demands for acetyl groups imposed by growth and spontaneous deacetylation (Hoare et al., 1967). The significance of the acetyl group in this sequence of intermediates is that it imparts some stability to acetylglutamyl phosphate, and it prevents cyclization of the aldehyde, a potentially competitive nonenzymic reaction normally seen in proline synthesis (Thompson, 1980). Thus acetylation of intermediates metabolically isolates proline and arginine syntheses at points of equivalent chemical changes.

The enzymes of ornithine synthesis in both fungi are all located in the mitochondria (Weiss and Davis, 1973; Cybis and Davis, 1975; Wolf and Weiss, 1980; Jauniaux et al., 1978) (Fig. 5.2). At one point, it was thought that acetylglutamate kinase might be extramitochondrial (Cybis and Davis, 1975). Later studies showed that there had been confusion between the assay product of the kinase (acetylglutamyl hydroxamate) and the final product (pyroglutamate) of an undefined, extramitochondrial activity in crude extracts (Wolf and Weiss, 1980).

In *Neurospora,* there is an acetylornithinase that yields ornithine by deacetylation (Vogel and Vogel, 1963). The enzyme requires Co^{2+} and has not been characterized. It is apparently different from the transacetylase, but no mutants are known to lack the activity. No physiological role can be assigned to it because its specificity, regulation, and cellular location are not known. It may account in part, however, for the slight "leakiness" of transacetylase-deficient *Neurospora* mutants.

B. Carbamoyl Phosphate Synthetases

In the fungi, as in bacteria and many other organisms, carbamoyl phosphate is synthesized by carbamoyl phosphate synthetase, which requires glutamate, 2 ATP-Mg, and HCO_3^-. In addition, free Mg^{2+} and K^+ are required for activity. In both *Neurospora* and yeast, there are two carbamoyl phosphate synthetases, designated P and A, for pyrimidine and arginine biosynthesis, respectively (Lacroute et al., 1965; Davis, 1967, 1972). The arginine-specific enzyme (M_r = 170,000) consists of two different polypeptides (M_r = 125,000 and 45,000 for the *Neurospora* enzyme), which are products of unlinked genes (Davis et al., 1980; Piérard and Schroter, 1978). The large subunit is capable of carbamoyl phosphate synthesis from ATP-Mg, HCO_3^-, and NH_3. The small subunit imparts to the holoenzyme the ability to use glutamine but has no detectable enzyme activity, such as glutaminase, by itself. The ammonium-dependent reaction does not support the growth of mutants lacking the small subunit. Thus the glutamine-dependent reaction is the physiological activity of the enzyme.

In both fungi, carbamoyl phosphate synthetase A is negatively controlled by arginine, mainly by control of subunit synthesis, and is indifferent to pyrimidines. The reverse is true of carbamoyl phosphate synthetase P (Williams and Davis, 1970; Williams et al., 1970; Lue and Kaplan, 1970). In *Neurospora,* the carbamoyl phosphate synthesized by the arginine-specific enzyme is normally destined strictly for arginine synthesis. This is not true of the yeast system; carbamoyl phosphate made by the arginine-specific enzyme can be used by aspartate transcarbamylase of the pyrimidine pathway. The difference between the fungi arises because, in *Neurospora,* the enzyme (and presumably its product) is located within the mitochondrion (Davis, 1972; Weiss and Davis, 1973), while in yeast, the enzyme is cytosolic (Urrestarazu et al., 1977) (Fig. 5.2). Thus *Neurospora* mutants lacking

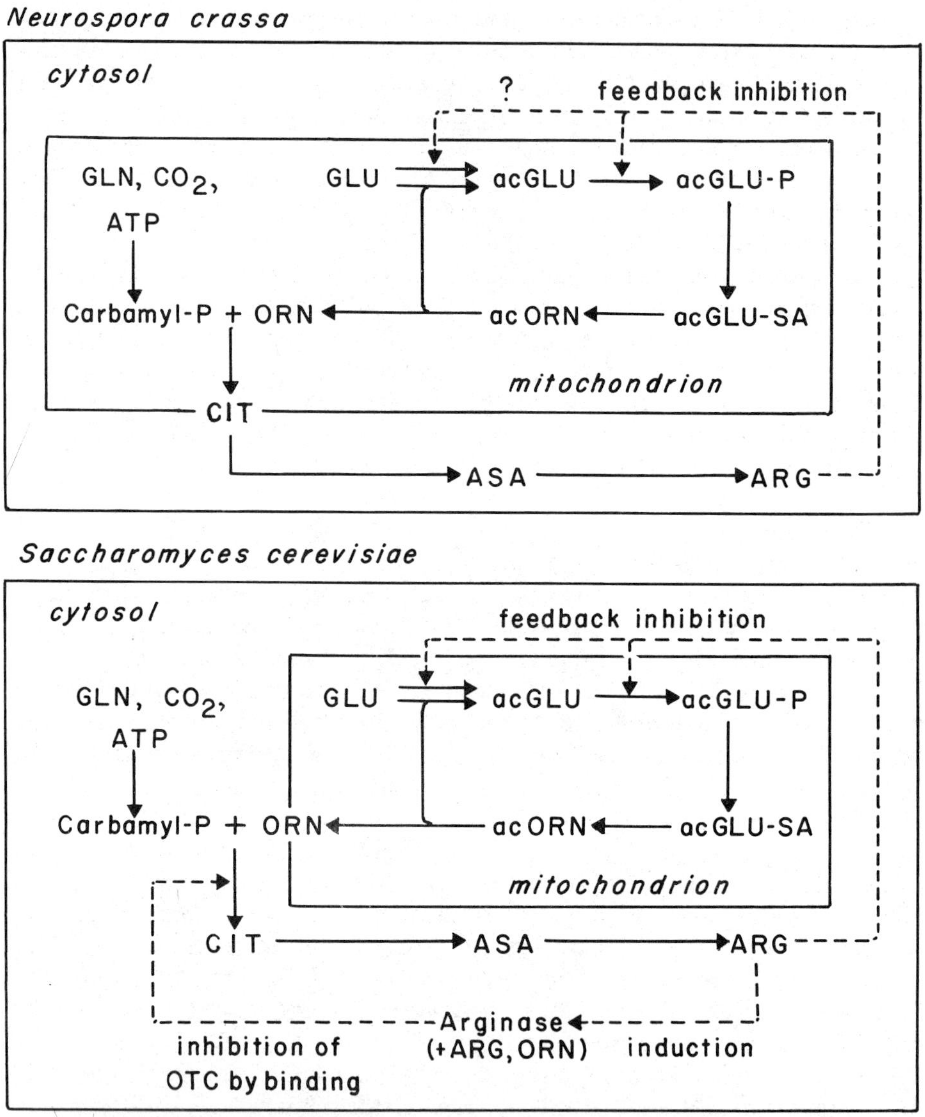

FIGURE 5.2 Compartmentation of enzymes and control of their activities in *Neurospora* and yeast. Feedback inhibition of acetylglutamate synthase has not been tested directly in *Neurospora*. The cytosolic location of ornithine transcarbamylase and carbamoyl phosphate synthetase A as seen in *S. cerevisiae* is a relatively rare phenomenon, even among yeasts; it is not seen among obligate aerobes (Urrestarazu et al., 1977). Enzymes controlled by the *argR* system of *S. cerevisiae* are the two cytosolic enzymes, ornithine transcarbamylase and argininosuccinate synthetase, and all mitochondrial enzymes except the transacetylase. OTC stands for ornithine transcarbamylase; all other abbreviations can be inferred from Figs. 5.1, 5.3, and 5.4, where metabolic intermediates are spelled out.

carbamoyl phosphate synthetase P are pyrimidine auxotrophs; in yeast, they are prototrophs (unless excess arginine is present to repress carbamoyl phosphate synthetase A). As might be expected, carbamoyl phosphate synthetase A mutants are arginine auxotrophs in *Neurospora* and weak or vigorous prototrophs in yeast (Davis, 1972).

C. Arginine Synthesis from Ornithine

As in bacteria and in other eukaryotes, the fungi need three enzymes to make arginine from ornithine (Fig. 5.3). The first is ornithine transcarbamylase, which transfers the carbamoyl group of carbamoyl phosphate to the δ-amino group of ornithine. The yeast enzyme is made of three identical subunits (Penninckx et al., 1974) and has high affinity for carbamoyl phosphate. The equilibrium constant is strongly in favor of citrulline. In *Neurospora,* the enzyme is mitochondrial (Bernhardt and Davis, 1972); in yeast, it is cytosolic (Urrestarazu et al., 1977). The significance of this will be discussed below. The second enzyme is argininosuccinate synthetase, which condenses citrulline and aspartate, with energy from ATP, to form argininosuccinate. The reaction, described in detail for the mammalian liver enzyme (Ratner, 1973), actually involves formation of pyrophosphate and an enzyme-bound AMP-citrulline adduct. The latter breaks down in the presence of aspartate to form free enzyme, AMP, and argininosuccinate. The pyrophosphate is hydrolyzed independently, a step that renders the argininosuccinate synthetase reaction physiologically irreversible. The third enzyme in arginine synthesis from ornithine is argininosuccinate lyase, which catalyzes a quite reversible hydrolysis of argininosuccinate to arginine and fumarate. Mutants for all three steps are known in both organisms. In both organisms, the last two enzymes of the pathway are cytosolic (Weiss and Davis, 1973; Jauniaux et al., 1978) and never become rate-limiting under normal circumstances (Flint et al., 1980).

D. The Vacuole

Large amounts of arginine and ornithine are stored in the vacuoles of *Neurospora* and yeast (Weiss, 1973; Wiemken and Nurse, 1973; Nurse and Wiemken, 1974). In steady state cultures of *Neurospora,* the maintenance of these pools (approximately 30 mmoles per milligram of dry weight) diverts approximately 15% of the glutamate carbon entering the pathway (Bowman and Davis, 1977a). The mechanisms by which these pools are used for amino nitrogen (in catabolism) and for regulation of cytosolic pool size are becoming better understood (see Section II.B) (Weiss, 1976; Weiss and Davis, 1977; Legerton and Weiss, 1979; Davis et al., 1978a, b; Nurse and Wiemken, 1974; Messenguy et al., 1980).

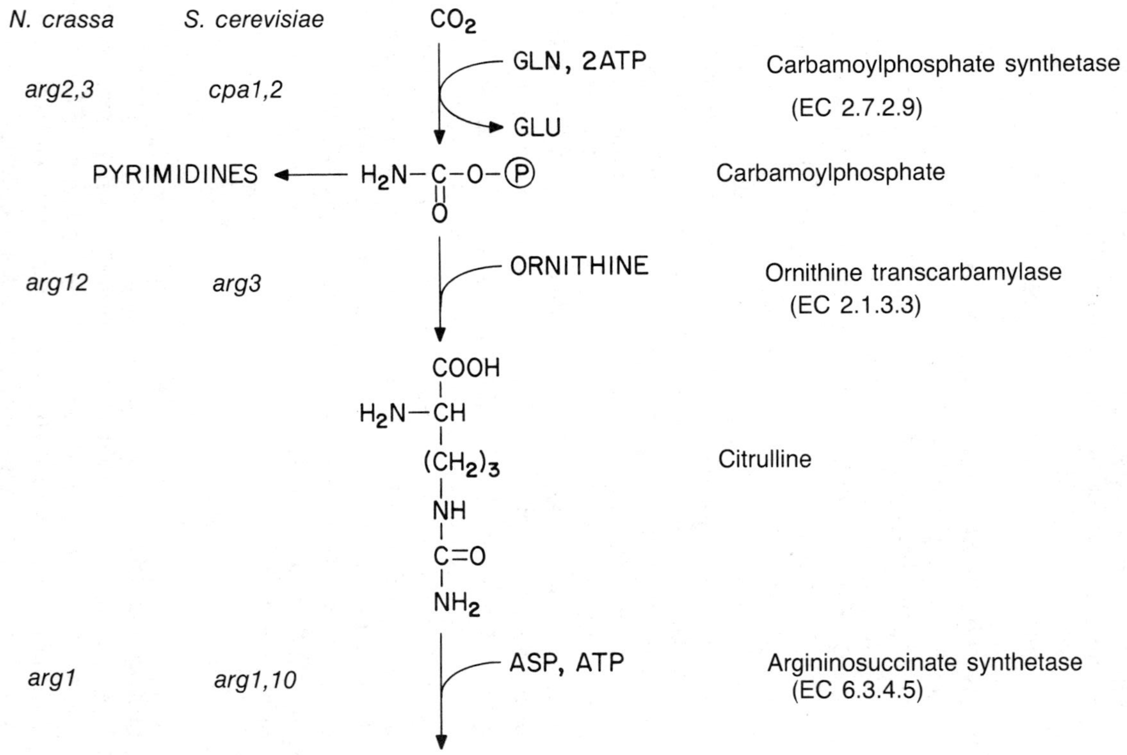
N. crassa
S. cerevisiae
CO2
GLN, 2ATP
GLU
arg2,3
cpa1,2
Carbamoylphosphate synthetase
(EC 2.7.2.9)
PYRIMIDINES
H2N-C-O-P
O
Carbamoylphosphate
ORNITHINE
arg12
arg3
Ornithine transcarbamylase
(EC 2.1.3.3)
COOH
H2N—CH
(CH2)3
NH
C=O
NH2
Citrulline
ASP, ATP
arg1
arg1,10
Argininosuccinate synthetase
(EC 6.3.4.5)

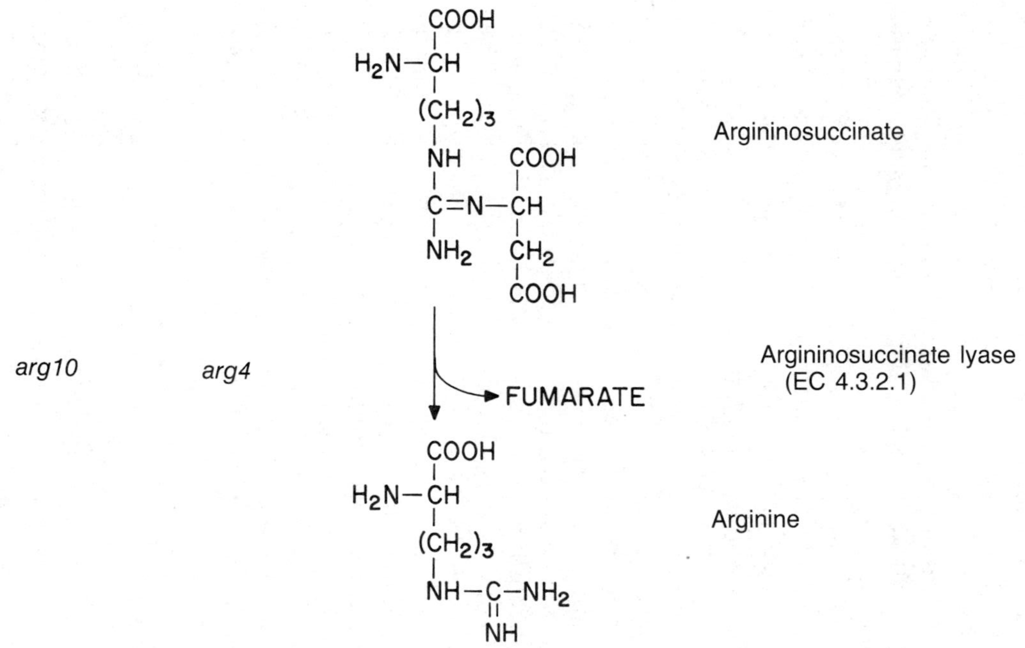

FIGURE 5.3 Biosynthesis of arginine from ornithine.

E. Arginine Catabolism and Polyamine Synthesis

In *Neurospora* and yeast, arginine is broken down to urea and ornithine by arginase (Davis et al., 1970; Middelhoven, 1964). Ornithine is converted to glutamate-γ-semialdehyde by ornithine transaminase (Davis, 1968; Davis and Mora, 1968; Middelhoven, 1964). The spontaneous cyclization of the semialdehyde to Δ^1pyrroline-5-carboxylate follows, and the compound is then enzymatically reduced to proline (Karlin et al., 1976; Brandriss and Magasanik, 1980). All of these reactions are cytosolic (Weiss and Davis, 1973; Jauniaux et al., 1978). In yeast, it has been shown recently that proline is an obligatory intermediate in the conversion of ornithine to glutamate; the proline derived from ornithine is oxidized to a separate intramitochondrial pool of Δ^1-pyrroline-5-carboxylate, which in turn is converted to glutamate in an NAD-linked dehydrogenase reaction (Brandriss and Magasanik, 1980). No critical evidence is available on this pathway in *Neurospora* except that ^{14}C-ornithine catabolism yields proline (at low levels of ornithine) and, at higher levels, glutamate as well (Karlin et al., 1976).

Urea, the second product of the arginase reaction, is broken down in *Neurospora* by a constitutive urease reaction to CO_2 and NH_3. This reaction is also the last step of purine breakdown (Davis, 1970). In yeast, however, disposal of urea takes place in a two-step sequence in which urea is first carboxylated by urea carboxylase, an ATP-requiring, biotin enzyme. The product, allophanate, is then hydrolyzed to 2 moles each of CO_2 and NH_3. This mode of urea catabolism is rare but occurs in *Chlorella* and *Chlamydomonas,* as well as in other yeasts (Roon and Levenberg, 1970; Whitney and Cooper, 1972).

Both fungi derive the putrescine moiety of polyamines from ornithine by decarboxylation (Fig. 5.4) (Paulus and Davis, 1981; Whitney and Morris, 1978). Ornithine decarboxylase is a cytosolic, pyridoxal phosphate-requiring enzyme with quite low activity and uses about 8% of the ornithine biosynthesized de novo in *Neurospora* (Karlin et al., 1976). In conditions of arginine catabolism, when ornithine biosynthesis is feedback-inhibited, the ornithine used by ornithine decarboxylase arises in the arginase reaction.

II. MECHANISMS OF CONTROL

In a study of enzyme activities in relation to the dosage of the corresponding genes in yeast tetraploids, Hilger et al. (1973) identified acetylglutamate kinase (encoded by *arg6*) and the small subunit of carbamoyl phosphate synthetase A (encoded by *cpa1*) as limiting factors in arginine synthesis, whereas ornithine transcarbamylase and argininosuccinate lyase were not. While other enzymes, if studied, might also be identified as limiting factors, the kinase and carbamoyl phosphate synthetase A subunit are appropriate controlling reactions. In *Neurospora,* carbamoyl phosphate pool determinations reveal that, at steady state, there is enough of this compound for about one second of arginine synthesis in exponential growth (Williams et al., 1971).

A. Control of Enzyme Catalysis

In yeast and *Neurospora,* two enzymes of ornithine synthesis are inhibited by arginine. These are acetylglutamate synthase and acetylglutamate kinase (Wipf and Leisinger, 1979; Hilger et al., 1973; Wolf and Weiss, 1980; Weiss and Lee, 1980). Both must be inhibited if the synthesis of acetylglutamyl phosphate is to be stopped quickly. The K_i of the yeast synthase for arginine is about 20 μM; that of the kinase in both fungi is 50 μM. The kinase of *Neurospora* is much less sensitive to ornithine, citrulline, or lysine (Wolf and Weiss, 1980). The problem posed by the feedback control of ornithine synthesis in both organisms is how it is exerted by arginine across the mitochondrial membrane. Presumably, the mitochondrial arginine concentration is related to the cytosolic concentration, but this may not always be the case (I. Goodman and R. L. Weiss, unpublished). Passage of arginine across the membrane is not known to involve a specific carrier; at higher concentrations, arginine is able to permeate nonspecifically if the mitochondrion has a negative-inside membrane potential (T. J. Paulus and R. H. Davis, unpublished).

Carbamoyl phosphate synthetase A is wholly indifferent to arginine, to any of its precursors, or to related compounds (Davis et al., 1980). No arginine-sensitive carbamoyl phosphate synthetase has ever been found in any organism. While the negative control of carbamoyl phosphate synthetase A evidently occurs at the level of enzyme synthesis, a short-term economic problem remains. As long as carbamoyl phosphate synthetase A is active, the possibility of a wasteful ornithine cycle prevails, because in conditions of excess arginine, ornithine will be made via arginase and—with the exception of weak arginine sensitivity of argininosuccinate synthetase (Wampler and Fairley, 1967)—none of the other enzymes of the cycle is inhibited by arginine. This economic problem is met in quite different ways by *Neurospora* and yeast.

In yeast, arginase has low activity in cells grown in an ammonium-based minimal medium, and the enzyme is induced by arginine (Wiame, 1971). Arginase, in the presence of relatively high concentrations of ornithine and arginine, makes a stoichiometric aggregate with ornithine transcarbamylase, thus inhibiting this cytosolic enzyme (Messenguy and Wiame, 1969; Penninckx, 1975). Therefore when arginine is added to cells, rapid induction of arginase effects a blockage of preexisting ornithine transcarbamylase activity. This antienzyme, or "epiarginase," role does not save the cell the ATP used in carbamoyl phosphate synthesis, but it diverts ornithine to the catabolic route and prevents consumption of two high-energy bonds in the argininosuccinate synthetase reaction. In yeast, the rate of carbamoyl phosphate production after addition of excess arginine is not known.

N. crassa	S. cerevisiae	
		Arginine
aga	*car1*(= *cargA*)	Arginase (EC 3.5.3.1)
		Ornithine
spe-1(= put-1)	*spe1*	Ornithine decarboxylase (EC 4.1.1.17)
		Putrescine

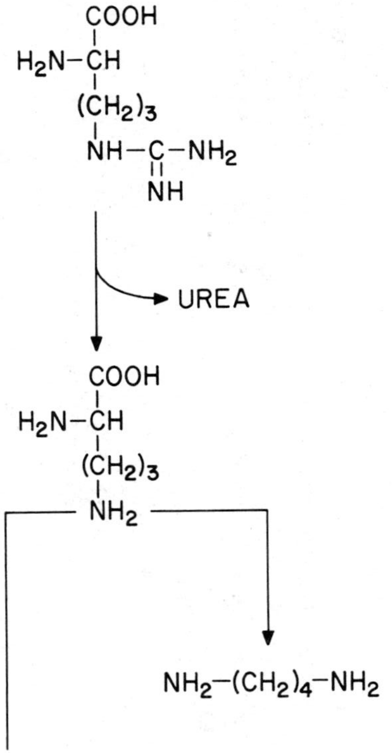

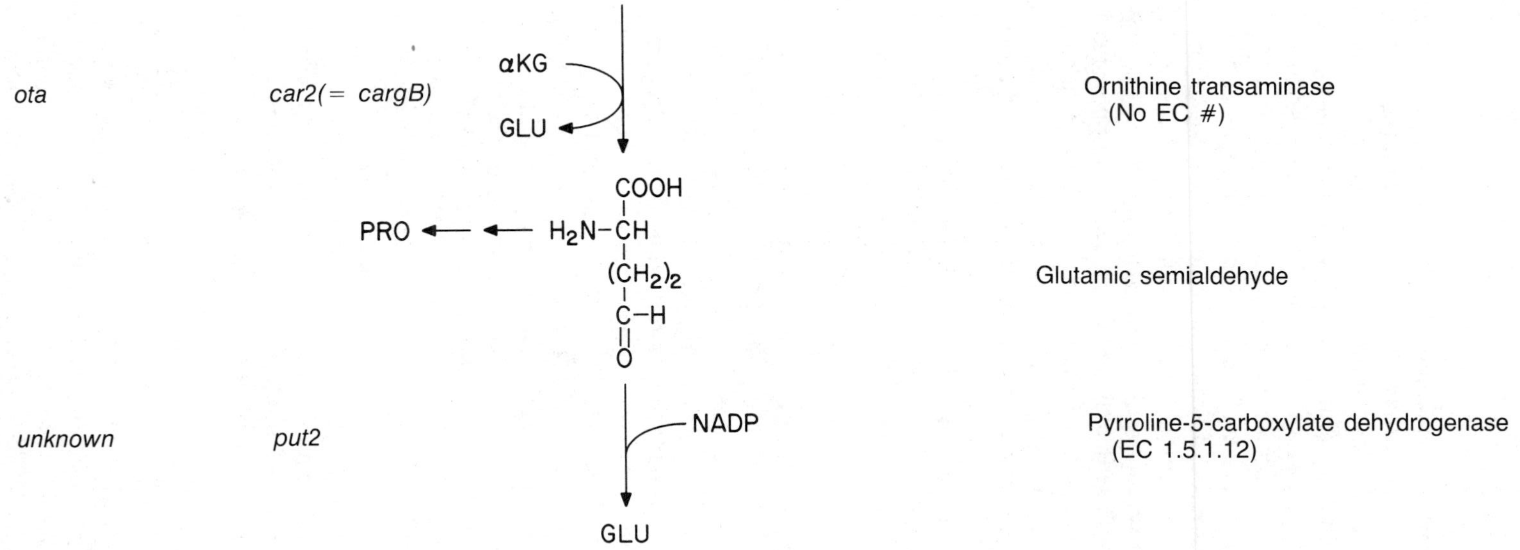

FIGURE 5.4 Arginine degradation and putrescine biosynthesis.

In *Neurospora,* a compartmental phenomenon is the basis of interrupting the ornithine cycle. Ornithine transcarbamylase is mitochondrial in this organism, and therefore arginase, a cytosolic enzyme, cannot act as an ornithine transcarbamylase antienzyme. Our studies have shown that minutes after arginine is added to minimal-grown cells, the rate of citrulline production falls to approximately 20% of its normal rate (Bowman and Davis, 1977a, b). The mechanism proposed for this is as follows: Excess arginine feedback inhibits the intramitochondrial source of ornithine and provides, via arginase action, cytosolic ornithine. The movement of cytosolic ornithine across the mitochondrial membrane to the site of ornithine transcarbamylase is probably the limiting factor in the cycle (T. J. Paulus and R. H. Davis, unpublished). It is not known whether arginine inhibits ornithine transport into mitochondria. However, the 20% "cycling" of ornithine continues for many hours until carbamoyl phosphate synthetase A reaches its repressed level of almost zero (Bowman and Davis, 1977a). The control of the ornithine cycle in *Neurospora* therefore appears to be somewhat inefficient. With a carbon-rich diet, however, *Neurospora* does not need to conserve ATP very efficiently (Drainas and Weiss, 1980).

B. Control of Cytosolic Ornithine and Arginine Pools

In the fungi and algae, vacuoles contain considerable amounts of arginine and ornithine, retained in part by interaction with polyphosphate and perhaps other fixed anions (Cramer et al., 1980; Vaughn and Davis, 1981; Dürr et al., 1979). In the fungi under consideration, the arginine in the vacuole is the major short-term determinant of cytosolic arginine levels. When arginine is in short supply, it moves out of the vacuole to support protein synthesis. When arginine is in excess, it is taken up by the vacuole, as well as being catabolized (Weiss and Davis, 1977). Thus the arginine concentration in the cytosol is affected more immediately by the interaction of the vacuole and its cellular milieu than by feedback inhibition or repression (Davis et al., 1978a).

C. Control of Enzyme Levels

Three control systems affecting arginine biosynthetic enzymes have been defined by mutation in yeast. Their characteristics and interactions will be described with comparative comments on the *Neurospora* system at each point.

1. The *cpa1*-Specific System. Carbamoyl phosphate synthesis appears to be controlled largely through variations in the level of the small subunit of carbamoyl phosphate synthetase A; the enzyme is not feedback-inhibited by arginine (Piérard and Schröter, 1978; Piérard et al., 1979). On a minimal medium with ammonium as the nitrogen source, arginine effects a fourfold to sixfold reduction in the level of the polypeptide. Upon starvation of auxotrophs for arginine, the level increases about tenfold. The amplitude of regulation is thus about 50-fold. Two mutations render the polypeptide's synthesis largely constitutive (Thuriaux et al., 1972). One, *cpa81* (originally, *cpaR*), is recessive and unlinked to *cpa1*, the structural gene for the small carbamoyl phosphate synthetase A polypeptide. The other, *cpa80* (originally, *cpa1O*), is *cis*-dominant and tightly linked to *cpa1*. While no molecular studies on the regulatory system have been done, a negative control model is favored: *cpa81* codes for a regulatory protein, which, in the presence of arginine or a derivative, blocks transcription of the *cpa1* gene at the *cpa80* (operator) site. Neither of the mutations appears to influence the level of the large polypeptide directly. In *Neurospora,* the regulatory phenomena are the same, but no regulatory mutants have been discovered (Davis et al., 1981).

2. The Arginine-specific "*argR*" System. This system (Wiame, 1971; Messenguy, 1979) is defined in yeast by three unlinked mutations, *arg80, arg 81,* and *arg82* (originally designated *argRI, argRII,* and *argRIII*). Strains carrying these mutations have derepressed levels of six enzymes, the products of the *arg2, arg5/6, arg8, arg3,* and *arg1/10* loci. All three types of regulatory mutation are recessive, and they cannot be qualitatively distinguished. The original model of regulation (Wiame, 1971) held that the three regulatory loci contribute to a heteromultimeric, negative regulator (aporepressor) that exerts repression when bound by arginine or arginyl-tRNA. The mutations do not appear to affect the levels of the three arginyl-tRNA species, nor of arginyl-tRNA synthetase (Hoet and Wiame, 1974). The targets for the putative repressor are presumed to be operator-like segments adjacent to the five relevant structural genes. Support for this model comes from the isolation of *cis*-dominant, constitutive mutations, one very tightly linked to *arg3* (ornithine transcarbamylase) (Messenguy, 1976) and the other to *arg5/6* (acetylglutamate kinase and acetylglutamyl phosphate reductase) (Jacobs et al., 1980). The latter "0^c" mutation has been mapped to the left of the leftmost mutant of the *arg6* segment, which is in turn to the left of *arg5.* Polarity studies involving nonsense mutations imply that the two genes are transcribed in the direction from *arg6* through *arg5* (Minet et al., 1979).

In cultures grown in a minimal medium, repression by arginine of the six enzymes controlled by the *argR* system is about tenfold. The remaining biosynthetic enzymes are not repressed under these conditions. Similarly, mutations of the *argR* system (*arg80, arg81,* and *arg82*) cause a twofold elevation of the six controlled enzymes and render them nonrepressible. These mutations have no effect on the other arginine biosynthetic enzymes. However, all nine biosynthetic enzymes can be derepressed by starvation for arginine, even the six *argR*-regulated enzymes in *arg80, arg81,* or *arg82* mutants (Delforge et al., 1975; Messenguy, 1979). The further derepression of *arg* genes in *argR* mutants is due to the contribution of the "general" system discussed below.

Perhaps the most interesting feature of the *argR* system is that all three regulatory mutations cause *noninducibility* of the catabolic enzymes, arginase and ornithine transaminase. An early analysis has yielded a regulatory model (Wiame, 1971) whose form rests wholly on the behavior of a set of mutants affecting the regulation of catabolism. Briefly, a recessive, pleiotropic *car82* (originally, *cargR*) mutation leads to constitutivity of arginase and ornithine transaminase, even in *car82 arg80* double mutants. Moreover, *cis*-dominant, constitutive mutations lying adjacent to the arginase (*car1*) and ornithine transaminase (*car2*) structural genes have been isolated. The model is that *car82* specifies a negative repressor that binds to operators of the *car1* and *car2* genes. The wild-type *argR* system, in the presence of arginine, interferes with *car82*$^+$ function (at the gene or protein level). When the *argR* system is inactivated by mutation, however, *car82* function (negative control) cannot be relieved, and the catabolic enzymes become noninducible. The *car82* mutation will relieve repression of the catabolic enzyme loci regardless of the status of the *argR* system.

A serious deficiency in our knowledge of this system is that few molecular studies have been reported. Despite one report (Messenguy and Cooper, 1977), it is not certain whether the *argR* system controls the synthesis of mRNA, or enzyme processing, or protein activation. The significance of operator mutations of any kind in this system has been diminished by the discovery of transposable elements (for example, *ty1*) and their effects. These DNA elements are capable of inserting adjacent to genes that are then under the control of transposable element promoters (that is, immune to arginine regulatory signals). Finally, the nature of the *arg80, arg81,* and *arg82* products is not known, nor is the significance of the fact that three unlinked genes have similar and nonadditive effects.

Neurospora appears to have no counterpart of the yeast *argR* system. All enzymes except carbamoyl phosphate synthetase A have similar amplitudes of regulation (Barthelmess et al., 1974; Cybis and Davis, 1975). All tested enzymes appear to be controlled by the "general" system, and no mutations corresponding phenotypically to *arg80, arg81, arg82,* or *car82* are known.

3. The General Control System. A broad system of control of amino acid biosynthetic enzymes was recognized in *Neurospora* (Carsiotis and Jones, 1974; Carsiotis et al., 1974) and yeast (Schürch et al., 1974) when starvation for one amino acid was found to cause derepression of enzymes in many other pathways. While it is not known how many pathways are involved, this mutual, cross-pathway control system embraces some or all enzymes of the aromatic, basic, and aspartate family amino acids as well as serine and valine (Niederberger et al., 1981). Reversal of derepression can be effected only by adding the amino acid of the deficient pathway. In the case of the arginine pathway of yeast, all arginine enzymes—including the two separate polypeptides of carbamoyl phosphate synthetase A—are affected by the system (Messenguy, 1979; Piérard et al., 1979).

Both constitutive (*tra3, tra5*) and several nonallelic, nonderepressible (*ndr1, ndr2, aas1, aas2, aas3*) mutants for general control have been isolated in various regimes. Working mainly in the histidine system, Wolfner et al. (1975) found the *tra3* and *tra5* mutants on the basis of resistance to 1,2,4-triazole-3-alanine. When grown at 30°C, the mutants show high levels of amino acid biosynthetic enzymes, which are not diminished by amino acid supplements. When grown at 37°C, the *tra3* mutant behaves as a cell division cycle variant, arresting at the cycle "start" point in G1. In the same work, the *aas1, aas2,* and *aas3* mutants were selected as strains sensitive to amino acid analogs. These mutants failed to derepress amino acid biosynthetic enzymes upon starvation for any of several amino acids. The *tra aas* double mutants all have constitutive enzyme levels. This would indicate that aas^+ function opposes the tra^+ function, which in turn would be a negative regulator of the structural genes. It is thus implied that amino acid sufficiency opposes the aas^+ function.

Another mutation, Gen^c, has an unusual phenotype (Messenguy, 1979). Enzyme levels are normal in cells grown on a minimal medium but derepressed rather than repressed in cells grown on a rich medium. Arginine by itself has no effect. This mutation, like *tra3*, is epistatic to the two tested *aas* mutations *aas2* and *aas3*. Its effect upon the cell cycle is not known.

Recently, Niederberger et al. (1981) reported a comprehensive study of *ndr1* and *ndr2* mutations, the latter allelic to *aas1*. Both strains are nonderepressible in the general control circuit. The findings confirm previous demonstrations that enzyme derepression is not coordinate, nor are all enzymes of a given pathway affected. Amino acid imbalances imposed by additions to or omissions from the medium caused demonstrable derepression of certain enzymes in wild-type cells. The inability of *ndr* strains to effect such derepression was correlated with a limitation of an amino acid pool in the cell and a reduction of growth rate. The pool and growth limitations could be reversed by suitable supplementation. Outright limitation of nitrogen, phosphorus, carbon, or sulfur, however, did not cause derepression of the controlled enzymes in wild-type cells. Therefore the general control system appears to respond to amino acid balance rather than to the broader nutritional status.

The general control system, when manipulated by growth of bradytrophic strains, appears to work at the level of transcription (Messenguy and Cooper, 1977). The conclusion is based only upon indirect evidence concerning accumulation and turnover rates of enzyme synthetic capacity for ornithine transcarbamylase. In *Neurospora,* derepression appears to be controlled by the general system, for which nonderepressible mutations have been found (Davis, 1979; Barthelmess, 1982). A systematic study of all enzymes affected by this mutation has not been done.

4. Interactions of Control Mechanisms. The *cpa81* and *argR* systems do not interact in any clear way; this is in keeping with their different specificities. Both, however, interact in similar ways with the general system. In order to reveal the full derepression of the general system, the action of the more specific systems must be impaired (Delforge et al., 1975). This may be done by mutation *cpa81, arg80, arg81,* or *arg82* or by withdrawing arginine itself. Thus while arginine starvation brings out the full potential for derepression of arginine enzymes controlled by the *argR* system, histidine starvation does so only in mutants.

For the arginine loci without a specific control, derepression is rather small, and it is not limited by the *argR* products (Messenguy, 1979). While the enzyme loci not affected by *argR* or *cpa81* are not repressible by arginine, they are greatly repressed by a rich medium. This response is presumably mediated by the "general" system. If so, a rich medium might be expected to repress an *argR*-controlled enzyme even in an *arg80* strain. The fact that, in *argR* mutants, ornithine transcarbamylase is not repressible by a rich medium (Messenguy, 1979) indicates an interaction of the systems that has not been fully illuminated.

III. ARGININE SYNTHESIS IN OTHER EUKARYOTES

A. Plants (including Algae)

The synthesis of arginine in plants follows the chemical steps found in *Neurospora,* including the acetylglutamate cycle (Morris and Thompson, 1975, 1977; McKay and Shargool, 1977; Thompson, 1980). In *Chlorella* and in higher plants, as in fungi, control of activity is exerted by arginine on acetylglutamate synthase and acetylglutamate kinase activities, but not on the acetylornithine–glutamate transacetylase. In *Chlorella,* acetylglutamate synthase and the transacetylase copurify over a 180-fold purification, but the activity of the synthase is more labile. The question of whether the two activities are catalyzed by one entity is thus open.

Two differences in the pathway have been discovered. First, only one carbamoyl phosphate synthetase has so far been detected in plants. It has the general regulatory properties of the *E. coli* enzyme, namely, inhibition by UMP and activation by ornithine (Ong and Jackson, 1972; O'Neal and Naylor, 1976). This suggests its role in both pyrimidine and arginine synthesis. Second, this enzyme and ornithine transcarbamylase appear, in soybean, to be associated with the plastids, rather than with mitochondria or the cytosol (Shargool et al., 1978). There is also some evidence for the cytosolic location of argininosuccinate synthetase and lyase (Shargool et al., 1978).

In view of this rudimentary knowledge, one may speculate that arginine enzymes that are mitochondrial in *Neurospora* will be found in plant plastids.

B. Animals

Animals vary greatly in their patterns of arginine metabolism. Because few types of animals have been studied in detail, it is not possible to generalize with confidence. Many animals (from protozoa to vertebrates) can synthesize arginine from glutamate; arginine is commonly not an essential amino acid (Campbell, 1970; Meister, 1965). Recent studies are beginning to confirm the impression of early tracer work in mammals (Sallach et al., 1951) that ornithine may originate via ornithine transaminase from the proline precursor pyrroline-5-carboxylate (Ross et al., 1978; Smith et al., 1980). No evidence suggests an acetylated pathway among animals, but critical tests on invertebrates are lacking.

The ability to synthesize arginine is lost entirely in many insects and reptiles and in all birds, although some of these organisms retain the ability to convert citrulline to arginine (Tamir and Ratner, 1963a, b; Mora et al., 1965; Cohen, 1976).

ACKNOWLEDGMENTS

I thank Drs. T. G. Cooper, E. W. Jones, and M. E. Jones for valuable discussion; Dr. T. Paulus, Dr. R. L. Weiss, and Ms. C. L. Cramer for critical reading of the manuscript; and Ms. Virginia Baker for typing. The author's research has been supported by the National Institute for Arthritis, Metabolic and Digestive Diseases and the National Science Foundation.

REFERENCES

Barthelmess, I. B. (1982) *Genet. Res. 39,* 169–185.

Barthelmess, I. B., Curtis, C. F., and Kacser, H. (1974) *J. Mol. Biol. 87,* 303–316.

Bernhardt, S. A., and Davis, R. H. (1972) *Proc. Nat. Acad. Sci. U.S.A. 69,* 1868–1872.

Bowman, B. J., and Davis, R. H. (1977a) *J. Bacteriol. 130,* 274–284.

Bowman, B. J., and Davis, R. H. (1977b) *J. Bacteriol. 130,* 285–291.

Brandriss, M. C., and Magasanik, B. (1980) *J. Bacteriol. 143,* 1403–1410.

Campbell, J. W. (ed.) (1970) *Comparative Biochemistry of Nitrogen Metabolism* Vol. I, *The Invertebrates*; Vol. II, *The Vertebrates*; Academic Press, New York.

Carsiotis, M., and Jones, R. F. (1974) *J. Bacteriol. 119,* 889–892.

Carsiotis, M., Jones, R. F., and Wesseling, A. C. (1974) *J. Bacteriol. 119,* 893–898.

Cohen, P. P. (1976) in *The Urea Cycle* (Grisolia, S., Baguena, R., and Mayor, F., eds.) pp. 21–38, John Wiley, New York.

Cramer, C. L., Vaughn, L. E., and Davis, R. H. (1980) *J. Bacteriol. 142,* 945–952.
Cybis, J. J., and Davis, R. H. (1974) *Biochem. Biophys. Res. Comm. 60,* 629–634.
Cybis, J., and Davis, R. H. (1975) *J. Bacteriol. 123,* 196–202.
Davis, R. H. (1967) in *Organizational Biosynthesis* (Vogel, H. J., Lampen, J. O., and Bryson, V., eds.) pp. 303–322, Academic Press, New York.
Davis, R. H. (1968) *J. Bacteriol. 96,* 389–395.
Davis, R. H. (1970) *Biochim. Biophys. Acta 215,* 412–414.
Davis, R. H. (1972) *Science 178,* 835–840.
Davis, R. H. (1979) *Genetics 93,* 557–575.
Davis, R. H., and Mora, J. (1968) *J. Bacteriol. 96,* 383–388.
Davis, R. H., Lawless, M. B., and Port, L. A. (1970) *J. Bacteriol. 102,* 299–305.
Davis, R. H., Bowman, B. J., and Weiss, R. L. (1978a) *J. Supramol. Struc. 9,* 473–488.
Davis, R. H., Weiss, R. L., and Bowman, B. J. (1978b) in *Microenvironments and Metabolic Compartmentation,* (Srere, P. A., and Estabrook, R. W., eds.) pp. 197–210, Academic Press, New York.
Davis, R. H., Ristow, J. L., and Hanson, B. A. (1980) *J. Bacteriol. 141,* 144–155.
Davis, R. H., Ristow, J. L., and Ginsburgh, C. L. (1981) *Mol. Gen. Genet. 181,* 215–221.
DeDeken, R. H. (1963) *Biochim. Biophys. Acta 78,* 606–616.
Delforge, J., Messenguy, F., and Wiame, J.-M. (1975) *Eur. J. Biochem. 57,* 231–239.
Drainas, C., and Weiss, R. L. (1980) *J. Bacteriol. 141,* 205–212.
Dürr, M., Urech, K., Boller, T., Wiemken, A., Schwencke, J., and Nagy, M. (1979) *Arch. Microbiol. 121,* 169–175.
Flint, H. J., Porteous, D. J., and Kacser, H. (1980) *Biochem. J. 190,* 1–15.
Hilger, F., Culot, M., Minet, M., Piérard, A., Grenson, M., and Wiame, J.-M. (1973) *J. Gen. Microbiol. 75,* 33–41.
Hoare, D. S., Hoare, S. L., and Brame, J. (1967) *J. Bacteriol. 94,* 782–783.
Hoet, P. P., and Wiame, J.-M. (1974) *Eur. J. Biochem. 43,* 87–92.
Jacobs, P., Jauniaux, J.-C., and Grenson, M. (1980) *J. Mol. Biol. 139,* 691–704.
Jauniaux, J.-C., Urrestarazu, L. A., and Wiame, J.-M. (1978) *J. Bacteriol. 133,* 1096–1107.
Karlin, J. N., Bowman, B. J., and Davis, R. H. (1976) *J. Biol. Chem. 251,* 3948–3955.
Lacroute, F., Piérard, A., Grenson, M., and Wiame, J.-M. (1965) *J. Gen. Microbiol. 40,* 127–142.
Legerton, T. L., and Weiss, R. L. (1979) *J. Bacteriol. 138,* 909–914.
Lue, P. F., and Kaplan, J. G. (1970) *Biochim. Biophys. Acta 220,* 365–372.
McKay, G., and Shargool, P. D. (1977) *Plant Sci. Lett. 9,* 189–193.
Meister, A. (1965) *Biochemistry of the Amino Acids* Vols. I and II (2nd ed.), Academic Press, New York.
Messenguy, F. (1976) *J. Bacteriol. 128,* 49–55.
Messenguy, F. (1979) *Mol. Gen. Genet. 169,* 85–95.
Messenguy, F., and Cooper, T. G. (1977) *J. Bacteriol. 130,* 1253–1261.
Messenguy, F., and Wiame, J.-M. (1969) *FEBS Lett. 3,* 47–49.
Messenguy, F., Colin, D., and ten Have, J.-P. (1980) *Eur. J. Biochem. 108,* 439–447.
Middelhoven, W. J. (1964) *Biochim. Biophys. Acta 93,* 650–652.

Minet, M., Jauniaux, J.-C., Thuriaux, P., Grenson, M., and Wiame, J.-M. (1979) *Mol. Gen. Genet. 168,* 299–308.
Mora, J., Martuscelli, J., Ortiz-Pineda, J., and Soberon, G. (1965) *Biochem. J. 96,* 28–35.
Morris, C. J., and Thompson, J. F. (1975) *Plant Phys. 55,* 960–967.
Morris, C. J., and Thompson, J. F. (1977) *Plant Phys. 59,* 684–687.
Niederberger, P., Miozzari, G., and Hutter, R. (1981) *Mol. Cell. Biol. 1,* 584–593.
Nurse, P., and Wiemken, A. (1974) *J. Bacteriol. 117,* 1108–1116.
O'Neal, T. D., and Naylor, A. W. (1976) *Plant Phys. 57,* 23–38.
Ong, B. L., and Jackson, J. F. (1972) *Biochem. J. 129,* 583–593.
Paulus, T. J., and Davis, R. H. (1981) *J. Bacteriol. 145,* 14–20.
Penninckx, M. (1975) *Eur. J. Biochem. 58,* 533–538.
Penninckx, M., Simon, J.-P., and Wiame, J.-M. (1974) *Eur. J. Biochem. 49,* 429–442.
Piérard, A., and Schröter, B. (1978) *J. Bacteriol. 134,* 167–176.
Piérard, A., Messenguy, F., Feller, A., and Hilger, F. (1979) *Mol. Gen. Genet. 174,* 163–171.
Ratner, S. (1973) *Adv. Enzymol. 39,* 1–90.
Roon, R. J., and Levenberg, B. (1970) *J. Biol. Chem. 245,* 4593–4595.
Ross, G., Dunn, D., and Jones, M. E. (1978) *Biochem. Biophys. Res. Comm. 85,* 140–147.
Sallach, H. J., Koeppe, R. E., and Rose, W. C. (1951) *J. Amer. Chem. Soc. 73,* 4500.
Schürch, A., Miozzari, J., and Hütter, R. (1974) *J. Bacteriol. 117,* 1131–1140.
Shargool, P. D., Steeves, T., Weaver, M., and Russell, M. (1978) *Can. J. Biochem. 56,* 273–279.
Smith, R. J., Downing, S. J., Phang, J. M., Lodato, R. F., and Aoki, T. T. (1980) *Proc. Nat. Acad. Sci. U.S.A. 77,* 5221–5225.
Tamir, H., and Ratner, S. (1963a) *Arch. Biochem. Biophys. 102,* 249–258.
Tamir, H., and Ratner, S. (1963b) *Arch. Biochem. Biophys. 102,* 259–269.
Thompson, J. F. (1980) in *The Biochemistry of Plants* Vol. V. (Mifflin, B. J., ed.) pp. 375–402, Academic Press, New York.
Thuriaux, P., Ramos, F., Piérard, A., Grenson, M., and Wiame, J.-M. (1972) *J. Mol. Biol. 67,* 277–287.
Urrestarazu, L. A., Vissers, S., and Wiame, J.-M. (1977) *Eur. J. Biochem. 79,* 473–481.
Vaughn, L. E., and Davis, R. H. (1981) *Mol. Cell Biol. 1,* 797–806.
Vogel, R. H., and Vogel, H. J. (1963) *Genetics 48,* 914.
Wampler, D. E., and Fairley, J. L. (1967) *Arch. Biochem. Biophys. 121,* 580–586.
Weiss, R. L. (1973) *J. Biol. Chem. 248,* 5409–5413.
Weiss, R. L. (1976) *J. Bacteriol. 126,* 1173–1179.
Weiss, R. L., and Davis, R. H. (1973) *J. Biol. Chem. 248,* 5403–5408.
Weiss, R. L., and Davis, R. H. (1977) *J. Bacteriol. 129,* 866–873.
Weiss, R. L., and Lee, C. A. (1980) *J. Bacteriol. 141,* 1305–1311.
Whitney, P. A., and Cooper, T. G. (1972) *J. Biol. Chem. 247,* 1349–1353.
Whitney, P. A., and Morris, D. R. (1978) *J. Bacteriol. 134,* 214–220.
Wiame, J.-M. (1971) *Curr. Topics Cell. Reg. 4,* 1–38.
Wiemken, A., and Nurse, P. (1973) *Planta 109,* 293–306.
Williams, L. G., and Davis, R. H. (1970) *J. Bacteriol. 103,* 335–341.

Williams, L. G., Bernhardt, S. A., and Davis, R. H. (1970) *Biochemistry 9,* 4329–4335.
Williams, L. G., Bernhardt, S. A., and Davis, R. H. (1971) *J. Biol. Chem. 246,* 973–978.
Wipf, B., and Leisinger, T. (1977) *FEMS Microbiol. Lett. 2,* 239–242.
Wipf, B., and Leisinger, T. (1979) *J. Bacteriol. 140,* 874–880.
Wolf, E. C., and Weiss, R. L. (1980) *J. Biol. Chem. 255,* 9189–9195.
Wolfner, M., Yep, D., Messenguy, F., and Fink, G. R. (1975) *J. Mol. Biol. 96,* 273–290.

CHAPTER 6

Regulation of Serine, Glycine, and One-Carbon Biosynthesis

George V. Stauffer

I. INTRODUCTION

The biosynthesis of serine and glycine serves a greater need than simply providing these two amino acids for protein synthesis. Serine is used in cysteine, tryptophan, and phospholipid biosynthesis, and glycine is a precursor for purines and the heme tetrapyrrole ring. The conversion of serine to glycine also produces most of the one-carbon units needed for purine, thymine, methionine, and histidine biosynthesis, and numerous methylated products. Thus the serine–glycine pathway can be considered a highly branched pathway with a central physiological role.

The pathway for serine and glycine biosynthesis is shown in Fig. 6.1 (Ichihara and Greenberg, 1957; Umbarger and Umbarger, 1962; Pizer, 1963; Umbarger et al., 1963). The final step in the pathway, the interconversion of serine and glycine, is catalyzed by a single enzyme, serine hydroxymethyltransferase (Scrimgeour and Heunnekens, 1962; Stauffer and Brenchley, 1974). Mutations producing enzymatic deficiencies in the biosynthetic pathway to serine impose a nutritional requirement that can be satisfied by either serine or glycine (Pizer, 1963; Umbarger et al., 1963). Mutants deficient in serine hydroxymethyltransferase have a requirement for glycine that is not

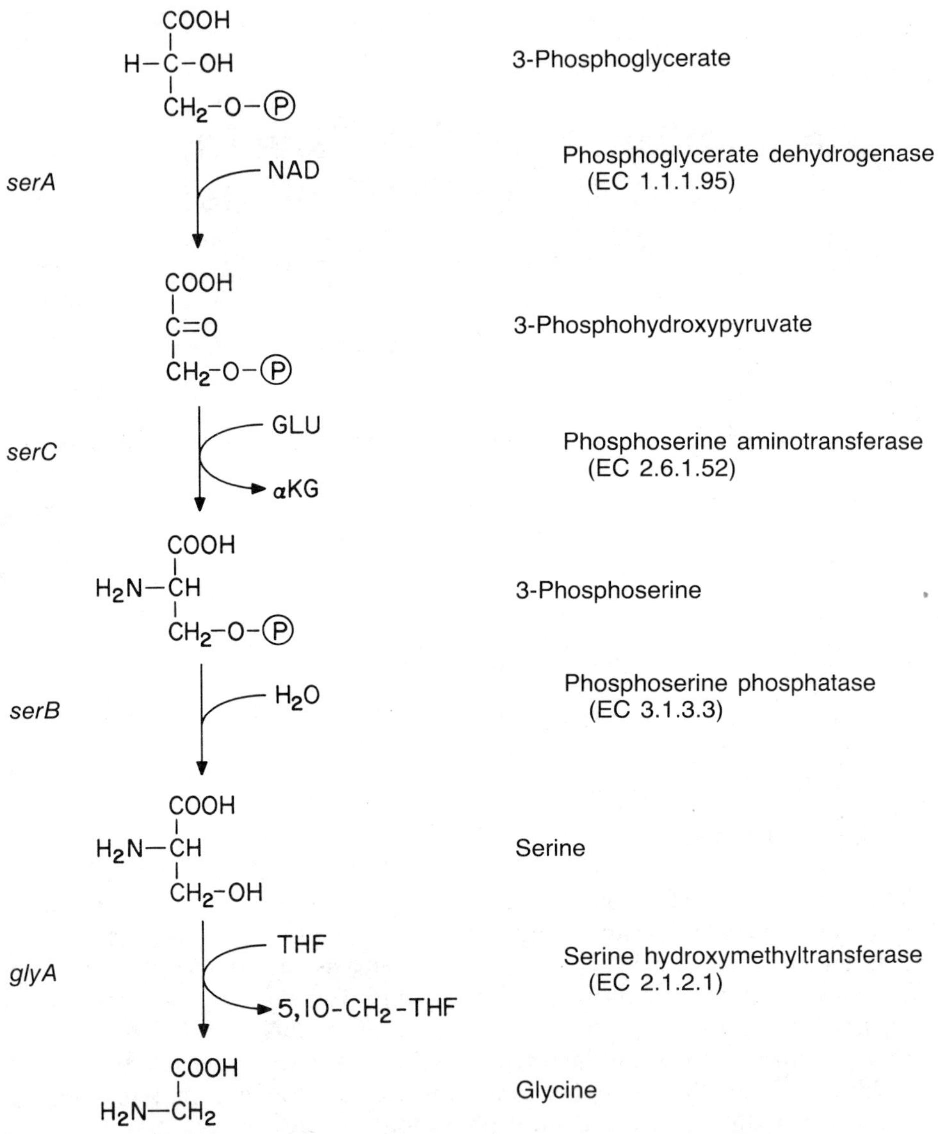

FIGURE 6.1 **Pathways for serine and glycine biosynthesis.**

satisfied by serine, directly proving that serine is the precursor of glycine (Pizer, 1965). The frequent occurrence of single mutations resulting in a deficiency satisfied by either serine or glycine favors the notion that only a single serine–glycine biosynthetic pathway is operative in bacteria (Umbarger and Umbarger, 1962; Umbarger et al., 1963). This review summarizes present knowledge of the regulation of serine, glycine, and one-carbon biosynthesis, obtained primarily through work in *Escherichia coli* and *Salmonella typhimurium*.

II. FEEDBACK INHIBITION OF SERINE BIOSYNTHESIS

The flow of carbon into the serine–glycine pathway is controlled primarily through inhibition by serine of 3-phosphoglycerate dehydrogenase, which catalyzes the first reaction in the pathway (Umbarger et al., 1963; Pizer, 1963; Sugimoto and Pizer, 1968a; Tosa and Pizer, 1971a). The enzyme consists of four identical subunits of about 41,000 daltons each (Rosenbloom et al., 1968; Winicov and Pizer, 1974). Serine is a noncompetitive inhibitor. Half-maximal inhibition occurs at 0.04 mM. D-Serine and glycine are the only other recognized inhibitors with 50% inhibition at 0.85 mM and 4.8 mM, respectively (Pizer, 1963). Serine inhibition causes a conformational change in the enzyme (Sugimoto and Pizer, 1968b; Dubrow and Pizer, 1977a). Fluorescence-quenching titrations indicate that two serine molecules are necessary to inhibit enzyme activity 80–90%, while saturating amounts inhibit more than 99% (Dubrow and Pizer, 1977b). The actual number of serine binding sites has not been unambiguously determined.

Glycine inhibition of 3-phosphoglycerate dehydrogenase is also the result of a conformational change in the enzyme (Dubrow and Pizer, 1977a). Whether serine and glyine bind to the enzyme at different sites is not known.

Inhibition of 3-phosphoglycerate dehydrogenase occurs at serine levels that are 1000-fold lower than the levels that inhibit 3-phosphoserine phosphatase, a later enzyme in the pathway (Fig. 6.1). There is no evidence for inhibition of the 3-phosphohydroxypyruvate–glutamate transaminase (Umbarger et al., 1963; Pizer, 1963). An *E. coli* mutant resistant to the analog serine hydroxamate has a 3-phosphoglycerate dehydrogenase that is insensitive to end product inhibition. This mutant overproduces serine (Tosa and Pizer, 1971b). Inhibition of 3-phosphoglycerate dehydrogenase by serine constitutes a major mode of regulation of serine bioysnthesis in both *E. coli* and *S. typhimurium*.

The enzyme from *Saccharomyces* is also inhibited by serine (Ulane and Ogur, 1972). In mammalian systems, however, the final enzyme, phosphoserine phosphatase, is inhibited by serine, with 50% inhibition at 2.5 $\times$ 10^{-4} *M* (Neuhaus and Byrne, 1958; Borkenhagen and Kennedy, 1959; Pizer, 1964; Bridgers, 1965). It is possible that the intermediates of this pathway, 3-phosphohydroxypyruvate and 3-phosphoserine, are precursors of other compounds in mammalian systems. The different location of the control point in the mammalian cell assures a supply of the intermediates even in the presence of high serine concentrations. A specific role for the intermediates has not been described. *E. coli* and *S. typhimurium serC* mutants, deficient in 3-phosphohydroxypyruvate–glutamate transaminase, require both serine and pyridoxine for growth (Dempsey, 1969; Dempsey and Itoh, 1970). This could suggest another role for intermediates of serine biosynthesis.

III. FEEDBACK INHIBITION OF GLYCINE BIOSYNTHESIS

Serine hydroxymethyltransferase of *S. typhimurium* is not inhibited by D-serine, α-methylserine, methionine, threonine, S-adenosylmethionine, purines, or thymine (Stauffer, 1976). Neither methionine nor S-adenosylmethionine affects the enzyme from *E. coli* (Silber and Mansouri, 1971; Mansouri et al., 1972). Thus the bacterial enzyme appears to be unregulated. However, regulation of the activity has been observed in some species. The folate coenzymes 5-methyltetrahydrofolate and 5-formyltetrahydrofolate inhibit the rabbit liver enzyme (Schirch and Ropp, 1967), while methionine and, to a limited extent, S-adenosylmethionine inhibit the enzyme in *Saccharomyces cerevisiae* (Botsford and Parks, 1969). Guanosine, guanine, thymine, thymidine, adenine, adenosine, and adenosine diphosphate or triphosphate, alone or in combination with methionine, have no effect on the yeast enzyme (Nakamura et al., 1973).

The partially purified enzyme from *E. coli,* having a molecular weight of 170,000 (Mansouri et al., 1972), is slightly smaller than the enzyme from rabbit liver (215,000 daltons), bovine liver (180,000 daltons), *S. cerevisiae* (180–200,000 daltons), and a facultative methylotroph (200,000 daltons) (Martinez-Carrion et al., 1972; Fujioka, 1969; Rowe and Lewis, 1973; Nakamura et al., 1973; O'Connor and Hanson, 1975). The *E. coli* structural gene for serine hydroxymethyltransferase, *glyA,* has been cloned into a multicopy plasmid. Using the plasmid DNA as template in a cell-free transcription–translation system, the *glyA* gene product was identified by SDS polyacrylamide gel electrophoresis as a 46,500-dalton polypeptide (Plamann et al., 1980; Stauffer et al., 1981). Thus the *E. coli* enzyme is most likely a tetramer consisting of identical subunits. Strains bearing multicopy plasmids containing the *glyA* gene overproduce the enzyme 17-fold to 26-fold (Plamann et al., 1980; Stauffer et al., 1981).

IV. TRANSCRIPTIONAL CONTROL OF SERINE BIOSYNTHESIS

The enzymes of serine biosynthesis are synthesized constitutively in both *E. coli* and *S. typhimurium* (Umbarger et al., 1963; Pizer, 1963; McKitrick and Pizer, 1980). The enzyme levels are not increased by starvation for serine or decreased by excess serine in the growth medium.

In several *E. coli* mutants with either high or low seryl-tRNA synthetase activity (Tosa and Pizer, 1971b; Pizer et al., 1972), the levels of the serine biosynthetic enzymes appear to be unchanged. Furthermore, the presence of L-serine hydroxamate in the growth medium does not derepress 3-phosphoglycerate dehydrogenase synthesis, even though the activity of seryl-tRNA synthetase is reduced (Tosa and Pizer, 1971b; McKitrick and Pizer, 1980). If either seryl-tRNA or seryl-tRNA synthetase were involved in regulation, altered levels of the serine biosynthetic enzymes might be expected under these conditions.

Although the enzymes of serine biosynthesis appear to be constitutively synthesized, 3-phosphoglycerate dehydrogenase levels in *E. coli* are reduced by growth on different carbon sources (McKitrick and Pizer, 1980). Although there appears to be an inverse relationship between generation time and the levels of this enzyme, the other two enzymes involved in serine biosynthesis remain constant under similar growth conditions (Pizer and Potochny, 1964). The 3-phosphoglycerate dehydrogenase levels are also reduced by growth in the presence of threonine, methionine, leucine, and isoleucine, amino acids not related in any obvious way to serine biosynthesis (Pizer and Potochny, 1964; McKitrick and Pizer, 1973). These results suggest an unusual type of regulation for the enzymes of serine biosynthesis.

It is also possible that 3-phosphoglycerate dehydrogenase levels are negatively modulated in part by cAMP. Exogenous cAMP at 3 mM reduces the enzyme levels in glucose-grown *E. coli* K12. The effect is abolished in a strain defective in the cAMP-binding protein (McKitrick and Pizer, 1980). In addition, an adenylate cyclase mutant grown on glucose has higher levels of the enzyme than the wild-type strain.

Interestingly, the concentrations of 3-phosphoglyceric acid and serine are constant in cells with marked differences in 3-phosphoglycerate dehydrogenase levels (McKitrick and Pizer, 1980). Thus in vivo the enzyme activity must be strongly inhibited by serine, a condition indicating that end product inhibition rather than the amount of the enzyme is the major factor controlling carbon flow in serine biosynthesis. Carbon flow in the shikimate pathway seems to be regulated in the same way (Chapter 17).

On the other hand, in mammalian systems, nutritional and hormonal studies suggest that the rate of synthesis of the biosynthetic enzymes may control intracellular serine concentrations (Pizer, 1964; Fallon et al., 1966, 1968; Fallon and Byrne, 1967; Davis and Fallon, 1970; Mauron et al., 1973; Hayashi et al., 1975).

V. TRANSCRIPTIONAL CONTROL OF GLYCINE BIOSYNTHESIS

Serine hydroxymethyltransferase is the enzyme responsible for the synthesis of glycine and 5,10-methylenetetrahydrofolate from serine (Fig. 6.1). This reaction is the only recognized route to glycine in bacteria (Pizer, 1965), and it accounts for a major source of one-carbon units (Mudd and Cantoni, 1964). The formation of the enzyme involves a complex repression mechanism.

Several products of one-carbon metabolism (serine, glycine, methionine, purines, thymine) repress enzyme synthesis in *S. typhimurium* (Stauffer et al., 1974) and *E. coli* (Taylor et al., 1966; Folk and Berg, 1970a; Mansouri et al., 1972; Miller and Newman, 1974; Greene and Radovich, 1975).

Some mutants with an altered glycyl-tRNA synthetase (*glyS* mutants) are glycine auxotrophs (Folk and Berg, 1970a, b). When such mutants are deprived of glycine, the level of charged $tRNA^{Gly}$ drops to about 4% of the level in a control strain, but serine hydroxymethyltransferase levels remain unchanged. In addition, serine hydroxymethyltransferase does not increase when a temperature-sensitive *glyS* mutant is shifted to the nonpermissive temperature. Thus serine hydroxymethyltransferase activity is not regulated by the functional state of glycyl-tRNA synthetase or glycyl-tRNA. The experiments do not rule out the involvement of a minor species of glycyl-tRNA in regulation of this enzyme.

There is some disagreement as to whether the enzyme is derepressed during methionine-limited growth. In *E. coli* (Silber and Mansouri, 1971; Mansouri et al., 1972; Greene and Radovich, 1975), but not *S. typhimurium* (Taylor et al., 1966; Stauffer et al., 1974) *metE* and *metF* mutants, the enzyme is derepressed. The differences may be due to the mode of methionine limitation. In *E. coli,* growth of cells to methionine exhaustion has only a modest effect on enzyme synthesis, and prolonged methionine-limited growth is necessary for derepression to occur (Greene and Radovich, 1975). Derepression does occur in *metE* and *metF* mutants of *S. typhimurium* during prolonged methionine-limited growth on D-methionine sulfoxide (L. Stauffer and G. Stauffer, unpublished results). These results support a role for methionine in the overall production of glycine and one-carbon units in both *E. coli* and *S. typhimurium* by controlling serine hydroxymethyltransferase synthesis.

Two other growth conditions, purine limitation and trimethoprim addition, also derepress the enzyme in both *E. coli* and *S. typhimurium* (Miller and Newman, 1974; Stauffer et al., 1974; Stauffer et al., 1981). It is not known if these effects are additive. The components that serve as the regulatory signal and the mechanism of this regulation are unknown.

Depression of serine hydroxymethyltransferase is not seen during thymine-, serine-, or glycine-limited growth (Folk and Berg, 1970a; Stauffer et al., 1974). Furthermore, serine, methionine, purines, and thymine repress the enzyme even during a condition of glycine-limited growth; evidence is thus provided that the effect of these compounds on enzyme synthesis may be indirect.

Two classes of mutants isolated as methionine regulatory mutants (*metK* and *metJ* mutants) display altered regulation of serine hydroxymethyltransferase synthesis. The *metK* gene codes for S-adenosylmethionine synthetase (Ahmed, 1973; Lawrence, 1972; Hobson, 1974; Stauffer and Brenchley, 1977). The *metJ* gene codes for a repressor protein of methionine biosynthesis (Lawrence, 1972; Ahmed, 1973; Greene et al., 1973; Hobson and Smith,

1973). In *S. typhimurium, metK* and *metJ* mutations prevent complete repression of serine hydroxymethyltransferase (Stauffer and Brenchley, 1977). In *E. coli, metK* mutants, but not *metJ* mutants, have elevated levels of this enzyme (Meedel and Pizer, 1974; Greene and Radovich, 1975).

Using a procedure devised to select for mutants with altered levels of serine hydroxymethyltransferase, mutants with *metK* characteristics have been isolated: ethionine, norleucine, and α-methylmethionine resistance; reduced S-adenosylmethionine synthetase activity; and cotransducibility with the *serA* gene (Stauffer and Brenchley, 1977). These mutants also have elevated serine hydroxymethyltransferase levels, and enzyme synthesis is not as sensitive to repression in them as in the parent strain. Thus serine hydroxymethyltransferase is partially regulated as a methionine enzyme. S-Adenosylmethionine synthetase, or an altered S-adenosylmethionine pool, is part of a regulatory signal for both methionine and glycine biosynthesis. The signal compound could act in conjunction with the *metJ* gene product (Chapter 11). One factor that impedes a clear analysis is that altered S-adenosylmethionine pools would affect a number of methylated compounds, one of which could be part of the regulatory signal for serine hydroxymethyltransferase synthesis.

Although the glycine and methionine pathways evidently share common regulatory elements, there are differences in the regulation of these two pathways. First, in *metJ* mutants, the methionine biosynthetic enzymes are elevated in both minimal and supplemented media. The serine hydroxymethyltransferase is not elevated in a minimal medium, but it is not repressed in a supplemented medium (Greene and Radovich, 1975; Stauffer and Brenchley, 1977). If the *metJ* product is the sole repressor for both pathways, increased serine hydroxymethyltransferase levels would be expected. Second, prolonged methionine-limited growth of either *metE, metF,* or *metA* mutants results in elevated levels of the methionine biosynthetic enzymes, but elevated levels of serine hydroxymethyltransferase occur only in the *metE* and *metF* mutants (Greene and Radovich, 1975). Finally, *metK* mutants have elevated serine hydroxymethyltransferase levels when grown in a minimal medium; but when they are grown in a medium supplemented with glycine, purines, and thymine, the enzyme levels are reduced. The methionine biosynthetic enzymes are elevated under both growth conditions (Greene and Radovich, 1975). It appears that the methionine component may be only one factor of a complicated regulatory scheme involving several elements.

Two additional mutants have been isolated with altered regulation of serine hydroxymethyltransferase synthesis (Stauffer and Brenchley, 1978). One mutant with sevenfold elevated levels of the enzyme is repressed normally when grown in a supplemented medium. The second mutant has twofold elevated enzyme levels, which are incompletely repressed in a supplemented medium. Both mutations are cotransducible with the *glyA* gene and are distinct from the *metK* and *metJ* mutations. Although the phenotypes of these two mutants are distinct from each other, suggesting two new control sites for *glyA* expression, both mutations could lie within the same gene. Whether these mutations identify a DNA control site or a structural gene for a control element is unknown.

In summary, four different loci affect *glyA* gene expression. One control appears to involve the methionine pathway, but additional control elements not involving the methionine pathway are also found.

VI. REGULATION OF ONE-CARBON BIOSYNTHESIS

The 5,10-methylenetetrahydrofolate formed during the synthesis of glycine from serine is essential for the synthesis of several compounds. However, it has been estimated that only about 75% of the total cellular need for one-carbon units can be accounted for by this reaction. An alternative route to one-carbon units is the enzymatic cleavage of glycine to a one-carbon unit, CO_2, and NH_3 (Sagers and Gunsalus, 1961). Such an enzyme, inducible by exogenous glycine, has been demonstrated in both *E. coli* (Meedel and Pizer, 1974) and *S. typhimurium* (M. Plamann and G. Stauffer, unpublished results). Since serine auxotrophs can grow on exogenous glycine, the glycine cleavage to 5,10-methylenetetrahydrofolate, CO_2, and NH_3 could account for the one-carbon units in serine auxotrophs (Pizer, 1965). Support for this hypothesis is the finding that loss of the glycine cleavage pathway in mutants blocked in serine biosynthesis leads to the inability to use glycine as a serine source (Newman et al., 1974; Pizer et al., 1977).

Glycine cleavage may act to balance the glycine and one-carbon unit requirements of the cell (Umbarger, 1977). If the one-carbon needs exceed glycine needs, some of the glycine made from serine would be converted to additional one-carbon units. In support of this hypothesis is the observation that mutants of both *E. coli* and *S. typhimurium,* with no detectable glycine cleavage activity, excrete glycine and can feed glycine auxotrophs (M. Plamann and G. Stauffer, unpublished results).

Most of the one-carbon units are generated during the synthesis of glycine from serine (Pizer and Potochny, 1964). However, the glycine cleavage enzymes have only one tenth of the activity of serine hydroxymethyltransferase, a result posing the question of whether glycine is the only source of one-carbon units in *E. coli* and *S. typhimurium* serine auxotrophs. Such mutants obtain about 46–89% of their one-carbon units from glycine (Pitts et al., 1961; Pizer, 1965), and strains defective in both the serine hydroxymethyltransferase reaction and the glycine cleavage system still have an unknown source of one-carbon units (Newman et al., 1974).

Regulation of the glycine cleavage system is not well understood. Growth of serine–glycine auxotrophs of *E. coli* in a medium containing serine and a purine base results in a temporary inability of the cells to grow upon transfer to a medium containing only glycine (Newman and Magasanik, 1963). An excess of one-carbon units could repress the synthesis of the enzymes responsible for one-carbon and serine formation. The presence of glycine, serine, and a purine base in the medium allows growth immediately upon transfer to a medium containing only glycine. The absence of a growth lag suggests that glycine induces the enzyme system necessary for its utilization as a one-carbon source. One-carbon units appear not to repress the enzyme system. S-Adenosylmethionine and methionine have no effect on the inducibility of the glycine cleavage enzymes (Meedel and Pizer, 1974).

ACKNOWLEDGMENTS

This work was supported by Public Health Service grant GM-26878 from the National Institute of General Medical Sciences.

REFERENCES

Ahmed, A. (1973) *Mol. Gen. Genet. 123,* 299–324.
Borkenhagen, L. F., and Kennedy, E. P. (1959) *J. Biol. Chem. 234,* 849–853.
Botsford, J. L., Jr., and Parks, L. W. (1969) *J. Bacteriol. 97,* 1176–1183.
Bridgers, W. F. (1965) *J. Biol. Chem. 240,* 4591–4597.
Davis, J. L., and Fallon, H. J. (1970) *J. Biol. Chem. 245,* 5838–5846.
Dempsey, W. B. (1969) *Biochem. Biophys. Res. Comm. 37,* 89–93.
Dempsey, W. B., and Itoh, H. (1970) *J. Bacteriol. 104,* 658–667.
Dubrow, R., and Pizer, L. I. (1977a) *J. Biol. Chem. 252,* 1527–1538.
Dubrow, R., and Pizer, L. I. (1977b) *J. Biol. Chem. 252,* 1539–1551.
Fallon, H. J., and Byrne, W. L. (1967) *Endocrinology 80,* 847–850.
Fallon, H. J., Hackney, E. J., and Byrne, W. L. (1966) *J. Biol. Chem. 241,* 4157–4167.
Fallon, H. J., Davis, J. L., and Goyer, R. A. (1968) *J. Nutrition 96,* 220–226.
Folk, W. R., and Berg, P. (1970a) *J. Bacteriol. 102,* 193–203.
Folk, W. R., and Berg, P. (1970b) *J. Bacteriol. 102,* 204–212.
Fujioka, M. (1969) *Biochim. Biophys. Acta 185,* 338–349.
Greene, R. C., and Radovich, C. (1975) *J. Bacteriol. 124,* 269–278.
Greene, R. C., Hunter, J. S. V., and Coch, E. H. (1973) *J. Bacteriol. 115,* 57–67.
Hayashi, S., Tanaka, T., Naito, J., and Suda, M. (1975) *J. Biochem. 77,* 207–219.
Hobson, A. C. (1974) *Mol. Gen. Genet. 131,* 263–273.
Hobson, A. C., and Smith, D. A. (1973) *Mol. Gen. Genet. 126,* 7–18.
Ichihara, A., and Greenberg, D. M. (1957) *J. Biol. Chem. 224,* 331–340.
Lawrence, D. A. (1972) *J. Bacteriol. 109,* 8–11.
Mansouri, A., Decter, J. B., and Silber, R. (1972) *J. Biol. Chem. 247,* 348–352.
Martinez-Carrion, M., Critz, W., and Quashnock, J. (1972) *Biochemistry 11,* 1613–1615.
Mauron, J., Mottu, F., and Spohr, G. (1973) *Eur. J. Biochem. 32,* 331–342.
McKitrick, J. C., and Pizer, L. I. (1973) Abstracts of the Annual Meeting of the American Society for Microbiology, p. 164.

McKitrick, J. C., and Pizer, L. I. (1980) *J. Bacteriol. 141,* 235–245.
Meedel, T. H., and Pizer, L. I. (1974) *J. Bacteriol. 118,* 905–910.
Miller, B. A., and Newman, E. B. (1974) *Can. J. Microbiol. 20,* 41–47.
Mudd, S. H., and Cantoni, G. L. (1964) in *Comprehensive Biochemistry* Vol. XV (Florkin, M., and Stotz, E. H., eds.) pp. 1–47, Elsevier, Amsterdam.
Nakamura, K. D., Trewyn, R. W., and Parks, L. W. (1973) *Biochim. Biophys. Acta 327,* 328–335.
Neuhaus, F. C., and Byrne, W. L. (1958) *Biochim. Biophys. Acta 28,* 223–224.
Newman, E. B., and Magasanik, B. (1963) *Biochim. Biophys. Acta 78,* 437–448.
Newman, E. B., Miller, B., and Kapoor, V. (1974) *Biochem. Biophys. Acta 338,* 529–539.
O'Connor, M. L., and Hanson, R. S. (1975) *J. Bacteriol. 124,* 985–996.
Pitts, J. D., Stewart, J. A., and Crosbie, G. W. (1961) *Biochim. Biophys. Acta 50,* 361–363.
Pizer, L. I. (1963) *J. Biol. Chem. 238,* 3934–3944.
Pizer, L. I. (1964) *J. Biol. Chem. 239,* 4219–4226.
Pizer, L. I. (1965) *J. Bacteriol. 89,* 1145–1150.
Pizer, L. I., and Potochny, M. L. (1964) *J. Bacteriol. 88,* 611–619.
Pizer, L. I., McKitrick, J., and Tosa, T. (1972) *Biochem. Biophys. Res. Comm. 49,* 1351–1357.
Pizer, F. L., McKitrick, J. C., Stern, J. L., Share, D., and Pizer, L. I. (1977) Abstracts of the Annual Meeting of the American Society for Microbiology, p. 220.
Plamann, M. D., Stauffer, L. T., and Stauffer, G. V. (1980) *Genetics 94,* S82.
Rosenbloom, J., Sugimoto, E., and Pizer, L. I. (1968) *J. Biol. Chem. 243,* 2099–2107.
Rowe, P. B., and Lewis, G. P. (1973) *Biochemistry 12,* 1962–1969.
Sagers, R. D., and Gunsalus, I. C. (1961) *J. Bacteriol. 81,* 541–549.
Schirch, L., and Ropp, M. (1967) *Biochemistry 6,* 253–257.
Scrimgeour, K. G., and Huennekens, F. M. (1962) in *Methods in Enzymology* Vol. V (Colowick, S. P., and Kaplan, N. O., eds.) pp. 838–843, Academic Press, New York.
Silber, R., and Mansouri, A. (1971) *Ann. N.Y. Acad. Sci. 186,* 55–69.
Stauffer, G. V. (1976) "Characterization of *S. typhimurium* mutants with altered regulation of the methionine and glycine biosynthetic enzymes" Ph.D. thesis, The Pennsylvania State University, University Park, Pennsylvania.
Stauffer, G. V., and Brenchley, J. E. (1974) *Genetics 77,* 185–198.
Stauffer, G. V., and Brenchley, J. E. (1977) *J. Bacteriol. 129,* 740–749.
Stauffer, G. V., and Brenchley, J. E. (1978) *Genetics 88,* 221–233.
Stauffer, G. V., Baker, C. A., and Brenchley, J. E. (1974) *J. Bacteriol. 120,* 1017–1025.
Stauffer, G. V., Plamann, M. D., and Stauffer, L. T. (1981) *Gene 14,* 63–72.
Sugimoto, E., and Pizer, L. I. (1968a) *J. Biol. Chem. 243,* 2081–2089.
Sugimoto, E., and Pizer, L. I. (1968b) *J. Biol. Chem. 243,* 2090–2098.
Taylor, R. T., Dickerman, H., and Weissbach, H. (1966) *Arch. Biochem. Biophys. 117,* 405–412.
Tosa, T., and Pizer, L. I. (1971a) *J. Bacteriol. 106,* 966–971.
Tosa, T., and Pizer, L. I. (1971b) *J. Bacteriol. 106,* 972–982.

Ulane, R., and Ogur, M. (1972) *J. Bacteriol. 109,* 34–43.
Umbarger, H. E. (1977) *Biochem. Educ. 5,* 67–71.
Umbarger, H. E., and Umbarger, M. A. (1962) *Biochim. Biophys. Acta 62,* 193–195.
Umbarger, H. E., Umbarger, M. A., and Siu, P. M. L. (1963) *J. Bacteriol. 85,* 1431–1439.
Winicov, I., and Pizer, L. I. (1974) *J. Biol. Chem. 249,* 1348–1355.

CHAPTER

7

Regulation of Cysteine Biosynthesis in *Escherichia coli* and *Salmonella typhimurium*

Nicholas M. Kredich

I. INTRODUCTION

The biosynthesis of cysteine is of interest not only as an aspect of amino acid metabolism, but also because it involves a key step in the sulfur cycle, that is, the covalent binding, or "fixation," of reduced inorganic sulfur to carbon (Siegel, 1975). This process together with the actual reduction of inorganic sulfate to sulfide occurs mainly in plants and microorganisms, which then provide reduced organic sulfur compounds for all other life forms. In plants and many fungi, sulfide can be directly fixed either by reaction with an O-acyl derivative of homoserine to give homocysteine or by reaction with O-acetylserine to give cysteine (Giovanelli and Mudd, 1967; Flavin, 1975). In *Escherichia coli* and *Salmonella typhimurium,* direct homocysteine synthesis from sulfide is very inefficient and demonstrable only in certain mutants (Flavin and Slaughter, 1967). In these two organisms, sulfide fixation first generates cysteine, which then donates a sulfur moiety via transsulfuration to generate homocysteine (Flavin, 1975). Reduced sulfur in methionine, thiamine, biotin, lipoic acid, pantetheine, and other compounds originates from cysteine in these bacteria. This chapter deals exclusively with cysteine biosynthesis in *E. coli* and *S. typhimurium*, the two microorganisms for which the genetics and regulation of cysteine biosynthesis are best understood.

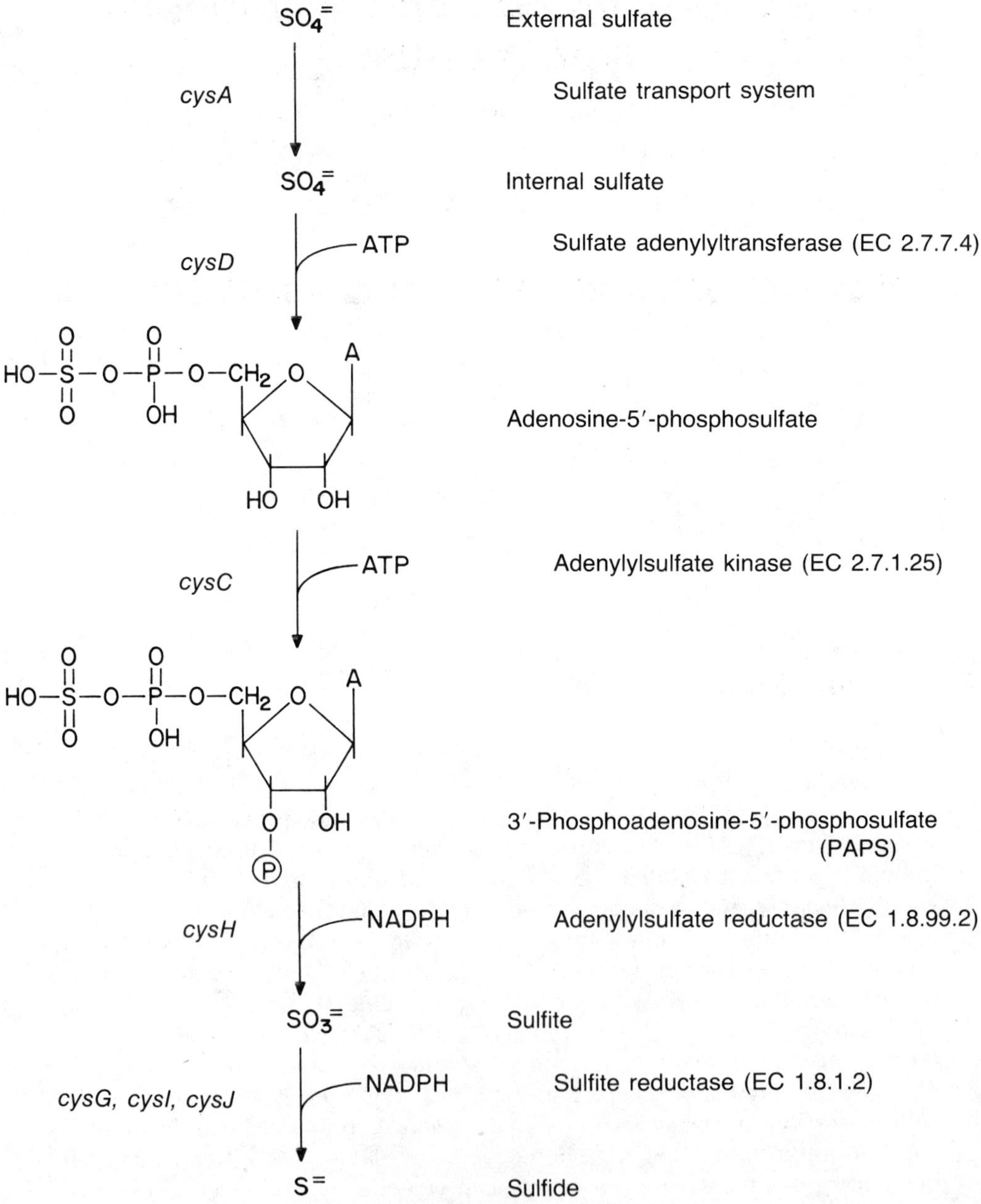

FIGURE 7.1 Pathway of sulfate reduction in *E. coli* and *S. typhimurium*.

II. THE PATHWAY

Cysteine biosynthesis constitutes a branched, convergent pathway. Along one arm, sulfate is transported into the cell and reduced to sulfide (Fig. 7.1). On the other arm, serine is converted to O-acetylserine, and the two arms converge at the point where O-acetylserine and sulfide react to give cysteine (Fig. 7.2). Sulfate reduction in these bacteria is "assimilatory" in the sense that it provides reduced sulfur for biosynthetic purposes, rather than supporting respiration by using sulfate as a terminal electron acceptor (Postgate, 1959). The latter form of sulfate reduction is termed "dissimilatory" and differs from the assimilatory process in several ways (Peck, 1962; Siegel, 1975).

A. Sulfate Transport

The uptake of inorganic sulfate requires an active transport system, which is specified, at least in part, by the products of the *cysA* region (Mizobuchi et al., 1962; Dreyfuss, 1964; Jones-Mortimer, 1968a). This system also mediates the transport of thiosulfate (Dreyfuss, 1964) and anions such as selenate, molybdate, and chromate (Pardee et al., 1966). Thus *cysA* mutants can utilize neither sulfate nor thiosulfate as sulfur sources. In addition,

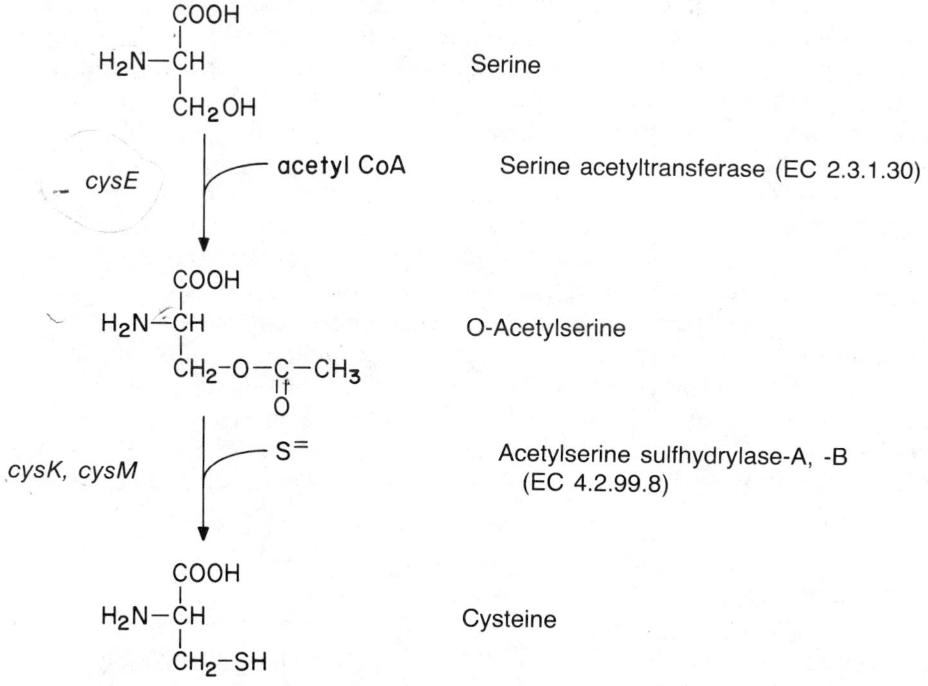

FIGURE 7.2 Biosynthesis of cysteine from serine and sulfide in *E. coli* and *S. typhimurium*.

they are resistant to chromate, a phenotype useful for the positive selection of such mutants.

Pardee and his associates have extensively characterized a sulfate-binding protein from *S. typhimurium,* which may play a role in sulfate transport (Pardee, 1966; Pardee et al., 1966; Pardee and Watanabe, 1968). This protein is located near the cell surface, is released by osmotic shock, has a molecular weight of 32,000, and binds one molecule of sulfate with a K_D of about 0.1 μM. These properties and the fact that it is coregulated with other *cys* genes imply that sulfate-binding protein is involved in sulfate uptake or utilization (Ohta et al., 1971). More conclusive evidence might be obtained by studying mutants specifically lacking sulfate-binding protein. Attempts to isolate such mutants have been unsuccessful, showing only that this protein is not encoded by any of the three cistrons of *cysA* (Ohta et al., 1971).

B. Sulfate Activation

Intracellular sulfate must be activated prior to its reduction to sulfite and sulfide. This activation involves a two-step sequence of reactions leading to 3′-phosphoadenosine-5′-phosphosulfate (PAPS) (Pasternak, 1962; Dreyfuss and Monty, 1963a). In the first reaction, catalyzed by sulfate adenylyltransferase, ATP and sulfate combine to give the mixed anhydride adenosine-5′-phosphosulfate and inorganic pyrophosphate (Bandurski et al., 1956). The equilibrium constant of this reaction in the physiological direction is 10 mM (Robbins and Lipmann, 1958). Thus a net flow of sulfate past this step depends on further metabolism of adenosine-5′-phosphosulfate and pyrophosphate.

In the second step, adenosine-5′-phosphosulfate is phosphorylated at the 3′ position to give PAPS (Robbins and Lipmann, 1956) in a reaction catalyzed by adenylylsulfate kinase. In many organisms, PAPS is the sulfate donor for the synthesis of sulfate-containing polysaccharides and other sulfate esters (De Meio, 1975). Such esters are not found in *E. coli* and *S. typhimurium*. The normal growth on cysteine of mutant strains unable to synthesize PAPS indicates that this compound is only required as an intermediate in sulfate reduction. Two closely linked genes, *cysD* and *cysC*, code for sulfate adenylyltransferase and adenylylsulfate kinase, respectively (Dreyfuss and Monty, 1963a; Jones-Mortimer, 1968a; Kredich, 1971).

C. Sulfate Reduction

There are two discrete reductive steps in the pathway: the first converts PAPS to sulfite, and the second reduces sulfite to sulfide. PAPS reduction in *E. coli* (Tsang and Schiff, 1976a) is similar to that in yeast (Wilson et al., 1961; Asahi et al., 1961; Wilson and Bierer, 1976). The reaction sequence is initiated by a specific sulfotransferase that transfers the sulfonyl group of PAPS to a carrier molecule (R—SH) to form an organic thiosulfate (R—S—SO_3^-). This enzyme is probably the adenylylsulfate reductase encoded by the *cysH* gene (Jones-Mortimer, 1968a; Kredich, 1971). Thioredoxin, an essential component of this system, may be the carrier molecule. The exact fate of the R—S—SO_3^- intermediate is unknown. Two general mechanisms have been considered (Tsang and Schiff, 1976a; Schiff, 1980). The first mechanism postulates reductive release of the sulfonyl moiety to generate free sulfite. Further reduction to free sulfide is catalyzed by sulfite reductase. The electron donor in the release of sulfite could be either NADPH or a thiol such as glutathione, which would in turn be reduced by NADPH. The second mechanism, analogous to the *Chlorella* system (Schmidt et al., 1974; Tsang and Schiff, 1976b, 1978), requires an organic thiosulfate reductase, presumably NADPH-dependent, that acts directly on R—S—SO_3^- to give R—S—S^- without the release of free sulfite. This R—S—S^- persulfide could then be further reduced, either to free sulfide or directly to cysteine in a concerted mechanism involving O-acetylserine and acetylserine sulfhydrylase.

Extracts of *E. coli* do contain organic thiosulfate reductase activity of unknown significance (Tsang and Schiff, 1976a). Since well-defined mutants lacking sulfite reductase activity cannot grow aerobically on either sulfate or sulfite, either the organic thiosulfate reductase activity is not involved in sulfate reduction in *E. coli* and *S. typhimurium,* or it is an activity of sulfite reductase itself. Significant quantities of free sulfite are found in sulfate-grown, wild-type *E. coli* (Ellis, 1966). Large amounts of sulfite are excreted by sulfate-grown sulfite reductase–deficient mutants (Dreyfuss and Monty, 1963a). This indicates that free sulfite is an intermediate in sulfate reduction.

Assimilatory sulfate reduction requires the action of sulfite reductase (Mager, 1960; Dreyfuss and Monty, 1963a; Kemp et al., 1963). The enzyme catalyzes an NADPH-dependent, 6-electron reduction of free sulfite to free sulfide, without release of intermediates (Siegel, 1975). Purified *E. coli* sulfite reductase has a molecular weight of 670,000 and contains 4 FMN, 4 FAD, 16 nonheme irons, 14–15 labile sulfides, and 4 molecules of a novel heme, siroheme (Siegel et al., 1973; Murphy et al., 1973; Murphy and Siegel, 1973). The enzyme is composed of two different polypeptide chains, α and β, and has the subunit structure $\alpha_8\beta_4$ (Siegel and Davis, 1974). The α subunit is approximately 60,000 daltons, contains FMN and FAD, and as an α_8 species has NADPH–cytochrome *c* reductase activity. The β subunit binds the nonheme iron, acid labile sulfide, and siroheme moieties and has reduced methyl viologen–sulfite reductase activity (Siegel and Davis, 1974). Electron flow is believed to occur from NADPH, via FAD, FMN, and heme, to sulfite (Siegel et al., 1974). The properties of the *S. typhimurium* sulfite reductase are very similar to those of the *E. coli* enzyme (Siegel et al., 1971).

Studies of *S. typhimurium* mutants indicate that *cysJ* codes for the α subunit and *cysI* codes for the β subunit (Siegel and Kamin, 1971; Siegel et al., 1971). The product of *cysG* is believed to be required for synthesis of the prosthetic group siroheme (Siegel and Davis, 1974; Siegel et al., 1971).

D. Cysteine Synthase

The direct precursor of cysteine is O-acetylserine, formed from serine and acetyl-CoA in a reaction catalyzed by serine acetyltransferase (Kredich and Tomkins, 1966; Kredich et al., 1969). In the final step of the pathway, acetylserine sulfhydrylase catalyzes the reaction of O-acetylserine and sulfide to give cysteine and acetate (Kredich and Tomkins, 1966; Becker et al., 1969). Two different acetylserine sulfhydrylases, designated -A and -B, have been described in *S. typhimurium* (Becker et al., 1969; Becker and Tomkins, 1969). Either isoenzyme alone is sufficient for cysteine prototrophy under conditions of aerobic growth on a minimal-salts, sulfate medium (Hulanicka et al., 1974, 1979; Kredich et al., 1980).

Acetylserine sulfhydrylase-A is the major isoenzyme. Depending on the regulatory status of the biosynthetic pathway, this enzyme comprises 60–95% of the total activity in a crude extract (Becker and Tomkins, 1969; Hulanicka et al., 1979). A portion of the acetylserine sulfhydrylase-A and all of the serine acetyltransferase are found in a multienzyme complex designated cysteine synthase. This multimeric species (molecular weight: 309,000) consists of one molecule of serine acetyltransferase (molecular weight: 160,000) and two molecules of acetylserine sulfhydrylase-A (molecular weight: 68,000) (Kredich et al., 1969). Serine acetyltransferase, encoded by *cysE* (Kredich and Tomkins, 1966; Kredich, 1971; Jones-Mortimer et al., 1968), is probably a tetramer, but the exact subunit structure has not been determined. Acetylserine sulfhydrylase-A, encoded by *cysK* (Hulanicka et al., 1974; Fimmel and Loughlin, 1977a), is a homodimer (subunit molecular weight: 34,000) and contains one mole of pyridoxal phosphate per monomer (Becker et al., 1969).

An intriguing property of cysteine synthase is its rapid dissociation into serine acetyltransferase and acetylserine sulfhydrylase-A in the presence of O-acetylserine (Kredich et al., 1969). This process is readily reversible with a $K_{0.5}$ for O-acetylserine of 20–40 μM and is essentially complete at 1.0 mM. Dissociation of the complex does not occur in the presence of sulfide even at O-acetylserine concentrations as high as 0.1 M.

Acetylserine sulfhydrylase-B has not been extensively purified but is known to differ from acetylserine sulfhydrylase-A with respect to chromatographic behavior and antigenicity (Becker and Tomkins, 1969; Hulanicka et al., 1979). Either isoenzyme alone is sufficient for normal aerobic growth on sulfate. For unknown reasons, mutants lacking acetylserine sulfhydrylase-B are cysteine bradytrophs under anaerobic conditions (Kredich et al.,

1980). Acetylserine sulfhydrylase-B is encoded by *cysM* and is part of the cysteine regulon (Hulanicka et al., 1979).

III. REGULATION OF ENZYME ACTIVITIES

Of the cysteine biosynthetic enzymes, only serine acetyltransferase is end product inhibited. Early observations in *E. coli* (Ellis, 1964) suggested that cysteine inhibited sulfate transport. Subsequent studies in *S. typhimurium* failed to confirm these results (Dreyfuss, 1964). Thiosulfate, sulfite, and other sulfate analogs do inhibit sulfate uptake, probably by direct competition for some component of the transport system (Pardee et al., 1966). Ellis et al. (1964) showed that sulfide and cysteine do not inhibit the conversion of sulfate to PAPS in vitro. Sulfide, though a known product inhibitor of sulfite reductase, has very little effect on this enzyme at concentrations likely to be present in vivo (L. M. Siegel, personal communication).

Mutants blocked in sulfite reduction accumulate large amounts of sulfite when grown in the presence of sulfate (Dreyfuss and Monty, 1963a; Ellis, 1966). Sulfate-grown mutants that are resistant to repression of *cys* enzymes excrete large amounts of sulfide (Borum and Monty, 1976). These observations indicate that control of gene expression rather than feedback inhibition is the most significant factor regulating sulfate assimilation.

The aggregation of serine acetyltransferase and acetylserine sulfhydrylase-A into the multienzyme complex cysteine synthase has invited speculation about the functional significance of this association (Kredich et al., 1969; Cook and Wedding, 1977a, 1978). One obvious advantage would be to allow direct intracomplex transfer of O-acetylserine from its product site on serine acetyltransferase to its substrate site on acetylserine sulfhydrylase-A. Kinetic studies indicate that O-acetylserine is actually released prior to its utilization by the sulfhydrylase moiety of the complex. It is reasonable to suggest that the complex affords a kinetic advantage by decreasing the distance *O*-acetylserine must travel from the one active center to the other (Cook and Wedding, 1977a).

Detailed kinetic and physicochemical studies of cysteine synthase and its component parts have demonstrated substrate and product inhibition of both enzymes (Cook and Wedding, 1976, 1977a,b, 1978). The kinetic parameters are such, however, that neither substrate nor product inhibition of acetylserine sulfhydrylase-A is likely to play a role in regulating the rate of cysteine biosynthesis in vivo.

Serine acetyltransferase, either as cysteine synthase or as the free enzyme, is inhibited by cysteine (K_i approximately 11 μM) in a manner competitive with acetyl-CoA (Kredich and Tomkins, 1966; Kredich et al., 1969; Cook and Wedding, 1978). This feedback inhibition is probably the most important regulatory property of any enzyme in the cysteine biosynthetic pathway. Through its effects on O-acetylserine synthesis, cysteine can control not only the level of its own carbon skeleton precursor, but also the expression of those genes required for sulfate reduction (see below).

IV. REGULATION OF GENE EXPRESSION

A. Regulatory Components of the Cysteine Regulon

Regulation of gene expression in cysteine biosynthesis requires three factors: sulfur limitation, availability of O-acetylserine, and the *cysB* gene product (Spencer et al., 1967; Jones-Mortimer et al., 1968; Jones-Mortimer, 1968b; Kredich, 1971). The cysteine regulon comprises all the genes that are so regulated and, in addition, genes *cysE* and *cysB*.

1. Sulfur source. Growth of wild-type *E. coli* or *S. typhimurium* in the presence of cysteine, cystine, or sulfide severely represses sulfate transport and sulfate-binding protein (Dreyfuss, 1964; Pardee et al., 1966); sulfate adenylyltransferase, adenylylsulfate kinase, adenylylsulfate reductase, and sulfite reductase (Mager, 1960; Pasternak, 1962; Dreyfuss and Monty, 1963b; Pasternak et al., 1965; Kredich, 1971); acetylserine sulfhydrylase-A (Kredich and Tomkins, 1966; Jones-Mortimer et al., 1968; Kredich, 1971); acetylserine sulfhydrylase-B (Hulanicka et al., 1979); and the cystine transport system designated CTS-1 (Baptist and Kredich, 1977). Conversely, growth on a limiting sulfur source such as glutathione or djenkolate results in high levels of these enzyme activities. Thus these polypeptides of the cysteine regulon are repressed by cysteine and sulfide and derepressed by sulfur limitation. Serine acetyltransferase is the only enzyme of the pathway that is not so regulated (Jones-Mortimer et al., 1968; Kredich, 1971). Control of this enzyme activity appears to be mediated only through feedback inhibition by cysteine.

2. *O*-Acetylserine. Mutations in *cysE*, the structural gene for serine acetyltransferase, pleiotropically affect the regulation of the entire cysteine biosynthetic pathway (Spencer et al., 1967; Jones-Mortimer et al., 1968; Kredich, 1971) as well as that of cystine transport (Baptist and Kredich, 1977). Lack of the *cysE* gene product prevents derepression of the cysteine regulon. Derepression is restored by adding O-acetylserine to the medium. Therefore O-acetylserine is considered to be an internal inducer of the cysteine regulon.

A second type of *cysE* mutation (*trz-312*), described in *S. typhimurium*, is also pleiotropic. The mutation lies at one extremity of *cysE*, probably within the promoter. It increases the level of serine acetyltransferase sevenfold (Hulanicka and Kredich, 1976). Expression of the cysteine regulon is normal in a *trz-312* strain except in response to 1,2,4-triazole. This compound inhibits cysteine biosynthesis in wild type through its conversion to 1,2,4-triazole-1-alanine. This substance is nontoxic to *S. typhimurium* (Kredich et al., 1975), but its formation from O-acetylserine depletes the levels of this metabolite below those necessary for induction of cysteine biosynthetic enzymes (Hulanicka et al., 1972; Kredich et al., 1975). Growth is inhibited unless a sulfur source such as sulfite, sulfide, or cysteine is supplied. The high level of serine acetyltransferase found in *trz-312* strains enables O-acetylserine synthesis to proceed at a rate sufficient to prevent its depletion by reaction with 1,2,4-triazole. Induction of the cysteine regulon and resistance to this inhibitor are the outcome.

3. The *cysB* gene. Most mutations in *cysB*, including all known deletions, give a Cys^- phenotype. Such *cysB* mutants cannot be derepressed for the biosynthetic pathway by any nutritional manipulation, even though they have normal levels of serine acetyltransferase. Introduction of an F′*cysB*$^+$ restores derepression capability and gives a Cys^+ phenotype (Jones-Mortimer, 1968b; Jagura et al., 1978). Therefore *cysB*$^+$ is dominant to *cysB*, and this gene must code for an element of positive control that is required for expression of the *cys* genes.

Other mutations in *cysB* result in derepressed levels of one or more cysteine biosynthetic enzyme whether or not cells are grown on limiting sulfur. Alleles that allow such constitutive expression of the cysteine regulon are designated *cysB*^{c}. The best studied of these, *cysB1352*, also eliminates the need for the inducer O-acetylserine in derepression (Spencer et al., 1967; Kredich, 1971) and gives a Cys^+ phenotype that is resistant to growth inhibition by 1,2,4-triazole (Hulanicka and Klopotowski, 1972). The *cysB482* mutation gives a Cys^- phenotype that cannot be derepressed for enzymes of the sulfate-reducing pathway but is constitutive for acetylserine sulfhydrylase-A. Therefore this allele is both *cysB* and *cysB*^{c} (Kredich, 1971).

Early genetic studies in *S. typhimurium* employing abortive transduction indicated that *cysB* consists of three distinct complementation groups, designated *cysBa, cysBb,* and *cysBc* (Mizobuchi et al., 1962). Subsequently, a fine structure map of this gene in *S. typhimurium* showed that mutations of the types *cysBa* and *cysBc* are flanked on both sides by *cysBb* mutations, suggesting that complementation is intracistronic rather than intercistronic (Cheney and Kredich, 1975). Similar studies in *E. coli* demonstrated that mutations at the two extreme ends of *cysB* do not complement each other in stable merodiploids (Tully and Yudkin, 1977). Therefore on the basis of current genetic evidence, *cysB* is a single cistron.

The isolation of amber and temperature-sensitive *cysB* mutations proved that the active product of this gene is a protein (Tully and Yudkin, 1975). More recently, Mascarenhas and Yudkin (1980a) have identified a 39,000-dalton polypeptide specified by λ*cysB* that is the *cysB* gene product. Two-dimensional electrophoretic gels of crude extracts from wild-type and various *cysB*. strains of *S. typhimurium* have also shown that *cysB* codes for a polypeptide with a molecular weight of 36,000 (E. W. Baptist and N. M. Kredich, unpublished). Physicochemical data (Mascarenhas and Yudkin, 1980a) and genetic evidence (Jagura et al., 1978) suggest an oligomeric structure for the native *cysB* protein.

B. Genetic Organization

The genes of the cysteine regulon are dispersed among a number of noncontiguous sites on the chromosome (Fig. 7.3) (Jones-Mortimer, 1973; Sanderson and Hartman, 1978). A minimum of four and a maximum of eight transcriptional units must be postulated for those genes that are regulated. Another two transcriptional units for *cysE* and *cysB* are not linked to any other *cys* gene and must exert their regulatory effects in *trans*.

In *S. typimurium, cysK* and *cysA* are separated by 5–10 kb (Hulanicka et al., 1974) and may be controlled by the same promoter (Wiater and

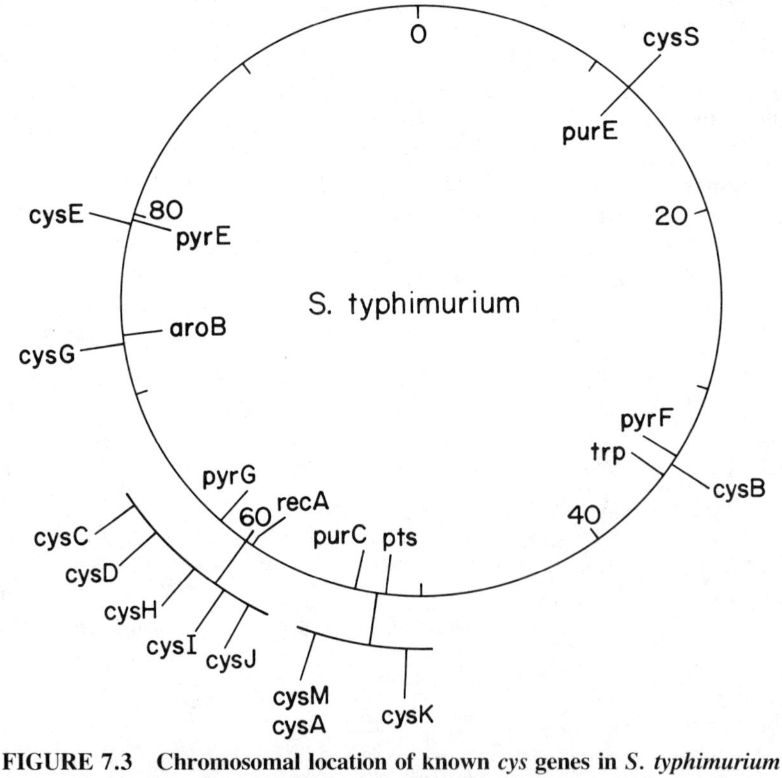

FIGURE 7.3 Chromosomal location of known *cys* genes in *S. typhimurium* (Sanderson and Hartman, 1978). A similar arrangement exists in *E. coli*. The order of *cysA* and *cysM* is not known, but indirect evidence suggests *purC cysM cysA cysK*. The map is calibrated to 100 units.

Hulanicka, 1978). The *cysA* region and *cysM* are closely linked; but since *cysK* and *cysM* belong to different transcriptional units (Hulanicka et al., 1979), one can infer that *cysA* and *cysM* are not cotranscribed.

The *cysHIJ* cluster is a single operon, which is transcribed in the direction *cysJ* to *cysH* (Loughlin, 1975). Starting with a *cysB* deletion mutant, Ostrowski and Hulanicka (1979) have isolated a strain constitutive for *cysHIJ* expression, which carries a mutation in or near *cysJ*, presumably in the promoter region for this transcriptional unit. The presence of normal sulfite reductase activity in the double mutant indicates that *cysG*, which is believed to be required for siroheme synthesis, is not tightly controlled by *cysB* and in fact may not be regulated at all. The *cysC* and *cysD* genes are separated from *cysHIJ* by 15–20 kb (Demerec et al., 1963) and do not belong to this operon (Loughlin, 1975). The contiguity of *cysC,D* suggests that they may share a common promoter. The structural genes for sulfate-binding protein and the CTS-1 cystine transport system have not been identified.

C. Interactions Between O-Acetylserine, Cysteine, and Sulfide

The extent of derepression of the cysteine regulon varies with the sulfur source used for growth. Derepression is greatest with glutathione and djenkolate; sulfate and sulfite are intermediate; both sulfide and cysteine (or cystine) give full repression (Dreyfuss and Monty, 1963b; Ellis et al., 1964; Pasternak et al., 1965; Jones-Mortimer et al., 1968; Kredich, 1971). Wheldrake (1967) has shown that intracellular cysteine levels correlate well with the degree of repression of sulfate-activating enzymes in *E. coli*. This effect might be explained by varying degrees of feedback inhibition of serine acetyltransferase by cysteine, which would control levels of the internal inducer O-acetylserine. However, the addition of O-acetylserine to cultures grown on either sulfide or cystine does not overcome repression, even though the inducer seems to be taken up readily under these conditions (Jones-Mortimer et al., 1968; Kredich, 1971). Both sulfur limitation and O-acetylserine are required for derepression, a condition suggesting a dual system of control where the repressing effect of cysteine or sulfide is epistatic to the inducing effect of O-acetylserine.

It is very difficult to distinguish experimentally between the effects of sulfide and cysteine on induction by O-acetylserine, owing to the presence (at least in *S. typhimurium*) of a cysteine-induced cystathionine lyase (Kredich et al., 1972; Collins et al., 1973). Furthermore, addition of O-acetylserine to sulfide-grown cells increases intracellular cysteine to a level sufficient to induce cystathionine lyase (Kredich, 1971). Thus growth either on cysteine or on sulfide plus O-acetylserine is equivalent to growth on a mixture of both sulfur sources, so it is impossible to determine whether sulfide or cysteine is the metabolite that prevents induction.

Sulfide or cysteine could interfere with induction by competing with exogenously supplied O-acetylserine for the same target, for example, a single binding site on the *cysB* protein. Alternatively, they might act independently at a second site on the *cysB* protein or on some other regulatory element. In either case, gene expression would depend on the balance between these two kinds of effectors rather than on the level of just one. A second possibility is that sulfide, either added to the culture directly or derived from the action of cystathionine lyase on cysteine, reacts so readily with O-acetylserine that levels of the latter fall below those required for induction. The fact that cystathionine lyase is induced by a combination of O-acetylserine and sulfide indicates that even repressed cells have sufficient acetylserine sulfhydrylase to make this explanation feasible. Thus sulfide would serve simply to deplete the inducer and not interact directly with any other regulatory component. Conversely, one might argue that sulfide is the true repressor and that the only regulatory role of O-acetylserine is to lower the intracellular concentration of this metabolite. This seems unlikely because *cysE* strains require O-acetylserine for derepression even when grown on sulfur sources that should provide virtually no free sulfide at all.

D. A Model of the Cysteine Regulon

The exact mechanism by which the *cysB* protein modulates the expression of the genes is not known, but genetic and biochemical evidence is sufficient to formulate a plausible model. One possible function of the *cysB* protein would be to produce or interact with another factor, which itself mediates the regulatory effect, as for instance serine acetyltransferase does by catalyzing the synthesis of O-acetylserine. However, the regulatory properties of the *cysB484* allele show that the *cysB* protein is multifunctional and can be modified by a single mutation to be both *cysB*^{c} for acetylserine sulfhydrylase-A expression and *cysB* for other *cys* genes. One could postulate multiple diffusible mediators through which the *cysB* protein exerts its effects, but it seems more likely that different portions of this regulatory molecule interact directly with the various genes under its control.

The roles of sulfur limitation and O-acetylserine could then be explained through their effects on two different but interconvertible configurations of the *cysB* protein, one an active form and the other inactive. O-Acetylserine would then be necessary for the formation of an active *cysB* protein, which is required for derepression; sulfide and/or cysteine would shift the equilibrium to the inactive form (Fig. 7.4).

Derepression of different enzymes of the cysteine biosynthetic pathway is not always coordinate (Pasternak et al., 1965; Jones-Mortimer et al., 1968; Kredich, 1971). Sensitivity to derepression appears to be greater for acetylserine sulfhydrylase-A and the reductive enzymes than for the activating enzymes. These findings could be explained by assuming different affinities of the active *cysB* protein for the control regions of the various genes it

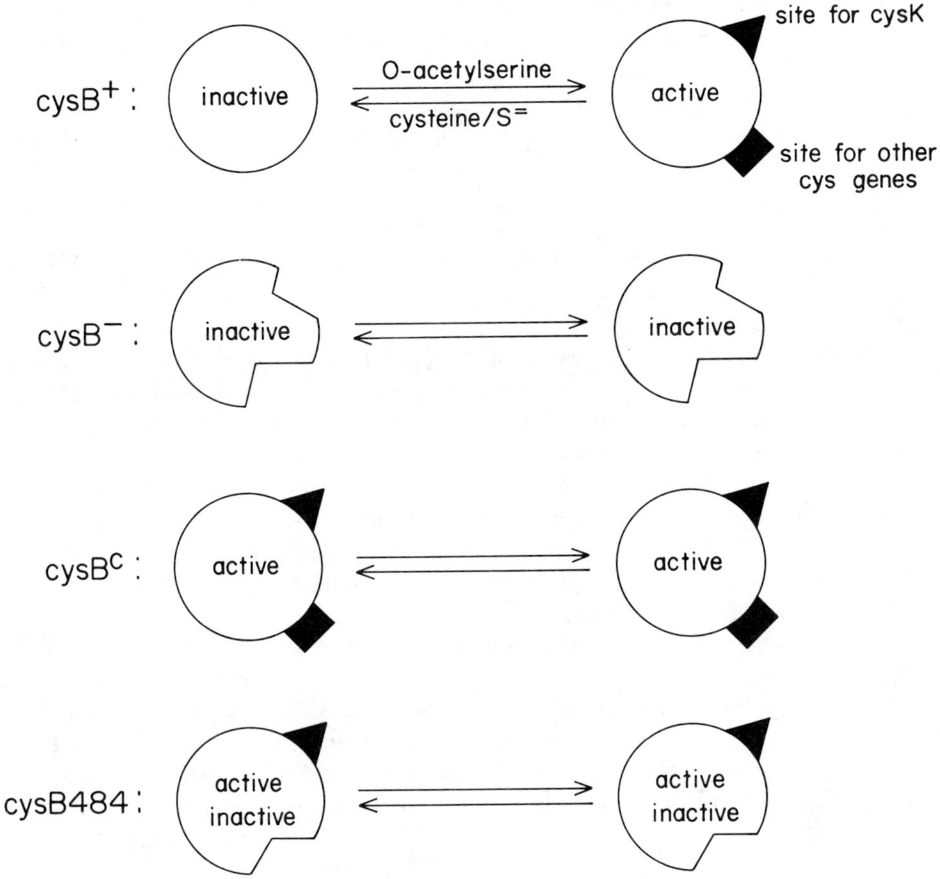

FIGURE 7.4 The *cysB* product as regulator of the cysteine regulon. In the presence of O-acetylserine, the *cysB* protein assumes an active configuration that can interact with the promoter regions of *cys* genes. At least two DNA binding sites are postulated, one for the *cysK* promoter to regulate acetylserine sulfhydrylase-A expression and a second for the other *cys* genes. The inactive configuration is favored in the presence of cysteine and/or sulfide. Mutant *cysB* alleles give proteins that are not affected by ligands and that can be either inactive (*cysB*), active (*cysB*^{c}), or active for one regulatory site while inactive for the other (*cysB484*).

regulates. Maximum derepression of the entire regulon would occur when all of the *cysB* protein is in the active form; intermediate levels of active *cysB* protein would preferentially activate those genes with the greatest affinities for this regulatory element, such as *cysK* and *cysHIJ*. This model would apply equally well whether active *cysB* protein has a single site for all the genes under its control or separate sites specific for different genes.

The behavior of *cysB484* indicates that there should be at least two sites on the active *cysB* protein, one for acetylserine sulfhydrylase-A regulation, the other for the rest of the pathway.

According to this model, a *cysB* mutation would result in either a *cysB* protein that cannot assume an active form or no protein at all; a pure $cysB^c$ mutation like *cysB1352* would give a protein that is active even in the absence of O-acetylserine and resistant to the effects of sulfide and/or cysteine; the effect of *cysB484* would be the same as that of *cysB1352* for that portion of the molecule involved in acetylserine sulfhydrylase-A regulation, but it would also give a protein incapable of stimulating expression of other *cys* genes (Fig. 7.4).

Regulation of the expression of the *cysDHIJ* region in *E. coli* has been shown to occur at the level of transcription (Fimmel and Loughlin, 1977b). Given the finding that constitutive expression of *cysHIJ* in a *cysB* deletion strain can be caused by a mutation in or near the promoter end of *cysJ* (Ostrowski and Hulanicka, 1979), it seems likely that the active *cysB* protein interacts with promoter regions, thereby facilitating transcription.

This model of the cysteine regulon is analogous to that proposed and demonstrated for the arabinose operon of *E. coli* (Sheppard and Englesberg, 1966). One important distinction deserves mention. The product of the *araC* regulatory gene exerts not only positive control but also negative control. Arabinose induction requires both the presence of the former and the absence of the latter. Thus an $araC^c/araC^+$ merodiploid still requires arabinose for induction because the $araC^+$ protein retains its capability for negative control. In contrast, a $cysB^c/cysB^+$ merodiploid is constitutively derepressed, an indication that the *cysB* protein itself is capable only of positive control (Jagura et al., 1978).

E. Attenuation and Autoregulation

There is no experimental basis for postulating attenuation in the cysteine regulon. Temperature-sensitive strains carrying mutations in *cysS*, the structural gene for cysteinyl-tRNA synthetase, have been isolated in *E. coli* (Bohman and Isaksson, 1979), but no studies on the regulatory behavior of these mutants have been reported. Gene *cysS* is not linked to any of the other known genes of the cysteine regulon.

Jagura-Burdzy and Hulanicka (1981) have recently constructed several *E. coli* strains in which a promoter-deficient fragment of *lacZ* is fused to the *cysB* promoter, thus allowing *cysB* expression to be monitored by measuring levels of β-galactosidase. Introduction of an episome or plasmid carrying an intact wild-type *cysB* gene gives a 50–90% decrease in β-galactosidase activity in these strains, an indication that *cysB* can regulate its own expression through a diffusible intermediate, presumably the *cysB* protein. Autoregulation of *cysB* differs from regulation of the biosynthetic pathway in that it is independent of the sulfur source used for growth. These studies also localized the *cysB* promoter to the *trp* side of *cysB* in *E. coli*.

F. Other Factors Influencing the Cysteine Regulon

Selection for cysteine auxotrophs often yields mutants that can grow on either cysteine or methionine. Some of these prove to be leaky mutants, which can synthesize sufficient cysteine for growth only if none of it need be diverted for methionine biosynthesis. Others appear to have no capacity at all for cysteine biosynthesis unless they are grown in the presence of methionine (Qureshi et al., 1975). These mutants carry mutations, provisionally designated Cym, that map in either *cysA*, the *cysCD* region, *cysG, cysH, cysI,* or *cysJ;* where tested, they result in enzyme deficiencies identical to those expected for mutations in these genes. Radiolabeled precursor studies have shown that the sulfur moiety of cysteine in methionine-grown Cym strains is obtained from sulfate in the medium and not from methionine (Qureshi et al., 1975). The mechanism by which methionine allows expression of an otherwise nonfunctional *cys* gene in a Cym mutant is not understood. There is evidence, however, that Cym mutations tend to be located at one particular end of certain *cys* genes (Qureshi et al., 1975; Kingsman, 1977; Kingsman et al., 1978; Kingsman and Smith, 1978), a finding suggesting that they alter *cys* regulatory regions that are sensitive to methionine or to a metabolite derived from this amino acid.

Major differences exist in the enzymology and regulation of cysteine biosynthesis during aerobic and anaerobic growth. One of these is the bradytrophic phenotype of *cysM* $cysK^+$ mutants under anaerobic but not aerobic conditions, mentioned earlier. Another is the observation of Barrett and Chang (1979) that *cysB, cysI,* and *cysJ* auxotrophs of *S. typhimurium* can grow on sulfate as a sole sulfur source in an atmosphere of 95% hydrogen and 5% carbon dioxide. Strains carrying mutations in *cysA, cysC, cysD, cysE, cysG,* and *cysH* remain auxotrophs under these conditions. These results indicate that there is an alternate route of sulfite reduction that requires siroheme, is distinct from the sulfite reductase specified by *cysI* and *cysJ,* and is inactive during aerobic growth. Furthermore, the response of *cysB* strains implies that, during anaerobic growth, the cysteine biosynthetic pathway can be regulated by a system that is independent of the *cysB* protein. Further biochemical and genetic studies of these differences should be most interesting. One possibility is that the silent DNA regions between *cysD* and *cysH* and between *cysH* and *cysI* (Demerec et al., 1963), as well as the 5–10 kb region between *cysK* and *cysMA,* may specify *cys* genes that are required only for anaerobic growth.

G. Cloned cys Genes

Lambda phage have been isolated that carry the *cysDHIJ* region (Fimmel and Loughlin, 1977b) and the *cysB* gene (Mascarenhas and Yudkin, 1980b), both of *E. coli*. In addition, a 5-kb fragment of the *E. coli* chromosome carrying *cysB* has been cloned in a ColE1 vector (Jagura-Burdzy and Hulanicka, 1981). Several *cys* genes from *S. typhimurium* have been cloned (G. Jagura-Burdzy, D. Hulanicka, and N. M. Kredich, unpublished), including *cysB* as a 2.7-kb *Sal*I fragment, *cysMA* as contiguous 3.8- and 9.1-kb *Eco*RI fragments, and *cysHIJ* as a 14-kb *Sal*I fragment. The availability of these cloned genes should facilitate efforts to purify and characterize the *cysB* protein and to elucidate the manner in which this regulatory element interacts with the genes under its control.

ACKNOWLEDGMENT

The author's research has been supported by Grant AM12828 from the National Institute for Arthritis, Diabetes, Digestive and Kidney Diseases.

REFERENCES

Asahi, T., Bandurski, R. S., and Wilson, L. G. (1961) *J. Biol. Chem. 236,* 1830–1835.
Bandurski, R. S., Wilson, L. G., and Squires, C. L. (1956) *J. Amer. Chem. Soc. 78,* 6408–6409.
Baptist, E. W., and Kredich, N. M. (1977) *J. Bacteriol. 131,* 111–118.
Barrett, E. L., and Chang, G. W. (1979) *J. Gen. Microbiol. 115,* 513–516.
Becker, M. A., and Tomkins, G. M. (1969) *J. Biol. Chem. 244,* 6023–6030.
Becker, M. A., Kredich, N. M., and Tomkins, G. M. (1969) *J. Biol. Chem. 244,* 2418–2427.
Bohman, K., and Isaksson, L. A. (1979) *Mol. Gen. Genet. 176,* 53–55.
Borum, P. R., and Monty, K. J. (1976) *J. Bacteriol. 125,* 94–101.
Cheney, R. W., Jr., and Kredich, N. M. (1975) *J. Bacteriol. 124,* 1273–1281.
Collins, J. M., Wallenstein, A., and Monty, K. J. (1973) *Biochim. Biophys. Acta 313,* 156–162.
Cook, P. F., and Wedding, R. T. (1976) *J. Biol. Chem. 251,* 2023–2029.
Cook, P. F., and Wedding, R. T. (1977a) *Arch. Biochem. Biophys. 178,* 293–302.
Cook, P. F., and Wedding, R. T. (1977b) *J. Biol. Chem. 252,* 3459.
Cook, P. F., and Wedding, R. T. (1978) *J. Biol. Chem. 253,* 7874–7879.
De Meio, R. H. (1975) in *Metabolic Pathways* Vol. VII, *Metabolism of Sulfur Compounds* (Greenberg, D. M., ed.) pp. 287–358, Academic Press, New York.
Demerec, M., Gillespie, D. H., and Mizobuchi, K. (1963) *Genetics 48,* 997–1009.
Dreyfuss, J. (1964) *J. Biol. Chem. 239,* 2292–2297.
Dreyfuss, J., and Monty, K. J. (1963a) *J. Biol. Chem. 238,* 1019–1024.
Dreyfuss, J., and Monty, K. J. (1963b) *J. Biol. Chem. 238,* 3781–3783.
Ellis, R. J. (1964) *Biochem. J. 93,* 20 p.
Ellis, R. J. (1966) *Nature 211,* 1266–1268.
Ellis, R. J., Humphries, S. K., and Pasternak, C. A. (1964) *Biochem. J. 92,* 167–172.
Fimmel, A. L., and Loughlin, R. E. (1977a) *J. Gen. Microbiol. 103,* 37–43.
Fimmel, A. L., and Loughlin, R. E. (1977b) *J. Bacteriol. 132,* 757–763.
Flavin, M. (1975) in *Metabolic Pathways* Vol. VII, *Metabolism of Sulfur Compounds* (Greenberg, D. M., ed.) pp. 457–503, Academic Press, New York.
Flavin, M., and Slaughter, C. (1967) *Biochim. Biophys. Acta 132,* 400–405.

Giovanelli, J., and Mudd, S. H. (1967) *Biochem. Biophys. Res. Comm. 27,* 150–156.
Hulanicka, D., and Klopotowski, T. (1972) *Acta Biochim. Pol. 19,* 251–260.
Hulanicka, M. D., and Kredich, N. M. (1976) *Mol. Gen. Genet. 148,* 143–148.
Hulanicka, D., Klopotowski, T., and Smith, D. A. (1972) *J. Gen. Microbiol. 72* 291–301.
Hulanicka, M. D., Kredich, N. M., and Treiman, D. M. (1974) *J. Biol. Chem. 249,* 867–872.
Hulanicka, M. D., Hallquist, S. G., Kredich, N. M., and Mojica-A, T. (1979) *J. Bacteriol. 140,* 141–146.
Jagura, G., Hulanicka, D., and Kredich, N. M. (1978) *Mol. Gen. Genet. 165,* 31–38.
Jagura-Burdzy, G., and Hulanicka, D. (1981) *J. Bacteriol. 147,* 744–751.
Jones-Mortimer, M. C. (1968a) *Biochem. J. 110,* 589–595.
Jones-Mortimer, M. C. (1968b) *Biochem. J. 110,* 597–602.
Jones-Mortimer, M. C. (1973) *Heredity 31,* 213–221.
Jones-Mortimer, M. C., Wheldrake, J. F., and Pasternak, C. A. (1968) *Biochem. J. 107,* 51–53.
Kemp, J. D., Atkinson, D. E., Ehret, A., and Lazzarini, R. A. (1963) *J. Biol. Chem. 238,* 3466–3471.
Kingsman, A. J. (1977) *Mol. Gen. Genet. 156,* 327–332.
Kingsman, A. J., and Smith, D. A. (1978) *Genetics 89,* 439–451.
Kingsman, A. J., Smith, D. A., and Hulanicka, M. D. (1978) *Genetics 89,* 419–437.
Kredich, N. M. (1971) *J. Biol. Chem. 246,* 3474–3484.
Kredich, N. M., and Tomkins, G. M. (1966) *J. Biol. Chem. 241,* 4955–4965.
Kredich, N. M., Becker, M. A., and Tomkins, G. M. (1969) *J. Biol. Chem. 244,* 2428–2439.
Kredich, N. M., Keenan, B. S., and Foote, L. J. (1972) *J. Biol. Chem. 247,* 7157–7162.
Kredich, N. M., Foote, L. J., and Hulanicka, M. D. (1975) *J. Biol. Chem. 250,* 7324–7331.
Kredich, N. M., Hulanicka, M. D., and Hallquist, S. G. (1980) in *Sulphur in Biology: Ciba Foundation Symposium 72* pp. 87–99, Elsevier/North Holland, New York.
Loughlin, R. E. (1975) *J. Gen. Microbiol. 86,* 275–282.
Mager, J. (1960) *Biochim. Biophys. Acta 41,* 553–555.
Mascarenhas, D. M., and Yudkin, M. D. (1980a) *Mol. Gen. Genet. 177,* 535–539.
Mascarenhas, D. M., and Yudkin, M. D. (1980b) *J. Gen. Microbiol. 118,* 405–410.
Mizobuchi, K., Demerec, M., and Gillespie, D. H. (1962) *Genetics 47,* 1617–1627.
Murphy, M. J., and Siegel, L. M. (1973) *J. Biol. Chem. 248,* 6911–6919.
Murphy, M. J., Siegel, L. M., Kamin, H., and Rosenthal, D. (1973) *J. Biol. Chem. 248,* 2801–2814.
Ohta, N., Galsworthy, P. R., and Pardee, A. B. (1971) *J. Bacteriol. 105,* 1053–1062.
Ostrowski, J., and Hulanicka, D. (1979) *Mol. Gen. Genet. 175,* 145–149.
Pardee, A. B. (1966) *J. Biol. Chem. 241,* 5886–5892.
Pardee, A. B., and Watanabe, K. (1968) *J. Bacteriol. 96,* 1049–1054.
Pardee, A. B., Prestidge, L. S., Whipple, M. B., and Dreyfuss, J. (1966) *J. Biol. Chem. 241,* 3962–3969.

Pasternak, C. A. (1962) *Biochem. J. 85,* 44–49.
Pasternak, C. A., Ellis, R. J., Jones-Mortimer, M. C., and Crichton, C. E. (1965) *Biochem. J. 96,* 270–275.
Peck, H. D., Jr. (1962) *Bacteriol. Rev. 26,* 67–94.
Postgate, J. (1959) *Ann. Rev. Microbiol. 13,* 505–520.
Qureshi, M. A., Smith, D. A., and Kingsman, A. J. (1975) *J. Gen. Microbiol. 89,* 353–370.
Robbins, P. W., and Lipmann, JF. (1956) *J. Amer. Chem. Soc. 78,* 6409–6410.
Robbins, P. W., and Lipmann, F. (1958) *J. Biol. Chem. 233,* 686–690.
Sanderson, K. E., and Hartman, P. E. (1978) *Microbiol. Rev. 42,* 471–519.
Schiff, J. A. (1980) in *Sulphur in Biology: Ciba Foundation Symposium 72* pp. 49–69, Elsevier/North Holland, New York.
Schmidt, A., Abrams, W. R., and Schiff, J. A. (1974) *Eur. J. Biochem. 47,* 423–434.
Sheppard, D., and Englesberg, E. (1966) *Cold Springs Harbor Symp. Quant. Biol. 31,* 345–347.
Siegel, L. M. (1975) in *Metabolic Pathways* Vol. III, *Metabolism of Sulfur Compounds* (Greenberg, D. M., ed.) pp. 217–286, Academic Press, New York.
Siegel, L. M., and Davis, P. S. (1974) *J. Biol. Chem. 249,* 1587–1598.
Siegel, L. M., and Kamin, H. (1971) *Fed. Proc. 30,* 1261.
Siegel, L. M., H., Rueger, D. C., Presswood, R. P., and Gibson, Q. H. (1971) in *Flavins and Flavoproteins* (Kamin, H., ed.) pp. 523–553, University Park Press, Baltimore, MD.
Siegel, L. M., Murphy, M. J., and Kamin, H. (1973) *J. Biol. Chem. 248,* 251–264.
Siegel, L. M., Davis, P. S., and Kamin, H. (1974) *J. Biol. Chem. 249,* 1572–1586.
Spencer, H. T., Collins, J., and Monty, K. J. (1967) *Fed. Proc. 26,* 677.
Tsang, M. L.-S., and Schiff, J. A. (1976a) *J. Bacteriol. 125,* 923–933.
Tsang, M. L.-S., and Schiff, J. A. (1976b) *Plant Cell Physiol. 17,* 1209–1220.
Tsang, M. L.-S., Schiff, J. A. (1978) *Plant Sci. Lett. 11,* 177–183.
Tully, M., and Yudkin, M. D. (1975) *Mol. Gen. Genet. 136,* 181–183.
Tully, M., and Yudkin, M. D. (1977) *J. Bacteriol. 131,* 49–56.
Wheldrake, J. F. (1967) *Biochem. J. 105,* 697–699.
Wiater, A., and Hulanicka, D. (1978) *Acta Biochim. Pol. 25,* 281–287.
Wilson, L. G., and Bierer, D. (1976) *Biochem. J. 158,* 255–270.
Wilson, L. G., Asahi, T., and Bandurski, R. S. (1961) *J. Biol. Chem. 236,* 1822–1829.

CHAPTER 8

Aspartate and Asparagine Biosynthesis

Lawrence J. Reitzer

I. INTRODUCTION

Aspartate and asparagine represent metabolic extremes. Aspartate is a precursor for other amino acids, purines, and pyrimidines and is used directly for protein biosynthesis. Asparagine is generally destined for protein biosynthesis, although there are reports of purified enzymes that can use asparagine as an amide donor (Wilcox and Nirenberg, 1968; Cooper, 1977).

This review deals primarily with the genetics and biochemistry of aspartate and asparagine biosynthesis in *Escherichia coli*. An imposing literature on aspartate transaminases dominates the study of aspartate biosynthesis (Braunstein, 1973) (Chapter 2). The genetics and biochemistry of asparagine synthesis have been elucidated only recently. Two recent reviews on asparagine focus on the enzymology (Meister, 1974) and on asparagine biosynthesis in plants (Lea and Fowden, 1975a).

Figure 8.1 shows the reactions leading to aspartate and asparagine. Aspartate arises through transamination: oxaloacetate and glutamate serve as substrates generating aspartate and α-ketoglutarate. Other amino donors, such as alanine, may function in vivo (Chapter 2). A potential pathway in bacteria, independent of transamination, involves the reversal of aspartase

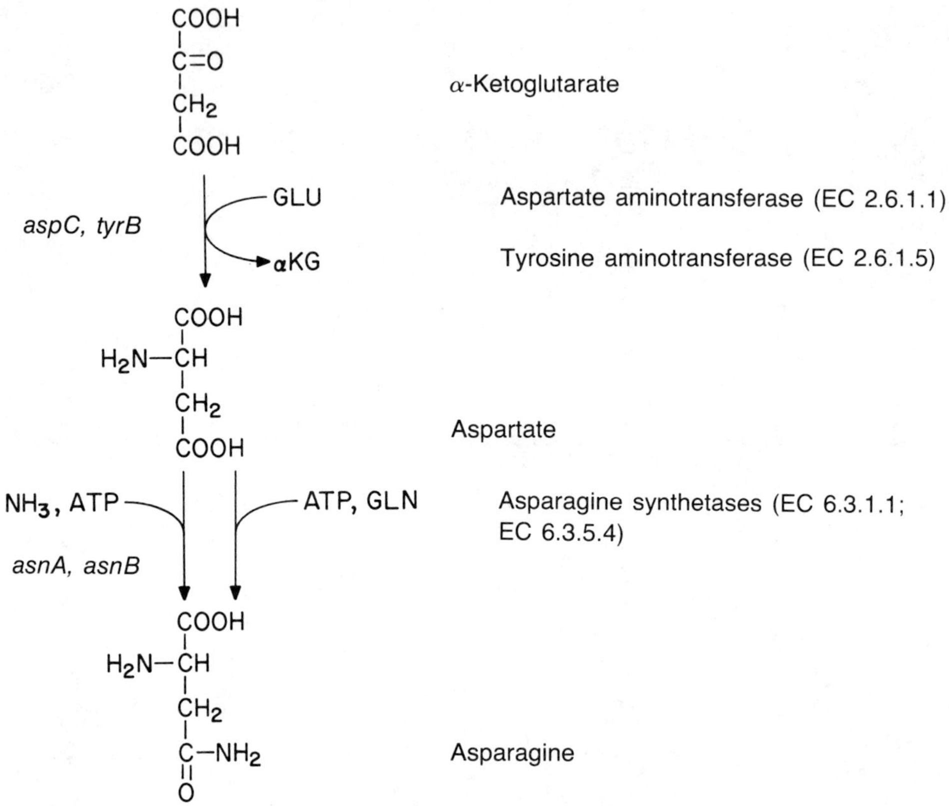

FIGURE 8.1 Pathway for asparagine biosynthesis.

by a reaction between ammonia and fumarate. As we will see, this does not happen in vivo.

There are two routes of asparagine biosynthesis. Asparagine synthetase catalyzes an ATP-dependent transfer of the glutamine amide (in all organisms) or ammonia (in bacteria only) to aspartate, liberating AMP and pyrophosphate in the process. A second pathway, involving cyanide fixation, is considered mainly a detoxification mechanism in plants.

II. ASPARTATE

Two routes of aspartate synthesis in enteric bacteria have been proposed. The universal route involves transamination of oxaloacetate with glutamate as the amino donor (Fig. 8.1). The author is not aware of a wild-type organism that lacks an aspartate transaminase. A second potential route involves ammonia addition to fumarate, a reversal of the aspartase reaction. Aspartase, encoded by the *aspA* gene in *E. coli,* appears in many bacteria

(Virtanen and Ellfolk, 1955; Hanson and Havir, 1974). The equilibrium constant would allow the aspartase reaction to go in either direction in vivo (Bada and Miller, 1968).

However, several lines of evidence argue that aspartase is a degradative enzyme only. If aspartase could assimilate ammonia, that is, synthesize aspartate, then all glutamate requiring strains would have to be *aspA* mutants. This is not the case. Two enzymes are responsible for glutamate symthesis in enteric bacteria in minimal media: glutamate synthase and glutamate dehydrogenase (Tyler, 1978) (Chapter 1). Only the double mutant is a glutamate auxotroph. Glutamate auxotrophs require glutamate or a nitrogen source readily transaminated to glutamate, such as aspartate. These strains do not grow with ammonia as a nitrogen source. Therefore ammonia cannot be assimilated by reversal of aspartase. Furthermore, the genes responsible for glutamate synthesis do not lie near *aspA*.

Evidence for aspartate formation via transamination in vivo has been difficult to obtain. Aspartate transaminase mutants have been isolated in *E. coli* (Marcus and Halpern, 1969; Gelfand and Steinberg, 1977), *Brevibacterium flavum* (Shiio and Ujigawa, 1978), *Bacillus subtilis* (Hoch, 1972), and *Saccharomyces cerevisiae* (de Robichon-Szulmajster et al., 1966). The minimum number of mutations necessary for aspartate auxotrophy in *E. coli* is two (Gelfand and Steinberg, 1977). The genes involved are *aspC*, coding for a high-affinity aspartate transaminase, and *tyrB*, coding for a transaminase of tyrosine and phenylalanine biosynthesis (Chapter 2). The double mutant, *aspC tyrB*, requires only tyrosine and aspartate for growth, since phenylalanine can be synthesized via a branched-chain amino acid transaminase, the *ilvE* gene product. An *aspC* $tyrB^+$ strain has no phenotype, since the *tyrB* gene product has asparatate transaminase activity. Tyrosine repression of *tyrB* should generate aspartate auxotrophy in such a strain. This was not tested. Future mutant searches for aspartate transaminase mutants should include tyrosine in the medium. The *aspC* gene lies at 20 min of the *E. coli* linkage map (Gelfand and Steinberg, 1977). However, the aspartate transaminase mutant of Marcus and Halpern (1969), though it eliminated the same electrophoretic species of aminotransferase, did not map in the same location as the mutation of Gelfand and Steinberg (1977). Thus the map location of the aspartate transaminase structural gene has not been established unambiguously.

Despite the multiplicity of aspartate transaminases in *E. coli*, there is good agreement concerning the biochemistry of these enzymes. Rudman and Meister (1953) defined three transaminase fractions: A for aspartate, tryptophan, tyrosine, and phenylalanine; B for leucine, isoleucine, and valine; and C for valine, alanine, and α-aminobutyrate. This nomenclature will be retained here because of its frequent use in the literature. In this discussion, aspartate activity means activity in a crude extract. Aspartate transaminase refers to a specific protein for which aspartate is the substrate with the highest affinity, unless qualified.

Transaminase A consists of two enzymes. Aspartate activity of transaminase A was heat stable, while the tyrosine and phenylalanine activity of transaminase A was not (Collier and Kohlhaw, 1972). Aspartate activity was less subject to inactivation by dialysis than the aromatic amino acid activity (Collier and Kohlhaw, 1972). This means that the cofactor dissociated more rapidly for the latter. Aspartate did not repress either activity, but the aromatic amino acid activity was tyrosine repressible (Collier and Kohlhaw, 1972; Mavrides and Orr, 1974). The purified aspartate transaminase had activity with aromatic amino acids as substrates (Chesne and Pelmont, 1973; Mavrides and Orr, 1975). The aromatic amino acid transaminase does not have activity with aspartate but does have activity with oxaloacetate in the other direction (Mavrides and Orr, 1975). This agrees remarkably well with the genetic studies, which convincingly demonstrated the use of a tyrosine repressible transaminase, the *tyrB* gene product, for aspartate biosynthesis and, conversely, the use of the aspartate transaminase for aromatic amino acid biosynthesis (Gelfand and Steinberg, 1977). Mavrides and Orr (1975) reported a list of K_m's for the purified transaminase components of transaminase A. The assays of Collier and Kohlhaw (1972) were performed well below these K_m's; any differences between these two reports as to substrate specificity are easily understood.

Mavrides and Orr (1974, 1975) purified the two components of transminase A. Both were dimers of identical subunits. The molecular weights were 88,000 and 82,000 for the aromatic amino acid and the aspartate transaminase, respectively. Subtilisin cleaved the tyrosine-repressible transaminase to generate "aspartate transaminase" (Mavrides and Orr, 1975). This observation is not of physiological significance, since both transaminases can use oxaloacetate as a substrate, but it suggests an evolutionary relationship between the two enzymes. Paris and Magasanik (1981) purified the high-affinity aspartate transaminase of *Klebsiella aerogenes*. The properties were similar to the 82,000 molecular weight aspartate transaminase of *E. coli*.

The predominant aspartate transaminase activity in vitro is constitutive. Nineteen amino acids (singly) did not repress aspartate activity (Collier and Kohlhaw, 1972). Gelfand and Steinberg (1977) and Mavrides and Orr (1974) obtained similar results. One report of repression exists, but the conditions of growth were so unusual that the repression may not occur in more normal media (Urm et al., 1973).

Is glutamate the amino donor for aspartate synthesis in vivo? This is a subtle point. The major alternative amino donor is alanine. In the early transaminase literature, mention is made of alanine–aspartate transaminases. Rudman and Meister (1953) did not look for such a possibility. The work of Lowry et al. (1971) allows a reasonable guess as to the cellular alanine versus glutamate levels. Assuming that the only route of alanine synthesis is transamination by glutamate and assuming further that the equilibrium constant for this transamination (Bulos and Handler, 1965) determines the maximal alanine level possible, then the ratio of glutamate to alanine internally is at least 3 to 1. This favors glutamate as an amino donor, if the kinetic constants are assumed to be similar. Furthermore, alanine induces its own degradation (Raunio and Jenkins, 1973). This means that intracellular levels of alanine are strictly controlled. Finally, if alanine could aminate oxaloacetate, then a glutamate auxotroph should grow with alanine in place of aspartate or glutamate; this is not the case (Tyler, 1978). This does not imply the nonexistence of a glutamate–pyruvate or an alanine–oxalacetate transaminase, but it does imply that these reactions are not particularly reversible in enteric bacteria in vivo. In conclusion, alanine does not appear to be a likely amino donor for aspartate synthesis (Chapter 2).

III. ASPARAGINE

Asparagine auxotrophs have been isolated in *E. coli, K. aerogenes, S. cerevisiae,* and mammalian cell lines. In the first three organisms, auxotrophs are double mutants. Many mammalian cell lines and tumors require asparagine (Eagle et al., 1966). Of clinical significance is the fact that high-affinity asparaginases cause some leukemias and tumors to regress.

Virtually all asparagine synthetase assays involve the separation of asparagine from aspartate. Aspartyl–hydroxamate formation is unsatisfactory in crude systems, since aspartatekinases react with the asparagine synthetase substrates. The assay of Prusiner and Milner (1970) is rapid and convenient. Recent studies of asparagine synthesis have included efforts to eliminate the effect of asparaginases on product detection. The use of asparaginase mutants and the chemical inactivation of asparaginase have both been effective (Cedar and Schwartz, 1969a; Humbert and Simoni, 1980; Ramos and Wiame, 1979; Reitzer and Magasanik, 1982).

A. Physiology and Genetics of Bacterial Asparagine Synthetases

Cedar and Schwartz (1969a) isolated the first bacterial asparagine auxotroph. Their mutant *E. coli* strain could be transduced to prototrophy by phage P1, propagated on a wild-type strain, but not vice versa. Auxotrophy resulted from mutations in two unlinked genes designated *asnA* and *asnB* (Felton et al., 1980; Humbert and Simoni, 1980). These loci lie at 84 and 15 min, respectively, on the revised *E. coli* linkage map (Bachmann and Low, 1980). The *asnA* gene product was the previously characterized ammonia-dependent enzyme (Cedar and Schwartz, 1969a, b) with respect to molecular weight, substrate preference, and thermal stability (Humbert and Simoni, 1980). The *asnB* gene product was cold labile, more heat stable, and able to use glutamine or ammonia as the nitrogen donor. The *asnA* product was specific for ammonia. Strains with only one wild-type allele, *asnA*$^+$ *asnB*, or *asnA* *asnB*$^+$, did not have a detectable phenotype, even though *asnB* mutants had 24 times more enzyme activity than *asnA* mutants (Humbert and Simoni, 1980).

The function of the multiple asparagine synthetases was recently elucidated. Enteric bacteria, grown in a minimal medium, can ordinarily use a variety of compounds as their sole nitrogen source (Tyler, 1978). Enterics assimilate the preferred nitrogen source, ammonia, through the combined action of glutamate dehydrogenase and glutamine synthetase (Chapter 1). On slowly metabolized nitrogen sources, such as proline or aspartate, glutamine synthetase and glutamate synthase assimilate ammonia into glutamine and glutamate, and glutamate synthase aminates α-ketoglutarate with glutamine as the nitrogen donor. *K. aerogenes asnB* mutants do not grow on a variety of nitrogen compounds (Reitzer and Magasanik, 1982). The *asnB* mutant strains had a phenotype that is different from *glnG* and *gltB* strains, the previously isolated pleiotropic nitrogen nonutilizing strains. The *glnG* gene product is a regulatory protein essential for growth on secondary nitrogen sources (Gaillardin and Magasanik, 1978; Kustu et al., 1979; Pahel and Tyler, 1979). The *gltB* gene encodes glutamate synthase (Tyler, 1978). The *K. aerogenes asnB* strains assimilated low levels of ammonia, although with glutamine as the sole nitrogen source the growth rate was radically altered. The levels of the nitrogen assimilatory enzymes also differed from the levels in the isogenic parent. Asparagine reversed these effects in *asnB* strains missing a high-affinity asparaginase.

Asparagine-requiring strains were isolated from an *asnB*::Tn*5* parental strain after further mutagenesis. The orginial Tn*5* insertion mutation conferred kanamycin resistance. If the asparagine requirement were the result of a double mutation, then two classes of asparagine-independent strains should and did result from transduction to prototrophy, distinguishable by kanamycin resistance or sensitivity. The two *asn* genes lay in chromosomal positions analogous to those of the *E. coli* genes. Only the kanamycin resistant *asnA*$^+$ *asnB* strains had the pleiotropic nitrogen phenotype. The kanamycin sensitive *asnA asnB*$^+$ strains had no growth phenotype. Therefore in *K. aerogenes,* as in *E. coli,* asparagine auxotrophy requires a double mutation (Reitzer and Magasanik, 1982).

The *asnA*$^+$ *asnB* strains contained an ammonia-dependent asparagine synthetase, while *asnA asnB*$^+$ strains had a glutamine-dependent enzyme. At pH 7.0, using 2.5 mM ammonium chloride or 2 mM glutamine, the substrate preference was absolute. Above pH 7.5, *asnA asnB*$^+$ strains will also use ammonia for asparagine biosynthesis, a common property of glutamine amidotransferases. Purification of the asparagine synthetases verified the substrate preference (Section II.D.).

The *asnA*$^+$ *asnB* strains fail to grow on secondary nitrogen sources. The glutamine-dependent enzyme mediates asparagine synthesis on poor nitrogen sources, that is, when the intracellular level of ammonia is low. In excess ammonia, asparagine synthesis proceeds via the ammonia-dependent enzyme. However, when the nitrogen source is poor and the intracellular ammonia levels are low, the ammonia-dependent enzyme cannot function. Suppressors of *asnA*$^+$ *asnB* mutations, which allow growth on secondary nitrogen sources, have altered nitrogen assimilation. Almost all such strains have a less active glutamine synthetase, presumably causing ammonia accumulation sufficient for the ammonia-dependent enzyme. Analysis of the suppressor strains verified the postulated physiological function of the two asparagine synthetases (Reitzer and Magasanik, 1982).

The asparagine synthetases of *K. aerogenes* are regulated by the nitrogen source of the medium (Reitzer and Magasanik, 1982). Ammonia derepresses the ammonia-dependent asparagine synthetase and represses the glutamine-dependent enzyme. The reverse is true for a poor nitrogen source. The analysis of asparagine synthetase regulation in strains having various combinations of *asnA* and *asnB* with two nitrogen assimilatory mutations gave the following results:

1. Glutamine-dependent asparagine synthetase levels are controlled solely by intracellular asparagine pools; 0.7 mM asparagine represses this enzyme. An increase in the glutamine-dependent enzyme invariably accompanies a reduction of the ammonia-dependent enzyme. When ammonia is present in the medium, the asparagine formed via the ammonia-dependent enzyme represses the glutamine-dependent enzyme.
2. A combination of asparagine repression (0.7 mM is sufficient) and the cellular levels of the *glnG* gene product controls the ammonia-dependent enzyme. Whenever the *glnG* gene product is low, ammonia-dependent activity is high, and vice versa (Reitzer and Magasanik, 1982).

There are two rationales for the interplay of asparagine synthetase regulation and nitrogen assimilation. The repression of an ammonia-consuming enzyme during nitrogen limitation has advantages for nitrogen assimilation because then only glutamine synthetase consumes ammonia. This is especially critical because asparagine is a dead-end metabolite, since there is no transfer of the asparagine nitrogen to other metabolites. The second rationale is based on energy expenditure. Nitrogen excess can sometimes be equated with energy deficiency. Under these conditions, the glutamine enzyme is repressed, thus saving the energy required to synthesize glutamine.

Low levels of asparagine (0.3 mM) repress the *Lactobacillus arabinosus* asparagine synthetase (Ravel et al., 1962). In *E. coli,* significant repression of the enzyme is only seen with high levels of asparagine (Cedar and Schwartz, 1969a; Humbert and Simoni, 1980). In this organism, asparagine may not be readily transported, or alternatively, it may be actively degraded.

The physiological outline of asparagine synthetase regulation in enteric bacteria does not define the molecular mechanism. The control mechanism may involve attenuation or repression or both. Because there are two *asn* genes, the possibility of more than one repressor gene is intriguing. Each asparagine synthetase can accumulate to the point at which it constitutes a few percent of the soluble protein of the cell (Reitzer and Magasanik, 1982). The *asn* promoters may have some value in recombinant DNA research. The *asnA* gene has been cloned on a bacteriophage lambda vector (von Meyenburg et al., 1978).

B. Fungal Asparagine Synthesis

There are multiple asparagine synthetase isoenzymes in *S. cerevisiae*. Jones (1978) isolated asparagine auxotrophs by counterselection. The genetic data indicated two cistrons, *ASN1* and *ASN2*. Three independent mutant strains were recessive to wild type. Heteroallelic diploids were auxotrophs. The loss of only one gene did not result in a detectable phenotype. Ramos and Wiame (1979, 1980) isolated 16 asparagine auxotrophs. Fifteen were asparagine synthetase deficient and were double mutants. The sixteenth was an asparaginyl-tRNA synthetase mutant. The genes for the asparagine synthetases were called *asnA* and *asnB*. The *asnA* $asnB^+$ strains had less total asparagine synthetase activity than the wild type or $asnA^+$ *asn*B, which had essentially the same activity. The two activities were separable by ion exchange chromatography. Unlike bacteria, they possessed activity with glutamine and ammonia, but glutamine was the preferred substrate. Each enzyme had a molecular weight of 150,000. Asparagine repressed slightly; variation of nitrogen source was not tried.

C. Plant Asparagine Synthesis

Asparagine synthesis in plants (Lea and Fowden, 1975a; Miflin and Lea, 1977) and the role of asparagine in nitrogen transport (Pate, 1980) have been reviewed. Asparagine can accumulate in seeds and seedlings (Chibnall, 1939), and it can be the major nitrogenous product in xylum (Pate, 1980).

Plants synthesize asparagine by aspartate amidation or by a cyanide-dependent route (Lea and Fowden, 1975a; Miflin and Lea, 1977; Streeter, 1973; Meister, 1974). The cyanide route involves condensation of cyanide with serine or cysteine to form β-cyanoalanine, which is cleaved by a nitrilase to give asparagine. However, the primary physiological pathway appears to be synthesis from aspartate (Streeter, 1973; Lees et al., 1968; Ting and Zschoche, 1970). Purified plant asparagine synthetases have biosynthetic capacities similar to the bacterial and mammalian enzymes (Lea and Fowden, 1975b; Rognes, 1975, 1980). The plant enzymes can definitely satisfy the asparagine requirements of normal growth.

The enzyme activities vary during plant development (Dilworth and Dure, 1978; Stulen and Oaks, 1977; Stulen et al., 1979). The similarity between bacterial and plant nitrogen assimilation (Miflin and Lea, 1976) suggests an interaction between nitrogen assimilation and asparagine synthetase regulation (Lea and Fowden, 1975b).

D. Enzymology

Asparagine synthetases have been purified from several sources (Table 8.1). Pure glutamine-dependent asparagine synthetase has an estimated turnover number of 400–500 nmoles/min·mg.

Several catalytic properties of the asparagine synthetases are preserved between species. All enzymes cleave ATP to AMP and pyrophosphate and catalyze a pyrophosphate–ATP exchange reaction. Glutamine is the source of amide nitrogen in all organisms, including bacteria. Bacteria also possess an ammonia-dependent enzyme. Even when glutamine is the preferred amide donor, high levels of ammonia can replace glutamine. The glutamine-dependent enzymes exhibit a substrate-independent glutaminase activity.

The glutaminase activity of more highly purified enzyme preparations can exceed the biosynthetic capacity (Horowitz and Meister, 1972; Milman and Cooney, 1979; Rognes, 1975, 1980; Reitzer and Magasanik, 1982). Asparagine inhibits the glutaminase activity, the implication being that the glutaminase is an integral part of the asparagine synthetase. In the plant *Lupinus luteus,* glutaminase and glutamine-dependent asparagine synthetase activities are stimulated by Cl^- and Br^-, while the ammonia-dependent activity is unaffected (Rognes, 1980). Glutaminase activity may be a consequence of damage to asparagine synthetase during purification (Horowitz and Meister, 1972; Reitzer and Magasanik, 1982). Enzyme preparations are highly susceptible to oxidation, and crude and purified asparagine synthetases differ with respect to stability and kinetics; multiple species of the enzyme are found by two-dimensional gel electrophoresis (Reitzer and Magasanik, 1982). Damage to another amidotransferase, carbamoylphosphate synthetase of *E. coli,* results in generation of a glutaminase activity (Wellner and Meister, 1975). While glutaminase is not unusual in glutamine amidotransferases (Horowitz and Meister, 1972; Buchanan, 1973), the ratio of glutaminase to asparagine synthetase appears to be unusual.

Two ammonia-dependent (Ravel et al., 1962; Cedar and Schwartz, 1969b) and three glutamine-dependent asparagine synthetases (Horowitz and Meister, 1972; Rognes, 1975; Milman and Cooney, 1979) catalyze a pyrophosphate–ATP exchange reaction. In all but one case, the exchange is aspartate-dependent. In the exception, the exchange reaction is unaffected by any substrate and exceeds the synthetic reaction fivefold to sevenfold (Rognes, 1975). The substrate-independent exchange is a potential ATP degradation uncoupled from any useful work, and thus it probably is not physiological.

All asparagine synthetases are inhibited by asparagine (Table 8.1). Despite the asparagine accumulation, plant enzymes appear to be only slightly more resistant to asparagine than the bacterial enzymes. Asparagine inhibits the glutaminase and the synthetase activity of the glutamine-dependent enzymes (Horowitz and Meister, 1972; Milman and Cooney, 1979; Reitzer and Magasanik, 1982). However, the ammonia-dependent activity of the *K. aerogenes* glutamine enzyme is less well inhibited. Asparagine may inhibit both as a glutamine analog and as a product inhibitor. The evidence favors the former

TABLE 8.1 Properties of Purified Asparagine Synthetases

				K_m/s (mM)						
	Preferred Substrate	*Specific Activity*[1]	*Estimated Purity (%)*	*Assay pH*	*asp*	*ATP*	NH_3	*gln*	*Asparagine Inhibition*	*Reference*
Bacteria										
Streptococcus bovis	NH_3	380	—	7.2	26	4	4	—	2.2 mM (55%)	Burchall et al. (1964)
Lactobacillus arabinosus	NH_3	525	—	6.8	4.2	2	—	—	10 mM (99%)	Ravel et al. (1962)
Escherichia coli	NH_3	454	7	78	0.8	0.5	0.1	—	0.12 mM (50%)	Cedar and Schwartz (1969a, b)
Klebsiella aerogenes	NH_3	2400	40	7.0	0.9	0.27	0.33	—	1 mM (79%)	Reitzer and Magasanik (1982)
				8.0	—	—	0.06	—		
Klebsiella aerogenes	glutamine	430	95	7.0	0.5	0.24	22	0.57	0.1 mM (48%)	
Plant										
Lupinus albus	glutamine	194	20	8.0	0.8	0.15	2.1	0.04	2 mM (50%)	Lea and Fowden (1975b)
Lupinus luteus	glutamine	225	20–30	7.8	1.3	0.14	—	0.16	—	Rognes (1975, 1980)

1. nmoles/min · mg.

explanation. Certain kinetic parameters are strikingly similar in all asparagine synthetases; pH profile and K_m's for asparate and ATP. The K_m's for glutamine vary over a 15-fold range, plant enzymes having lower K_m's. The glutamine analogs 6-diazo-5-oxonorleucine and azaserine inhibit glutamine-dependent asparagine synthetases (Horowitz and Meister, 1972; Patterson and Orr, 1968; Milman and Cooney, 1979; Reitzer and Magasanik, 1982). Sensitivity to such inhibitors is a characteristic of amidotransferases (Buchanan, 1973).

The subunit structures of few asparagine synthetases are known. The glutamine-dependent enzyme from *K. aerogenes* is a tetramer of 57,000-dalton subunits (Reitzer and Magasanik, 1982). This type of subunit organization is also seen in the glutamine amidotransferases CTP synthetase and phosphoribosyl–pyrophosphate amidotransferase (Buchanan, 1973). The asparagine synthetase tetramer dissociates to a monomer in the absence of glutamine. The monomer has glutaminase and synthetase activity, but the latter only with high ammonia. The plant enzyme has a molecular weight of 320,000 in the presence of ATP and 160,000 in the absence of ATP (Rognes, 1975). This is similar to CTP synthetase, in which nucleotides are involved in aggregation (Buchanan, 1973).

Studies on the mechanism of the catalytic reaction show that hydroxylamine can replace ammonia or glutamine as the nitrogen donor (Ravel et al., 1962; Burchall et al., 1964; Cedar and Schwartz, 1969a; Horowitz and Meister, 1972; Rognes, 1975). This implies that hydroxylamine reacts with an activated aspartate, possibly an enzyme–aspartyl–AMP complex. Consistent with this hypothesis are the findings that aspartate and ATP bind first and randomly to the enzyme, then pyrophosphate is released, then ammonia binds, and the products are released randomly (Cedar and Schwartz, 1969b). Also, β-aspartyl-adenylate can participate in the pyrophosphate–ATP exchange reaction (Horowitz and Meister, 1972). For the glutamine-dependent asparagine synthetases, glutamine appears to be the first substrate to bind to the enzyme (Rognes, 1975; Milman et al., 1980). However, the possibility that glutaminase activity may be the result of damage was not considered. An altered asparagine synthetase may indeed bind glutamine first. Thus the reaction mechanisms proposed for the glutamine-dependent asparagine synthetases should be considered to be tentative.

ACKNOWLEDGMENTS

The work of the author was supported by Public Health Service research grants from the National Institute of General Medical Sciences (GM-07446), the National Institute of Arthritis, Metabolism and Digestive Diseases (AM-13894), and the National Science Foundation (PCM-7808576) to Dr. Boris Magasanik, in whose laboratory the work of the author was performed.

REFERENCES

Bachmann, B. J., and Low, K. B. (1980) *Microbiol. Rev. 44,* 1–56.

Bada, J. L., and Miller, S. L. (1968) *Biochemistry 7,* 3403–3408.

Braunstein, A. E. (1973) in *The Enzymes* (Boyer, P. D., ed.) Vol. IX, pp. 379–481, Academic Press, New York.

Buchanan, J. M. (1973) *Adv. Enzymol. 39,* 91–183.

Bulos, B., and Handler, P. (1965) *J. Biol. Chem. 240,* 3283–3294.

Burchall, J. J., Reichelt, E. C., and Wolin, M. J. (1964) *J. Biol. Chem. 239,* 1794–1798.

Cedar, H., and Schwartz, J. H. (1969a) *J. Biol. Chem. 244,* 4112–4121.

Cedar, H., and Schwartz, J. H. (1969b) *J. Biol. Chem. 244,* 4122–4127.

Chesne, S., and Pelmont, J. (1973) *Biochimie 55,* 237–244.

Chibnall, A. C. (1939) *Protein Metabolism in the Plant,* Yale University Press, New Haven, Conn.

Collier, R. H., and Kohlhaw, G. (1972) *J. Bacteriol. 112,* 365–371.

Cooper, A. J. L. (1977) *J. Biol. Chem. 252,* 2032–2038.

de Robichon-Szulmajster, H., Surdin, Y., and Mortimer, R. K. (1966) *Genetics 53,* 609–619.

Dilworth, M. F., and Dure, L. (1978) *Plant Physiol. 61,* 698–702.

Eagle, H., Washington, C. L., Levy, M., and Cohen, L. (1966) *J. Biol. Chem. 241,* 4994–4999.

Felton, J., Michaelis, S., and Wright, A. (1980) *J. Bacteriol. 142,* 221–228.

Gaillardin, C. M., and Magasanik, B. (1978) *J. Bacteriol. 133,* 1329–1338.

Gelfand, D. H., and Steinberg, R. A. (1977) *J. Bacteriol. 130,* 429–440.

Hanson, K. R., and Havir, E. A. (1972) in *The Enzymes* (Boyer, P. D., ed.) Vol. VII, pp. 75–166, Academic Press, New York.

Hoch, J. (1972) in *Spores V* (Halvorson, H. O., Hanson, R., and Campbell, L. L., eds.) pp. 113–116, American Society for Microbiology, Washington, D.C.

Horowitz, B., and Meister, A. (1972) *J. Biol. Chem. 247,* 6708–6719.

Humbert, R., and Simoni, R. D. (1980) *J. Bacteriol. 142,* 212–220.

Jones, G. E. (1978) *J. Bacteriol. 134,* 200–207.

Kustu, S., Burton, D., Garcia, E., McCarter, L., and McFarland, N. (1979) *Proc. Nat. Acad. Sci. U.S.A. 76,* 4576–4580.

Lea, P. J., and Fowden, L. (1975a) *Biochem. Physiol. Pflanzen 168,* S.3–14.

Lea, P. J., and Fowden, L. (1975b) *Proc. Roy. Soc. London B 192,* 13–26.

Lees, E. M., Farnden, K. J. F., and Elliott, W. H. (1968) *Arch. Biochem. Biophys. 126,* 639–546.

Lowry, O. H., Carter, J., Ward, J. B., and Glaser, L. (1971) *J. Biol. Chem. 246,* 6511–6521.

Marcus, M., and Halpern, Y. S. (1969) *Biochim. Biophys Acta 177,* 314–320.

Mavrides, C., and Orr, W. (1974) *Biochim. Biphys. Acta 336,* 70–78.

Mavrides, C., and Orr, W. (1975) *J. Biol. Chem. 250,* 4128–4133.

Meister, A. (1974) in *The Enzymes* (Boyer, P. D., ed.) Vol. X, pp. 561–580, Academic Press, New York.

Miflin, B. J., and Lea, P. J. (1976) *Phytochemistry 15,* 873–885.

Miflin, B. J., and Lea, P. J. (1977) *Ann. Rev. Plant Physiol. 28,* 299–329.

Milman, H. A., and Cooney, D. A. (1979) *Biochem. J. 181,* 51–59.

Milman, H. A., Cooney, D. A., and Huang, C. Y. (1980) *J. Biol. Chem. 255,* 1862–1866.
Pahel, G., and Tyler, B. (1979) *Proc. Nat. Acad. Sci. U.S.A. 76,* 4544–4548.
Paris, C. G., and Magasanik, B. (1981) *J. Bacteriol. 145,* 266–271.
Pate, J. S. (1980) *Ann. Rev. Plant Physiol. 31,* 313–340.
Patterson, M. K., and Orr, G. R. (1968) *J. Biol. Chem. 243,* 376–380.
Prusiner, S., and Milner, L. (1970) *Anal. Biochem. 37,* 429–438.
Ramos, F., and Wiame, J.-M. (1979) *Eur. J. Biochem. 94,* 4309–417.
Ramos, F., and Wiame, J.-M. (1980) *Eur. J. Biochem. 108,* 373–377.
Raunio, R. P., and Jenkins, W. T. (1973) *J. Bacteriol. 115,* 560–566.
Ravel, J. M., Norton, S. J. Humphreys, J. S., and Shive, W. (1962) *J. Biol. Chem. 237,* 2845–2849.
Reitzer, L. J., and Magasanik, B. (1982) *J. Bacteriol. 151,* 1299–1313.
Rognes, S. E. (1975) *Phytochemistry 14,* 1975–1982.
Rognes, S. E. (1980) *Phytochemistry 19,* 2287–2293.
Rudman, D., and Meister, A. (1953) *J. Biol. Chem. 200,* 591–604.
Shiio, I., and Ujigawa, K. (1978) *J. Biochem.* (Tokyo) *84,* 647–657.
Streeter, J. G. (1973) *Arch. Biochem. Biophys. 157,* 613–624.
Stulen, I., and Oaks, A. (1977) *Plant Physiol. 60,* 680–683.
Stulen, I., Israelstam, G. F., and Oaks, A. (1979) *Planta 146,* 237–241.
Ting, I. P., and Zschoche, W. C. (1970) *Plant Physiol. 45,* 429–434.
Tyler, B. (1978) *Ann. Rev. Biochem. 47,* 1127–1162.
Urm, E., Leisinger, T., and Vogel, H. J. (1973) *Biochem. Biophys. Acta 302,* 249–260.
Virtanen, A. I., and Ellfolk, N. (1955) *Methods in Enzymology 2,* 386–388.
von Meyenburg, K., Hansen, F. G., Nielsen, L. D., and Riise, E. (1978) *Mol. Gen. Genet. 160,* 287–295.
Wellner, V. P., and Meister, A. (1975) *J. Biol. Chem. 250,* 3261–3266.
Wilcox, M., and Nirenberg, M. (1968) *Proc. Nat. Acad. Sci. U.S.A. 61,* 229–236.

CHAPTER
9

The Common Pathway to Lysine, Methionine, and Threonine

Georges N. Cohen

I. INTRODUCTION

In *Escherichia coli,* diaminopimelate, lysine, methionine, threonine, and isoleucine derive part or all of their carbon atoms from aspartate (Fig. 9.1). The first reaction of the common pathway is catalyzed by three distinct aspartate kinases, which differ in the way their synthesis and activity are regulated. Aspartate kinase I is inhibited by threonine. Its synthesis is repressed by threonine *plus* isoleucine. The synthesis of aspartate kinase II is repressed by methionine. Lysine inhibits the activity and represses the synthesis of aspartate kinase III. The reduction of aspartate semialdehyde to homoserine is catalyzed by two distinct homoserine dehydrogenases. Homoserine dehydrogenase I synthesis is repressed by threonine *plus* isoleucine, and its activity is inhibited by threonine. Methionine represses the synthesis of homoserine dehydrogenase II. The regulatory features are summarized in Table 9.1.

This chapter reviews the reactions of *E. coli* aspartate kinase I–homoserine dehydrogenase I, aspartate kinase II–homoserine dehydrogenase II, and aspartate semialdehyde dehydrogenase. Aspartate kinase I–homoserine dehydrogenase I is encoded by *thrA*, which is part of the *thr* operon. Regulation

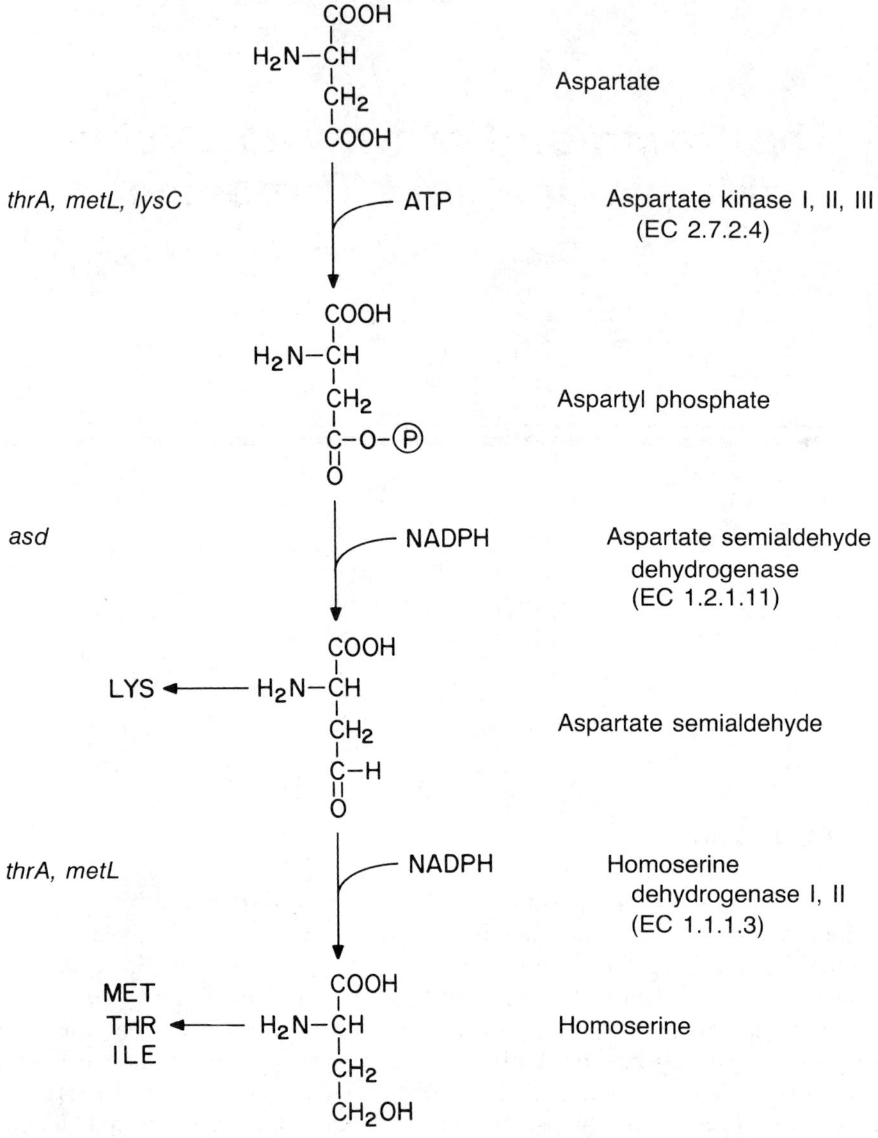

FIGURE 9.1 The common biosynthetic pathway to lysine, methionine, and threonine.

TABLE 9.1 Regulation of the Common Pathway to Lysine, Methionine, and Threonine

Enzyme	*Activity Inhibited by*	*Corresponding Structural Gene*	*Gene Expression Regulated by*
Aspartate kinase I	threonine	*thrA*	threonine, isoleucine
Aspartate kinase II	—	*metL*	methionine
Aspartate kinase III	lysine	*lysC*	lysine
Aspartate semialdehyde dehydrogenase	—	*asd*	lysine, threonine, methionine
Homoserine dehydrogenase I	threonine	*thrA*	threonine, isoleucine
Homoserine dehydrogenase II	—	*metL*	methionine

of *thrA* expression is discussed in Chapter 10. The specific pathway from homoserine to methionine is described in Chapter 11. Aspartate kinase III is covered under lysine biosynthesis in Chapter 12.

II. ASPARTATE KINASE I–HOMOSERINE DEHYDROGENASE I

A. Structure

A threonine-regulated aspartate kinase was first recognized by Stadtman et al. (1961), and a threonine-regulated homoserine dehydrogenase was recognized two years later (Patte et al., 1963). These two enzymes were considered to be separate entities. Early evidence that the two enzyme activities were part of the same protein (Cohen et al., 1965a, b; Patte et al., 1966) was confirmed by Truffa-Bachi et al. (1968).

The molecular weight of the native protein determined by several methods is 355,000 ± 27,000 (Truffa-Bachi et al., 1968; Falcoz-Kelly et al., 1972; Starnes et al., 1972). A variety of chemical and genetic studies indicate that the enzyme has an α_4 structure with subunit molecular weights of 86,000 ± 4,000 (Truffa-Bachi et al., 1969; Clark and Ogilvie, 1972; Starnes et al., 1972; Wampler, 1972; Falcoz-Kelly et al., 1972; Veron et al., 1972).

Functionally active fragments isolated from a nonsense mutant (aspartate kinase I fragment) or from native enzyme by limited proteolysis (homoserine dehydrogenase I fragment) possess only one or the other activity of the intact polypeptide. The tetrameric aspartate kinase I fragment (M_r about 4 × 53,000) is located in the N-terminal part of the polypeptide, and it is still inhibited by threonine. The dimeric homoserine dehydrogenase fragment (M_r = 2 × 57,000) is in the C-terminal part of the polypeptide chain, and it is insensitive to threonine. There is thus an overlap of about 21,000 daltons between the two fragments. The existence of these two fragments has helped in the early positioning of some tryptic and cyanogen bromide peptides (Sibilli et al., 1977; Sibilli, 1977; Cossart, 1977; Briley et al., 1978; Cossart-Gheerbrant et al., 1978). About 75% of the protein sequence was known when *thrA* was cloned (Cossart et al., 1979). DNA sequence analysis completed the primary structure of the *thrA* product (Katinka et al., 1980).

Figure 9.2 shows the nucleotide sequence of *thrA* and the corresponding amino acid sequence of its product, aspartate kinase I–homoserine dehydrogenase I. A predicted secondary structure for the enzyme is shown in Fig. 9.3. No salient features emerge from a rough examination of the protein sequence, although three tetrapeptides occur twice (Leu–Arg–Thr–Leu, positions 406 and 811; Leu–Ala–Arg–Glu, positions 659 and 686; Ala–Ala–Arg–Val, positions 332 and 730), and one tetrapeptide, Ile–Ser–Val–Val, is present three times at positions 344, 399, and 439. Codon usage within *thrA* is highly nonrandom and similar to that in *trpA* (Nichols and Yanofsky, 1979).

Aspartate kinase I–homoserine dehydrogenase I contains 22 methionine residues per polypeptide chain. Nine of the 22 are centrally clustered in a region corresponding to 15% of the protein, starting at Met-249. The ATG corresponding to Met-249 is preceded by a possible ribosome-binding site GAGGT (Shine and Dalgarno, 1975). The RNA sequence encoded by this region of *thrA* could potentially form a stem and loop structure similar to functional intergenic ribosome binding sites identified in the *trp* operon of *E. coli* (Selker and Yanofsky, 1979). The calculated ΔG value for formation of this structure is about -13 kcal (-54 kJ/mol). Thus Met-249 could represent a start codon for the dehydrogenase domain of the polypeptide chain. The ATG encoding Met-249 is immediately preceded by TTG TTG AAG TCG. Each of these four triplets could be converted into amber termination codons by a single base change. One could speculate that two separate base changes could have created a bifunctional polypeptide chain by abolishing two termination codons frequently found at the end of genes. The subtilisin fragment carrying only the dehydrogenase activity starts at Ser-298 (Briley et al., 1978). Limited proteolysis by a series of proteases with widely different specificities all cleave within a short segment between Thr-293 and Asp-300 (Sibilli et al., 1981). These findings are compatible with the idea that a gene fusion occurred during evolution (Veron et al., 1972). The presumptive ancestral kinase would have had a molecular weight of about 25,000. Coordinate control of both activities at the transcription and translation level, in addition to common allosteric control of both activities by threonine, could be the evolutionary advantage of a fused protein (Patte et al., 1967; Cohen et al., 1967; Veron and Cohen, 1974; Cohen and Dautry-Varsat, 1980).

The native aspartate kinase I–homoserine dehydrogenase I and the isolated fragments refold and regain their enzyme activities after total denaturation by guanidine hydrochloride (Garel and Dautry-Varsat, 1980a, b; Dautry-Varsat and Garel, 1981). The reappearance of the kinase activity is a monomolecular process, much faster for the fragment than for the whole protein. The dehydrogenase fragment renatures according to biphasic kinetics, corresponding to the folding of the monomer followed by an association reaction that limits the appearance of the threonine-sensitive activity. The native enzyme refolds in three steps. A monomolecular step leads to a monomeric species with kinase activity. An association step then leads to a dimeric species with kinase and threonine-sensitive dehydrogenase activities. Finally, a second association step leads to a tetrameric species with the two activities, both sensitive to threonine. The folding of this large protein appears as a sequential process during which the functional properties are regained successively as the protein structure becomes more complex. During this process, the two regions of each polypeptide chain, respectively responsible for the kinase and the dehydrogenase activities, seem to acquire their native conformation rather independently of each other.

Aspartate kinase I–homoserine dehydrogenase I forms tetragonal crystals

ATG CGA GTG TTG AAG TTC GGC GGT ACA TCA GTG GCA AAT GCA GAA CGT TTT CTG CGT GTT GCC GAT ATT CTG GAA AGC AAT GCC AGG CAG$_{90}$
Met Arg Val Leu Lys Phe Gly Gly Thr Ser Val Ala Asn Ala Glu Arg Phe Leu Arg Val Ala Asp Ile Leu Glu Ser Asn Ala Arg Gln$_{30}$

GGG CAG GTG GCC ACC GTC CTC TCT GCC CCT GCC AAA ATC ACC AAC CAC CTG GTG GCG ATG ATT GAA AAA ACC ATT AGC GGC CAG GAT GCT$_{180}$
Gly Gln Val Ala Thr Val Leu Ser Ala Pro Ala Lys Ile Thr Asn His Leu Val Ala Met Ile Glu Lys Thr Ile Ser Gly Gln Asp Ala$_{60}$

TTA CCC AAT ATC AGC GAT GCC GAA CGT ATT TTT GCC GAA CTT TTG ACG GGA CTG GCC GCC GCC CAG CCG GGG TTC CCG TTG GCG CAA TTG$_{270}$
Leu Pro Asn Ile Ser Asp Ala Glu Arg Ile Phe Ala Glu Leu Leu Thr Gly Leu Ala Ala Ala Gln Pro Gly Phe Pro Leu Ala Gln Leu$_{90}$

AAA ACT TTC GTC GAT CAG GAA TTT GCC CAA ATA AAA CAT GTC CTG CAT GGC ATT AGT TTG TTG GGG CAG TGC CCG GAT AGC ATC AAC GCT$_{360}$
Lys Thr Phe Val Asp Gln Glu Phe Ala Gln Ile Lys His Val Leu His Gly Ile Ser Leu Leu Gly Gln Cys Pro Asp Ser Ile Asn Ala$_{120}$

GCG CTG ATT TGC CGT GGC GAG AAA ATG TCG ATC GCC ATT ATG GCC GGC GTA TTA GAA GCG CGC GGT CAC AAC GTT ACT GTT ATC GAT CCG$_{450}$
Ala Leu Ile Cys Arg Gly Glu Lys Met Ser Ile Ala Ile Met Ala Gly Val Leu Glu Ala Arg Gly His Asn Val Thr Val Ile Asp Pro$_{150}$

GTC GAA AAA CTG CTG GCA GTG GGG CAT TAC CTC GAA TCC ACC GTC GAT ATT GCC GAG TCC ACC CGC CGT ATA GCG GCA AGC CGC ATT CCG$_{540}$
Val Glu Lys Leu Leu Ala Val Gly His Tyr Leu Glu Ser Thr Val Asp Ile Ala Glu Ser Thr Arg Arg Ile Ala Ala Ser Arg Ile Pro$_{180}$

GCT GAT CAC ATG GTG CTG ATG GCA GGT TTC ACC GCC GGT AAT GAG AAA GGC GAA CTG GTG GTG CTT GGA CGC AAC GGT TCC GAC TAC TCT$_{630}$
Ala Asp His Met Val Leu Met Ala Gly Phe Thr Ala Gly Asn Glu Lys Gly Glu Leu Val Val Leu Gly Arg Asn Gly Ser Asp Tyr Ser$_{210}$

GCT GCG GTG CTG GCT GCC TGT TTA CGC GCC GAT TGT TGC GAG ATT TGG ACG GAC GTT AAC GGG GTC TAT ACC TGC GAC CCG CGT CAG GTG$_{720}$
Ala Ala Val Leu Ala Ala Cys Leu Arg Ala Asp Cys Cys Glu Ile Trp Thr Asp Val Asn Gly Val Tyr Thr Cys Asp Pro Arg Gln Val$_{240}$

CCC GAC GCG AGG TTG TTG AAG TCG ATG TCC TAC CAG GAA GCG ATG GAG CTT TCC TAC TTC GGC GCT AAA GTT CTT CAC CCC CGC ACC ATT$_{810}$
Pro Asp Ala Arg Leu Leu Lys Ser Met Ser Tyr Gln Glu Ala Met Glu Leu Ser Tyr Phe Gly Ala Lys Val Leu His Pro Arg Thr Ile$_{270}$

ACC CCC ATC GCC CAG TTC CAG ATC CCT TGC CTG ATT AAA AAT ACC GGA AAT CCT CAA GCA CCA GGT ACG CTC ATT GGT GCA AGC CGT GAT$_{900}$
Thr Pro Ile Ala Gln Phe Gln Ile Pro Cys Leu Ile Lys Asn Thr Gly Asn Pro Gln Ala Pro Gly Thr Leu Ile Gly Ala Ser Arg Asp$_{300}$

GAA GAC GAA TTA CCG GTC AAG GGC ATT TCC AAT CTG AAT AAC ATG GCA ATG TTC AGC GTT TCT GGT CCG GGG ATG AAA GGG ATG GTC GGC$_{990}$
Glu Asp Glu Leu Pro Val Lys Gly Ile Ser Asn Leu Asn Asn Met Ala Met Phe Ser Val Ser Gly Pro Gly Met Lys Gly Met Val Gly$_{330}$

ATG GCG GCG CGC GTC TTT GCA GCG ATG TCA CGC GCC CGT ATT TCC GTG GTG CTG ATT ACG CAA TCA TCT TCC GAA TAC AGC ATC AGT TTC$_{1080}$
Met Ala Ala Arg Val Phe Ala Ala Met Ser Arg Ala Arg Ile Ser Val Val Leu Ile Thr Gln Ser Ser Ser Glu Tyr Ser Ile Ser Phe$_{360}$

TGC GTT CCA CAA AGC GAC TGT GTG CGA GCT GAA CGG GCA ATG CTG GAA GAG TTC TAC CTA GAA CTG AAA GAA GGC TTA CTG GAG CCG CTG$_{1170}$
Cys Val Pro Gln Ser Asp Cys Val Arg Ala Glu Arg Ala Met Leu Glu Glu Phe Tyr Leu Glu Leu Lys Glu Gly Leu Leu Glu Pro Leu$_{390}$

GCA GTG GCG GAA CGG CTG GCC ATT ATC TCG GTG GTA GGT GAT GGT TTG CGC ACC TTG CGT GGG ATC TCG GCG AAA TTC TTT GCC GCA CTG$_{1260}$
Ala Val Ala Glu Arg Leu Ala Ile Ile Ser Val Val Gly Asp Gly Leu Arg Thr Leu Arg Gly Ile Ser Ala Lys Phe Phe Ala Ala Leu$_{420}$

```
GCC CGC GCC AAT ATC AAC ATT GTC GCC ATT GCT CAG GGA TCT TCT GAA CGC TCA ATC TCT GTC GTG GTA AAT AAC GAT GAT GCG ACC ACT 1350
Ala Arg Ala Asn Ile Asn Ile Val Ala Ile Ala Gln Gly Ser Ser Glu Arg Ser Ile Ser Val Val Val Asn Asn Asp Asp Ala Thr Thr 450

GGC GTG CGC GTT ACT CAT CAG ATG CTG TTC AAT ACC GAT CAG GTT ATC GAA GTG TTT GTG ATT GGC GTC GGT GGC GTT GGC GGT GCG CTG 1440
Gly Val Arg Val Thr His Gln Met Leu Phe Asn Thr Asp Gln Val Ile Glu Val Phe Val Ile Gly Val Gly Gly Val Gly Gly Ala Leu 480

CTG GAG CAA CTG AAG CGT CAG CAA AGC TGG CTG AAG AAT AAA CAT ATC GAC TTA CGT GTC TGC GGT GTT GCC AAC TCG AAG GCT CTG CTC 1530
Leu Glu Gln Leu Lys Arg Gln Gln Ser Trp Leu Lys Asn Lys His Ile Asp Leu Arg Val Cys Gly Val Ala Asn Ser Lys Ala Leu Leu 510

ACC AAT GTA CAT GGC CTT AAT CTG GAA AAC TGG CAG GAG GAA CTG GCG CAA GCC AAA GAG CCG TTT AAT CTC GGG CGC TTA ATT CGC CTC 1620
Thr Asn Val His Gly Leu Asn Leu Glu Asn Trp Gln Glu Glu Leu Ala Gln Ala Lys Glu Pro Phe Asn Leu Gly Arg Leu Ile Arg Leu 540

GTG AAA GAA TAT CAT CTG CTG AAC CCG GTC ATT GTT AAC TGC ACT TCC AGC CAG GCA GTG GCG GAT CAA TAT GCC GAC TTC CTG CGC GAA 1710
Val Lys Glu Tyr His Leu Leu Asn Pro Val Ile Val Asn Cys Thr Ser Ser Gln Ala Val Ala Asp Gln Tyr Ala Asp Phe Leu Arg Glu 570

GGT TTC CAC GTT GTC ACG CCG AAC AAA AAG GCC AAC ACC TCG TCG ATG GAT TAC TAC CAT CAG TTG CGT TAT GCG GCG GAA AAA TCG CGG 1800
Gly Phe His Val Val Thr Pro Asn Lys Lys Ala Asn Thr Ser Ser Met Asp Tyr Tyr His Gln Leu Arg Tyr Ala Ala Glu Lys Ser Arg 600

CGT AAA TTC CTC TAT GAC ATC AAC GTT GGG GCT GGA TTA CCG GTT ATT GAG AAC CTG CAA AAT CTG CTC AAT GCA GGT GAT GAA TTG ATG 1890
Arg Lys Phe Leu Tyr Asp Ile Asn Val Gly Ala Gly Leu Pro Val Ile Glu Asn Leu Gln Asn Leu Leu Asn Ala Gly Asp Glu Leu Met 630

AAA TTC TCC GGC ATT CTT TCT GGT TCG CTT TCT TAT ATC TTC GGC AAG TTA GAC GAA GGC ATG AGT TTC TCC GAG GCG ACC CGG CTG GCG 1980
Lys Phe Ser Gly Ile Leu Ser Gly Ser Leu Ser Tyr Ile Phe Gly Lys Leu Asp Glu Gly Met Ser Phe Ser Glu Ala Thr Arg Leu Ala 660

CGG GAA ATG GGT TAT ACC GAA CCG GAC CCG CGA GAT GAT CTT TCT GGT ATG GAT GTG GCG CGT AAA CTA TTG ATT CTC GCT CGT GAA ACG 2070
Arg Glu Met Gly Tyr Thr Glu Pro Asp Pro Arg Asp Asp Leu Ser Gly Met Asp Val Ala Arg Lys Leu Leu Ile Leu Ala Arg Glu Thr 690

GGA CGT GAA CTG GAG CTG GCG GAT ATT GAA ATT GAA CCT GTG CTG CCC GCA GAG TTT AAC GCC GAG GGT GAT GTT GCC GCT TTT ATG GCG 2160
Gly Arg Glu Leu Glu Leu Ala Asp Ile Glu Ile Glu Pro Val Leu Pro Ala Glu Phe Asn Ala Glu Gly Asp Val Ala Ala Phe Met Ala 720

AAT CTG TCA CAA CTC GAC GAT CTC TTT GCC GCG CGC GTG GCG AAG GCC CGT GAT GAA GGA AAA GTT TTG CGC TAT GTT GGC AAT ATT GAT 2250
Asn Leu Ser Gln Leu Asp Asp Leu Phe Ala Ala Arg Val Ala Lys Ala Arg Asp Glu Gly Lys Val Leu Arg Tyr Val Gly Asn Ile Asp 750

GAA GAT GGC GTC TGC CGC GTG AAG ATT GCC GAA GTG GAT GGT AAT GAT CCG CTG TTC AAA GTG AAA AAT GGC GAA AAC GCC CTG GCC TTC 2340
Glu Asp Gly Val Cys Arg Val Lys Ile Ala Glu Val Asp Gly Asn Asp Pro Leu Phe Lys Val Lys Asn Gly Glu Asn Ala Leu Ala Phe 780

TAT AGC CAC TAT TAT CAG CCG CTG CCG TTG GTA CTG CGC GGA TAT GGT GCG GGC AAT GAC GTT ACA GCT GCC GGT GTC TTT GCT GAT CTG 2430
Tyr Ser His Tyr Tyr Gln Pro Leu Pro Leu Val Leu Arg Gly Tyr Gly Ala Gly Asn Asp Val Thr Ala Ala Gly Val Phe Ala Asp Leu 810

CTA CGT ACC CTC TCA TGG AAG TTA GGA GTC TGA 2463
Leu Arg Thr Leu Ser Trp Lys Leu Gly Val 820
```

FIGURE 9.2 **The DNA sequence of *thrA* and the corresponding amino acid sequence of the encoded aspartate kinase I–homoserine dehydrogenase I.**

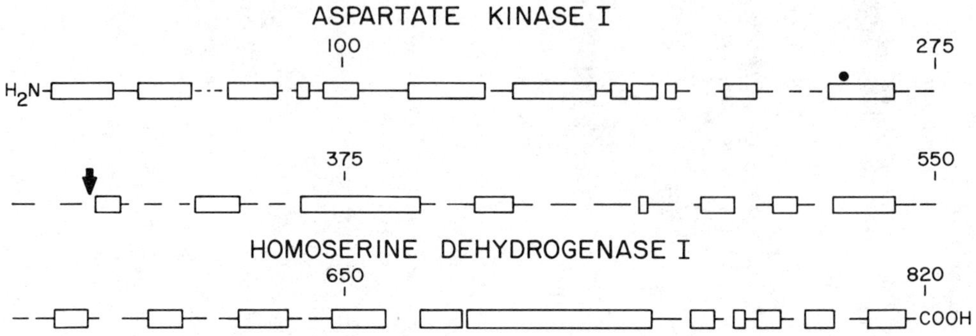

FIGURE 9.3 Predicted secondary structure of the aspartate kinase I–homoserine dehydrogenase I. The boxes indicate α-helix; the lines indicate extended regions of the protein. The amino acid residues of the polypeptide chains are numbered from 1 to 820. The dot indicates Met-249. The arrow indicates the position of limited proteolysis cleavage generating an active homoserine dehydrogenase fragment.

in the presence of threonine and NADPH. The asymmetric unit is $10^6 Å^3$; the space group is 1422 or 14122. If it is assumed that the solvent occupies 53% of the total volume, the asymmetric unit contains one tetrameric molecule (Janin, 1974). The symmetry of the crystal form does not express the presumed symmetrical arrangement of the four identical subunits. Attempts to obtain crystals suitable for high resolution X-ray crystallographic analysis have been unsuccessful.

B. Ligands Other than Threonine

Direct determination of ATP binding gives 4.6 ± 1.6 sites per tetramer with a dissociation constant of 2.8 ± 1.0 mM (Ehrlich and Takahashi, 1973). The ATP analog, 6-mercapto-9-β-D-ribofuranosylpurine-5′-triphosphate, is a substrate and an inactivator of aspartate kinase I. It binds covalently to the protein with a stoichiometry of one equivalent per monomer, without affecting the homoserine dehydrogenase activity or the tetrameric state of the protein (Truffa-Bachi and Heck, 1971).

Pyridine nucleotide binding has been studied by equilibrium dialysis, fluorescence spectroscopy (Janin and Cohen, 1969; Janin and Iwatsubo, 1969) and circular dichroism (Heck and Truffa-Bachi, 1970). The resulting data show that there are four sites per tetramer with dissociation constants of about 12 μM for NADP and 0.3 μM for NADPH. These values are independent of homoserine, threonine, or K^+ ion concentrations. The binding is strictly Michaelian.

The affinity of the protein for homoserine is so low and aspartate semialdehyde is so unstable that their binding stoichiometry cannot be determined by equilibrium dialysis. Indirect measurements employing an affinity label analog of homoserine, 2-amino-4-oxo-5-chloropentanoate, suggest one equivalent per monomer. Complete NADP-dependent inactivation of the dehydrogenase occurs without affecting aspartate kinase or the oligomeric state of the protein. The analog forms a reversible complex with the protein prior to irreversible binding (Hirth et al., 1975).

C. Regulation at the Enzyme Level

The allosteric inhibition by threonine had been studied long before the discovery that the same polypeptide chain carries two activities. Both enzyme activities are inhibited by threonine in a sigmoidal fashion. The inhibition of the two activities can be affected simultaneously in a parallel fashion by a single mutational event (Cohen et al., 1965a, b).

1. Threonine Binding. Binding of ^{14}C-threonine was studied by equilibrium dialysis and gel filtration (Janin and Cohen, 1969; Janin and Iwatsubo, 1969). In the presence of 150 mM KCl, saturation of the binding sites is reached at 0.3–0.5 mM threonine (Fig. 9.4); 8.1 ± 0.4 moles of threonine are bound to 1 mole of protein. Threonine binding depends on the K^+ concentration. High concentrations of KCl diminish the binding of threonine and the inhibition of homoserine dehydrogenase (Cohen et al., 1965a, 1965b; Wampler and

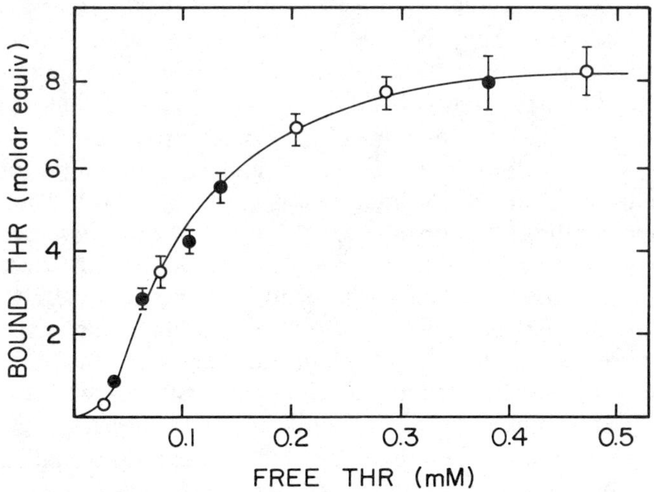

FIGURE 9.4 Binding of threonine to aspartate kinase I–homoserine dehydrogenase I (from Truffa-Bachi et al., 1974).

Westhead, 1968). However, threonine binding is cooperative only at high K^+ concentrations (Fig. 9.4); the sigmoidicity of the binding curves disappears below 50 mM K^+ (Vernon et al., 1973), where a Scatchard plot shows eight equal affinity sites for threonine, a result that is not readily interpretable (Veron et al., 1973). Threonine inhibition of aspartate kinase I is competitive with aspartate (Stadtman et al., 1961), and threonine decreases the binding of aspartate (Janin and Cohen, 1969; Janin and Iwatsubo, 1969).

Treatment with -SH reagents that desensitize the homoserine dehydrogenase activity and destroy the aspartate kinase activity (Truffa-Bachi et al., 1966, 1968) yields a modified protein that no longer binds threonine. Also, the homoserine dehydrogenase fragment does not bind threonine (Truffa-Bachi et al., 1974), and only four binding sites for threonine are found on the tetrameric aspartate kinase I fragment (Costrejean, 1978).

2. The Catalytic Sites of Aspartate Kinase I–Homoserine Dehydrogenase I: Identification of Four of the Eight Threonine-binding Sites. Aspartate kinase requires Mg^{2+} for enzyme activity. Nuclear magnetic resonance studies using Mn^{2+} instead of Mg^{2+} indicate that the metal serves as a bridge between the enzyme and the ATP (Ehrlich and Takahashi, 1973). Mg^{2+} can be replaced by Co^{2+} (Ryzewski and Takahashi, 1975). Co^{2+} is added to the enzyme and oxidized in situ to Co^{3+} with H_2O_2. The Co^{3+} is not removable by dialysis, yet it is released by incubation with dithiothreitol. The Co^{3+}-containing protein has no aspartate kinase activity, yet it retains 60% of the original dehydrogenase activity. The kinetics of aspartate kinase inactivation parallel those of Co^{3+} incorporation, which saturates at one Co^{3+} per subunit. The residual homoserine dehydrogenase remains threonine-sensitive. This suggests that the integrity of the aspartate kinase active site is not obligatory for threonine binding that inhibits homoserine dehydrogenase activity. Limited proteolysis studies show that Co is physically attached to the aspartate kinase moiety of the polypeptide chain (Tilak et al., 1976).

The ternary complexes enzyme–Co^{3+}–ATP, enzyme–Co^{3+}–aspartate, and enzyme–Co^{3+}–threonine have been prepared. Threonine binds in the first coordination sphere of the metal (Wright et al., 1976). The metal and one threonine binding site are adjacent, with 0.44 nm between the metal and the carboxyl carbon of threonine (Tilak et al., 1976). In quaternary adducts, enzyme–Co^{3+}–ATP–aspartate and enzyme–Co^{3+}–ATP–threonine, the sum of the number of aspartate and threonine molecules per active site is one (Wright et al., 1976), a result confirming that aspartate and threonine cannot occupy the kinase active site simultaneously, as was suggested by earlier steady state kinetic studies (Stadtman et al., 1961).

The ternary enzyme–Co^{3+}–threonine adduct has four additional threonine binding sites per tetramer (Wright and Takahashi, 1977a), as shown by equilibrium dialysis. The binding of threonine to the enzyme–Co^{3+}–threonine adduct causes inhibition of the residual homoserine dehydrogenase activity, changes in the circular dichroism, protection of -SH groups, and protection against proteolytic attack. All changes qualitatively resemble the behavior of the native enzyme (Truffa-Bachi et al., 1974), a finding supporting the existence of four additional nonderivatized threonine binding sites in the enzyme–Co^{3+}–threonine complex.

In addition, the inhibition of dehydrogenase by kinase substrates (Patte et al., 1966) was shown to be due to the binding of aspartate and ATP within the dehydrogenase domain of the enzyme (Wright and Takahashi, 1977a). When the fluorescent probe 1,N(6)ethenoadenosine-5′-triphosphate (ε-ATP), an ATP analog, is attached to the kinase active site through Co^{3+} and the dehydrogenase site is saturated with the NADPH analog, thionicotinamide adenine dinucleotide phosphate, the latter quenches the fluorescence of ε-ATP. The distance between ε-ATP and the quencher is 2.9 ± 0.4 nm. In the presence of threonine, the distance increases to 3.6 ± 0.5 nm. Threonine binding either increases the intersite distance by 0.7 nm or causes a reorientation of the probe at the dehydrogenase site (Wright and Takahashi, 1977b).

3. Association of Regulation by Threonine with Conformational Changes of the Protein. UV difference spectroscopy of protein solutions having ±0.5 nM threonine shows differences in the 250 to 310-nm range, attributable to perturbations of the environment of several aromatic residues. Two protein conformations are postulated: a T form with buried aromatic residues and an R form with more exposed aromatic residues (Janin and Cohen, 1969). The equilibrium is shifted back to the R form by aspartate.

The change in the absorption spectrum of the protein upon threonine addition is associated with a quenching and a slight shift of the protein fluorescence (Janin and Cohen, 1969; Janin and Iwatsubo, 1969). Excitation below 280 nm shows quenching by threonine only if aspartate or potassium ions are present. The significance of these findings is discussed below. Through excitation of the enzyme–NADPH complex at 290 nm, energy transfer from an aromatic residue of the protein to NADPH produces an emission at 450 nm. Threonine strongly quenches this transfer, a condition indicating a distance increase between the excited aromatic residue and NADPH. Threonine affects neither the affinity of NADPH nor the total number of equivalents of the coenzyme that is bound (Janin et al., 1969; Janin and Cohen, 1969).

The circular dichroism spectrum of the protein shows a complex overlap of negative bands from aromatic transitions. Threonine binding results in a decrease in ellipticity that is indicative of conformational changes (Heck and Truffa-Bachi, 1970). A plot of the ellipticity at 280 nm versus the threonine concentration is sigmoidal and identical to that obtained by using other physicochemical methods (Janin and Cohen, 1969; Janin and Iwatsubo, 1969). The circular dichroism at 280 nm is used to measure the state function.

Protein hydrogen exchangeable with the solvent decreases with increasing threonine concentrations, this decrease indicating a tighter structure with the hydrophobic interior of the protein becoming less accessible to the solvent (Takahashi and Westhead, 1971).

The spectral data summarized above define two conformations of the protein, an allosteric equilibrium (Janin and Cohen, 1969; Janin and Iwatsubo, 1969). The T form is stable in the absence of ligands or in the presence of threonine. The R form is stable in the absence of threonine or in the presence of aspartate or K^+. The ligands act by shifting an equilibrium between conformations rather than by merely modifying the environment of their respective binding sites. The experimental facts may be formalized by a two-state model:

$$R \underset{\text{aspartate, } K^+}{\overset{\text{threonine}}{\rightleftharpoons}} T$$

A strong correlation exists between the catalytic and regulatory properties of the protein and the conformational states (Table 9.2). The competitive inhibition of the kinase activity by threonine can be accounted for by the preferential binding of aspartate to the R form. In contrast, the inhibition of the dehydrogenase activity and the analogous activation by K^+ (Patte et al., 1963; Cohen et al., 1965a) do not seem to be due to a conformational change upon binding of the substrates. They may be due to an actual requirement for K^+, or more likely to an improper configuration of the dehydrogenase active site in the T form.

Fast kinetic experiments (stopped flow) to study the relaxation of the allosteric equilibrium demonstrate a single rate-determining step in the R to T conformational change (Barber and Bright, 1968; Janin and Iwatsubo, 1969; Bearer and Neet, 1978a). The more powerful temperature-jump technique reveals two relaxations that are independent of the protein concentration. The two relaxations indicate that the R to T equilibrium goes through two isomerizations with an intermediate conformation, S, of the protein that equilibrates fast with one and slowly with the other conformation (Janin and Iwatsubo, 1969). Further studies distinguish between threonine inhibition of the kinase and the dehydrogenase activity on the basis of different initial velocities (Bearer and Neet, 1978b).

D. A Model of Aspartate Kinase I–Homoserine Dehydrogenase I

The two catalytic regions of aspartate kinase I–homoserine dehydrogenase I are profoundly different. The kinase is quite susceptible to disruption by numerous agents, including heat (Patte et al., 1966), SH-reagents (Truffa-Bachi et al., 1966, 1974), and proteolysis (Veron et al., 1972). The dehy-

TABLE 9.2 The Conformational States of Aspartate Kinase I–Homoserine Dehydrogenase I

Property	*T Form*	*R Form*	*Remarks*
Binding of ligands:			
(Dissociation constants)			Shifts the equilibrium toward:
Threonine	$K_T = 30\ \mu M$	$K_R \geqslant 1$ mM	T form
Aspartate	$K_T \geqslant 20$ mM	$K_R = 3$ mM	R form
K^+ ions	Low affinity	High affinity	R form
NADPH	$K_T = 0.3\ \mu M$	$K_R = 0.3\ \mu M$	No detectable effect
Spectrophotometric properties			
Relative fluorescence intensity	100	116	Excitation at 280 nm, emission at 335 nm
Tryptophan and tyrosine residues	Buried	Exposed	Difference spectrum
Enzymatic activities:			
Aspartate kinase	None	Active	T does not bind the substrate
Homoserine dehydrogenase	10–15%	100%	Different configurations of the active sites

Used with permission of the European Journal of Biochemistry.

drogenase is little affected by these agents. The existence of separate domains, although inferred by indirect and circumstantial evidence, explains the ability to obtain independent fragments (Veron et al., 1972) and the intersubunit and intrasubunit interactions in the native enzyme (Veron et al., 1973).

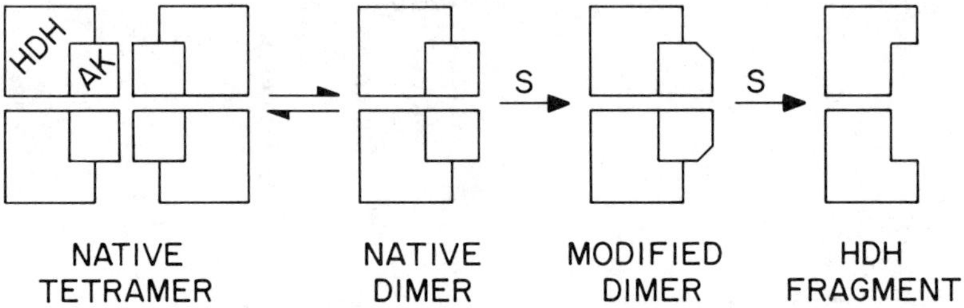

FIGURE 9.5 Model for the subunit organization in aspartate kinase I–homoserine dehydrogenase I. AK and HDH indicate the aspartate kinase and homoserine dehydrogenase regions, respectively, within the same polypeptide chain. The kinase region buried in the tetramer appears more exposed in the dimer. Sublisin (S) cleavage within the kinase region is indicated by a notch (from Veron et al., 1973, with permission of the European Journal of Biochemistry).

The model shown in Fig. 9.5 is consistent with both the dimeric nature of the dehydrogenase fragment and the tetrameric nature of the kinase fragment (Veron et al., 1972). The model is compatible with immunological evidence and with the effects of dissociation on the accessibility of the sulfhydryl groups and the rate of proteolysis (Veron et al., 1973). The Hill coefficient for threonine inhibition for the kinase fragment argues that the fragment tetramer retains the allosteric properties of the native tetrameric enzyme. The major conformational changes characteristic of the native enzyme may be largely confined to this region. Using a quite different approach, Mackall and Neet (1974) have proposed a model that is essentially identical to the one in Fig. 9.5. On the basis of refined proteolysis experiments, the hybrid nature of the modified dimer has been established as one native chain and one dehydrogenase fragment. Isolation of other active intermediates has led to a triglobular model of the protein with the third domain responsible for subunit contacts (A. M. Fazel et al., in preparation).

E. Regulation at the Transcriptional Level, the *thr* Operon

The structural gene for aspartate kinase I–homoserine dehydrogenase I, *thrA*, lies in the same region of the *E. coli* chromosome as *thrB* and *thrC*, encoding homoserine kinase and threonine synthase, respectively. The gene order is *serB thrA thrB thrC pyrA* (Thèze et al., 1974). Polar effects of phase Mu–induced nonsense mutations are evidence for the existence of a *thrA,B,C* operon transcribed in that order (Thèze and Saint-Girons, 1974). Saint-Girons (1978) has provided a fine structure analysis of this operon.

Selecting for resistance for α-amino-β-hydroxyvalerate, the C-5 homolog of threonine (Cohen and Patte, 1963), threonine-excreting mutants have been isolated that express the *thr* operon constitutively (Saint-Girons and Margarita, 1975). The mutations are closely linked to the *thrA* side of the *thr* cluster and lead in merodiploids to derepression only when located in the *cis* position. By a different method, Gardner and Smith (1975) have isolated similar mutants. The DNA region preceding *thrA* has been cloned and sequenced (Gardner, 1979), as has the corresponding DNA for one of the mutants (Parsot et al., 1983). The exact nature of the *cis*-dominant constitutive mutations is discussed in Chapter 10.

Bacteria constitutive for *thr* operon expression due to mutations unlinked to the cluster have been isolated by several stratagems. One mutation, within *thrS*, results in an apparent affinity of threonyl-tRNA synthetase for threonine 200-fold lower than that of the wild-type enzyme (Johnson et al., 1977). A *thrS* mutation causing a tenfold decreased affinity of the enzyme for threonine (Nass and Thomale, 1974) does not lead to derepression of the *thr* operon, nor do the *thrS* mutations isolated during this study. However, some of the mutants are derepressed in the presence of borrelidin, which inhibits threonyl-tRNA synthetase (Nass et al., 1969). Thus threonyl-tRNA synthetase affects regulation of the *thr* operon either directly or via charged threonyl-tRNA.

Another class of constitutive mutants has been isolated starting from *thr–lac* fusions selecting for growth on a lactose minimal medium containing high concentrations of homoserine. In addition to the *cis*-dominant constitutive mutations described above, some constitutive mutants have been obtained that are resistant to α-amino-β-hydroxyvalerate, do not excrete threonine, and are not cotransducible with *thrA* or *thrS*. This new type of regulatory mutation, which has so far not been biochemically characterized, is called *thrX* (Saint-Girons, 1978).

III. ASPARTATE KINASE II–HOMOSERINE DEHYDROGENASE II

A. Structure

Aspartate kinase II–homoserine dehydrogenase II is also a bifunctional protein (Patte et al., 1967). The two activities remain associated during a 400-fold purification. The molecular weight of the protein, determined by equilibrium sedimentation (Falcoz-Kelly et al., 1969; Dautry-Varsat et al., 1977), is 185,000 $\pm$ 10,000. The protein is a dimer, composed of two identical subunits of M_r = 88,000. The N-terminus is Ser–Val–Ile–Ala–Gln–Ala–Gly–Ala–Lys, and the C-terminus is Leu–Leu–COOH (Dautry-Varsat et al., 1977). The homo-dimeric nature of this protein implies that the two catalytic activities are carried by the same polypeptide chain.

Limited proteolysis of aspartate kinase II–homoserine dehydrogenase II yields two dimeric fragments. One has dehydrogenase activity (M_r = 2 × 35,000). The other is devoid of catalytic activity (M_r = 2 × 25,000). The smaller fragment, although inactive, remains dimeric. It originates from a different part of the polypeptide chain than the dehydrogenase fragment, and one can regard it as part of the kinase domain. A model for the quaternary structure of aspartate kinase II–homoserine dehydrogenase II (Fig. 9.6) correlates the fragmentation pattern to the structure of the native enzyme (Dautry-Varsat and Cohen, 1977).

The native enzyme and the dehydrogenase fragment refold and renature after denaturation in 6 M guanidine hydrochloride (Dautry-Varsat and Garel, 1978). When the intact polypeptide is denatured, the reappearance of both enzyme activities follows biphasic kinetics with the reassociation of the two chains to form a dimer as rate-limiting step. The two activities reappear in parallel. The homoserine dehydrogenase II fragment also refolds and regains activity after complete denaturation, albeit 20 times faster than the entire protein. Thus the compact region corresponding to the dehydrogenase II fragment is an independent folding unit.

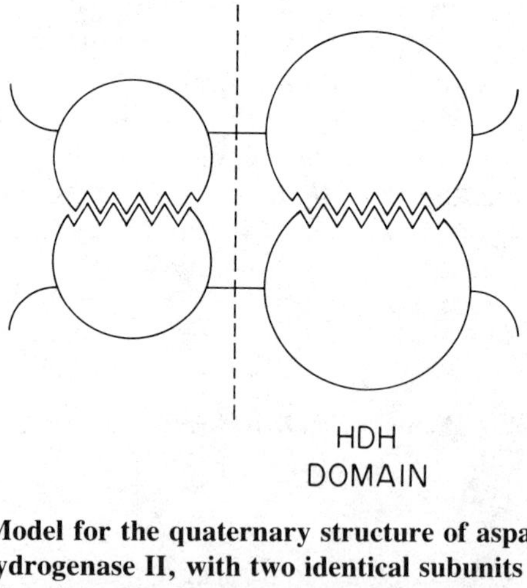

FIGURE 9.6 Model for the quaternary structure of aspartate kinase II–homoserine dehydrogenase II, with two identical subunits each containing two nonoverlapping domains. Proteolytic cleavage, indicated by the dashed line, generates two dimeric fragments. The homoserine dehydrogenase fragment retains enzyme activity. The relative positions of the two fragments with respect to the amino and carboxy termini of the polypeptide chain are not known (from Dautry-Varsat and Cohen, 1977, with permission of the American Society of Biological Chemists).

Neither aspartate kinase II nor homoserine dehydrogenase II is inhibited by methionine, S-adenosylmethionine, or threonine or by combinations, two by two, of methionine, threonine, isoleucine, and lysine (Patte et al., 1967; Falcoz-Kelly et al., 1969). There is no cooperative binding of the substrates for either of the activities.

B. Regulation at the Transcriptional Level

In *E. coli,* the synthesis of the methionine biosynthetic enzymes is repressed by methionine (Cohn et al., 1953; Wijesundera and Woods, 1953; Rowbury, 1965; Rowbury and Woods, 1966; Patte et al., 1967) (Chapter 11). The genes encoding aspartate kinase II and homoserine dehydrogenase II have been designated *metL* and *metM* (Thèze et al., 1974). Since the two activities reside on a single polypeptide chain (Dautry-Varsat and Cohen, 1977; Dautry-Varsat et al., 1977), only *metL* will be used. The genes encoding methionine biosynthetic enzymes are not organized as a single operon but are dispersed on the bacterial chromosome (Bachmann and Low, 1980). However, the four genes *metJ, metB, metL,* and *metF,* coding for a pleiotropic regulatory gene (Cohen and Jacob, 1959; Su and Greene, 1971; Patte et al., 1967), cystathionine synthase, aspartate kinase II–homoserine dehydrogenase II, and 5,10-methylenetetrahydrofolate reductase, respectively, are clustered at 88 mm (Su and Greene, 1971; Thèze et al., 1974), suggesting a possible operon. The knowledge of the genetic structure of the cluster is a prerequisite for the understanding of the regulation of its genetic expression. The *metJBLF* cluster has been cloned in plasmid vectors, the physical map has been determined, and sequence studies are in progress (Zakin, et al., 1982).

IV. COMPARISON OF THE TWO BIFUNCTIONAL PROTEINS

E. coli produces two different bifunctional enzymes carrying the same two activities. The discovery of aspartate kinase II–homoserine dehydrogenase II raises questions regarding the similarity of the two proteins and their evolutionary origin (Patte et al., 1967). Several attempts have been made to answer these questions.

A. Chemical Comparison

Both proteins possess sites for the same substrates. In addition, enzyme I binds threonine, its allosteric inhibitor. Although the turnover numbers are different, the apparent affinity constants of the two proteins for their substrates are quite similar. The subunit molecular weights of the two proteins are also quite similar, but one is a tetramer and the other is a dimer. In both cases, modification of cysteinyl groups leads to the loss of the kinase activity with considerable retention of the dehydrogenase activity (Truffa-Bachi et al., 1974; Falcoz-Kelly, 1971).

AGT GTG ATT GCG CAG GCA GGG GCG AAA GGT CGT CAG CTG CAT AAA TTT GGT GGC AGT AGT CTG GCT GAT GTG AAG TGT TAT TTG CGT GTC 90
Ser Val Ile Ala Gln Ala Gly Ala Lys Gly Arg Gln Leu His Lys Phe Gly Gly Ser Ser Leu Ala Asp Val Lys Cys Tyr Leu Arg Val 30

GCG GGC ATT ATG GCG GAG TAC TCT CAG CCT GAC GAT ATG ATG GTG GTT TCC GCC GCC GGT AGC ACC ACT AAC CGG TTG ATT AGC TGG TTG 180
Ala Gly Ile Met Ala Glu Tyr Ser Gln Pro Asp Asp Met Met Val Val Ser Ala Ala Gly Ser Thr Thr Asn Arg Leu Ile Ser Trp Leu 60

AAA CTA AGC CAG ACC GAT CGT CTC TCT GCG CAT CAG GTT CAA CAA ACG CTG CGT CGC TAT CAG TGC GAT CTG ATT AGC GGT CTG CTA CCC 270
Lys Leu Ser Gln Thr Asp Arg Leu Ser Ala His Gln Val Gln Gln Thr Leu Arg Arg Tyr Gln Cys Asp Leu Ile Ser Gly Leu Leu Pro 90

GCT GAA GAA GCC GAT AGC CTC ATT AGC GCT TTT GTC AGC GAC CTT GAG CGC CTG GCG GCG CTG CTC GAC AGC GGT ATT AAC GAC GCA GTG 360
Ala Glu Glu Ala Asp Ser Leu Ile Ser Ala Phe Val Ser Asp Leu Glu Arg Leu Ala Ala Leu Leu Asp Ser Gly Ile Asn Asp Ala Val 120

TAT GCG GAA GTG GTG GGC CAC GGG GAA GTA TGG TCG GCA CGT CTG ATG TCT GCG GTA CTT AAT CAA CAA GGG CTG CCA GCG GCC TGG CTT 450
Tyr Ala Glu Val Val Gly His Gly Glu Val Trp Ser Ala Arg Leu Met Ser Ala Val Leu Asn Gln Gln Gly Leu Pro Ala Ala Trp Leu 150

GAT GCC CGC GAG TTT TTA CGC GCT GAA CGC GCC GCA CAA CCG CAG GTT GAT GAA GGG CTT TCT TAC CCG TTG CTG CAA CAG CTG CTG GTG 540
Asp Ala Arg Glu Phe Leu Arg Ala Glu Arg Ala Ala Gln Pro Gln Val Asp Glu Gly Leu Ser Tyr Pro Leu Leu Gln Gln Leu Leu Val 180

CAA CAT CCG GGC AAA CGT CTG GTG GTG ACC GGA TTT ATC AGC CGC AAC AAC GCC GGT GAA ACG GTG CTG CTG GGG CGT AAC GGT TCC GAC 630
Gln His Pro Gly Lys Arg Leu Val Val Thr Gly Phe Ile Ser Arg Asn Asn Ala Gly Glu Thr Val Leu Leu Gly Arg Asn Gly Ser Asp 210

TAT TCC GCG ACA CAA ATC GGT GCG CTG GCG GGT GTT TCT CGC GTA ACC ATC TGG AGC GAC GTC GCC GGG GTA TAC AGT GCC GAC CCG GGT 720
Tyr Ser Ala Thr Gln Ile Gly Ala Leu Ala Gly Val Ser Arg Val Thr Ile Trp Ser Asp Val Val Gly Val Tyr Ser Ala Asp Pro Arg 240

AAA GTG AAA GAT GCC TGC CTG CTG CCG TTG CTG CGT CTG GAT GAG GCC AGC GAA CTG GCG CGC CTG GCG GCT CCC GTT CTT CAC GCC CGT 810
Lys Val Lys Asp Ala Cys Leu Leu Pro Leu Leu Arg Leu Asp Glu Ala Ser Glu Leu Ala Arg Leu Ala Ala Pro Val Leu His Ala Arg 270

ACT TTA CAG CCG GTT TCT GGC AGC GAA ATC GAC CTG CAA CTG CGC TGT AGC TAC ACG CCG GAT CAA GGT TCC ACG CGC ATT GAA CGC GTG 900
Thr Leu Gln Pro Val Ser Gly Ser Glu Ile Asp Leu Gln Leu Arg Cys Ser Tyr Thr Pro Asp Gln Gly Ser Thr Arg Ile Glu Arg Val 300

CTG GCC TCC GGT ACT GGT GCG CGT ATT GTC ACC AGC CAC GAT GAT GTC TGT TTG ATT GAG TTT CAG GTG CCC GCC AGT CAG GAT TTC AAA 990
Leu Ala Ser Gly Thr Gly Ala Arg Ile Val Thr Ser His Asp Asp Val Cys Leu Ile Glu Phe Gln Val Pro Ala Ser Gln Asp Phe Lys 330

CTG GCG CAT AAA GAG ATC GAC CAA ATC CTG AAA CGC GCG CAG GTA CGC CCG CTG GCG GTT GGC GTA CAT AAC GAT CGC CAG TTG CTG CAA 1080
Leu Ala His Lys Glu Ile Asp Gln Ile Leu Lys Arg Ala Gln Val Arg Pro Leu Ala Val Gly Val His Asn Asp Arg Gln Leu Leu Gln 360

TTT TGC TAC ACC TCA GAA GTG GCC GAC AGT GCG CTG AAA ATC CTC GAC GAA GCG GGA TTA CCT GGC GAA CTG CGC CTG CGT CAG GGG CTG 1170
Phe Cys Tyr Thr Ser Glu Val Ala Asp Ser Ala Leu Lys Ile Leu Asp Glu Ala Gly Leu Pro Gly Glu Leu Arg Leu Arg Gln Gly Leu 390

GCG CTG GTG GCG ATG GTC GGT GCA GGC GTC ACC CGT AAC CCG CTG CAT TGC CAC CGC TTC TGG CAG CAA CTG AAA GGC CAG CCG GTC GAA 1260
Ala Leu Val Ala Met Val Gly Ala Gly Val Thr Arg Asn Pro Leu His Cys His Arg Phe Trp Gln Gln Leu Lys Gly Gln Pro Val Glu 420

TTT ACC TGG CAG TCC GAT GAC GGC ATC AGC CTG GTG GGA GTA CTG CGC ACC GGC CCG ACC GAA AGC CTG ATT CAG GGG CTG CAT CAG TCC 1350
Phe Thr Trp Gln Ser Asp Asp Gly Ile Ser Leu Val Ala Val Leu Arg Thr Gly Pro Thr Glu Ser Leu Ile Gln Gly Leu His Gln Ser 450

GTC TTC CGC GCA GAA AAA CGC ATC GGC CTG GTA TTG TTC GGT AAG GGC AAT ATC GGT TCC CGT TGG CTG GAA CTG TTC GCC CGT GAG CAG 1440
Val Phe Arg Ala Glu Lys Arg Ile Gly Leu Val Leu Phe Gly Lys Gly Asn Ile Gly Ser Arg Trp Leu Glu Leu Phe Ala Arg Glu Gln 480

AGC ACG CTT TCG GCA CGT ACC GGC TTT GAG TTT GTG CTG GCA GGT GTG GTG GAC AGC CGC CGC AGC CTG TTG AGC TAT GAC GGG CTG GAC 1530
Ser Thr Leu Ser Ala Arg Thr Gly Phe Glu Phe Val Leu Ala Gly Val Val Asp Ser Arg Arg Ser Leu Leu Ser Tyr Asp Gly Leu Asp 510

GCC AGC CGC GCG TTA GCC TTC TTC AAC GAT GAA GCG GTT GAG CAG GAT GAA GAG TCG TTG TTC CTG TGG ATG CGC GCC CAT CCG TAT GAT 1620
Ala Ser Arg Ala Leu Ala Phe Phe Asn Asp Glu Ala Val Glu Gln Asp Glu Glu Ser Leu Phe Leu Trp Met Arg Ala His Pro Tyr Asp 540

GAT TTA GTG GTG CTG GAC GTT ACC GCC AGC CAG CAG CTT GCT GAT CAG TAT CTT GAT TTC GCC AGC CAC GGT TTC CAC GTT ATC AGC GCC 1710
Asp Leu Val Val Leu Asp Val Thr Ala Ser Gln Gln Leu Ala Asp Gln Tyr Leu Asp Phe Ala Ser His Gly Phe His Val Ile Ser Ala 570

AAC AAA CTG GCG GGA GCC AGC GAC AGC AAT AAA TAT CGC CAG ATC CAC GAC GCC TTC GAA AAA ACC GGG CGT CAC TGG CTG TAC AAT GCC 1800
Asn Lys Leu Ala Gly Ala Ser Asp Ser Asn Lys Tyr Arg Gln Ile His Asp Ala Phe Glu Lys Thr Gly Arg His Trp Leu Tyr Asn Ala 600

ACC GTC GGT GCG GGC TTG CCG ATC AAC CAC ACC GTG CGC GAT CTG ATC GAC AGC GGC GAT ACT ATT TTG TCG ATC AGC GGG ATC TTC TCC 1890
Thr Val Gly Ala Gly Leu Pro Ile Asn His Thr Val Arg Asp Leu Ile Asp Ser Gly Asp Thr Ile Leu Ser Ile Ser Gly Ile Phe Ser 630

GGC ACG CTC TCC TGG CTG TTC CTG CAA TTC GAC GGT AGC GTG CCG TTT ACC GAG CTG GTG GAT CAG GCG TGG CAG CAG GGC TTA ACC GAA 1980
Gly Thr Leu Ser Trp Leu Phe Leu Gln Phe Asp Gly Ser Val Pro Phe Thr Glu Leu Val Asp Gln Ala Trp Gln Gln Gly Leu Thr Glu 660

CCT GAC CCG CGT GAC GAT CTC TCT GGC AAA GAC GTG AGT CGC AAG CTG GTG ATT CTG GCG CGT GAA GCA GGT TAC AAC ATC GAA CCG GAT 2070
Pro Asp Pro Arg Asp Asp Leu Ser Gly Lys Asp Val Ser Arg Lys Leu Val Ile Leu Ala Arg Glu Ala Gly Tyr Asn Ile Glu Pro Asp 690

CAG GTA CGT GTG GAA TCG CTG GTG CCT GCT CAT TGC GAA GGC GGC AGC ATC GAC CAT TTC TTT GAA AAT GGC GAT GAA CTG AAC GAG CAG 2160
Gln Val Arg Val Glu Ser Leu Val Pro Ala His Cys Glu Gly Gly Ser Ile Asp His Phe Phe Glu Asn Gly Asp Glu Leu Asn Glu Gln 720

ATG GTG CAA CGG CTG GAA GCG GCC CGC GAA ATG GGG CTG GTG CTG CGC TAC GTG GCG CGT TTC GAT GCC AAC GGT AAA GCG CGT GTA GGC 2250
Met Val Gln Arg Leu Glu Ala Ala Arg Glu Met Gly Leu Val Leu Arg Tyr Val Ala Arg Phe Asp Ala Asn Gly Lys Ala Arg Val Gly 750

GTG GAA GCG GTG CGT GAA GAT CAT CCG TTG CGA TCA CTG CTG CCG TGC GAT AAC GTC TTT GCC ATC GAA AGC CGC TGG TAT CGC GAT AAC 2340
Val Glu Ala Val Arg Glu Asp His Pro Leu Ala Ser Leu Leu Pro Cys Asp Asn Val Phe Ala Ile Glu Ser Arg Trp Tyr Arg Asp Asn 780

CCT CTG GTG ATC CGC GGA CCT GGC GCT GGG CGC GAC GTC ACC GCC GGG GCG ATT CAG TCG GAT ATC AAC CGG CTG GCA CAG TTG TTG 2427
Pro Leu Val Ile Arg Gly Pro Gly Ala Gly Arg Asp Val Thr Ala Gly Ala Ile Gln Ser Asp Ile Asn Arg Leu Ala Gln Leu Leu 809

FIGURE 9.7 The DNA sequence of *metL* and the corresponding amino acid sequence of the encoded aspartate kinase II–homoserine dehydrogenase II.

Both proteins yield proteolytic fragments carrying the dehydrogenase activity, although the dehydrogenase II fragment is shorter than the dehydrogenase I fragment. This shows that, in both proteins, a large part, and in particular a large part of the kinase domain, is not necessary for the dehydrogenase activity. In the case of enzyme I, the presence of the kinase domain may be necessary for the inhibition of the dehydrogenase.

Thus for both proteins, the dehydrogenase domain is quite independent of the kinase domain. The two domains of enzyme I can refold independently. This is also true for the dehydrogenase II domain; it is not testable for the kinase II domain, since an active fragment is not yet available. In enzyme I, a large part of the dehydrogenase domain is dispensable for the activity and inhibition of the kinase. Whether this is also true for enzyme II remains an open question.

B. Immunochemical Comparison

The two bifunctional proteins share antigenic determinants when denatured proteins are used as antigens (Zakin et al., 1978). The native proteins do not exhibit such cross-reactivity (Kaminski et al., 1969; Zakin et al., 1978). This may indicate that common determinants are essentially sequential in nature and that the two proteins derive from a common ancestor. The DNA sequence of the *thrA* gene and the amino acid sequence of its product are known (Katinka et al., 1980) (Fig. 9.2). The *metL* gene has been sequenced (Fig. 9.7); homologies between *thrA* and *metL* in several sections of the corresponding proteins (Fig. 9.8) point to a common ancestor (Zakin et al., 1983).

V. ASPARTATE SEMIALDEHYDE DEHYDROGENASE

A. Structure

Aspartate semialdehyde dehydrogenase catalyzes the reversible substrate-dependent reduction of NADP in the presence of phosphate or arsenate (Fig. 9.1, reverse reaction). This reaction is formally similar to that catalyzed by glyceraldehyde 3-phosphate dehydrogenase. The aspartate semialdehyde dehydrogenase has been obtained in 90% purity from *E. coli* K-12 grown under conditions of lysine limitation (Hegeman et al., 1970). The preparation of the enzyme has been highly improved by starting with a phenotypically derepressed strain (Biellmann et al., 1980a). The molecular weight of the enzyme is 77,000. The enzyme is a dimer of identical subunits (M_r =

FIGURE 9.8 Comparison between the amino acid sequences of the *E. coli* aspartate kinase I–homoserine dehydrogenase I (top line) and the aspartate kinase II–homoserine dehydrogenase II. Identical residues are boxed; single base changes are marked by a black dot; crosses indicate conservative amino acid substitutions. ▶

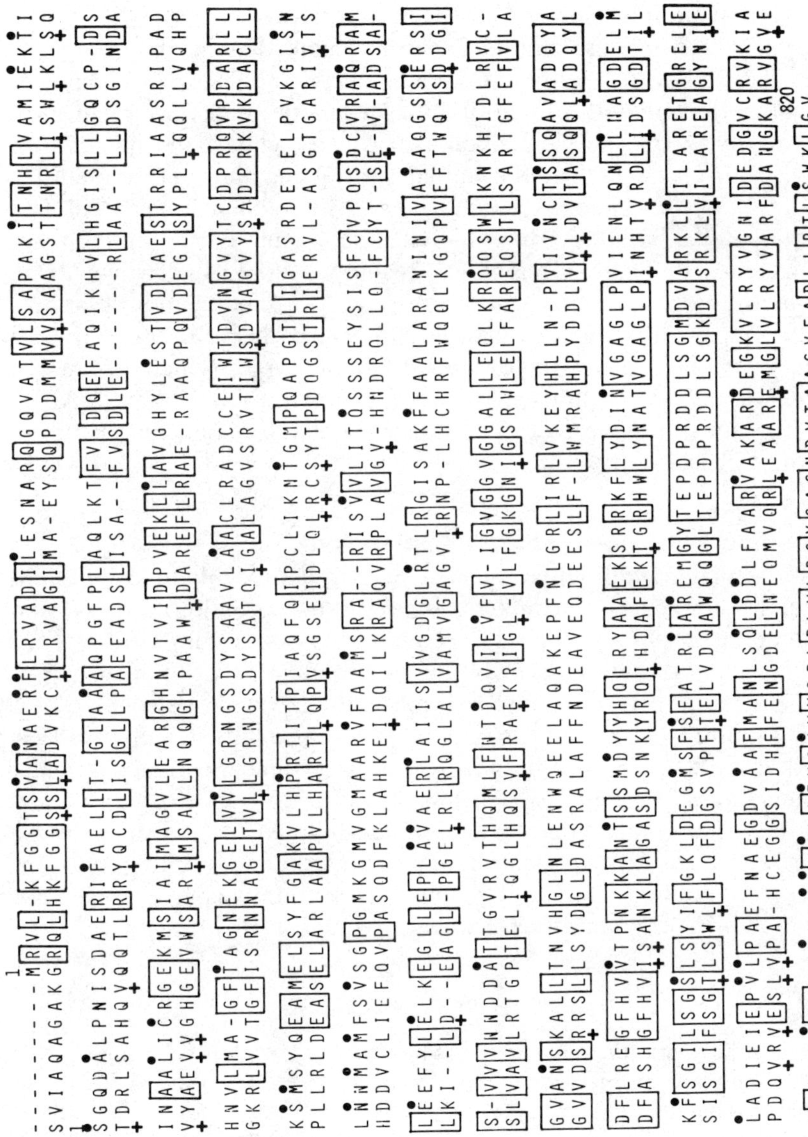

38,000 ± 2,000). The N-terminal sequence is Met–Lys–Asx–Val–Gly–. Each subunit contains three cysteine residues; two are reactive in the native enzyme, and one is partially protected by the substrate. Formation of an acyl-enzyme intermediate has been detected.

The substrate binding site of aspartate semialdehyde dehydrogenase has been labeled with 2-amino-4-oxo-5-chloropentanoate. This analog inactivates the enzyme with pseudo-first-order kinetics and with a half of the sites reactivity. Aspartate semialdehyde protects against the inactivation. A single group is labeled at the active site (Biellmann et al., 1980b). A peptic digest of the labeled enzyme yields the sequence Phe–Val–Gly–Gly–Asp-modified residue–Thr–Val–Ser. Biellmann et al. (1980b) suggested that the side chain of a histidine residue is modified; however, DNA sequencing indicates a cysteine to be the target of modification (Haziza et al., 1982). Aspartate semialdehyde dehydrogenase has also been alkylated with the coenzyme analog chloroacetylpyridine-ADP, which irreversibly inactivates the enzyme with pseudo-first-order kinetics. NADP and NADPH, but not the substrate, protect. The stoichiometry for total inactivation is again 1 mol of analog per dimer (Biellmann et al., 1980c).

B. Regulation at the Level of Gene Expression

Aspartate semialdehyde dehydrogenase activity is not inhibited by lysine; however, the synthesis of the enzyme is subject to repression. Excess lysine represses about 50%, and lysine limitation causes a 20-fold to 30-fold derepression. Thus the enzyme level of the cell can vary 40-fold to 60-fold (Cohen and Patte, 1963). Threonine and methionine may participate with lysine in a multivalent repression of aspartate semialdehyde dehydrogenase (Boy and Patte, 1972). Although the effect of lysine limitation is the dominant factor, limitation of the other two metabolites leads to a 1.5-fold derepression over the fully repressed level. An analysis of the regulation of the synthesis of aspartate semialdehyde dehydrogenase in mutants lacking phosphoglucoisomerase activity has led to indirect but convincing evidence that glucose-6-phosphate is an effector of this synthesis (Boy and Patte, 1979).

ACKNOWLEDGMENTS

Over the past 20 years, the work in the author's laboratory has been supported by the Centre National de la Recherche Scientifique, the Délégation Générale a la Recherche Scientifique et Technique, the Commissariat a l'Énergie Atomique, the National Science Foundation, and the Fondation pour la Recherche Médicale Française.

REFERENCES

Bachmann, B. J., and Low, K. B. (1980) *Microbiol. Rev., 44,* 1–56.

Barber, E. D., and Bright, H. J. (1968) *Proc. Nat. Acad. Sci. U.S.A., 60,* 1363–1370.

Bearer, C. F., and Neet, K. E. (1978a) *Biochemistry 17,* 3517–3522.
Bearer, C. F., and Neet, K. E. (1978b) *Biochemistry 17,* 3523–3530.
Biellmann, J. F., Eid, P., Hirth, C., and Jörnvall, H. (1980a) *Eur. J. Biochem. 104,* 53–58.
Biellmann, J. F., Eid, P., Hirth, C., and Jörnvall, H. (1980b) *Eur. J. Biochem. 104,* 59–64.
Biellmann, J. F., Eid, P., and Hirth, C., (1980c) *Eur. J. Biochem. 104,* 65–69.
Boy, E., and Patte, J. C. (1972) *J. Bacteriol. 112,* 84–92.
Boy, E., and Patte, J. C. (1979) *FEMS Microbiol. Lett. 6,* 189–192.
Briley, P. A., Sibilli, L., Chalvignac, M. A., Cossart, P., Le Bras, G., de Wolf, A., and Cohen, G. N. (1978) *J. Biol. Chem. 253,* 8867–8871.
Clark, R. B., and Ogilvie, J. W. (1972) *Biochemistry 11,* 1278–1282.
Cohen, G. N., and Dautry-Varsat, A. (1980) in *Multifunctional Proteins* (Bisswanger, H., and Schmincke-Ott, E., eds.) pp. 49–121, John Wiley, New York.
Cohen, G. N., and Jacob, F. (1959) *Comptes Rendus 248,* 3490–3492.
Cohen, G. N., and Patte, J. C. (1963) *Cold Spring Harbor Symp. Quant. Biol. 28,* 513–516.
Cohen, G. N., Patten, J. C., and Truffa-Bachi, P. (1965a) *Biochim. Biophys. Res. Comm. 19,* 546–550.
Cohen, G. N., Patte, J. C., Truffa-Bachi, P., Sawas, C., and Doudoroff, M. (1965b) in *Mécanismes de régulation des activités cellulaires chez les microorganismes* pp. 243–253, CNRS, Paris.
Cohen, G. N., Patte, J. C., Truffa-Bachi, P., and Janin, J. (1967) in *Regulation of Nucleic Acid and Protein Biosynthesis* (Koningsberger, V. V., and Bosch, L., eds.) pp. 357–365, Elsevier, New York.
Cohn, M., Cohen, G. N., and Monod, J. (1953) *Comptes Rendus 236,* 746–748.
Cossart, P. "Contribution à la détermination de la structure primaire de l'aspartokinase I-homosérine déshydrogénase I : distribution des méthionines et des peptides contenant de la cystéine et du tryptophane. Isolement et caractérisation des peptides obtenus par action du bromure de cyanogène. Séquence des deux plus grands fragments." Ph.D. thesis, University of Paris, 1977.
Cossart, P., Katinka, M., Yaniv, M., Saint-Girons, I., and Cohen, G. N. (1979) *Mol. Gen. Genet 175,* 39–44.
Cossart-Gheerbrant, P., Sibilli-Weill, L., Briley, P. A., Chalvignac, M. A., Le Bras, G., and Cohen, G. N. (1978) *Biochim. Biophys. Acta 535,* 206–215.
Costrejean, J. M. "Etude de l'aspartokinase I-homosérine déshydrogénase I d'*Escherichia coli* K12. Localisation et fonction des deux sites de fixation pour la thréonine." Ph.D. thesis, University of Paris, 1978.
Dautry-Varsat, A., and Cohen, G. N. (1977) *J. Biol. Chem. 252,* 7685–7689.
Dautry-Varsat, A., and Garel, J. R. (1978) *Proc. Nat. Acad. Sci. U.S.A. 75,* 5979–5982.
Dautry-Varsat, A., and Garel, J. R. (1981) *Biochemistry 20,* 1396–1401.
Dautry-Varsat, A., Sibilli-Weill, L., and Cohen, G. N. (1977) *Eur. J. Biochem. 76,* 1–6.
Ehrlich, R. S., and Takahashi, M. (1973) *Biochemistry 12,* 4309–4315.
Falcoz-Kelly, F. "Etude comparée des aspartokinases homosérine déshydrogénases de *E.coli* K12" Ph.D. thesis, University of Paris, 1971.
Falcoz-Kelly, F., van Rapenbusch, R., and Cohen, G. N. (1969) *Eur. J. Biochem. 8,* 146–152.

Falcoz-Kelly, F., Janin, J., Saari, J. C., Veron, M., Truffa-Bachi, P., and Cohen, G. N. (1972) *Eur. J. Biochem. 28,* 507–519.

Fazel, A., Müller, K., Le Bras, G., Garel, J. R., Veron, M., and Cohen, G. N. (1983) *Biochemistry 22,* 158–165.

Gardner, J. F. (1979) *Proc. Nat. Acad. Sci. U.S.A. 76,* 1706–1710.

Gardner, J. F., and Smith, O. H. (1975) *J. Bacteriol. 124,* 161–166.

Garel, J. R., and Dautry-Varsat, A. (1980a) in *Protein Folding* (Jaenicke, R., ed.) pp. 485–499, Elsevier/North-Holland Biomedical Press, New York.

Garel, J. R., and Dautry-Varsat, A. (1980b) *Proc. Nat. Acad. Sci. U.S.A. 77,* 3379–3383.

Haziza, C., Stragier, P., and Patte, J. C. (1982) *EMBO Journal 1,* 379–384.

Heck, H., and Truffa-Bachi, P. (1970) *Biochemistry 9,* 2776–2783.

Hegeman, G. D., Cohen, G. N., and Morgan, R. (1970) *Meth. Enzymol. 17A,* 708–713.

Hirth, C. G., Veron, M., Villar-Palasi, C., Hurion, N., and Cohen, G. N. (1975) *Eur. J. Biochem. 50,* 425–430.

Janin, J. (1974) *FEBS Lett. 45,* 318–319.

Janin, J., and Cohen, G. N. (1969) *Eur. J. Biochem. 11,* 520–529.

Janin, J., and Iwatsubo, M. (1969) *Eur. J. Biochem. 11,* 530–540.

Johnson, E. J., Cohen, G. N., and Saint-Girons, I. (1977) *J. Bacteriol. 129,* 66–70.

Kaminski, M., Falcoz-Kelly, F., Truffa-Bachi, P., Patte, J. C., and Cohen, G. N. (1969) *Eur. J. Biochem. 11,* 278–282.

Katinka, M., Cossart, P., Sibilli, L., Saint-Girons, I., Chalvignac, M. A., Le Bras, G., Cohen, G. N., and Yaniv, M. (1980) *Proc. Nat. Acad. Sci. U.S.A. 77,* 5730–5733.

Mackall, J. C., and Neet, K. E. (1974) *Eur. J. Biochem. 42,* 275–282.

Nass, G., and Thomale, J. (1974) *FEBS Lett. 39,* 182–186.

Nass, G., Poralla, K., and Zahner, H. (1969) *Biochem. Biophys. Res. Comm. 34,* 84–91.

Nichols, B. P., and Yanofsky, C. (1979) *Proc. Nat. Acad. Sci. U.S.A. 76,* 5244–5248.

Parsot, C., Saint-Girons, I., and Cossart, P. (1983) *Mol. Gen. Genetics, 188,* 455–458.

Patte, J. C., Le Bras, G., Loviny, T., and Cohen, G. N. (1963) *Biochim. Biophys. Acta 67,* 16–30.

Patte, J. C., Truffa-Bachi, P., and Cohen, G. N. (1966) *Biochim. Biophys. Acta 128,* 426–439.

Patte, J. C., Le Bras, G., and Cohen, G. N. (1967) *Biochim. Biophys. Acta 136,* 245–257.

Rowbury, R. J. (1965) *Nature* (London) *206,* 962–963.

Rowbury, R. J., and Woods, D. D. (1966) *J. Gen. Microbiol. 42,* 155–163.

Ryzewski, C., and Takahashi, M. T. (1975) *Biochemistry 14,* 4482–4486.

Saint-Girons, I. (1978) *Mol. Gen. Genet. 162,* 95–100.

Saint-Girons, I., and Margarita, D. (1975) *J. Bacteriol. 124,* 1137–1141.

Selker, E., and Yanofsky, C. (1979) *J. Mol. Biol. 130,* 135–143.

Shine, J., and Dalgarno, L. (1975) *Nature* (London) *254,* 34–38.

Sibilli, L. "Contribution à la détermination de la structure primaire de l'aspartokinase I-homosérine déshydrogénase I par :

- l'étude de la distribution des résidus méthioninyles et des peptides tryptiques contenant de la cystéine ou du tryptophane;
- l'étude des fragments obtenus par clivage avec le bromure de cyanogène;
- l'étude des peptides tryptiques contenant de la méthionine

Ph.D. thesis, University of Paris, 1977.

Sibilli, L., Cossart, P., Chalvignac, M. A., Briley, P. A., Costrejean, J. M., Le Bras, G., and Cohen, G. N. (1977) *Biochimie 59,* 943–946.

Sibilli, L., Le Bras, G., Le Bras, G., and Cohen, G. N. (1981) *J. Biol. Chem. 256,* 10228–10230.

Stadtman, E. R., Cohen, G. N., Le Bras, G., and de Robichon-Szulmajster, H. (1961) *J. Biol. Chem. 236,* 2033–2038.

Starnes, W. L., Munk, P., Maul, S. B., Cunningham, G. N., Cox, D. J., and Shive, W. (1972) *Biochemistry 11,* 677–687.

Su, C. H., and Greene, R. C. (1971) *Proc. Nat. Acad. Sci. U.S.A. 68,* 367–371.

Takahashi, M., and Westhead, E. W. (1971) *Biochemistry 10,* 1700–1705.

Thèze, J., and Saint-Girons, I. (1974) *J. Bacteriol. 118,* 990–998.

Thèze, J., Margarita, D., Cohen, G. N., Borne, F., and Patte, J. C. (1974) *J. Bacteriol. 117,* 133–143.

Tilak, A., Wright, K., Damle, S., and Takahashi, M. (1976) *Eur. J. Biochem. 69,* 249–255.

Truffa-Bachi, P., and Heck, H. (1971) *Biochemistry 10,* 2700–2706.

Truffa-Bachi, P., Le Bras, G., and Cohen, G. N. (1966) *Biochim. Biophys. Acta 128,* 440–449.

Truffa-Bachi, P., van Rapenbusch, R., Janin, J., Gros, C., and Cohen, G. N. (1968) *Eur. J. Biochem. 5,* 73–80.

Truffa-Bachi, P., van Rapenbusch, R., Janin, J., Gros, C., and Cohen, G. N. (1969) *Eur. J. Biochem. 7,* 401–407.

Truffa-Bachi, P., Veron, M., and Cohen, G. N. (1974) *CRC Critical Rev. Biochem. 2,* 379–415.

Veron, M., and Cohen, G. N. (1974) in *Metabolic Interconversion of Enzymes 1973* (Fischer, E. H., Krebs, E. G., Neurath, H., and Stadtman, E. R., eds.) pp. 335–347, Springer, Berlin.

Veron, M., Falcoz-Kelly, F., and Cohen, G. N. (1972) *Eur. J. Biochem. 28,* 520–527.

Veron, M., Saari, J. C., Villar-Palasi, C., and Cohen, G. N. (1973) *Eur. J. Biochem. 38,* 325–335.

Wampler, D. E. (1972) *Biochemistry 11,* 4428–4435.

Wampler, D. E., and Westhead, E. W. (1968) *Biochemistry 7,* 1661–1671.

Wijesundera, S., and Woods, D. D. (1953) *Biochem. J. 55,* viii.

Wright, J. K., and Takahashi, M. (1977a) *Biochemistry 16,* 1541–1548.

Wright, J. K., and Takahashi, M. (1977b) *Biochemistry 16,* 1548–1554.

Wright, J. K., Feldman, J., and Takahashi, M. (1976) *Biochemistry 15,* 3704–3710.

Zakin, M. M., Garel, J. R., Dautry-Varsat, A., Cohen, G. N., and Boulot, G. (1978) *Biochemistry 17,* 4318–4323.

Zakin, M. M., Greene, R. C., Dautry-Varsat, A., Duchange, N., Ferrara, P., Py, M. C., Margarita, D., and Cohen, G. N. (1982) *Molec. Gen. Genet. 187,* 101–106.

Zakin, M. M., Duchange, N., Ferrara, P., and Cohen, G. N. (1983) *J. Biol. Chem. 258,* 3028–3031.

CHAPTER

10

The Threonine Operon of *Escherichia coli*

Steven P. Lynn
Jeffrey F. Gardner

I. INTRODUCTION

Studies during the past 25 years have shown that amino acid biosynthesis in bacteria is regulated by controlling the catalytic proficiency for certain key enzymes and by controlling the enzyme level via gene expression (Umbarger, 1978). Threonine biosynthesis from aspartate consists of five enzymatic steps that appear to be the same in all bacteria (Fig. 10.1). The first three steps from aspartate to homoserine are common to the biosynthesis of threonine and methionine; the first two from aspartate to aspartate semialdehyde are also the initial steps in lysine biosynthesis. The enzymes catalyzing the first three steps are described in detail in Chapter 9. The aspartate kinase I and homoserine dehydrogenase I activities are found in one polypeptide chain (bifunctional enzyme), the product of the *thrA* gene (Truffa-Bachi et al., 1974). Aspartate kinase I–homoserine dehydrogenase I is inhibited by threonine (Patte et al., 1963). The structural genes encoding the enzymes that catalyze the two terminal steps in threonine biosynthesis, *thrB* and *thrC*, form together with *thrA* a transcriptional unit, the *thr* operon. Regulation of *thr* operon expression, at least in *Escherichia coli* and *Salmonella*

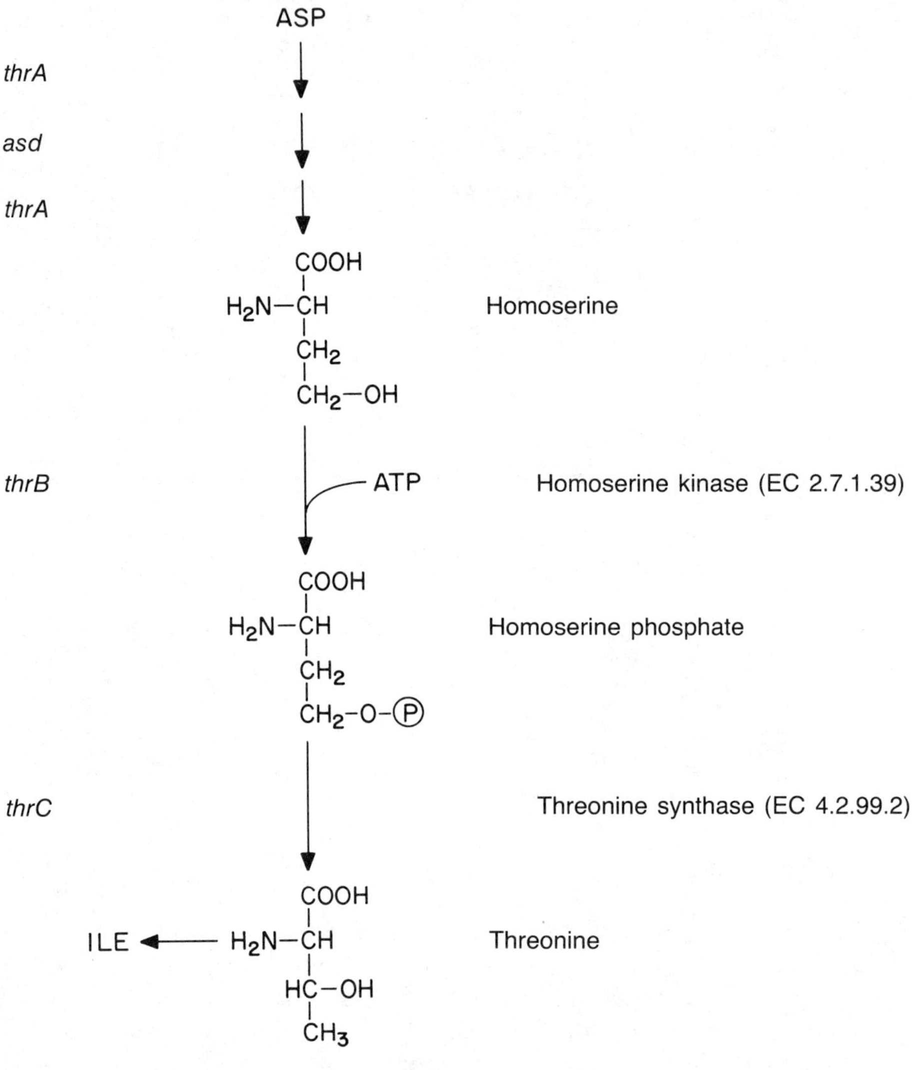

FIGURE 10.1 The threonine biosynthetic pathway.

typhimurium, is mediated jointly by the intracellular levels of both threonine and isoleucine through attenuation, a process involving transcription termination. Most of the information on *thr* operon attenuation has been obtained during the past five years through work with *E. coli* K-12. A broader view of attenuation, particularly as applied to the *trp* operon of *E. coli,* is found in a recent review by Yanofsky (1981).

II. ENZYMES AND REGULATION OF ENZYME ACTIVITY

A. Aspartate Kinase I–Homoserine Dehydrogenase I

The *thrA* gene encodes the threonine-sensitive bifunctional aspartate kinase I–homoserine dehydrogenase I (Cohen and Dautry-Varsat, 1980) (Chapter 9). *E. coli* K-12 contains two additional aspartate kinases and one additional homoserine dehydrogenase. The synthesis of the bifunctional aspartate kinase II–homoserine dehydrogenase II is regulated by methionine (Patte et al., 1967). The enzyme activity and the synthesis of aspartate kinase III is regulated by lysine (Stadtman et al., 1961).

B. Homoserine Kinase

The *thrB* gene encodes homoserine kinase, which catalyzes the phosphorylation of homoserine. The native protein consists of two identical 29,000-dalton subunits (Burr et al., 1976). The purified protein requires Mg^{2+} for activity, and K^+ stimulates the enzyme (Thèze et al., 1974a). The protein has been purified to homogeneity (Burr et al., 1976), and the first eight amino acids from the N-terminus have been determined. The entire DNA sequence of *thrB* is known (Cossart et al., 1981), from which the amino acid sequence of the protein has been deduced.

Homoserine kinase is inhibited by threonine with an apparent K_i of 0.6 mM (Burr et al., 1976). No product inhibition is observed, nor do the threonine analogs, D-threonine, glycyl-threonine, 3- or 4-hydroxy butyrate, 3-aminobutyrate, or O-methyl-threonine at concentrations of 10 mM inhibit the enzyme (Burr et al., 1976). Isoleucine inhibits partially purified preparations of homoserine kinase, but this inhibition is not additive with threonine inhibition (Thèze et al., 1974a). Threonine inhibition is competitive with respect to homoserine. A low level of cooperativity ($n = 1.4$) is observed with partially purified homoserine kinase but not with the pure enzyme (Thèze et al., 1974a). Since the intracellular homoserine level in *E. coli* is low, threonine may regulate homoserine kinase competitively rather than allosterically (Thèze et al., 1974a; Burr et al., 1976). The binding of ATP is strictly Michaelian (Burr et al., 1976).

C. Threonine Synthase

There are very few biochemical studies on the threonine synthase of *E. coli* (Wormser and Pardee, 1958; Szczesiul and Wampler, 1976). The enzyme catalyzes the rather complex conversion of homoserine phosphate to threonine. Like other threonine synthases, the *E. coli* enzyme requires pyridoxal phosphate (Flavin and Slaughter, 1960; Skarstedt and Greer, 1973). The apparent K_m for homoserine phosphate is 0.23 mM; the molecular weight as determined by gel filtration is 36,000 (Szczesiul and Wampler, 1976). No information about subunit composition or amino acid sequences is available. A possible amino terminus has been deduced by DNA sequencing (Cossart et al., 1981).

III. OPERON ORGANIZATION AND REGULATION OF GENE EXPRESSION

A. "Multivalent Repression," Gene Order, and Complementation Analysis

Coordinate regulation of the threonine biosynthetic enzymes was first demonstrated in *E. coli* and *S. typhimurium* by Freundlich (1963), who observed that both threonine and isoleucine were required for "repression." Since threonine is an intermediate in isoleucine biosynthesis, it appears that both amino acids, threonine and isoleucine, are involved in regulation of threonine operon expression. This form of regulation, termed "multivalent repression," is only observed when threonine and isoleucine are both present in excess. Today, the term "multivalent repression" appears to be a misnomer, since regulation of *thr* operon expression is by attenuation.

The structural genes encoding the enzymes of threonine biosynthesis from aspartate have all been identified and mapped (Thèze et al., 1974b; Thèze and Saint-Girons, 1974; Gardner and Smith, 1975; Bachmann and Low, 1980). The *thr* operon of *E. coli* consists of three structural genes—*thrA, thrB,* and *thrC* (Thèze and Saint-Girons, 1974)—and an analogous arrangement is found in *S. typhimurium* (Stuttard, 1973). The *thr* operon lies at 0 min on the standard *E. coli* linkage map. This cluster encodes four of the five enzymes of threonine biosynthesis. Gene *asd* encoding the fifth enzyme, aspartate semialdehyde dehydrogenase, lies at 75 min (Boy and Patte, 1972; Bachmann and Low, 1980). Its synthesis is regulated by threonine, lysine, and methionine (Chapter 12).

The order *thrA thrB thrC,* with transcription proceeding from *thrA* to *thrC,* has been established by analysis of phage Mu insertions and nonsense mutations in the *thr* cluster (Thèze and Saint-Girons, 1974) and by analysis of a phage λ insertion into the *thr* regulatory region (Gardner et al., 1974). The λ insertion lies to the left of *thrA* (that is, *trpR*-λ-*thrA*) and pleiotropically affects the expression of *thrA* and *thrB*. Deletion analysis indicates that *thrA* is adjacent to the site of λ insertion. The genetic fine structure of the *thr* operon has been analyzed by deletion mapping (Saint-Girons and Margarita, 1978).

Complementation analysis indicates that both *thrB* and *thrC* consist of single cistrons, while *thrA* is composed of two cistrons specified by a single polypeptide chain (Thèze and Saint-Girons, 1974). This is in accord with the bifunctional character of the *thrA* protein (Chapter 9).

B. The *thr* Operon Regulatory Region

1. Regulatory Mutants. The *thr* operon regulatory region was first located mutationally by analysis of a secondary λ lysogen inserted between the *thr* promoter and the first structural gene (Gardner et al., 1974; Gardner and Smith, 1975). The original lysogen is a threonine auxotroph, but it is possible to isolate λ lysogenic pseudorevertants from which nonlysogenic derivatives can be obtained. These derivatives show a fivefold elevation of the *thr* operon enzymes. Nonlysogenic derivatives of the original lysogen always display wild-type, multivalent regulation of the *thr*-operon enzymes. The pseudorevertants arise from mutations in the *thr* terminator located between the secondary λ insertion and the beginning of *thrA* (Gardner, 1979; Chapman and Gardner, 1981; Lynn et al., 1982).

Other regulatory mutants have been isolated on the basis of enhanced resistance to the threonine analog α-amino-β-hydroxyvalerate, which inhibits aspartate kinase I–homoserine dehydrogenase I (Cohen and Patte, 1963; Thèze et al., 1974b). Regulatory mutants are identified by their excretion of threonine into the media. Mutations of this type lie adjacent to *thrA*. The mutants are *cis*-dominant (Saint-Girons and Margarita, 1975), and at least one contains a large insertion or deletion in the *thr* regulatory region (Parsot et al., 1983).

Regulatory mutants have also been isolated from *thr–lac* fusions (Saint-Girons, 1978). Strains with elevated levels of β-galactosidase are identified and shown to act in *cis*. The mutations are localized upstream of *thrA* (Saint-Girons and Margarita, 1978).

These studies identify a regulatory region preceding the *thr* operon structural genes. Mutations in this region cause pleiotropic effects on *thr* operon expression. Originally, the mutants were thought to be classical operator mutants, analogous to those isolated in the *trp* and *lac* operon (Gardner and Smith, 1975; Saint-Girons and Margarita, 1975). However, because no apo-repressor mutants have been isolated and because all mutations analyzed to date lie in the *thr* attenuator, it does not appear that the *thr* operon of *E. coli* is regulated by a repressor–operator mechanism.

2. DNA Sequencing of the *thr* Regulatory Region. The sequence of the *thr* operon regulatory region has been determined (Gardner, 1979) by using DNA cloned from λ *thr* transducing phages (Gardner and Reznikoff, 1978). The sequence of the regulatory region (Fig. 10.2) has been identified by matching the DNA sequence with the known amino-terminal sequence of the *thrA* encoded polypeptide (Cohen and Veron, 1974). Several aspects of the DNA sequence are discussed below.

3. The Promoter. The promoter for the *thr* operon, identified from in vitro transcription studies, has been localized approximately 230 nucleotides up-

```
BstEII                                                                                   HhaI
      -250                                                   -200
CTGGTTACCTGCCGTGAGTAAATTAAAATTTTATTGACTTAGGTCACTAAATACTTTAACCAATATAGGCATAGCGCACAGACAGATAAAAATTACAGAG
                                      -35                    -10

      -150                                                   -100
TACACAACATCCATGAAACGCATTAGCACCACCATTACCACCACCATCACCATTACCACAGGTAACGGTGCGGGCTGACGCGTACAGGAAACACAGAAAA
           MetLysArgIleSerThrThrIleThrThrThrIleThrIleThrThrGlyAsnGlyAlaGly

       -50                                                     +1
AAGCCCGCACCTGACAGTGCGGGCTTTTTTTTTCGACCAAAGGTAACGAGGTAACAACCATGCGAGTGTTGAAGTTCGGCGGTACATCA
                                                               MetArgValLeuLysPheGlyGlyThrSer
```

FIGURE 10.2 The DNA sequence of the *thr* operon regulatory region and the amino acid sequences of the presumed leader peptide and the amino terminus of aspartate kinase I–homoserine dehydrogenase I. Position 1 marks the first base encoding the amino terminal methionine of aspartate kinase I. Positions -147 to -84 encode the presumed leader peptide. The wedge shapes indicate the two predominant in vitro transcription termination sites at -29 and -28. The -35 and -10 regions of the promoter are indicated by brackets. The arrow marks the ATP initiation of transcription at -189; brackets and dashed lines indicate dyad symmetries.

stream of *thrA* (Lynn et al., 1982). DNA between a BstEII site at position −254 and a HhaI site −184 must contain the *thr* promoter, since in vitro transcription from templates extending from *thrA* to the BstEII site produce a terminated transcript but those from templates extending only to the HhaI site do not. Sequencing of the transcript RNA, labeled with [γ-^{32}P]ATP, revealed the 5′ end of the transcript as pppAUAGCGC, which is complementary to the DNA of the coding strand starting at position −189 (Fig. 10.2). This transcript initiates with A, like a majority of transcripts (Rosenberg and Court, 1979), although a G start is also observed (Gardner, 1982). In the DNA sequence preceding the RNA initiation site, two versions of the *thr* promoter can be identified by comparison of the −35 and the −10 regions with the consensus promoter sequence of Rosenberg and Court (1979). One version shows a better match in the −35 region than in the −10 region; the other, obtained by shifting the sequence two base pairs upstream, shows a better match with the −10 consensus sequence but a less perfect match in the −35 region. Additional biochemical and genetic studies are required to precisely identify RNA polymerase recognition and initiation sites of the *thr* operon.

4. The Attenuator Region. Regulation at the level of transcription termination is a common regulatory mechanism in phage and bacteria (Adhya and Gottesman, 1978; Rosenberg and Court, 1979). Transcription terminators are either rho-independent or rho-dependent, depending upon in vitro termination efficiency. Rho-independent terminators contain a region of dyad symmetry, high in G–C content, followed by an A–T rich region of four to nine base pairs. When transcribed into RNA, the terminators give rise to RNA structures containing a stem and loop with a run of uridine residues at the 3′ end. Mutant and nucleotide analog studies indicate that both the stem of the RNA secondary structure (Rosenberg and Court, 1979; Farnham and Platt, 1980) and the run of uridine residues are necessary for efficient termination (Farnham and Platt, 1980). In addition, RNA-DNA interactions may be involved in the termination event (Farnham and Platt, 1980). The in vivo role of the rho factor in so-called rho-independent termination is not known, but rho may act as a release factor to increase the rate of dissociation of termination complexes from the DNA (Fuller and Platt, 1978; Farnham and Platt, 1980).

The *thr* regulatory region contains extensive dyad symmetry and a stretch of A–T base pairs characteristic of rho-independent terminators at positions −63 to −30 (Fig. 10.2). This region has been termed the *thr* attenuator by analogy with similar regions found in other amino acid biosynthetic operons (Kasai, 1974; Yanofsky, 1981). In vitro transcription of templates bearing the *thr* regulatory region including promoter and attenuator generates a terminated transcript that initiates at the promoter and terminates at the attenuator. The termination efficiency is 90% as judged by measurement of the amounts of terminated and readthrough transcripts produced (Lynn et al., 1982). A transcript identical in size is also formed in vivo (Lynn et al., 1982).

RNA sequencing studies have shown that in vitro termination at the attenuator is heterogeneous, as observed for several other terminators (Rosenberg and Court, 1979). The major RNAse T1 products, containing the 3′ terminus of the terminated RNA, are CU_7OH and CU_8OH (Gardner, 1982). This places the terminator event in the run of A–T base pairs at positions −29 and −28 (Fig. 10.2), similar to those in other rho-independent terminators where the 3′ end of the RNA has been sequenced (Rosenberg and Court, 1979).

All *thr* regulatory mutations that have been identified by DNA sequencing lie within the attenuator (Lynn et al., 1982). The mutational changes are shown in Fig. 10.3. The mutants show increased levels of *thr* operon expression, but not the multivalent regulation characteristic of the wild-type parent. The mutations are analogous to those found in the *trp* attenuator (Stauffer et al., 1978), in the *his* attenuator (Johnston and Roth, 1981), and at other terminators (Rosenberg and Court, 1979). The mutations decrease the efficiency of transcription termination in vitro (Stauffer et al., 1978) and in vivo (Stauffer et al., 1978; Johnston and Roth, 1981). A common feature

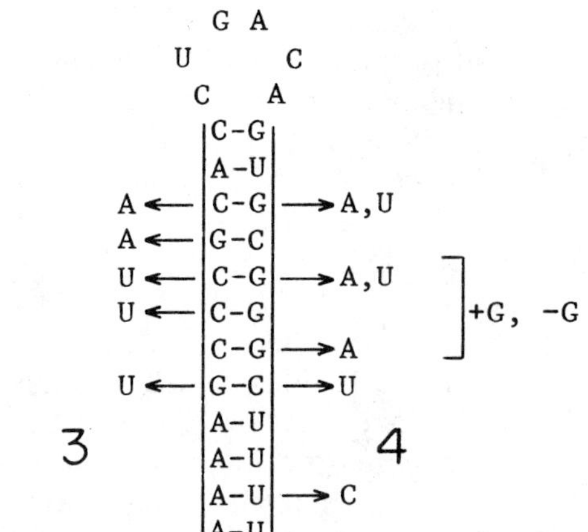

FIGURE 10.3 Termination relief mutants. The secondary structure of the DNA encoded by the *thr* regulatory region from -71 to -27 (Fig. 10.2) is presented in the most stable conformation to maximize base pairing in the RNA termination structure. Identified mutational base changes are indicated by arrows. The first nonlysogenic pseudorevertant isolated as described in Section III.B.1 is an insertion of a G between positions -40 and -36 as indicated by a bracket. A single base deletion has also been identified in this region.

of these mutations is the disruption of the dyad symmetry in the attenuator. In turn, the presumed RNA secondary structure, which constitutes part of the termination signal, is disrupted by base mismatches or bulge loops. Thus the increased expression of the *thr* operon in vivo in strains bearing such mutations is a result of increased transcription through the *thr* attenuator even under growth conditions of threonine and isoleucine excess.

In addition to the promoter and attenuator, the *thr* regulatory region contains two features relevant to attenuation: a sequence encoding a leader peptide and a pause site for RNA polymerase. Positions −147 to −82 encode a peptide of 21 amino acids. This putative leader peptide contains eight threonine and four isoleucine residues, the amino acids that regulate *thr* operon expression. Eleven of the 12 threonine plus isoleucine codons are consecutive.

The *thr* regulatory region contains several regions of overlapping rotational symmetry (Fig. 10.2); one is in the attenuator itself, 28 base pairs between positions −63 and −30. A second region of perfect rotational symmetry includes 20 base pairs between positions −93 and −49 and overlaps the attenuator. Since the sequence extending from −58 to −49 is part of the attenuator, RNA transcribed from −93 to −30 can form two mutually exclusive secondary structures. If the attenuator secondary structure is formed during transcription, a transcription termination signal is generated (Lee and Yanofsky, 1977; Rosenberg and Court, 1979). Alternatively, if the secondary structure of −93 to −49 is formed, the termination structure is excluded, RNA polymerase transcribes through the attenuator, and transcription proceeds into the structural genes. Two other regions of incomplete rotational symmetry include the sequence −99 to −90 with either −130 to −121 or −112 to −103, which are direct repeats (Fig. 10.2). Sequence −93 to −90 overlaps the second region of symmetry described above.

A pause site for RNA polymerase has been localized within the region −90 to −80 (Gardner, 1982). Pausing of RNA polymerase also occurs in the analogous region of the *trp* operon (Farnham and Platt, 1981; Winkler and Yanofsky, 1981) and of the phage T7 early genes (Kassavetis and Chamberlin, 1981). The significance of these sites is unknown. It is possible that pausing is used in vivo to couple transcription and translation of the leader region.

C. A Model for Control of the *thr* Operon by Attenuation

On the basis of the existing data, we propose a model for multivalent regulation of *thr* operon expression involving coupled transcription and translation of the leader region and the consequences of this coupling on RNA secondary structure formation and transcription termination (Gardner, 1979). The model assumes that transcription of the *thr* operon in vivo initiates at the *thr* promoter, that the *thr* leader RNA is translated into the leader peptide, and that formation of the RNA secondary structure of the

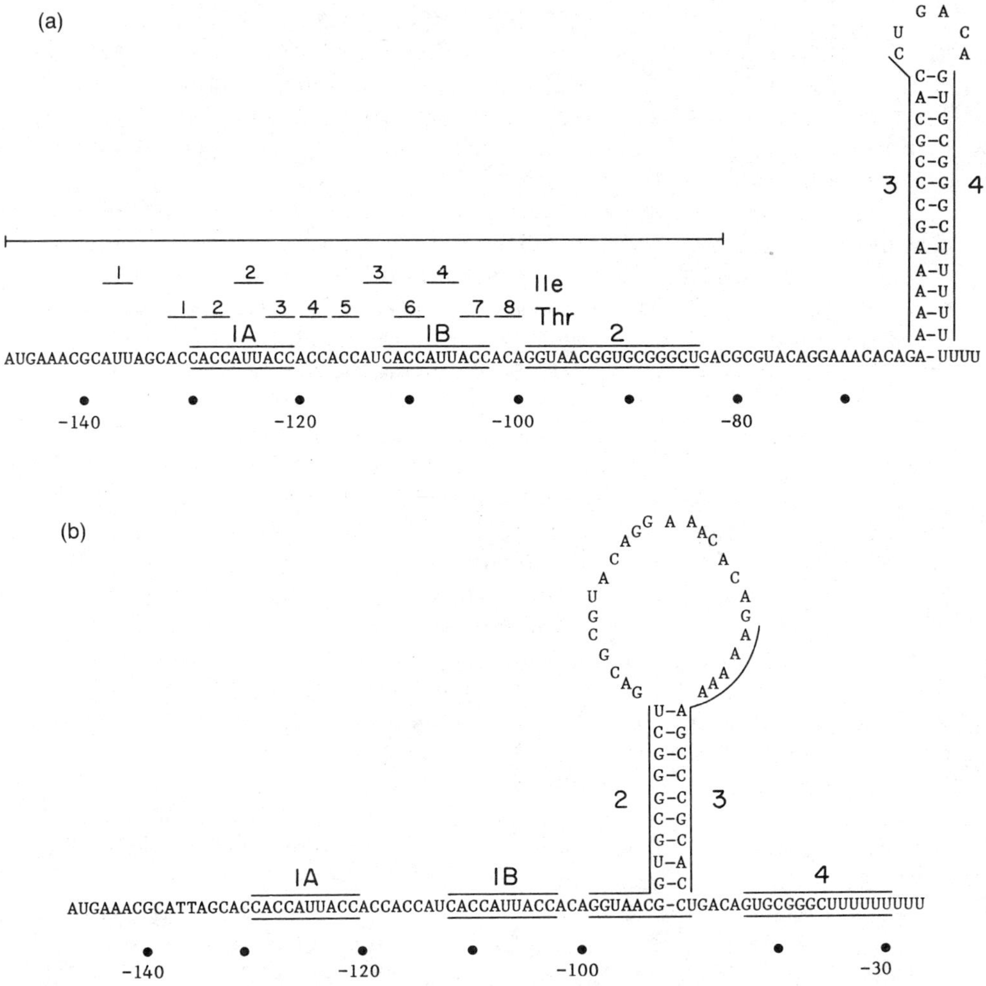

FIGURE 10.4 A model for control of *thr* operon expression by attenuation. The RNA encoded by the *thr* regulatory region from -147 to -27 (Fig. 10.2) is shown in four possible secondary structures reflecting (a) attenuation or transcription termination, (b) transcription through the attenuator, and (c, d) poor or no translation of the leader RNA. For details see Section III.C. The bar in (a) indicates the coding region for the leader peptide. The numbers above the sequence in (a) denote the positions of the threonine and isoleucine codons in the leader peptide. The boxed areas indicate regions involved in secondary structure formation.

(c)

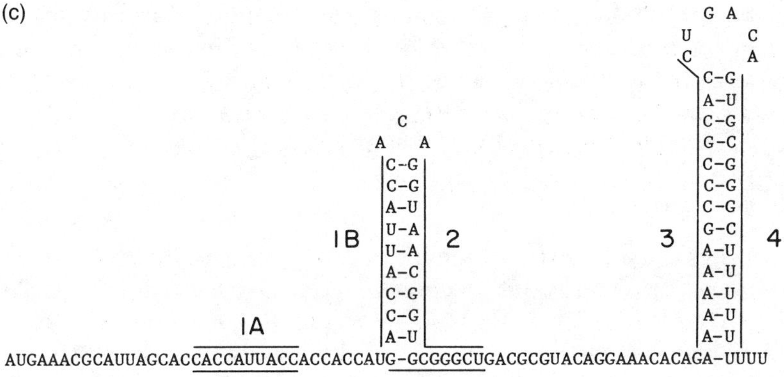

(d)

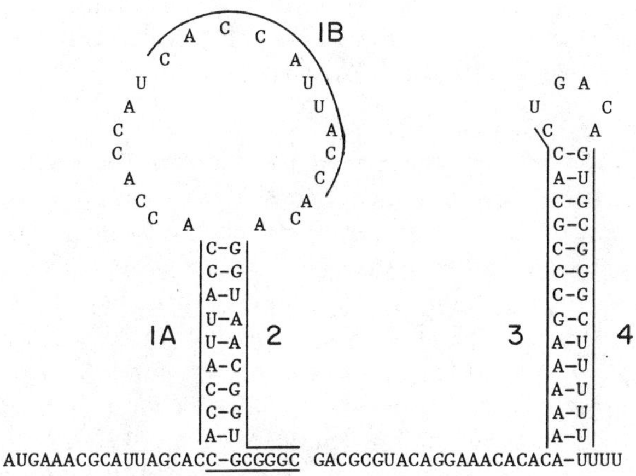

attenuator generates a transcription termination signal. This model is similar to those proposed for other amino acid biosynthetic operons (Lee and Yanofsky, 1977; DiNocera et al., 1978; Barnes, 1978; Zurawski et al., 1978; Gemmill et al., 1979; Keller and Calvo, 1979; Oxender et al., 1979; Johnston et al., 1980; Lawther and Hatfield, 1980; Nargang et al., 1980).

1. Attenuation or Transcription Termination. Under conditions of threonine and isoleucine excess, transcription initiating at the *thr* promoter is coupled to translation, and ribosomes translate the leader peptide and terminate translation at the UGA codon at position −82. The formation of the RNA termination structure (stem 3–4 of Fig. 10.4) is favored, since the ribosomes sterically block pairing of the −58 to −49 region with the proximal −93 to −84 region. Therefore the threonine and isoleucine excess signal results in transcription termination. Attenuator mutants synthesize elevated levels of threonine biosynthetic enzymes, regardless of the intracellular levels of threonine and isoleucine, because the stability of the RNA termination structure is decreased, and consequently, an effective termination signal cannot be generated.

2. Transcription Through the Attenuator. When intracellular threonine and/or isoleucine concentrations become limiting for growth, the ribosomes translating the leader RNA slow down or stall in the region of consecutive threonine and isoleucine codons, thereby uncoupling transcription and translation. As RNA polymerase continues transcription, an RNA secondary structure is formed in which positions −93 to −84 pair with positions −58 to −49 (stem 2–3 of Fig. 10.4). This, in turn, precludes the formation of the RNA termination structure, and RNA polymerase continues transcription into the structural genes of the *thr* operon.

3. Failure to Translate the Leader RNA. When the leader RNA is not translated, as in transcription experiments in vitro, transcription is efficiently terminated because RNA secondary structures form sequentially during transcription uncoupled from translation. The −99 to −90 region pairs with either the −130 to −121 (stem 1A–2 of Fig. 10.4) or the −112 to −103 region (stem 1B–2 of Fig. 10.4), thus favoring the formation of the termination structure. This suggests that no translation or poor translation of leader RNA in vivo leads to transcription termination. Evidence for this prediction has been obtained in both the *trp* (Stauffer et al., 1978) and *his* (Johnston and Roth, 1981) operons where mutations in the AUG codon of the leader peptide cause decreased expression of the structural genes.

Most aspects of the *thr* model presented here remain to be tested, and other factors may be involved in regulation of *thr* operon expression by attenuation in vivo. Several other amino acid biosynthetic operons in *E. coli* and *S. typhimurium* are regulated by a similar mechanism (Chapters 14, 15, 18, 20, 21). It appears that attenuation is a common mechanism for the control of expression in some, but not all, amino acid biosynthetic operons.

D. Other Factors Affecting *thr* Operon Expression

1. Amino Acyl-tRNA Synthetases. Growth of *E. coli* in the presence of the macrolide antibiotic borrelidin specifically increases the levels of the threonine biosynthetic enzymes threefold (Nass et al., 1969). Borrelidin inhibits threonyl-tRNA synthetase (Hütter et al., 1966), encoded by *thrS*. Mutations to borrelidin resistance are either *cis*-dominant and affect transcription of *thrS* and therefore increase theonyl-tRNA synthetase levels (Fröhler et al., 1980) or have altered tRNA synthetases with either increased threonine or decreased borrelidin affinities (Nass and Thomale, 1974; Paetz and Nass, 1973).

In one *thrS* mutant, the K_m for threonine is 200 times the wild-type value; in this mutant, *thr* operon expression is elevated tenfold and not regulated by threonine and isoleucine (Johnston et al., 1977). Isoleucyl-tRNA synthetase is encoded by *ilvS*. In *S. typhimurium*, an *ilvS* mutation increases the level of *thr* operon expression fourfold to fivefold (Blatt and Umbarger, 1972). From these studies, it is postulated that either the tRNA synthetases, charged tRNA's, or uncharged tRNA's are involved in the regulation of the *thr* operon. The attenuation model accommodates the data readily, since translation of the leader peptide is postulated to be involved in the regulation of the *thr* operon.

2. Other Factors. Besides the α-amino-β-hydroxyvalerate-resistant mutants (Section III.B.) that have mutations in the *thr* regulatory region and excrete threonine (Saint-Girons and Margarita, 1975), a second class of mutants resistant to this analog does not excrete threonine (Saint-Girons, 1978). In these mutants, homoserine dehydrogenase I activity is increased tenfold, and *thr* operon is not regulated by threonine and isoleucine. The mutations do not lie within the *thr* operon or the *thrS* gene. The exact positions of the lesions on the *E. coli* chromosome are unknown (*thrX*). It is speculated that *thrX* mutations cause alterations in tRNA modification.

IV. CLONING AND DNA SEQUENCING

Various regions of the *thr* operon have been cloned into the multicopy plasmid pBR322 (Cossart et al., 1979; Kozlov et al., 1980; Lynn et al., 1982). Cossart et al. (1979) cloned a HindIII-EcoRI fragment from a λ *thr* transducing phage. The *thr* operon expression in the plasmid-bearing strain is not elevated, most likely because transcription is not initiated at a *thr* promoter. Kozlov et al. (1980) cloned the *thr* operon by inserting a SalI-EcoRI fragment into pBR322. Strains containing the cloned *thr* operon exhibit normal regulation by threonine and isoleucine, and homoserine dehydrogenase I activity in the cell extract is elevated 30-fold.

Katinka et al. (1980) have determined the DNA sequence of *thrA* (Chapter 9). Cossart et al. (1981) have determined the sequence of the *thrB* gene and possibly the start of the *thrC* gene (Fig. 10.5). The *thrB* gene consists of 928 base pairs encoding 309 amino acids. The amino acid sequence predicted for homoserine kinase agrees with the protein sequence assuming proteolytic removal of the amino terminal methionine. There appears to be no obvious homology between *thrA* and *thrB*.

An amino acid sequence for the amino terminus of the *thrC* product

```
CGC GGA TAT GGT GCG GGC AAT GAC GTT ACA GCT GCC GGT GTC TTT GCT GAT CTG CTA CGT ACC CTC TCA TGG AAG TTA GGA GTC TGA  C
Arg Gly Tyr Gly Ala Gly Asn Asp Val Thr Ala Ala Gly Val Phe Ala Asp Leu Leu Arg Thr Leu Ser Trp Lys Leu Gly Val

ATG GTT AAA GTT TAT GCC CCG GCT TCC AGT GCC AAT ATG AGC GTC GGG TTT GAT GTG CTC GGG GCG GCG GTG ACA CCT GTT GAT GGT GCA 90
Met Val Lys Val Tyr Ala Pro Ala Ser Ser Ala Asn Met Ser Val Gly Phe Asp Val Leu Gly Ala Ala Val Thr Pro Val Asp Gly Ala 30

TTG CTC GGA GAT GTA GTC ACG GTT GAG GCG GCA GAG ACA TTC AGT CTC AAC AAC CTC GGA CGC TTT GCC GAT AAG CTG CCG TCA GAA CCA 180
Leu Leu Gly Asp Val Val Thr Val Glu Ala Ala Glu Thr Phe Ser Leu Asn Asn Leu Gly Arg Phe Ala Asp Lys Leu Pro Ser Glu Pro 60

CGG GAA AAT ATC GTT TAT CAG TGC TGG GAG CGT TTT TGC CAG GAA CTG GGT AAG CAA ATT CCA GTG GCG ATG ACC CTG GAA AAG AAT ATG 270
Arg Glu Asn Ile Val Tyr Gln Cys Trp Glu Arg Phe Cys Gln Glu Leu Gly Lys Gln Ile Pro Val Ala Met Thr Leu Glu Lys Asn Met 90

CCG ATC GGT TCG GGC TTA GGC TCC AGT GCC TGT TCG GTG GTC GCG GCG CTG ATG GCG ATG AAT GAA CAC TGC GGC AAG CCG CTT AAT GAC 360
Pro Ile Gly Ser Gly Leu Gly Ser Ser Ala Cys Ser Val Val Ala Ala Leu Met Ala Met Asn Glu His Cys Gly Lys Pro Leu Asn Asp 120

ACT CGT TTG CTG GCT TTG ATG GGC GAG CTG GAA GGC CGT ATC TCC GGC AGC ATT CAT TAC GAC AAC GTG GCA CCG TGT TTT CTC GGT GGT 450
Thr Arg Leu Leu Ala Leu Met Gly Glu Leu Glu Gly Arg Ile Ser Gly Ser Ile His Tyr Asp Asn Val Ala Pro Cys Phe Leu Gly Gly 150

ATG CAG TTG ATG ATC GAA GAA AAC GAC ATC ATC AGC CAG CAA GTG CAG GGT TTG ATG AGT GGC TGT GGG TGC TGG CGT ATC CGG GGA TTA 540
Met Gln Leu Met Ile Glu Glu Asn Asp Ile Ile Ser Gln Gln Val Gln Gly Leu Met Ser Gly Cys Gly Cys Trp Arg Ile Arg Gly Leu 180

AAG TCT CGA CGG CAG AAG CAG GGC TAT TTA CCG GCG CAG TAT CGC CGC CAG GAT TGC ATT GCG CAC GGG CGA CAT CTG GCA GGC TTC ATT 630
Lys Ser Arg Arg Gln Lys Gln Gly Tyr Leu Pro Ala Gln Tyr Arg Arg Gln Asp Cys Ile Ala His Gly Arg His Leu Ala Gly Phe Ile 210

CAC GCC TGC TAT TCC CGT CAG CCT GAG CTT GCC GCG AAG CTG ATG AAA GAT GTT ATC GCT GAA CCC TAC CGT GAA CGG TTA CTG CCA GGC 720
His Ala Cys Tyr Ser Arg Gln Pro Glu Leu Ala Ala Lys Leu Met Lys Asp Val Ile Ala Glu Pro Tyr Arg Glu Arg Leu Leu Pro Gly 240

TTC CGG CAG GCG CGG CAG GCG GTC GCG GAA ATC GGC GCG GTA GCG AGC GGT ATC TCC GGC TCC GGC CCG ACC TTG TTC GCT CTG TGT GAC 810
Phe Arg Gln Ala Arg Gln Ala Val Ala Glu Ile Gly Ala Val Ala Ser Gly Ile Ser Gly Ser Gly Pro Thr Leu Phe Ala Leu Cys Asp 270

AAG CCG GAA ACC GCC CAG CGC GTT GCC GAC TGG TTG GGT AAG AAC TAC CTG CAA AAT CAG GAA GGT TTT GTT CAT ATT TGC CGG CTG GAT 900
Lys Pro Glu Thr Ala Gln Arg Val Ala Asp Trp Leu Gly Lys Asn Tyr Leu Gln Asn Gln Glu Gly Phe Val His Ile Cys Arg Leu Asp 300

ACG GCG GGC GCA CGA GTA CTG GAA AAC TAA ATG AAA CTC TAC AAC TTG AAA GAT CAC AAC GAG CAG GTC AGC TTT GCG CAA GCC GTA ACC 60
Thr Ala Gly Ala Arg Val Leu Glu Asn 309 Met Lys Leu Tyr Asn Leu Lys Asp His Asn Glu Gln Val Ser Phe Ala Gln Ala Val Thr 20

CAG GGG TTG GGC AAA AAT CAG GGG CTG TTT TTT CCG CAC GAC CTG CCG GAA TTC AGC CTG 114
Gln Gly Leu Gly Lys Asn Gln Gly Leu Phe Phe Pro His Asp Leu Pro Gln Phe Ser Leu 38
```

has been deduced (Cossart et al., 1981). An initiator AUG codon, an open reading frame of 114 nucleotides to the end of the cloned insert, and a putative ribosome-binding site have been identified. Further DNA sequencing and protein sequencing studies are required to identify the exact location of the *thrC* gene.

The intracistronic region between *thrA* and *thrB,* TGACATG, contains only one base pair (G–C), while the putative boundary between *thrB* and *thrC, TAAATG,* has no gap or overlap.

V. CONCLUSION

Regulation of threonine biosynthesis is at the level of enzymatic activity and at the level of enzyme synthesis. The aspartate kinase I–homoserine dehydrogenase I has been studied extensively and is well characterized. The other enzymes, especially threonine synthase, have not been studied in great detail. DNA sequence analysis of the entire *thr* operon is nearly complete, and comparisons of the aspartate kinases and homoserine dehydrogenases should provide insights into the evolution of these enzymes.

The regulation of gene expression appears to be primarily at the level of transcription termination via attenuation. This type of regulation allows the cell to gauge the levels of threonine and isoleucine and thereby control the expression of the *thr* operon. One remarkable feature of this system of regulation is that control is mediated by the intracellular levels of two amino acids. Although the basic attenuation model is highly plausible, many of its aspects, including the roles of other possible regulatory proteins, remain to be tested at the molecular level.

ACKNOWLEDGMENT

Work in the author's laboratory was supported by Grant PCM 78-07931 from the National Science Foundation.

REFERENCES

Adhya, S., and Gottesman, M. (1978) *Ann. Rev. Biochem. 47,* 967–996.
Bachmann, B. J., and Low, B. K. (1980) *Microbiol. Rev. 44,* 1–56.

◀ **FIGURE 10.5 The DNA sequence of *thrB* and its flanking regions and the deduced amino acid sequences of the carboxy terminus of homoserine dehydrogenase I, of the homoserine kinase, and of the amino terminus of threonine synthase (from Cossart et al., 1981 as corrected by P. Cossart and I. Saint-Girons, personal communication). The amino acids and the corresponding bases are numbered starting with the amino terminal methionine of homoserine kinase as 1.**

Barnes, W. M. (1978) *Proc. Nat. Acad. Sci. U.S.A. 75,* 4281–4285.
Blatt, J. M., and Umbarger, H. E. (1972) *Biochem. Genet. 6,* 99–118.
Boy, E., and Patte, J. C. (1972) *J. Bacteriol. 112,* 84–92.
Burr, B., Walker, J., Truffa-Bachi, P., and Cohen, G. N. (1976) *Eur. J. Biochem. 62,* 519–526.
Chapman, J., and Gardner, J. F. (1981) *J. Bacteriol. 146,* 1046–1054.
Cohen, G. N., and Dautry-Varsat, A. (1980) in *Multifunctional Proteins* (Bisswanger, H., and Schmincke-Ott, E., eds.) pp. 49–121, John Wiley, New York.
Cohen, G. N., and Patte, J. C. (1963) *Cold Spring Harbor Symp. Quant. Biol. 28,* 513–516.
Cossart, P., Katinka, M., Yaniv, M., Saint-Girons, I., and Cohen, G. N. (1979) *Mol. Gen. Genet. 175,* 39–44.
Cossart, P., Katinka, M., and Yaniv, M. (1981) *Nucl. Acids Res. 9,* 339–347.
DiNocera, P. P., Blasi, F., DiLauro, R., Frunzio, R., and Bruni, C. B. (1978) *Proc. Nat. Acad. Sci. U.S.A. 75,* 4276–4280.
Farnham, P. J., and Platt, T. (1980) *Cell 20,* 739–748.
Farnham, P. J., and Platt, T. (1981) *Nucl. Acids Res. 9,* 563–577.
Flavin, M., and Slaughter, C. (1960) *J. Biol. Chem. 235,* 1103–1108.
Freundlich, M. (1963) *Biochem. Biophys. Res. Comm. 10,* 277–282.
Fröhler, J., Rechenmacher, A., Thomale, J., Nass, G., and Böck, A. (1980) *J. Bacteriol. 143,* 1135–1141.
Fuller, R. S., and Platt, T. (1978) *Nucl. Acids Res. 5,* 4613–4623.
Gardner, J. F. (1979) *Proc. Nat. Acad. Sci. U.S.A. 76,* 1706–1710.
Gardner, J. F. (1982) *J. Biol. Chem. 257,* 3896–3904.
Gardner, J. F., and Reznikoff, W. S. (1978) *J. Mol. Biol. 126,* 241–258.
Gardner, J. F., and Smith, O. H. (1975) *J. Bacteriol. 124,* 161–166.
Gardner, J. F., Smith, O. H., Fredricks, W. W., and McKinney, M. A. (1974) *J. Mol. Biol. 90,* 613–631.
Gemmill, R. M., Wessler, S. R., Keller, E. B., and Calvo, J. M. (1979) *Proc. Nat. Acad. Sci. U.S.A. 76,* 4941–4945.
Hütter, R., Poralla, K., Zachau, H. G., and Zähner, H. (1966) *Biochem. Z. 344,* 190–196.
Johnson, E. J., Cohen, G. N., and Saint-Girons, I. (1977) *J. Bacteriol. 129,* 66–70.
Johnston, H. M., and Roth, J. R. (1981) *J. Mol. Biol. 145,* 735–756.
Johnston, H. M., Barnes, W. M., Chumley, F. G., Bossi, L., and Roth, J. R. (1980) *Proc. Nat. Acad. Sci. U.S.A. 77,* 508–512.
Kasai, T. (1974) *Nature* (London) *249,* 523–527.
Kassavetis, G. A., and Chamberlin, M. J. (1981) *J. Biol. Chem. 256,* 2777–2786.
Katinka, M., Cossart, P., Sibilli, L., Saint-Girons, I., Chalvignac, M. A., Le Bras, G., Cohen, G. N., and Yaniv, M. (1980) *Proc. Nat. Acad. Sci. U.S.A. 77,* 5730–5733.
Keller, E. B., and Calvo, J. M. (1979) *Proc. Nat. Acad. Sci. U.S.A. 76,* 6186–6190.
Kozlov, Y. I., Kochetova, L. P., Livshits, V. A., Mashko, S. V., Moshenteva, V. N., Rebentish, B. A., Rozinov, M. N., Khurges, E. M., Yankovskii, N. K., and Debabov, V. G. (1980) *Genetika 16,* 66–77.
Lawther, R. P., and Hatfield, G. W. (1980) *Proc. Nat. Acad. Sci. U.S.A. 77,* 1862–1866.

Lee, F., and Yanofsky, C. (1977) *Proc. Nat. Acad. Sci. U.S.A. 74,* 4365–4369.

Lynn, S. P., Gardner, J. F., and Reznikoff, W. S. (1982) *J. Bacteriol. 152,* 363–371.

Nargang, F. E., Subrahmanyam, C. S., and Umbarger, H. E. (1980) *Proc. Nat. Acad. Sci. U.S.A. 77,* 1823–1827.

Nass, G., and Thomale, J. (1974) *FEBS Lett. 39,* 182–186.

Nass, G., Poralla, K., and Zähner, H. (1969) *Biochem. Biophys. Res. Comm. 34,* 84–91.

Oxender, D. L., Zurawski, G., and Yanofsky, C. (1979) *Proc. Nat. Acad. Sci. U.S.A. 76,* 5524–5528.

Paetz, W., and Nass, G. (1973) *Eur. J. Biochem. 35,* 331–337.

Parsot, C., Saint-Girons, I., and Cossart, P. (1983) *Mol. Gen. Genetics 188,* 455–458.

Patte, J. C., Le Bras, G., Loviny, T., and Cohen, G. N. (1963) *Biochim. Biophys. Acta 67,* 16–30.

Patte, J. C., Le Bras, G., and Cohen, G. N. (1967) *Biochim. Biophys. Acta 136,* 245–257.

Rosenberg, M., and Court, D. (1979) *Ann. Rev. Genet. 13,* 319–353.

Saint-Girons, I. (1978) *Mol. Gen. Genet. 162,* 95–100.

Saint-Girons, I., and Margarita, D. (1975) *J. Bacteriol. 124,* 1137–1141.

Saint-Girons, I., and Margarita, D. (1978) *Mol. Gen. Genet. 162,* 101–107.

Skarstedt, M. T., and Greer, S. B. (1973) *J. Biol. Chem. 248,* 1032–1044.

Stadtman, E. R., Cohen, G. N., Le Bras, G., and de Robichon-Szulmajster, H. (1961) *J. Biol. Chem. 236,* 2033–2038.

Stauffer, G. V., Zurawski, G., and Yanofsky, C. (1978) *Proc. Nat. Acad. Sci. U.S.A. 75,* 4833–4837.

Stuttard, C. (1973) *J. Bacteriol. 116,* 1–11.

Szczesiul, M., and Wampler, D. E. (1976) *Biochemistry 15,* 2236–2244.

Theze, J., and Saint-Girons, I. (1974) *J. Bacteriol. 118,* 990–998.

Theze, J., Kleidman, L., and Saint-Girons, I. (1974a) *J. Bacteriol. 118,* 577–581.

Theze, J., Margarita, D., Cohen, G. N., Borne, F., and Patte, J. C. (1974b) *J. Bacteriol. 117,* 133–143.

Truffa-Bachi, P., Veron, M., and Cohen, G. N. (1974) in *CRC Critical Reviews in Biochemistry* Vol. II (Fasman, G. R., ed.) pp. 379–415, Chemical Rubber Company Press, Cleveland, Ohio.

Umbarger, H. E. (1978) *Ann. Rev. Biochem. 47,* 533–606.

Veron, M., and Cohen, G. N. (1974) in *Metabolic Interconversions of Enzymes* (Fischer, E. H., Krebs, E. G., Neurath, H., and Stadtman, E. R., eds.) pp. 335–347, Springer, New York.

Winkler, M. E., and Yanofsky, C. (1981) *Biochemistry 20,* 3738–3744.

Wormser, E. H., and Pardee, A. B. (1958) *Arch. Biochem. Biophys. 78,* 416–432.

Yanofsky, C. (1981) *Nature* (London) *289,* 751–758.

Zurawski, G., Brown, K., Killingly, D., and Yanofsky, C. (1978) *Proc. Nat. Acad. Sci. U.S.A. 75,* 4271–4275.

CHAPTER

11

Methionine Biosynthesis and Its Regulation

Robin J. Rowbury

I. INTRODUCTION

Methionine, which is essential to man and other mammals, is synthesized de novo by most bacteria and fungi and by all plants. Most of the dietary methionine ultimately derives from the latter source. Although no one amino acid can be considered more important than any other, methionine fulfills other functions apart from its role as a protein constituent. Methionine participates in the initiation of mRNA translation as the universal N-terminal amino acid. The methionine derivative S-adenosylmethionine (SAM) not only is an important cellular methylating agent, but is also a precursor of the polyamine spermidine.

The pathway of methionine biosynthesis has many common features in most organisms, even though some species utilize more than one synthetic route. This chapter focuses upon methionine biosynthesis in bacteria and fungi with synthesis in plants to be briefly considered.

II. THE PATHWAYS OF METHIONINE BIOSYNTHESIS

A. Bacteria

1. *Escherichia coli* and *Salmonella typhimurium*. In *E. coli* and *S. typhimurium*, the carbon skeleton of methionine is provided by aspartate, the sulfur atom by cysteine, and the methyl group by serine. The intermediates are homoserine, O-succinylhomoserine, cystathionine, and homocysteine (Figs. 11.1 and 11.2).

Although cystathionine was a presumed intermediate in methionine biosynthesis, its mode of formation from homoserine remained unclear until the involvement of succinate was established. Methionine can partially replace the succinate requirement of certain *E. coli* mutants (Davis et al., 1959). Extracts of *E. coli metC* mutants form cystathionine upon incubation with homoserine, cysteine, succinate, and an energy source. Extracts of *metB* mutants convert homoserine and succinyl CoA to a succinylated derivative of homoserine. With cysteine, this derivative is converted to cystathionine by an enzyme in *metA* extracts (Rowbury, 1961, 1962; Rowbury and Woods, 1964b). The homoserine derivative is O-succinylhomoserine (Flavin et al., 1964; Rowbury and Woods, 1964b). Mutants (*metA*) blocked in the homoserine succinyltransferase reaction respond to cystathionine but not to O-succinylhomoserine.

Cystathionine is synthesized from O-succinylhomoserine and cysteine by cystathionine γ-synthase, which is absent from *metB* mutants (Flavin et al., 1964; Rowbury and Woods, 1964b). The enzyme from *S. typhimurium* has been purified and crystallized; it consists of four identical subunits, each carrying a tightly bound molecule of pyridoxal phosphate (Kaplan and Flavin, 1966). Cystathionine formation is undoubtedly essential for methionine biosynthesis, although the γ-synthase can catalyze the reaction of H_2S or methylsulfide with O-succinylhomoserine to give homocysteine or methionine directly (Kaplan and Flavin, 1966; Flavin and Slaughter, 1967). However, the obligate requirement for cystathionine β-lyase and a homocysteine methylating enzyme establishes that the formation of cystathionine and homocysteine cannot be bypassed in methionine biosynthesis. The conversion of cystathionine to homocysteine, pyruvate, and ammonia is catalyzed by cystathionine β-lyase, a pyridoxal phosphate–dependent enzyme encoded by *metC* (Wijesundera and Woods, 1964; Rowbury and Woods, 1964a).

The methylation of homocysteine is catalyzed by either of two enzymes in *E. coli*. For both, 5-methyltetrahydrofolate (or a polyglutamate derivative) is the methyl donor. This intermediate is formed from serine via 5,10-methylenetetrahydrofolate in two steps catalyzed by serinehydroxymethyltransferase and methylenetetrahydrofolate reductase (Fig. 11.2). The reductase, encoded by *metF*, is specific to methionine synthesis in *E. coli*. $FADH_2$ supplies the reducing power for this reaction; the enzyme can use either methylene tetrahydrofolate itself or a polyglutamate form as a substrate (Smith, 1971; Salem and Foster, 1972).

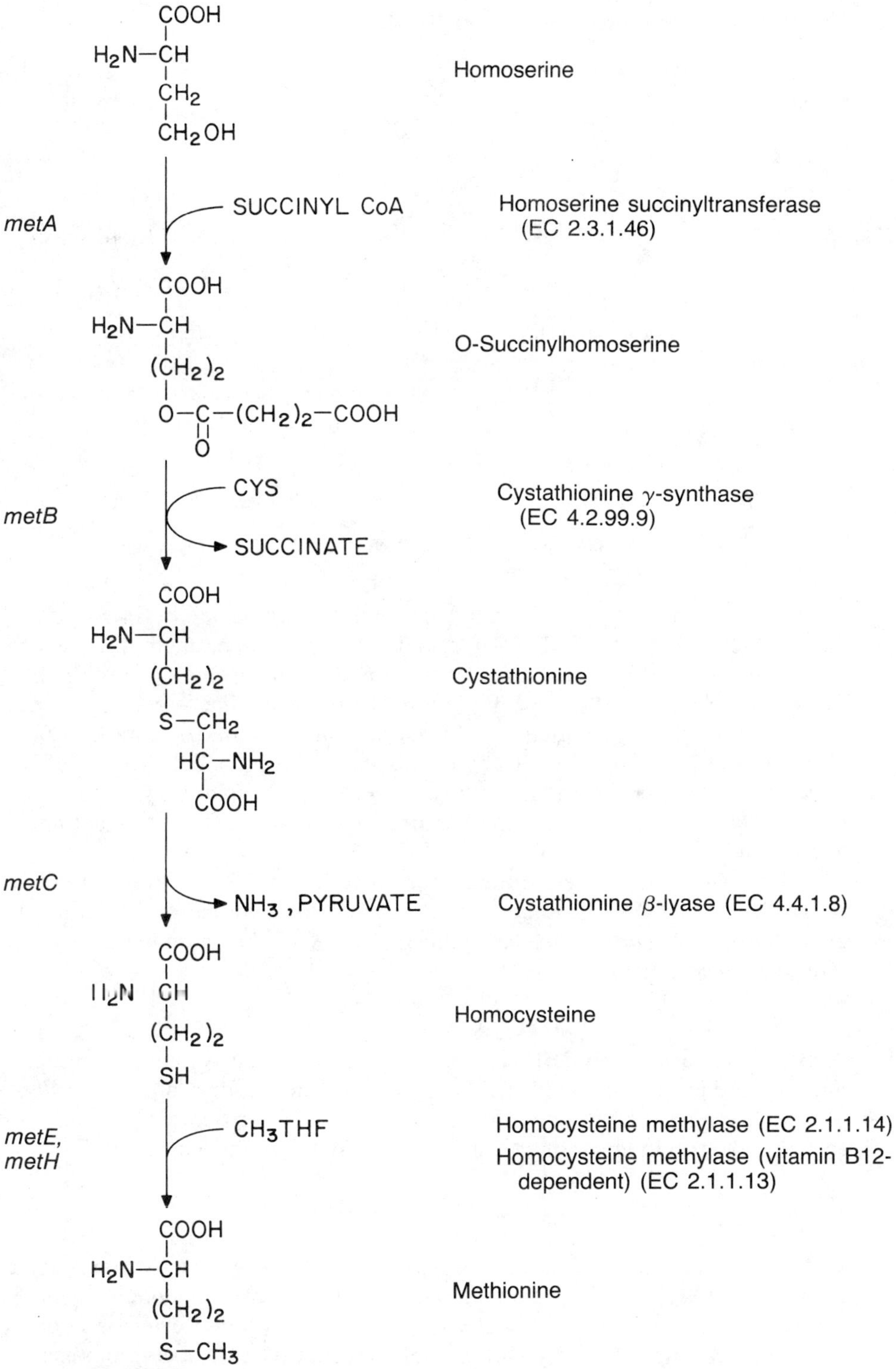

FIGURE 11.1 **The pathway of methionine biosynthesis in *E. coli* and *S. typhimurium*.**

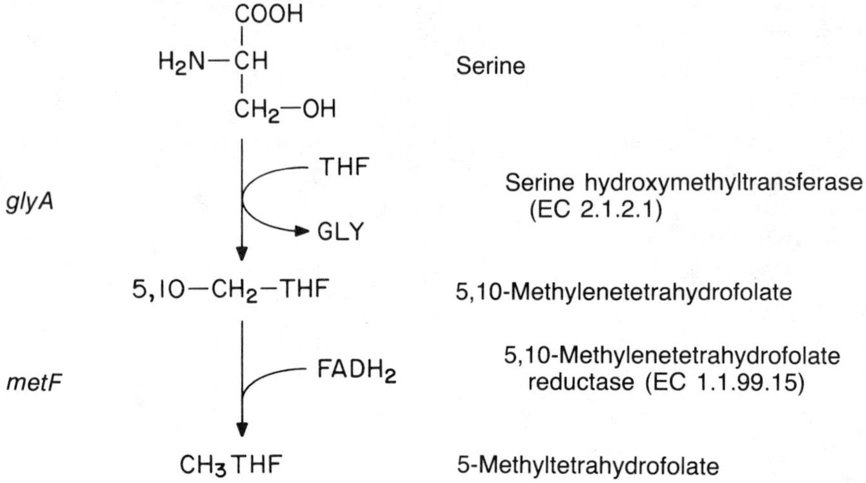

FIGURE 11.2 Generation of 5-methyltetrahydrofolate from serine.

One enzyme catalyzing methyl transfer to homocysteine in *E. coli* is the *metE*-encoded vitamin B12–independent homocysteine methylase, which utilizes only 5-methyl derivatives of the polyglutamate forms of tetrahydrofolate (Jones et al., 1961). Methylation of homocysteine is also catalyzed by a vitamin B12–dependent methylase encoded by *metH* (Smith, 1971). The enzyme uses 5-methyltetrahydrofolate and requires catalytic levels of SAM and a source of reducing power. Methyl cobalamin appears to be involved as an intermediate in the methyl transfer, with SAM acting in essentially a "priming" role.

Most bacteria, unlike the fungi, appear to be unable to catalyze reverse transsulfuration, that is, they cannot convert methionine to cysteine via cystathionine. *Pseudomonas aeruginosa* may be an exception in this respect (Gunther et al., 1979).

2. Other Bacteria. While the enterics utilize O-succinylhomoserine as an intermediate in methionine biosynthesis, other bacteria, for example, *Bacillus, Corynebacteria, Brevibacterium,* and *Arthrobacter,* resemble the fungi in forming O-acetylhomoserine (Flavin, 1975; Wyman and Paulus, 1975).

B. Fungi

In *Neurospora crassa,* O-acetylhomoserine replaces O-succinylhomoserine (Nagai and Flavin, 1967). Homoserine acetyltransferase, encoded by *met5,* catalyzes the synthesis of O-acetylhomoserine from homoserine and acetyl coenzyme A (Fig. 11.3). Cystathionine is formed from O-acetylhomoserine and cysteine in a reaction catalyzed by cystathionine γ-synthase (Kerr and

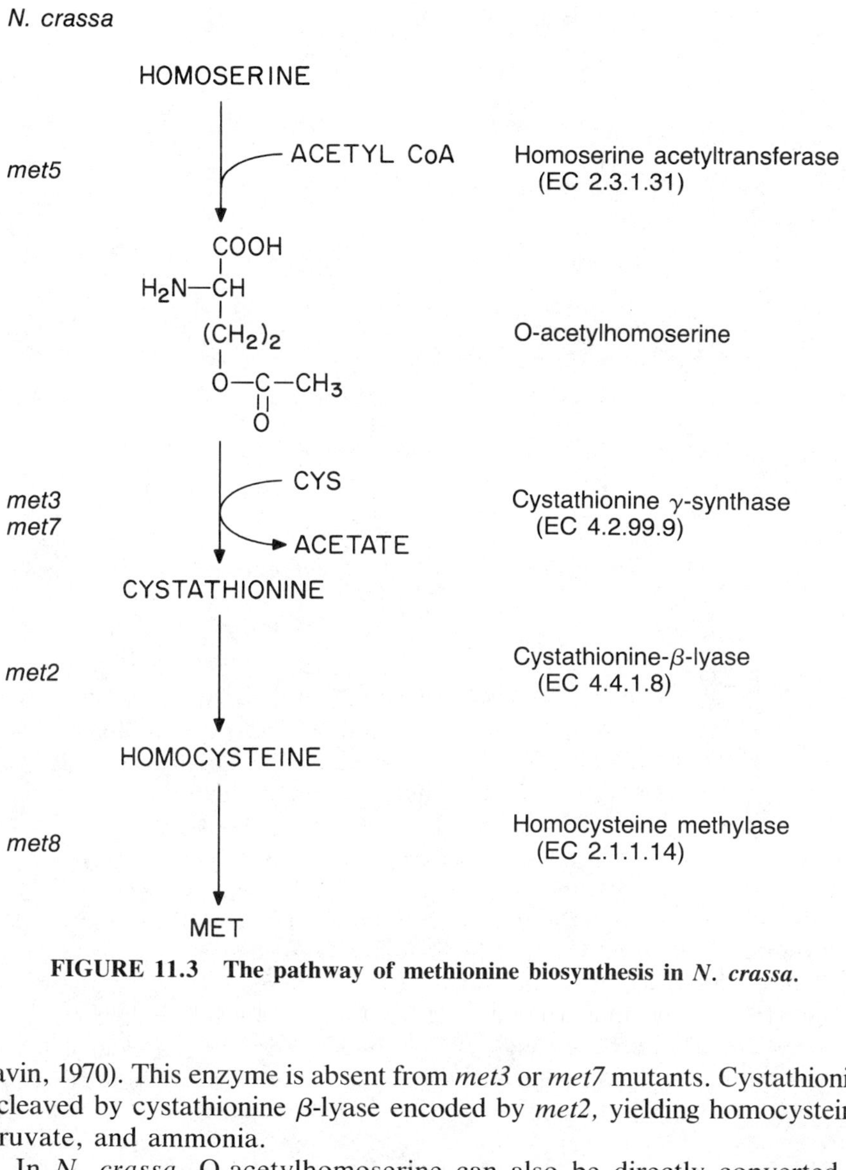

FIGURE 11.3 The pathway of methionine biosynthesis in *N. crassa*.

Flavin, 1970). This enzyme is absent from *met3* or *met7* mutants. Cystathionine is cleaved by cystathionine β-lyase encoded by *met2,* yielding homocysteine, pyruvate, and ammonia.

In *N. crassa,* O-acetylhomoserine can also be directly converted to homocysteine by reaction with H_2S; however, the obligate role of cystathionine γ-synthase and cystathionine β-lyase suggests that this conversion is not quantitatively important (Kerr and Flavin, 1970). *Saccharomyces cerevisiae* also synthesizes methionine from O-acetylhomoserine via a direct reaction with H_2S to give homocysteine. In contrast to *N. crassa,* this step, catalyzed by homocysteine synthase (Fig. 11.4), lies on the major synthetic route, although *S. cerevisiae* also contains a cystathionine γ-synthase (Savin and Flavin, 1972).

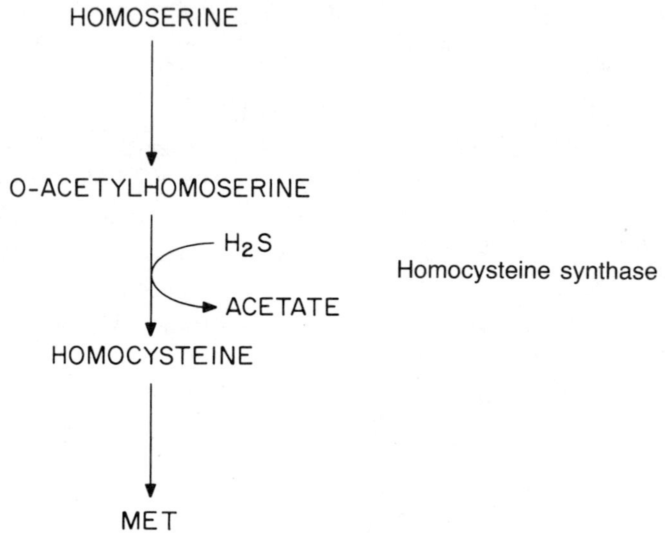

FIGURE 11.4 The pathway of methionine biosynthesis in *S. cerevisiae*.

In *N. crassa,* conversion of homocysteine to methionine involves methylenetetrahydrofolate reductase and a vitamin B12–independent homocysteine methylase, encoded by *met8* (Burton et al., 1969), which uses only polyglutamate forms of the C1 carrier. There is some evidence in *Candida* for vitamin B12–dependent methionine synthesis (Poston and Hemmings, 1979).

Aspergillus nidulans appears to be intermediate between *N. crassa* and *S. cerevisiae* in methionine biosynthesis. The major pathway is via O-acetylhomoserine, cystathionine, and homocysteine, and lesions in either cystathionine γ-synthase or cystathionine β-lyase cause a requirement for methionine. Such lesions can, however, be suppressed by mutations that derepress homocysteine synthase, generating strains in which methionine synthesis does not proceed via cystathionine. Direct conversion of O-acetylhomoserine to homocysteine also occurs during growth in limiting cysteine, a condition that derepresses homocysteine synthase (Paszewski and Grabski, 1975).

The fungi differ from most bacteria in that reverse transsulfuration can occur. This opens a pathway for the synthesis of cysteine from methionine (Fig. 11.5).

C. Higher Plants

Early studies using plant tissue culture cells implicated homoserine, cystathionine, and homocysteine as intermediates (Dougall and Fulton, 1967). The most probable pathway for plants is via cystathionine; the direct route to homocysteine constitutes at most a minor alternative (Fig. 11.6). For

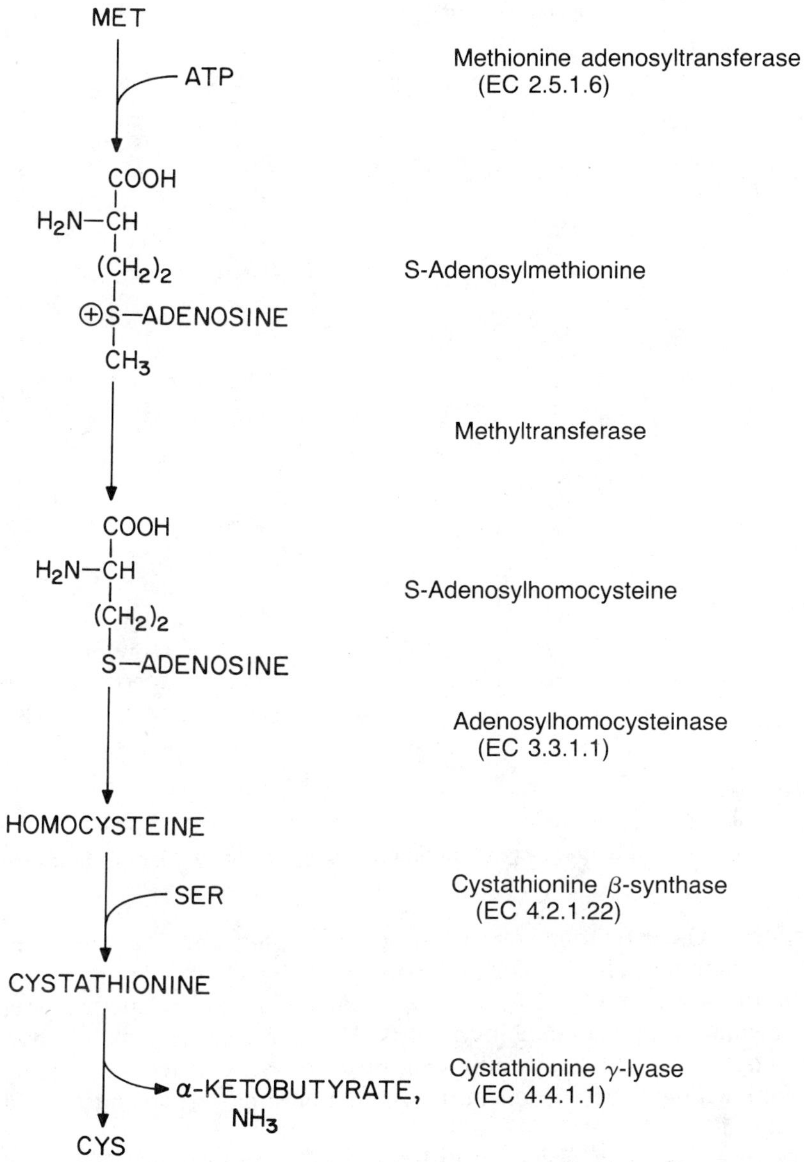

FIGURE 11.5 Synthesis of cysteine from methionine by reverse transsulfuration in fungi. The methyltransferase reaction is catalyzed by a variety of different enzymes.

most plant tissues, the only homoserine ester present is phosphohomoserine, and extracts of all the plants tested can synthesize this ester, then convert it to cystathionine in a reaction dependent upon cysteine (Giovanelli et al., 1974; Datko et al., 1974). Even plants that accumulate other homoserine esters (for example, *Pisum sativum* forms O-acetylhomoserine, and *Lathyrus*

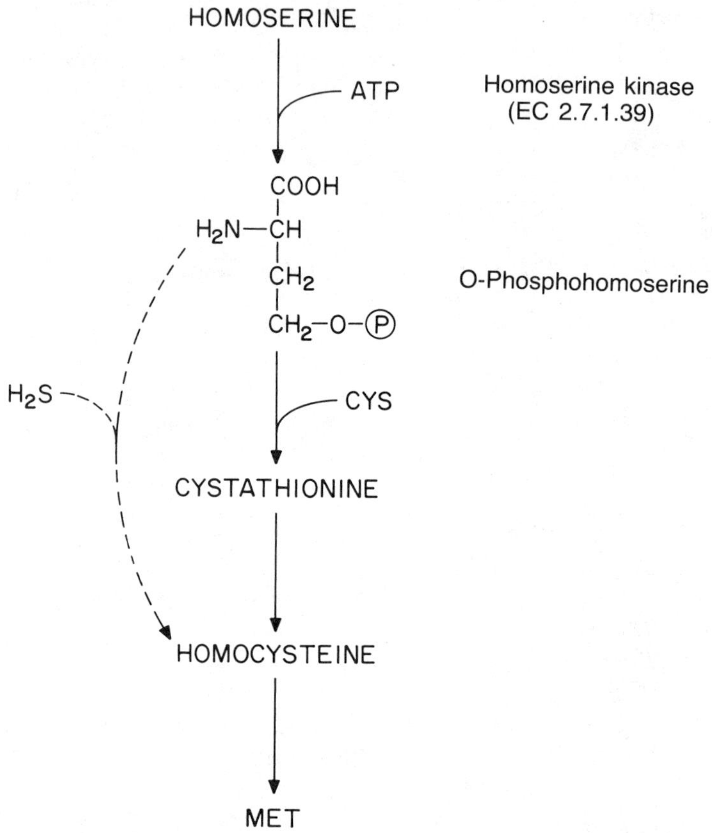

FIGURE 11.6 The pathway of methionine biosynthesis in higher plants.

sativus forms O-oxalylhomoserine) can form the phospho ester and convert it to cystathionine. The low rate of conversion of O-acetylhomoserine to cystathionine by extracts of *P. sativum* makes it unlikely that this ester is an intermediate in methionine biosynthesis. However, the O-oxalyl derivative is efficiently converted to cystathionine by *Lathyrus* extracts; accordingly, cystathionine might arise from both O-phosphohomoserine and O-oxalylhomoserine in *Lathyrus*.

Plant extracts also convert cystathionine to homocysteine. The reaction catalyzed by cystathionine β-lyase produces pyruvate and ammonia. There has been some doubt about the role of cystathionine in plants because most plant extracts can directly convert O-phosphohomoserine and H_2S to homocysteine, bypassing cystathionine γ-synthase and cystathionine β-lyase (Datko et al., 1977). However, cystathionine is an obligate intermediate in corn seedlings (Giovanelli et al., 1972), and labeling studies with ^{35}S-sulphate show that direct conversion of phosphohomoserine to homocysteine does not occur appreciably in *Chlorella* (Giovanelli et al., 1978).

The methylation of homocysteine to methionine in plants is catalyzed by a vitamin B12–independent homocysteine methylase using 5-methyltetrahydropteroyltriglutamate as the methyl donor (Clandinin and Cossins, 1974). Reverse transsulfuration apparently does not occur in plants.

III. CONTROL OF METHIONINE BIOSYNTHESIS AT THE LEVEL OF ENZYME ACTIVITY

A. Bacteria

In *E. coli* and *S. typhimurium,* the first specific enzyme of methionine biosynthesis is feedback-sensitive. Homoserine succinyltransferase is inhibited only by high levels of methionine; but if the enzyme is assayed in intact cells, the inhibition is seen at much lower levels (Rowbury, 1961; Rowbury and Woods, 1966). Evidently, methionine is converted to SAM in the whole cell system because methionine and SAM together produce synergistic inhibition (Lee et al., 1966). The analog α-methylmethionine mimics methionine as an inhibitor (Schlesinger, 1967; Rowbury, 1968). *S. typhimurium* mutants resistant to this analog, originally termed *metI* mutants (Lawrence et al., 1968), excrete methionine and contain homoserine succinyltransferase, which is resistant to the analogue and to methionine, but not to SAM. The levels of the enzyme are low in such mutants, possibly because the methionine overproduction causes partial repression (Lawrence, 1972). A series of such mutations all fall within *metA* (Chater and Rowbury, 1970); thus the catalytic and methionine regulatory sites are part of the same polypeptide. Homoserine succinyltransferase is also subject to product inhibition by O-succinylhomoserine (Savin et al., 1972).

SAM exerts a second effect on the activity of the bacterial methionine enzymes because it activates the vitamin B12–dependent homocysteine methylase (Rosenthal et al., 1965) of *E. coli.* SAM apparently serves to convert the vitamin B12 enzyme into a methylated derivative.

Synergistic feedback inhibition of the first enzyme of methionine biosynthesis is also found in *Bacilli.* Homoserine acetyltransferase of *Bacillus polymyxa* is inhibited by methionine and SAM; each inhibitor binds to a specific site distinct from the active site (Wyman and Paulus, 1975).

SAM inhibits homoserine dehydrogenase in *Chromatium;* in the presence of threonine, nearly 90% inhibition is observed. Growth inhibition by methionine might reflect threonine starvation attributable to the effect of SAM on the dehydrogenase. However, the growth inhibition is not overcome by homoserine (Sugimoto et al., 1976).

B. Fungi

In *S. cerevisiae,* the first specific enzyme of methionine biosynthesis, homoserine acetyltransferase, is inhibited by SAM alone rather than by methionine plus SAM (Cherest et al., 1969).

In *N. crassa,* the first specific enzyme of methionine biosynthesis, homoserine acetyltransferase, is not inhibited. The flow of intermediates through the methionine pathway is controlled at the second step catalyzed by cystathionine γ-synthase. This structurally complex enzyme is encoded by the *met3* and *met7* genes. The enzyme is unstable and prone to loss of activity on dilution (Kerr and Flavin, 1969, 1970). Extracts of *met3* or *met7* strains lack cystathionine γ-synthase activity, which is demonstrable on mixing extracts from such mutants. Extracts from wild-type cells grown in the presence of methionine contain virtually no synthase activity. This is not due to enzyme repression, but to the presence of inhibitory levels of SAM in the extracts (Kerr and Flavin, 1969, 1970). SAM also inhibits methylenetetrahydrofolate reductase of *N. crassa* (Burton and Metzenberg, 1975). Thus two *N. crassa* methionine biosynthetic enzymes are subject to feedback control by SAM. Strains altered in the synthesis of SAM overproduce methionine (Metzenberg, 1968), and extracts of such cells grown in the presence of methionine contain high levels of cystathionine γ-synthase (Kerr and Flavin, 1969).

Inhibition of the *N. crassa* γ-synthase by SAM is not immediate. Preincubation of SAM and enzyme is needed for the full effect (Kerr and Flavin, 1969, 1970). There is evidently a slow and readily reversible conformational change in the enzyme elicited by SAM. The *N. crassa* enzyme requires polyglutamate forms of reduced folates for activity. These not only activate the enzyme by a slow-acting process, but also reverse the inhibitory effect of SAM. The activator is apparently the 5-methyl derivative (Selhub et al., 1971). Because of this requirement, mutants unable to produce such folate derivatives, for example, *met1* and *met6,* behave as if they were synthase mutants.

C. Higher Plants

In plants, control mechanisms that affect enzyme activity are important, since enzyme repression does not appear to play an appreciable role. Two specific effects on methionine biosynthesis in plants have been documented. Methionine inhibits the aspartate kinase of barley, provided that lysine is also present. Inhibition occurs with lysine alone but not with methionine alone (Bright et al., 1979). SAM inhibits the homoserine kinase of peas, although only at high concentrations (Thoen et al., 1978a).

Whereas methionine and threonine originate in many organisms from distinct homoserine derivatives, O-phosphohomoserine is the common precursor in higher plants. Perhaps for this reason, plants can switch synthesis from methionine to threonine and vice versa. In several plants, these controls operate on threonine synthase, which is activated by SAM and inhibited by cysteine (Madison and Thompson, 1976; Thoen et al., 1978b). When methionine and hence SAM levels are high, O-phosphohomoserine is more actively converted to threonine than to cystathionine. High levels of cysteine, however, shift synthesis back to cystathionine and hence to methionine.

Low levels of methionine and lysine occur in most plant proteins. The levels of free amino acids in plant cells are elevated in mutants selected for resistance to threonine and lysine (Bright et al., 1979), but there is no evidence for any changes in protein amino acid levels in these mutants.

IV. REGULATION OF METHIONINE BIOSYNTHESIS AT THE LEVEL OF ENZYME SYNTHESIS

A. Bacteria

1. Repression by Methionine. Growth of *E. coli* strains with methionine greatly reduces their ability to convert homocysteine to methionine (Wijesundera and Woods, 1953, 1960). The other enzymes of the biosynthetic pathway are also subject to repression, although the control is not coordinate (Rowbury and Woods, 1966). This is not too surprising, since the structural genes encoding methionine biosynthetic enzymes in both *E. coli* and *S. typhimurium* are scattered on the chromosome (Fig. 11.7). Even *metB* and *metF,* which are closely linked, are not in the same operon in *E. coli* (Ahmed, 1973).

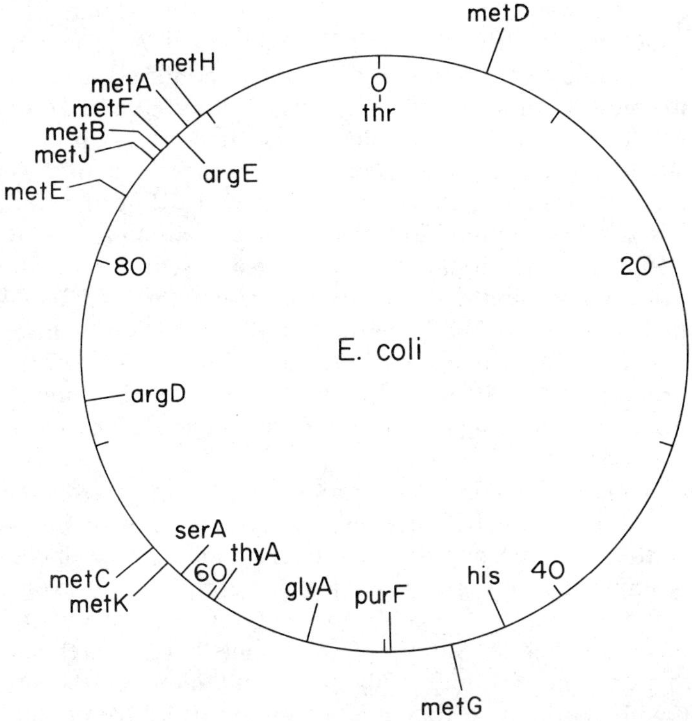

FIGURE 11.7 The positions of the *met* genes on the *E. coli* chromosome.

In addition to the enzymes of Fig. 11.1, some enzymes of homoserine synthesis and methionine adenosyltransferase are also controlled by methionine (Rowbury et al., 1968; Hobson and Smith, 1973). Furthermore, methionine plays a role in the control of serine hydroxymethyltransferase synthesis (Stauffer et al., 1974; Stauffer and Brenchley, 1977) (Chapter 6), and the high-affinity methionine permease appears to be repressed by methionine (Ayling et al., 1979).

2. Analog-resistant Mutants. In the early work on enzyme repression, exogenous methionine was added as an effector. To investigate the regulatory process further and to elucidate the nature of the corepressor, mutants resistant to α-methylmethionine, ethionine, and norleucine have been selected. Such mutants from *S. typhimurium* fall into three classes. Those mutants resistant only to α-methylmethionine are *metA* mutants altered in feedback control of homoserine succinyltransferase. The other two classes, termed *metJ* and *metK* (Lawrence et al., 1968), map at unit 88 and unit 64, respectively, on the standard *S. typhimurium* chromosome.

The *metJ* mutants, selected on the basis of resistance to ethionine, overproduce methionine, are derepressed for the methionine biosynthetic enzymes, and are poorly repressed by exogenous methionine. In *metJ* mutants, methionine adenosyltransferase is derepressed (Hobson and Smith, 1973), and serine hydroxymethyltransferase is not affected by the usual repressing metabolites (Stauffer and Brenchley, 1977). The derepression ratios for the various enzymes in *metJ* mutants confirm that the control of the *met* enzymes is noncoordinate (Lawrence et al., 1968). To explore the possibility that *metJ* mutants are altered in the conversion of methionine to a derivative that is the active corepressor, these mutants were tested for the ability to convert methionine to methionyl-tRNA and to SAM. The levels of $tRNA^{Met}$ and methionyl-tRNA synthetase are unaltered (Gross and Rowbury, 1971), and methionine adenosyltransferase is derepressed in *metJ* mutants (Hobson and Smith, 1973). Also, *metJ* mutants of *E. coli* are found to have no lesions in $tRNA^{Met}$ or methionyl-tRNA synthetase (Ahmed, 1973). The *metJ* product is a protein (Minson and Smith, 1972). For both *S. typhimurium* (Chater, 1970) and *E. coli* (Ahmed, 1973), *metJ*$^{+}$ is *trans*-dominant to *metJ,* indicating a diffusible product encoded by *metJ,* possibly a classical repressor.

The *metK* mutants fall into two groups. One group excretes methionine, derepresses *met* biosynthetic enzymes, and does not respond to exogenous methionine; the second group does not excrete methionine, has essentially normal *met* enzyme levels, and is subject to repression by exogenous methionine (Lawrence et al., 1968). Apparently, *metK* encodes methionine adenosyltransferase, and the various *metK* mutants have differentially reduced levels of this enzyme (Chater, 1970). SAM or a metabolite of it, alone or in concert with methionine (but not methionyl-tRNA), interacts with an aporepressor, possibly specified by *metJ*. The activated repressor supposedly acts on each of the *met* genes, whose operators have different affinities for the holorepressor.

In support of this suggestion, analog-resistant *metK* mutants of *E. coli* have methionine adenosyltransferase lesions; some of the mutants show derepression of the *met* biosynthetic enzymes, but others do not (Greene et al., 1970). Several *metK* mutants of *S. typhimurium* produce altered methionine adenosyltransferase with elevated K_m's for methionine (Hobson and Smith, 1973; Stauffer and Brenchley, 1977), while altered stability or temperature sensitivity is observed for the enzyme of some *E. coli metK* mutants (Greene et al., 1973; Hafner et al., 1977). This establishes firmly that *metK* is the structural gene for methionine adenosyltransferase in both *S. typhimurium* and *E. coli*. The few anomalous *metK* mutants of *S. typhimurium* with unaltered methionine adenosyltransferase presumably have alterations in some other regulatory element (Hobson and Smith, 1973).

Apparently, SAM is a corepressor for the *met* biosynthetic enzymes. Slight uncertainty about the role of SAM in repression arises, since several true *metK* mutants of both *S. typhimurium* and *E. coli* with alterations in methionine adenosyltransferase display relatively normal control of the biosynthetic enzymes. However, a detailed study shows that even nonexcreting *metK* mutants have abnormal methionine regulation, indicated by a slight derepression of the methylenetetrahydrofolate reductase (Hobson, 1974) and an increased pool. A slight excretion of methionine is detected, provided that a *metP* lesion (Ayling and Bridgland, 1972) is present. Since all normal *metK* mutants have lesions in methionine adenosyltransferase, which leads to altered methionine control, SAM or a derivative is considered to be the corepressor. For the *E. coli metK* mutants (Ahmed, 1973), the situation may be similar. Analog resistance of all the *metK* mutants of *E. coli* and *S. typhimurium* could stem from increased pools of methionine, and SAM could be the *met* corepressor.

There are, however, reasons for reserving judgment on this point, at least in *S. typhimurium*, which appears to have a pathway for SAM synthesis independent of methionine adenosyltransferase. *S. typhimurium metK* mutants, even those having undetectably low methionine adenosyltransferase levels, appear to be unimpaired in SAM synthesis, growth, and methylation reactions, and they have pool levels of SAM and its product spermidine that are very similar to those of wild-type strains (Hobson, 1976). Methionine adenosyltransferase is inhibited by cycloleucine, but this analog has little effect on the growth of *S. typhimurium metK* mutants. Extracts of *metK* mutants can convert labeled methionine to SAM. The conversion is not due to residual methionine adenosyltransferase activity because it is not inhibited by cycloleucine. Accordingly, *S. typhimurium* has a second system for synthesis of SAM independent of methionine adenosyltransferase (Hobson, 1976). *E. coli* probably has only one pathway for SAM synthesis because some *metK* mutants of *E. coli* have reduced pools of SAM (Greene et al., 1973a) and their growth is inhibited by cycloleucine. This inhibition is fully reversed by methionine. However, experiments to isolate conditional lethal mutants altered in methionine adenosyltransferase have not been successful (Hafner et al., 1977).

The paradox of *S. typhimurium metK* mutants is that, whereas SAM or a derivative is probably the corepressor for methionine biosynthesis, *metK* mutants with normal levels of SAM are derepressed. One possible explanation is that the SAM formed by the methionine adenosyltransferase is compartmentalized; possibly only SAM associated with the enzyme is active in repression, that is, the SAM–methionine adenosyltransferase complex might be the repressor, or the complex might interact with the *metJ* product to give the active repressor.

It seems unlikely that methionyl-tRNA is involved in control of methionine biosynthesis, since *metG* mutants of *S. typhimurium* (Gross and Rowbury, 1969, 1971) and *E. coli* (Ahmed, 1973) with altered methionyl-tRNA synthetase show normal regulation of *met* biosynthetic enzymes. The most likely model for control of *met* gene expression proposes that methionine is converted to SAM, which in association with methionine adenosyltransferase interacts with the diffusible *metJ* product to form the holorepressor. The holorepressor binds to the various *met* gene recognition sites with different affinities.

There is no direct evidence that the *metJ* gene product is an aporepressor. Indirect evidence comes from in vivo and in vitro studies with λd*met* phages carrying *metB, metF,* and *metJ* (Johnson et al., 1977; Krueger et al., 1978). In vivo measurements suggest that *metB* and *metF* are transcribed from *met* promoters because synthesis of the gene products is switched off by the *metJ* product. By using the λd*met* phages, an in vitro system for the synthesis of cystathionine γ-synthase was developed to test the presumptive repressors and repressing metabolites (Krueger et al., 1978). Cystathionine-γ-synthase formation in vitro, as in vivo (Somerville and Ahmed, 1977), is stimulated by guanosine tetraphosphate. However, none of the tested presumptive repressing metabolites—methionine, SAM, or spermidine—affect the in vitro enzyme synthesis whether the S30 fraction comes from wild-type or a *metJ* mutant. A metabolite other than those tested may be involved, or the metabolite may need to be in association with a specific protein, for example methionine adenosyltransferase, in addition to the *metJ* product. Since the rate of cystathionine γ-synthase formation is higher if the S30 fraction is from wild-type, the *metJ* product may be an activator, that is, converted to a repressor by the appropriate metabolites (Krueger et al., 1978). Guanosine tetraphosphate stimulates transcription of *met* structural genes in vivo (Somerville and Ahmed, 1977) and in vitro (Krueger et al., 1978). Introduction of *rel* lesions into partial revertants of *metG* mutants results in methionine auxotrophy, because *rel* mutants have lower levels of *met* biosynthetic enzymes.

The *metJ/metK* system of enteric bacteria does more than govern the levels of the enzymes that form methionine from homoserine and cysteine. The methionine-specific aspartate kinase–homoserine dehydrogenase of *S. typhimurium* is also derepressed in both *metJ* and *metK* mutants (Rowbury et al., 1968), and methionine adenosyltransferase is derepressed in *metJ* mutants (Hobson and Smith, 1973). The *metJ/metK* system may also control the high-affinity methionine permease, the *metP* product, because a *metBmetJ* strain has a higher specific transport activity than a *metB* strain grown with methionine (Ayling et al., 1979). The synthesis of serine hydroxymethyltransferase is controlled in part by the *metJ/metK* system. Serine hydroxymethyltransferase is not repressed by any single metabolite, but by a group of six metabolites including methionine (Stauffer et al., 1974) (Chapter 6). Methionine limitation, achieved by growing methionine auxotrophs with D-methionine, causes derepression of the enzyme (Greene and Radovich, 1975). Mutants selected for derepression of serine hydroxymethyltransferase are methionine analog–resistant and have lesions in *metK,* indicating that methionine acts via the *metJ/metK* system (Stauffer and Brenchley, 1977). None of the six repressing metabolites affect the synthesis of serine hydroxymethyltransferase in *metJ* or *metK* mutants.

There must be other specific regulatory effects of methionine or its metabolites that operate independently of the *metJ/metK* system to explain the behavior of certain biosynthetic enzymes in *metJ* mutants. For example, methionine induces the synthesis of the vitamin B12–independent homocysteine methylase in *metJ* mutants of *E. coli* (Green et al., 1973b).

3. Repression by Vitamin B12. Enteric bacteria have two pathways for homocysteine methylation (Fig. 11.1). One pathway uses vitamin B12. Organisms generally do not synthesize their own vitamin B12 but are dependent on exogenous supplies. When exogenous vitamin B12 is present, only the vitamin B12–dependent methylase operates. Vitamin B12 represses the synthesis of the vitamin B12–independent methylase and partially represses the methylenetetrahydrofolate reductase (Milner et al., 1969; Dawes and Foster, 1971).

Three observations suggest that this repression does not result from increased methionine synthesis: first, vitamin B12 is more effective as a repressor for the methylase than is methionine; second, addition of vitamin B12 does not alter the methionine pool level; third, vitamin B12, unlike methionine, does not repress enzymes for homocysteine synthesis (for example, cystathionine β-lyase), indicating that the vitamin and methionine activate independent repression systems. Furthermore, a mutant of *E. coli* derepressed and nonrepressible for the *met* biosynthetic enzymes is subject to normal repression by vitamin B12 (Dawes and Foster, 1971). Similarly, repression of the homocysteine methylase by vitamin B12 is not much affected by *metJ* or *metK* mutations (Kung et al., 1972).

Vitamin B12 induces the vitamin B12–dependent homocysteine methylase encoded by *metH;* the enzyme–vitamin B12 complex then functions as the repressor for the methylase encoded by *metE* (Kung et al., 1972; Greene et al., 1973b).

4. Regulation of the Synthesis of Methionyl-tRNA Synthetase. Expression of *metG* that encodes methionyl-tRNA synthetase is apparently not controlled by the *metJ/metK* system. Two independent controls are known for the formation of the synthetase. First, the enzyme is derepressed when the level of methionyl-tRNA falls. In *metG* mutants with methionyl-tRNA synthetases of reduced substrate affinity (Gross and Rowbury, 1969, 1971), levels of the synthetase parallel the extent of charging (Archibold and Williams, 1973). Also, growth with methionyl adenylate, which inhibits synthesis of methionyl-tRNA, derepresses the synthetase, supporting the view that methionyl-tRNA governs the synthesis of the enzyme (Cassio, 1975). Second, the synthesis of the synthetase is coupled to the overall cellular growth rate (Cassio, 1975).

B. Fungi

Because *S. cerevisiae* synthesizes methionine from H_2S (Fig. 11.4) and because cysteine can be formed by reverse transsulfuration from methionine (Fig. 11.5), it seems reasonable that methionine controls H_2S synthesis. Accordingly, methionine represses the synthesis of ATP sulfurylase, sulfite reductase, homoserine acetyltransferase, and homocysteine synthase, the so-called group I enzymes, which are controlled together although encoded by unlinked genes (Cherest et al., 1971). Two sulfate permeases are also regulated with this group (Breton and Surdin-Kerjan, 1977). While both methionine and SAM repress, the effect of SAM is not caused by conversion to methionine because the pool of free methionine remains low after addition of SAM (Cherest et al., 1973a).

Three observations show that repression by methionine is distinct from that caused by SAM. First, a number of ethionine-resistant mutants are insensitive to repression by methionine but remain responsive to SAM (Cherest et al., 1973b). Second, resynthesis of the group I enzymes occurs immediately upon the removal of SAM, but with a distinct lag when methionine is removed. Third, the addition of either effector during enzyme synthesis produces differing kinetic results: SAM causes enzyme synthesis to cease immediately, whereas residual synthesis occurs when methionine is added (Surdin-Kerjan and de Robichon-Szulmajster, 1975).

Since ethionine-resistant mutants have altered methionine adenosyltransferases (Cherest et al., 1978), the failure of methionine to repress in these mutants could reflect a reduced ability to form SAM. Methionine might act as a corepressor only after it has been converted to SAM. The kinetics of enzyme resynthesis would be in accord with this interpretation; however, measurements of SAM pool sizes are not (Cherest et al., 1973a). Apparently, methionine and SAM repress by different mechanisms, which act on different stages of enzyme synthesis. The RNA synthesis inhibitor lomofungin has been used to show that methionine acts on transcription, while SAM affects translation (Surdin-Kerjan and de Robichon-Szulmajster, 1975). Thus methionine does not repress through its conversion to SAM.

Repressibility by methionine is greatly decreased in a mutant with reduced ability to form methionyl-tRNA. The extent of repression of group I enzymes in this mutant parallels the extent of tRNA charging (Cherest et al., 1971). In contrast, repression by SAM is normal in this mutant strain. Repression by methionine evidently involves the activation of methionine to methionyl-tRNA, a requirement clearly not consistent with SAM's being the sole effector for both methionine-mediated and SAM-mediated repression. SAM represses by at least one mechanism independent of methionine or methionyl-tRNA because exogenous SAM does not alter the methionine pool level or the level of methionyl-tRNA.

However, methionyl-tRNA cannot be the only effector in the methionine-mediated repression mechanism, because the group I enzymes are not responsive to methionine in ethionine-resistant mutants, known to be altered in SAM synthesis. One model that accommodates the experimental observations presumes that SAM and methionyl-tRNA interact with an aporepressor to form a ternary complex that stops transcription of the group I genes. In addition, SAM alone or in association with a regulatory protein can prevent the translation of any mRNA for the group I enzymes. One proviso of this model is that exogenously added methionine alone generates only a limited amount of SAM, not enough to operate the second mechanism fully. Measurements of the SAM pool levels support this feature of the model (Cherest et al., 1973b).

Little is known about regulation of methionine biosynthesis in other fungi. Whereas in the fodder yeast *Saccharomycopsis lipolytica,* methionine and cysteine act independently on the synthetic enzymes (Morzycka and Paszewski, 1979), in *A. nidulans,* cysteine rather than methionine regulates the synthesis of the methionine pathway enzymes; for example, homocysteine methylase is derepressed by cysteine, whereas enzymes of reverse transsulfuration are repressed (Paszewski et al., 1977).

V. METHIONINE AND GROWTH CESSATION AT ELEVATED TEMPERATURES

Several bacterial species fail to grow at elevated temperatures in minimal media, whereas growth is unaffected in enriched media. *E. coli, Klebsiella,*

Serratia, and *Salmonella* show reduced growth rates at temperatures in the 40–44° range; growth is restored almost to normal if methionine is added (Ron, 1975). The elevated temperature inactivates homoserine succinyltransferase. The temperature sensitivity of this enzyme presumably evolved to limit the growth of mesophiles at elevated temperatures (Ron, 1975). Elevated temperatures might cause unbalanced growth and hence cell death, whereas inactivation of homoserine succinyltransferase blocks a wide range of metabolic functions, thereby stopping growth while preserving the viability of the organisms, which can rapidly recover if the temperature falls.

Homoserine acetyltransferase of *B. polymyxa* is also temperature sensitive and might limit growth in a minimal medium at elevated temperature (Wyman and Paulus, 1975). It would be of interest to isolate mutants of *E. coli* with temperature-resistant homoserine succinyltransferase (some α-methyl methionine–resistant *metA* mutants might fall in this class), then examine whether the mutants are killed at elevated temperature in minimal media, that is, to ascertain whether the instability of the enzyme has some protective function.

VI. NATURALLY OCCURRING METHIONINE ANALOGS

In a series on biotechnology, it is particularly appropriate to consider naturally occurring compounds that could be of technological or chemotherapeutic importance. Whereas many Gram-positive bacteria and fungi produce nonprotein antibiotics of relatively low molecular weight, such compounds have not generally been considered as common products of Gram-negative bacteria. Gram-negative bacteria commonly produce bacteriocins, but these large proteinaceous toxins are apparently ineffective chemotherapeutically (Hardy, 1975). Recently, however, several Gram-negative bacteria have been found to synthesize low-molecular-weight antibiotics, some of which are methionine analogs.

Early claims for bacterial production of methionine analogs involved plant pathogenic soil microorganisms. That toxins of *Pseudomonas tabaci* and *coronafaciens* function as methionine analogs is without foundation (Patil, 1974), but rhizobitoxine, a product of *Rhizobium japonicum,* is such an analog. This compound passes from the root nodule formed by this organism to cause lesions in the host plant. The analog inhibits cystathionine β-lyase of plant tissues (Giovanelli et al., 1971). The analog also forms during growth of *R. japonicum* in liquid culture. Rhizobitoxine might aid the competitive growth of the *Rhizobium* in the soil or rhizosphere, because the analog is a potent bacterial growth inhibitor, cystathionine β-lyase again being the target enzyme (Giovanelli et al., 1972).

Enteric bacteria also produce inhibitors of relatively low molecular weight (microcins) (Asensio et al., 1976). Apparently, some of these microcins are methionine analogs, since their inhibitory effects are fully reversed by methionine. These microcins shut off methionine biosynthesis by inhibiting homoserine succinyltransferase and methionyl-tRNA synthetase (Perez-Diaz and Clowes, 1980). This second inhibitory effect will not only reinforce inhibition in minimal media, that is, when methionine has to be synthesized, but would also presumably affect growth in rich media. There must be competition between the microcin and methionine for the synthetase because methionine reverses inhibition if given in high enough concentrations.

Strains of *E. coli* that produce the microcins are immune to their own microcin and to other similar compounds. The mechanism of this immunity is of considerable interest. Microcin-resistant mutants (Baquero et al., 1978) excrete methionine; presumably, such resistant mutants fall into one of the known classes of methionine analog–resistant mutants. Immunity for the producer strains is unlikely to result from methionine overproduction, since this would reduce or nullify the inhibitory effect of the agent on sensitive organisms in the vicinity. The mechanism of this immunity is worthy of investigation.

REFERENCES

Ahmed, A. (1973) *Mol. Gen. Genet. 123,* 299–324.

Archibold, E. R., and Williams, L. S. (1973) *J. Bacteriol. 114,* 1007–1013.

Asensio, C., Perez-Diaz, J. C., Martinez, M. C., and Baquero, F. (1976) *Biochem. Biophys. Res. Comm. 69,* 7–14.

Ayling, P. D., and Bridgland, E. S. (1972) *J. Gen. Microbiol. 73,* 127–141.

Ayling, P. D., Mojica, A. T., and Klopotowski, T. (1979) *J. Gen. Microbiol. 114,* 227–246.

Baquero, F., Bouanchaud, D., Martinez-Perez, M. C., and Fernandez, C. (1978) *J. Bacteriol. 135,* 342–347.

Breton, A., and Surdin-Kerjan, Y. (1977) *J. Bacteriol. 132,* 224–232.

Bright, S. W. J., Lea, P. J., and Mifflin, B. J. (1979) in *Sulphur in Biology: Ciba Foundation Symposium 72,* 101–114.

Burton, E. G., and Metzenberg, R. L. (1975) *Arch. Biochem. Biophys. 168,* 219–229.

Burton, E. G., Selhub, J., and Sakami, W. (1969) *Biochem. J. 111,* 793–795.

Cassio, D. (1975) *J. Bacteriol. 123,* 589–597.

Chater, K. F. (1970) *J. Gen. Microbiol. 63,* 95–109.

Chater, K. F., and Rowbury, R. J. (1970) *J. Gen. Microbiol. 63,* 111–120.

Cherest, H., Eichler, F., and de Robichon-Szulmajster, H. (1969) *J. Bacteriol. 97,* 328–336.

Cherest, H., Surdin-Kerjan, Y., and de Robichon-Szulmajster, H. (1971) *J. Bacteriol. 106,* 758–772.

Cherest, H., Surdin-Kerjan, Y., Antoniewski, J., and de Robichon-Szulmajster, H. (1973a) *J. Bacteriol. 114,* 928–933.

Cherest, H., Surdin-Kerjan, Y., Antoniewski, J., and de Robichon-Szulmajster, H. (1973b) *J. Bacteriol. 115,* 1084–1093.

Cherest, H., Surdin-Kerjan, Y., Exinger, F., and Lacroute, F. (1978) *Mol. Gen. Genet. 163,* 153–167.

Clandinin, M. T., and Cossins, E. A. (1974) *Phytochemistry 13,* 585–591.
Datko, A. H., Giovanelli, J., and Mudd, S. H. (1974) *J. Biol. Chem. 249,* 1139–1155.
Datko, A. H., Mudd, S. H., and Giovanelli, J. (1977) *J. Biol. Chem. 252,* 3436–3445.
Davis, B. D., Kornberg, H. L., Nagler, A., Miller, P., and Mingioli, E. (1959) *Fed. Proc. 18,* 211.
Dawes, J., and Foster, M. A. (1971) *Biochim. Biophys. Acta 237,* 455–464.
Dougall, D. K., and Fulton, M. M. (1967) *Plant Physiol. 42,* 941–945.
Flavin, M. (1975) in *Metabolism of Sulfur Compounds* (Greenberg, D. M., ed.) pp. 457–503, Academic Press, New York.
Flavin, M., and Slaughter, C. (1967) *Biochim. Biophys. Acta 132,* 400–405.
Flavin, M., Delavier-Klutchko, C., and Slaughter, C. (1964) *Science 143,* 50–52.
Giovanelli, J., Owens, L. D., and Mudd, S. H. (1971) *Biochim. Biophys. Acta 227,* 671–684.
Giovanelli, J., Owens, L. D., and Mudd, S. H. (1972) *Plant Physiol. 51,* 492–503.
Giovanelli, J., Mudd, S. H., and Datko, A. H. (1974) *Plant Physiol. 54,* 725–736.
Giovanelli, J., Mudd, S. H., and Datko, A. H. (1978) *J. Biol. Chem. 253,* 5665–5677.
Greene, R. C., and Radovich, C. (1975) *J. Bacteriol. 124,* 269–278.
Greene, R. C., Su, C.-H., and Holloway, C. T. (1970) *Biochem. Biophys. Res. Comm. 38,* 1120–1126.
Greene, R. C., Hunter, J. S. V., and Coch, E. H. (1973a) *J. Bacteriol. 115,* 57–67.
Greene, R. C., Williams, R. D., Kung, H. F., Spears, C., and Weissbach, H. (1973b) *Arch. Biochem. Biophys. 158,* 249–256.
Gross, T. S., and Rowbury, R. J. (1969) *Biochim. Biophys. Acta 184,* 233–236.
Gross, T. S., and Rowbury, R. J. (1971) *J. Gen. Microbiol. 65,* 5–21.
Gunther, E., Petruschka, L., and Herrmann, H. (1979) *Zeit. für Allg. Mikrobiol. 19,* 439–442.
Hafner, E. W., Tabor, C. W., and Tabor, H. (1977) *J. Bacteriol. 132,* 832–840.
Hardy, K. G. (1975) *Bacteriol. Rev. 39,* 464–515.
Hobson, A. C. (1974) *Mol. Gen. Genet. 131,* 263–273.
Hobson, A. C. (1976) *Mol. Gen. Genet. 144,* 87–95.
Hobson, A. C., and Smith, D. A. (1973) *Mol. Gen. Genet. 126,* 7–18.
Johnson, J. R., Greene, R. C., and Krueger, J. H. (1977) *J. Bacteriol. 131,* 795–800.
Jones, K. M., Guest, J. R., and Woods, D. D. (1961) *Biochem. J. 79,* 566–574.
Kaplan, M. M., and Flavin, M. (1966) *J. Biol. Chem. 241,* 5781–5789.
Kerr, D. S., and Flavin, M. (1969) *Biochim. Biophys. Acta 177,* 177–179.
Kerr, D. S., and Flavin, M. (1970) *J. Biol. Chem. 245,* 1842–1855.
Krueger, J. H., Johnson, J. R., and Greene, R. C. (1978) *J. Bacteriol. 133,* 1351–1357.
Kung, H. F., Spears, C., Greene, R. C., and Weissbach, H. (1972) *Arch. Biochem. Biophys. 150,* 23–31.
Lawrence, D. A. (1972) *J. Bacteriol. 109,* 8–11.
Lawrence, D. A., Smith, D. A., and Rowbury, R. J. (1968) *Genetics 58,* 473–492.
Lee, L. W., Ravel, J. M., and Shive, W. (1966) *J. Biol. Chem. 241,* 5479–5480.

Madison, J. T., and Thompson, J. F. (1976) *Biochem. Biophys. Res. Comm. 71,* 684–691.
Metzenberg, R. L. (1968) *Arch. Biochem. Biophys. 125,* 532–541.
Milner, L., Whitfield, C., and Weissbach, H. (1969) *Arch. Biochem. Biophys. 133,* 413–419.
Minson, A. C., and Smith, D. A. (1972) *J. Gen. Microbiol. 70,* 471–476.
Morzycka, E., and Paszewski, A. (1979) *Mol. Gen. Genet. 174,* 33–38.
Nagai, S., and Flavin, M. (1967) *J. Biol. Chem. 242,* 3884–3895.
Paszewski, A., and Grabski, J. (1975) *J. Bacteriol. 124,* 893–904.
Paszewski, A., Prazmo, W., and Landman-Balinska, M. (1977) *Mol. Gen. Genet. 155,* 109–112.
Patil, S. S. (1974) *Ann. Rev. Phytopath. 12,* 259–279.
Perez-Diaz, J. C., and Clowes, R. C. (1980) *J. Bacteriol. 141,* 1015–1023.
Poston, J. M., and Hemmings, B. A. (1979) *J. Bacteriol. 140,* 1013–1016.
Ron, E. Z. (1975) *J. Bacteriol. 124,* 243–246.
Rosenthal, S., Smith, L. C., and Buchanan, J. M. (1965) *J. Biol. Chem. 240,* 836–843.
Rowbury, R. J. (1961) *Biochem. J. 81,* 42p–43p.
Rowbury, R. J. (1962) *J. Gen. Microbiol. 28,* 5.
Rowbury, R. J. (1968) *J. Gen. Microbiol. 52,* 223–230.
Rowbury, R. J., and Woods, D. D. (1964a) *J. Gen. Microbiol. 35,* 145–158.
Rowbury, R. J., and Woods, D. D. (1964b) *J. Gen. Microbiol. 36,* 341–358.
Rowbury, R. J., and Woods, D. D. (1966) *J. Gen. Microbiol. 42,* 155–163.
Rowbury, R. J., Lawrence, D. A., and Smith, D. A. (1968) *J. Gen. Microbiol. 54,* 337–342.
Salem, A. R., and Foster, M. A. (1972) *Biochem. J. 127,* 845–853.
Savin, M. A., and Flavin, M. (1972) *J. Bacteriol. 112,* 299–303.
Savin, M. A., Flavin, M., and Slaughter, C. (1972) *J. Bacteriol. 111,* 547–556.
Schlesinger, S. (1967) *J. Bacteriol. 94,* 327–332.
Selhub, J., Savin, M. A., Sakami, W., and Flavin, M. (1971) *Proc. Nat. Acad. Sci. U.S.A. 68,* 312–314.
Smith, D. A. (1971) *Adv. Genet. 16,* 141–165.
Somerville, C. R., and Ahmed, A. (1977) *J. Mol. Biol. 111,* 77–81.
Stauffer, G. V., and Brenchley, J. E. (1977) *J. Bacteriol. 129,* 740–749.
Stauffer, G. V., Baker, C. A., and Brenchley, J. E. (1974) *J. Bacteriol. 120,* 1017–1025.
Sugimoto, Y., Nakatani, K., Shirakashi, T., Ohmori, H., Toraya, T., and Fukui, S. (1976) *Biochim. Biophys. Acta 437,* 333–344.
Surdin-Kerjan, Y., and de Robichon-Szulmajster, H. (1975) *J. Bacteriol. 122,* 367–374.
Thoen, A., Rognes, S. E., and Aarnes, H. (1978a) *Plant Sci. Lett. 13,* 103–113.
Thoen, A., Rognes, S. E., and Aarnes, H. (1978b) *Plant Sci. Lett. 13,* 113–119.
Wijesundera, S., and Woods, D. D. (1953) *Biochem. J. 55,* viii.
Wijesundera, S., and Woods, D. D. (1960) *J. Gen. Microbiol. 22,* 229–241.
Wijesundera, S., and Woods, D. D. (1962) *J. Gen. Microbiol. 29,* 353–366.
Wyman, A., and Paulus, H. (1975) *J. Biol. Chem. 250,* 3897–3903.

CHAPTER
12

Diaminopimelate and Lysine

Jean-Claude Patte

I. INTRODUCTION

There are two distinct pathways for lysine biosynthesis. The α-aminoadipate pathway is found in fungi (Chapter 13). Bacteria synthesize lysine by a pathway whose final step is the decarboxylation of diaminopimelate. This substance, like lysine, is an essential building block for peptidoglycan of the bacterial cell wall. Diaminopimelate and lysine fall within the aspartate family of amino acids. This chapter reviews only the biosynthesis of diaminopimelate and lysine in *Escherichia coli*.

The sequence of enzyme-catalyzed steps leading from aspartate semialdehyde to lysine and the structures of the metabolic intermediates are given in Fig. 12.1. Mutants blocked at almost every recognized enzymatic step have been isolated. They are either diaminopimelate or lysine auxotrophs. Most diaminopimelate auxotrophs undergo lysis in a complex medium unless diaminopimelate is present at low concentrations, presumably because of its requirement for cell wall synthesis. Some diaminopimelate auxotrophs do not lyse in a complex medium (Bukhari and Taylor, 1971a).

In addition to the biosynthetic enzymes (Fig. 12.1), three other enzymes play a role in lysine metabolism: aspartate kinase III, the lysine-sensitive isoenzyme that catalyzes the first reaction of the common pathway (Chapter

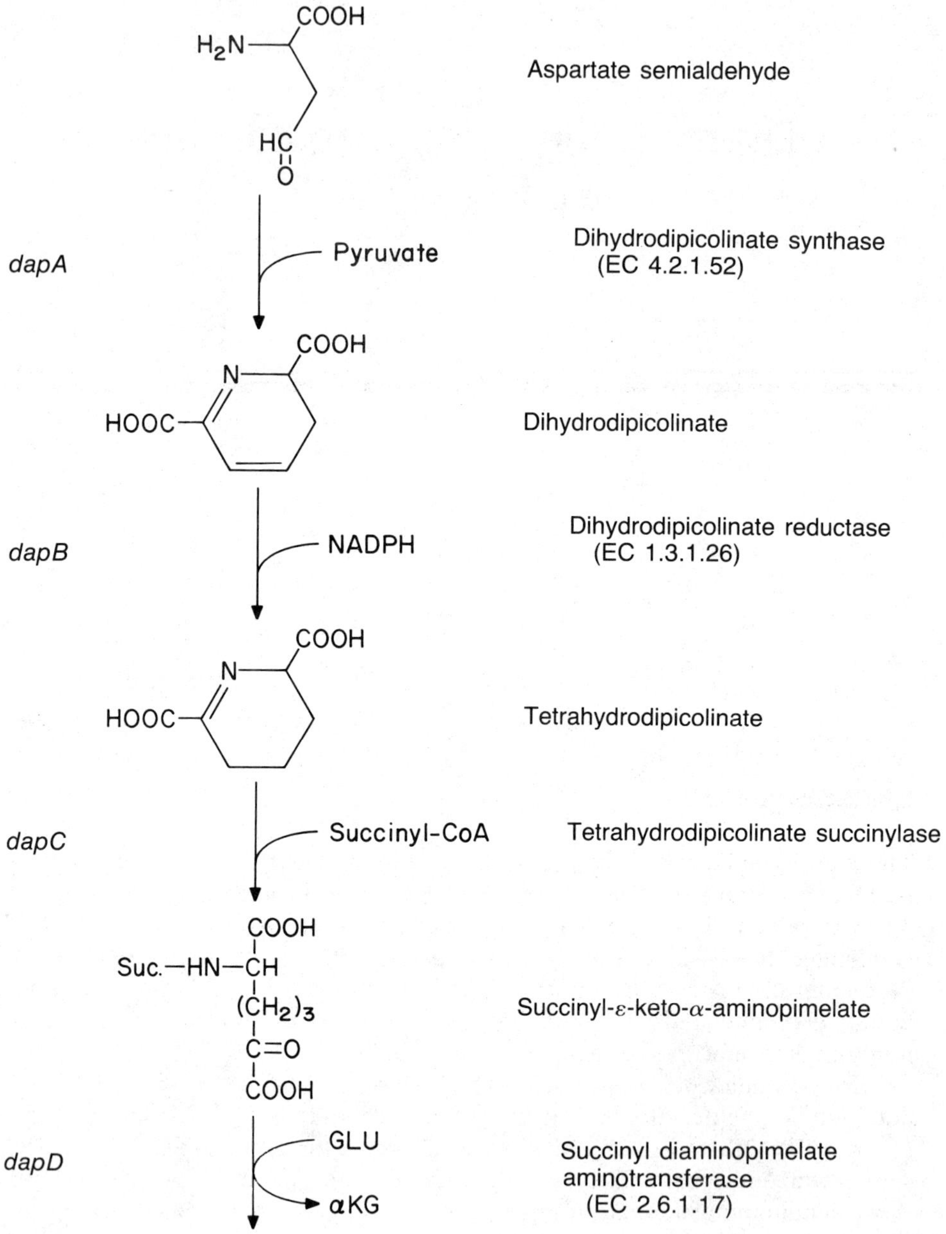
COOH
H_2N
HC
O
Aspartate semialdehyde
dapA
Pyruvate
Dihydrodipicolinate synthase
(EC 4.2.1.52)
COOH
N
HOOC
Dihydrodipicolinate
dapB
NADPH
Dihydrodipicolinate reductase
(EC 1.3.1.26)
COOH
N
HOOC
Tetrahydrodipicolinate
dapC
Succinyl-CoA
Tetrahydrodipicolinate succinylase
COOH
Suc.—HN—CH
$(CH_2)_3$
C=O
COOH
Succinyl-ε-keto-α-aminopimelate
dapD
GLU
αKG
Succinyl diaminopimelate
aminotransferase
(EC 2.6.1.17)

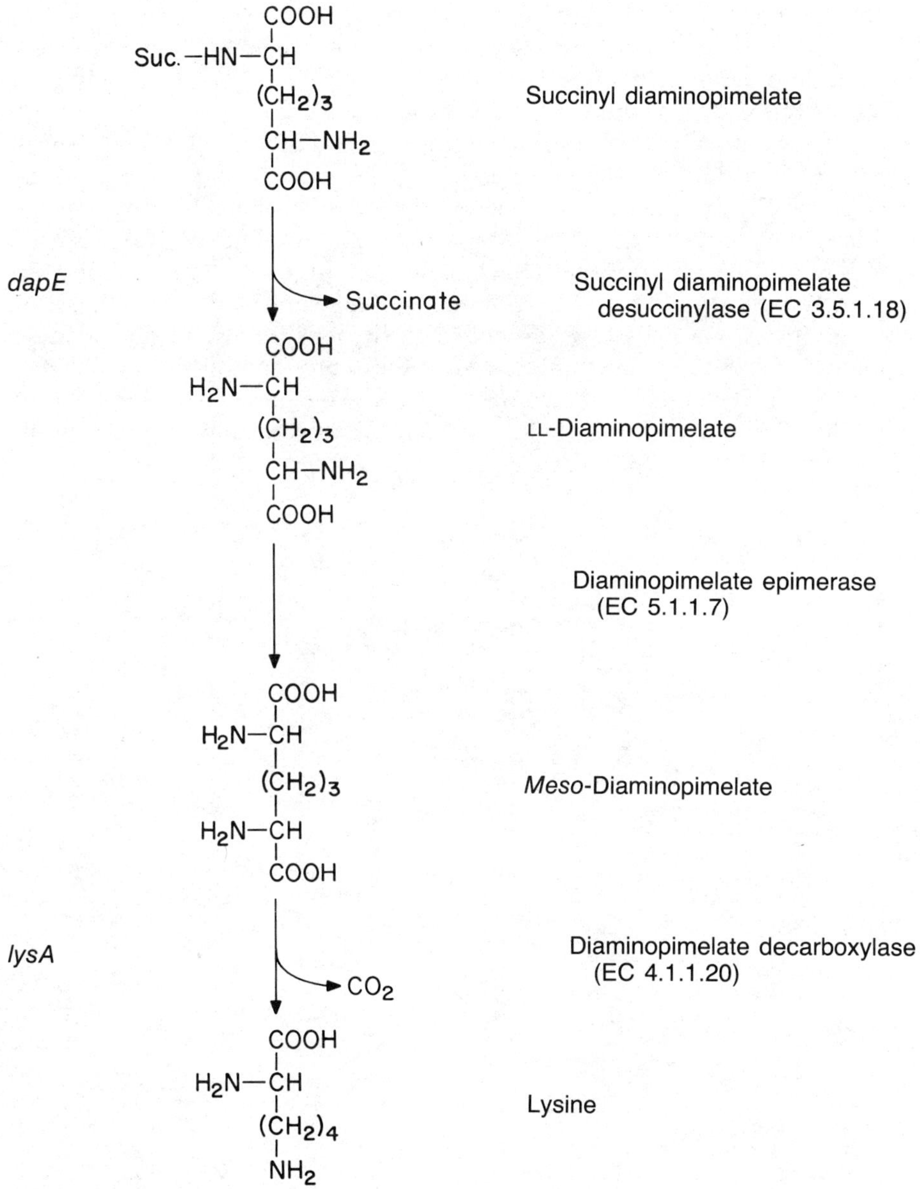

FIGURE 12.1 The biosynthetic pathway for diaminopimelate and lysine.

9); the catabolic lysine decarboxylase that converts lysine into cadaverine; and lysyl-tRNA synthetase. Enzymes that mediate diaminopimelate incorporation into peptidoglycan are not discussed here, although their activities could influence diaminopimelate and lysine pools.

The regulatory pattern of lysine biosynthesis described below applies only to *E. coli*. Many differences emerge when other bacterial species are considered (Umbarger, 1978). One of the most striking features of the diaminopimelate pathway is that the known structural genes are scattered over the *E. coli* chromosome (Bachmann and Low, 1980). There are no multigenic operons (Fig. 12.2). Although the regulation of the synthesis of some of the lysine enzymes has not been demonstrated and no common regulatory protein has been implicated in the repression of the *lys* and *dap* genes, the term *lys* regulon is used here to convey the notion of a physiological unit. The salient properties of the different enzymes of the *lys* regulon are discussed below.

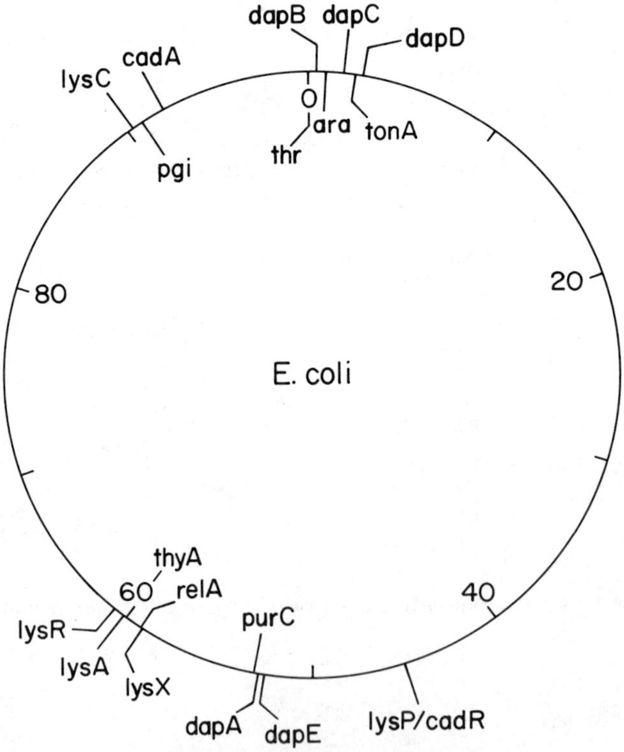

FIGURE 12.2 Chromosomal localization of the genes of lysine biosynthesis in *E. coli*.

II. ASPARTATE KINASE III

Aspartate kinase III is one of three isoenzymes that furnish aspartyl phosphate for further metabolism (Chapter 9). The enzyme activity is inhibited and its synthesis repressed by lysine (Stadtman et al., 1961). The enzyme is encoded by the *lysC* gene, a locus near *pgi* at min 90 of the *E. coli* linkage map (Thèze et al., 1974) (see Fig. 12.2).

A. Protein Structure

Aspartate kinase III has been purified to homogeneity (Truffa-Bachi and Cohen, 1966; Niles and Westhead, 1970; Lafuma et al., 1970; von Dippe et al., 1972). Contradictory assignments of molecular weights and subunit structures are due to different aggregation states of the enzyme under various conditions (Niles and Westhead, 1973a; Richaud et al., 1973). The native enzyme in the presence of 0.15 M KCl is a dimer of identical subunits with a molecular weight of 50,000. The dimeric protein is in equilibrium with both a tetrameric species, favored by high ionic strength in the presence of lysine, and a monomeric form, favored by low ionic strength in the absence of lysine. Only the dimeric form of the protein is enzymatically active (Richaud et al., 1974). However, lysine causes a conformational transition, leading to an inactive dimeric species (see below).

The N-terminal and C-terminal amino acids are serine and leucine, respectively (Richaud et al., 1973). Immunological studies reveal no homologies between the lysine-sensitive and the other two aspartate kinases, when native enzymes are analyzed (Kaminsky et al., 1969); however, cross-reactions occur with denatured proteins (Mouhli et al., 1980). DNA sequence comparisons may cast light on the possible common origin of the three kinase-encoding genes.

B. Regulation of Enzyme Activity

Aspartate kinase III is activated by ammonium and potassium ions (Wampler and Westhead, 1968).

1. Inhibitory Ligands. Feedback inhibition by lysine is noncompetitive with respect to aspartate (Stadtman et al., 1961). Lysine inhibition is cooperative (Patte and Cohen, 1964); 0.2 mM lysine, a concentration comparable to that of the intracellular lysine pool, inhibits about 50% of the enzyme activity (Truffa-Bachi and Cohen, 1966). Certain so-called "nonspecific" amino acids such as leucine, isoleucine, phenylalanine, and others at much higher concentrations also inhibit aspartate kinase III substantially (Patte et al., 1965). Synergism of inhibition by lysine and low concentrations of the nonspecific amino acids appears to be physiologically significant.

The dimeric protein binds four moles of lysine (Richaud et al., 1974). The four sites are nonequivalent; two have a K_d of 8 μM, while the remaining two have a K_d of 100 μM. Lysine binding is cooperative, but cooperativity disappears in the presence of leucine or the other nonspecific amino acids. This may explain the synergistic inhibition.

Lysine or leucine binding to aspartate kinase III leads to conformational changes at protein concentrations below those required for tetramerization (Richaud et al., 1974). Experimental kinetic data are in good agreement with theoretical values predicted (Mazat and Patte, 1976) for a concerted model of a class V allosteric enzyme (Monod et al., 1965). The kinetic parameters of the state and saturation functions have been determined (Mazat and Patte, 1976). The results support the hypothesis that aspartate kinase III is one of the best examples of a class V allosteric enzyme.

Several mutant proteins exhibit different degrees of resistance to inhibition by lysine and/or the nonspecific amino acids (Boy and Patte, 1972). All mutations lie within *lysC* without specific clustering (Thèze et al., 1974; J.-C. Patte, unpublished results).

2. Physiological Aspects. Two lines of evidence suggest that feedback inhibition of aspartate kinase III is the main point of regulation for lysine biosynthesis. First, most of the mutants selected on the basis of resistance to growth inhibition by lysine analogs, like aminoethylcysteine or lysine hydroxamate, are alleles of *lysC* encoding desensitized enzymes (Boy and Patte, 1972; J.-C. Patte and G. N. Cohen, unpublished results). Such mutants excrete lysine into the growth medium. Second, mutants with elevated expression of *lysC* (see below) are as sensitive to growth inhibition by lysine analogs as the nonconstitutive parental strain (Boy et al., 1979b).

Because lysine analog–resistant mutants excrete lysine but not threonine or methionine, it is relevant to inquire whether there is specific channeling of aspartyl phosphate synthesized by aspartate kinase III into diaminopimelate and lysine. Many attempts to demonstrate protein aggregates between aspartate kinases, aspartate semialdehyde dehydrogenase, and dihydrodipicolinate synthase have been unsuccessful (G. N. Cohen, unpublished results; E. W. Westhead, personal communication). It is thus likely that threonine inhibition of homoserine dehydrogenase is sufficient to direct aspartate semialdehyde toward diaminopimelate and lysine.

3. Enzyme Modifications. Tentatively proposed adenylylation of aspartate kinase III supposedly varies with the growth phase of the culture (Niles and Westhead, 1973b). Presumably, adenylylation also affects the aggregation state of the enzyme.

Aspartate kinase III is a good substrate for protease II, one of the *E. coli* cytoplasmic proteases (Pacaud and Richaud, 1975). Protease II copurifies with aspartate kinase III through several chromatographic steps. In the absence of lysine, protease II cleaves specific peptide bonds in aspartate kinase III, suggesting a physiological role for proteolysis in lysine biosynthesis.

C. Regulation of Enzyme Synthesis

The specific activity of aspartate kinase III varies in response to the intracellular lysine concentration. Approximately a tenfold repression is observed when lysine is added to the growth medium (Stadtman et al., 1961), whereas a 15-fold derepression is seen during lysine-limited growth in a chemostat (Patte et al., 1963). The number of active enzyme molecules, estimated to be 525 per wild-type cell growing in a minimal salts medium (Richaud et al., 1980), may thus vary about 150-fold. In *relA* strains derepression is only twofold (Patte et al., 1980), implicating guanosine tetraphosphate as an effector of transcription as proposed for general amino acid deficiency (Stephens et al., 1975).

Although lysine is a specific effector, arginine also plays a role in aspartate kinase III regulation. When both amino acids are present simultaneously, arginine prevents lysine from maximally repressing aspartate kinase III synthesis (Cassan et al., 1975). All identified regulatory mutants affecting aspartate kinase III synthesis are *cis*-dominant constitutive; no R-type mutants have yet been found (Cassan et al, 1975; Boy et al., 1979a). In a *lysC-lacZ* fusion, isolated by the method of Casadaban (1976), β-galactosidase levels reflect the intracellular levels of lysine and arginine (Richaud et al., 1980). This result provides strong support for regulation of aspartate kinase III at the transcriptional level. Although no derepression of the enzyme results when lysyl-tRNA charging is decreased, as in a K_m mutant of lysyl-tRNA synthetase (Boy et al., 1976), all other results are compatible with attenuation control (Artz and Broach, 1975; Lee and Yanofsky, 1977) of *lysC* expression. Since the *lysC–lacZ* fusion has been cloned on a multicopy plasmid (M. Cassan, unpublished results), DNA sequence analysis of the *lysC* region should help to clarify the issue.

III. DIHYDRODIPICOLINATE SYNTHASE

Dihydrodipicolinate synthase, encoded by the *dapA* gene at min 53 of the *E. coli* chromosome (Cohen et al., 1965; Bukhari and Taylor, 1971a), has been purified to homogeneity (Shedlarski and Gilvarg, 1970). The native protein has a molecular weight of 134,000. The subunit structure is not known. There is no methionine in the protein.

The enzyme is inhibited by lysine (Yugari and Gilvarg, 1962). Lysine inhibition is cooperative and noncompetitive toward the substrates (Truffa-Bachi et al., 1967). The physiological role of lysine inhibition is questionable. Mutants with desensitized aspartate kinase III excrete lysine but not threonine, an unexpected property of a cell with efficient regulation at the branch point. Mutants with desensitized dihydrodipicolinate synthase have not been found, despite the use of strong selective agents (J.-C. Patte, unpublished results).

However, dihydrodipicolinate synthase must catalyze the rate-limiting step in lysine biosynthesis after aspartate kinase III because increasing the *dapA* gene copy number in a *lysC* mutant with desensitized aspartate kinase III results in a parallel increase of dihydrodipicolinate synthase activity and lysine excretion (Dauce et al., 1982). Variations of the lysine pool have no effect on the synthesis of the apparently constitutive dihydrodipicolinate synthase (Butour et al., 1974). The *dapA* gene is part of a 2.8-kb *PstI* restriction fragment (Fig. 12.3) that has been identified in an *E. coli* library and transferred to pBR322 (Richaud et al., 1981).

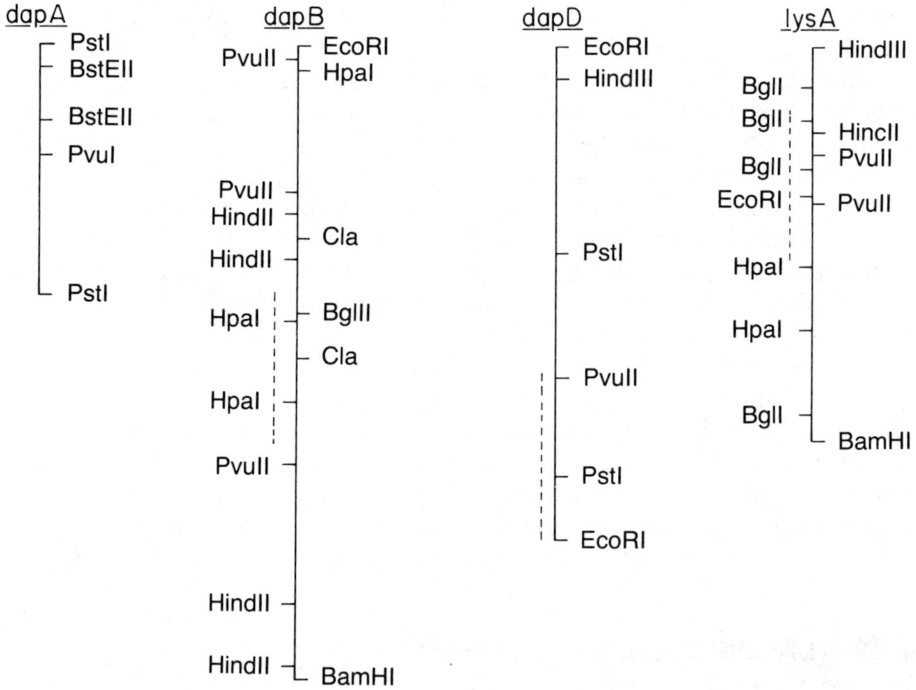

FIGURE 12.3 Restriction maps of endonucleolytic fragments encoding lysine biosynthetic genes. The fragments are drawn to scale; the coding regions (where known) are indicated by a dashed line. The 2.8-kb *PstI* fragment containing *dapA* has no restriction sites for *EcoRI, BamHI, HindIII, PvuII, BglI, II, AvaI,* or *HpaI*; the 7.1-kb *EcoRI-BamHI* fragment containing *dapB* has no restriction sites for *HindIII, SalI, KpnI,* or *SmaI;* the 5.5-kb *EcoRI* fragment containing *dapD* has no restriction sites for *BamHI, BglII, ClaI, AvaI,* or *KpnI* (F. Richaud, personal communication); the 4.4-kb *HindIII-BamHI* fragment containing *lysA* has no restriction sites for *SalI, BglII* or *KpnI* (Chenais, 1981).

IV. DIHYDRODIPICOLINATE REDUCTASE

Dihydrodipicolinate reductase, first described by Farkas and Gilvarg (1965), is encoded by *dapB* at min 0.5 on the *E. coli* chromosome between the *thr* and *ara* operons (Bukhari and Taylor, 1971a). The pure protein with a native molecular weight of 115,000 has a K_m for dihydrodipicolinate of 10^{-5} M; no specific inhibitors of the enzyme activity are known (Tamir and Gilvarg, 1974). A 32,000-dalton polypeptide has tentatively been identified as the *dapB* gene product, suggesting that the enzyme is a tetramer (Mackie, 1980).

Expression of *dapB* is regulated by lysine (Tamir and Gilvarg, 1974). Lysine represses threefold to fourfold, while growth limitation by diaminopimelate in a chemostat leads to a fourfold to sixfold derepression (Boy et al., 1978). Thus the specific activity of dihydrodipicolinate reductase can vary 15-fold to 20-fold. However, the range of variation is much lower in *relA* strains (Patte et al., 1980), indicating guanosine tetraphosphate to be a regulatory signal as discussed previously for aspartate kinase III synthesis. Charging of lysyl-tRNA also affects *dapB* expression (Boy et al., 1976, 1978). The *dapB* gene is part of a 7.1-kb *EcoRI-BamHI* restriction fragment (Fig. 12.3) that has been identified on λ-transducing phages and transferred to pBR322 (Mackie, 1980; Richaud et al., 1981).

V. TETRAHYDRODIPICOLINATE SUCCINYLASE AND N-SUCCINYL DIAMINOPIMELATE AMINOTRANSFERASE.

Tetrahydrodipicolinate succinylase from *E. coli* (Gilvarg, 1961) has not been studied in detail; however, the enzyme activity was recently demonstrated in Brevibacterium (Tosaka and Takinami, 1978). The aminotransferase from *E. coli* has been partially purified and characterized (Peterkofsky and Gilvarg, 1961). Nothing is known about the regulation of activity or synthesis of either the succinylase or the aminotransferase.

Mutants have been isolated that are blocked between aspartate semialdehyde and diaminopimelate but that do not fall into the *dapA, dapB,* or *dapE* class (Bukhari and Taylor, 1971a). The mutants affect one or the other of the two remaining metabolic steps. Two loci, designated *dapC* and *dapD,* have been identified at min 3.5 on the *E. coli* chromosome on either side of *tonA*. The *dapC* class is represented by only one mutant.

The aminotransferase encoding *dapD* gene is part of a 5.5-kb *EcoRI* restriction fragment (Fig. 12.3) that has been cloned on a λ-transducing phage and transferred to pBR322 (Bendiak and Friesen, 1981); Richaud et al., 1981).

Some auxotrophs simultaneously requiring lysine and methionine can grow on succinate in the absence of the amino acids (Davis et al., 1959). These auxotrophs were identified as *suc* mutants (Kaplan and Flavin, 1964) with impaired synthesis of succinyl-CoA, which participates in lysine and methionine biosynthesis through succinylation reactions.

VI. N-SUCCINYL DIAMINOPIMELATE DESUCCINYLASE

N-Succinyl diaminopimelate desuccinylase (Gilvarg, 1959) has been partially purified (Kindler and Gilvarg, 1960). The enzyme requires Co^{2+} for activity. The enzyme is encoded in *dapE* at min 52.5 on the *E. coli* chromosome (Cohen et al., 1963; Bukhari and Taylor, 1971a). Expression of *dapE* is repressed twofold by lysine (Cohen et al., 1963; J.-C. Patte, unpublished), suggesting that lysine could be a co-repressor for the synthesis of this enzyme.

VII. DIAMINOPIMELATE EPIMERASE

Diaminopimelate epimerase catalyzes the interconversion of L,L- and *meso*-diaminopimelate (Antia et al., 1957; Work, 1962a). Two mutants have been isolated that could grow on *meso*- but not on L,L-diaminopimelate; both mutants were unstable and lost (E. Work, personal communication; J.-C. Patte, unpublished results). To the best of my knowledge, no stable mutant lacking this epimerase has been described.

VIII. DIAMINOPIMELATE DECARBOXYLASE

A. The Structure of *lysA*

Diaminopimelate decarboxylase catalyzes the conversion of diaminopimelate to lysine (Work, 1962b). The enzyme is specific for *meso*-diaminopimelate; L,L-diaminopimelate is not a substrate. The purified protein has a native molecular weight of 200,000 (White and Kelly, 1965) as calculated from its sedimentation coefficient. Mutants devoid of this enzyme activity (Dewey and Work, 1952; Davis, 1952) accumulate diaminopimelate to an extent that is useful for diaminopimelate production.

Diaminopimelate decarboxylase is encoded in *lysA* at min 61 on the *E. coli* chromosome near *thyA* and *galR* (Jacob and Wollman, 1961). A second class of mutants with lesions at the same locus, formerly designated *lysB*, require lysine or pyridoxal phosphate for growth. These mutations are now also classified as *lysA* mutations, since they alter the affinity of the decarboxylase for pyridoxal phosphate in such a way that lysine cannot be synthesized unless a large excess of pyridoxal phosphate is supplied (Bukhari and Taylor, 1971b).

The *lysA* gene, identified on a λ-transducing phage (Shimada et al., 1972) is part of a 4.4-kb *HindIII-BamHI* restriction fragment (Fig. 12.3) that has been cloned into pBR322 (Chenais et al., 1981). The complete nucleotide sequence of *lysA* has been established (Fig. 12.4). The DNA sequence immediately preceding the coding sequence of *lysA* does not suggest attenuation control for *lysA* expression (P. Stragier, personal communication).

B. Expression of *lysA*

Lysine represses diaminopimelate decarboxylase (Patte et al., 1962). In *E. coli* W, the enzyme is induced by diaminopimelate, which may constitute the only effector if lysine acts indirectly by decreasing the diaminopimelate pool (White, 1976). However, growth limitation by lysine in a chemostat using *dap* mutants leads to a large increase in diaminopimelate decarboxylase activity (Boy et al., 1979b). Thus a multiple regulation involving repression by lysine and induction by diaminopimelate agrees well with the double metabolic role of this enzyme: anabolic for lysine biosynthesis and catabolic for diaminopimelate degradation.

Regulation of *lysA* expression is dependent upon *relA*. A large diaminopimelate-mediated induction is seen in guanosine tetraphosphate accumulating *relA*$^+$ strains but not in *relA* mutants (Patte et al., 1980). In addition, derepression of *lysA* under conditions of lysine limitation is partially abolished in *hisT* mutants (Boy et al., 1979a) where several tRNAs are not fully modified (Bruni et al., 1977; R. Cortese, personal communication).

Expression of *lysA* cloned into pBR322 does not parallel gene dosage; a 50-fold increase in gene copy number yields only a threefold increase in diaminopimelate decarboxylase activity (Chenais et al., 1981). This reflects the existence of an activator that is necessary for transcription of *lysA* and that is present in limiting amounts in the bacterial cell (P. Stragier, submitted for publication). A new class of *lys* mutants linked to *lysA* has been identified. These mutations appear to affect the proposed activator that is required for diaminopimelate decarboxylase synthesis. The mutations define a new gene, *lysR*, located on the *thyA*-distal side of *lysA*. The *lysR* product apparently does not affect the expression of the other genes of the *lys* regulon.

IX. LYSINE DECARBOXYLASE

Lysine decarboxylase catalyzes the conversion of lysine to cadaverine, one of the polyamines found in *E. coli*. No essential role is known for the enzyme activity. Mutants devoid of lysine decarboxylase show no obvious phenotype (Tabor et al., 1980); the corresponding gene, called *cadA*, lies at min 92.

The native lysine decarboxylase is a dimer of identical 80,000-dalton subunits that can aggregate into decamers (Sabo et al., 1974). Enzyme activity requires pyridoxal phosphate. Enzyme synthesis is induced by

```
ATG CCA CAT TCA CTG TTC AGC ACC GAT ACC GAT CTC ACC GCC GAA AAT CTG CTG CGT TTG CCC GCT GAA TTT GGC TGC CCG GTG TGG GTC 90
Met Pro His Ser Leu Phe Ser Thr Asp Thr Asp Leu Thr Ala Glu Asn Leu Leu Arg Leu Pro Ala Glu Phe Gly Cys Pro Val Trp Val 30

TAC GAT GCG CAA ATT ATT CGT CGG CAG ATT GCA GCG CTG AAA CAG TTT GAT GTG GTG CGC TTT GCA CAG AAA GCC TGT TCC AAT ATT CAT 180
Tyr Asp Ala Gln Ile Ile Arg Arg Gln Ile Ala Ala Leu Lys Gln Phe Asp Val Val Arg Phe Ala Gln Lys Ala Cys Ser Asn Ile His 60

ATT TTG CGC TTA ATG CGT GAG CAG GGC GTG AAA GTG GAT TCC GTC TCG TTA GGC GAA ATA GAG CGT GCG TTG GCG GCG GGT TAC AAT CCG 270
Ile Leu Arg Leu Met Arg Glu Gln Gly Val Lys Val Asp Ser Val Ser Leu Gly Glu Ile Glu Arg Ala Leu Ala Ala Gly Tyr Asn Pro 90

CAA ACG CAC CCC GAT GAT ATT GTT TTT ACG GCA GAT GTT ATC GAT CAG GCG ACG CTT GAA CGC GTC AGT GAA TTG CAA ATT CCG GTG AAT 360
Gln Thr His Pro Asp Asp Ile Val Phe Thr Ala Asp Val Ile Asp Gln Ala Thr Leu Glu Arg Val Ser Glu Leu Gln Ile Pro Val Asn 120

GCC GGT TCT GTT GAT ATG CTC GAC CAA CTG GGC CAG GTT TCG CCA GGG CAT CGG GTA TGG CTG CGC GTT AAT CCG GGG TTT GGT CAC GGA 450
Ala Gly Ser Val Asp Met Leu Asp Gln Leu Gly Gln Val Ser Pro Gly His Arg Val Trp Leu Arg Val Asn Pro Gly Phe Gly His Gly 150

CAT AGC CAA AAA ACC AAT ACC GGT GGC GAA AAC AGC AAG CAC GGT ATC TGG TAC ACC GAT CTG CCC GCC GCA CTG GAC GTG ATA CAA CGT 540
His Ser Gln Lys Thr Asn Thr Gly Gly Glu Asn Ser Lys His Gly Ile Trp Tyr Thr Asp Leu Pro Ala Ala Leu Asp Val Ile Gln Arg 180

CAT CAT CTG CAG CTG GTC GGC ATT CAC ATG CAC ATT GGT TCT GGC GTT GAT TAT GCC CAT CTG GAA CAG GTG TGT GGT GCT ATG GTG CGT 630
His His Leu Gln Leu Val Gly Ile His Met His Ile Gly Ser Gly Val Asp Tyr Ala His Leu Glu Gln Val Cys Gly Ala Met Val Arg 210

CAG GTC ATC GAA TTC GGT CAG GAT TTA CAG GCT ATT TCT GCG GGC GGT GGG CTT TCT GTT CCT TAT CAA CAG GGT GAA GAG GCG GTT GAT 720
Gln Val Ile Glu Phe Gly Gln Asp Leu Gln Ala Ile Ser Ala Gly Gly Gly Leu Ser Val Pro Tyr Gln Gln Gly Glu Glu Ala Val Asp 240

ACC GAA CAT TAT TAT GGT CTG TGG AAT GCC GCG CGT GAG CAA ATC GCC CGC CAT TTG GGC CAC CCT GTG AAA CTG GAA ATT GAA CCG GGT 810
Thr Glu His Tyr Tyr Gly Leu Trp Asn Ala Ala Arg Glu Gln Ile Ala Arg His Leu Gly His Pro Val Lys Leu Glu Ile Glu Pro Gly 270

CGC TTC CTG GTA GCG CAG TCT GGC GTA TTA ATT ACT CAG GTG CGG AGC GTC AAA CAA ATG GGG AGC CGC CAC TTT GTG CTG GTT GAT GCC 900
Arg Phe Leu Val Ala Gln Ser Gly Val Leu Ile Thr Gln Val Arg Ser Val Lys Gln Met Gly Ser Arg His Phe Val Leu Val Asp Ala 300

GGG TTC AAC GAT CTG ATG CGC CCG GCA ATG TAC GGT AGT TAC CAC CAT ATC AGT GCC CTG GCA GCT GAT GGT CGT TCT CTG GAA CAC GCG 990
Gly Phe Asn Asp Leu Met Arg Pro Ala Met Tyr Gly Ser Tyr His His Ile Ser Ala Leu Ala Ala Asp Gly Arg Ser Leu Glu His Ala 330

CCA ACG GTG GAA ACC GTC GTC GCC GGA CCG TTA TGT GAA TCG GGC GAT GTC TTT ACC CAG CAG GAA GGG GGA AAT GTT GAA ACC CGC GCC 1080
Pro Thr Val Glu Thr Val Val Ala Gly Pro Leu Cys Glu Ser Gly Asp Val Phe Thr Gln Gln Glu Gly Gly Asn Val Glu Thr Arg Ala 360

TTG CCG GAA GTG AAG GCA GGT GAT TAT CTG GTA CTG CAT GAT ACA GGG GCA TAT GGC GCA TCA ATG TCA TCC AAC TAC AAT AGC CGT CCG 1170
Leu Pro Glu Val Lys Ala Gly Asp Tyr Leu Val Leu His Asp Thr Gly Ala Tyr Gly Ala Ser Met Ser Ser Asn Tyr Asn Ser Arg Pro 390

CTG TTA CCA GAA GTT CTG TTT GAT AAT GGT CAG GCG CGG TTG ATT CGC CGT CGC CAG ACC ATC GAA GAA TTA CTG GCG CTG GAA TTG CTT TAA 1263
Leu Leu Pro Glu Val Leu Phe Asp Asn Gly Gln Ala Arg Leu Ile Arg Arg Arg Gln Thr Ile Glu Glu Leu Leu Ala Leu Glu Leu Leu 420
```

FIGURE 12.4 The DNA sequence of *lysA* and the corresponding amino acid sequence of the encoded diaminopimelate decarboxylase.

lysine (Sabo et al., 1974). Some aminoethylcysteine-resistant mutants synthesize this enzyme constitutively (Popkin and Maas, 1980; Tabor et al., 1980) and are called *lysP* or *cadR*. Genes *lysP* and *cadR* are located at min 46 and are presumably identical. In the *lysP* mutant, lysine transport decreases as lysine decarboxylase is derepressed. No modification of the regulation of the lysine biosynthetic enzymes is observed in the strain (P. S. Popkin and J.-C. Patte, unpublished results).

X. LYSYL-tRNA SYNTHETASE

Lysyl-tRNA synthetase has a molecular weight of 135,000 (Hirshfield et al., 1976). A previously described polypeptide of 100,000 daltons (Stern and Peterkofsky, 1969) is a proteolytic fragment of the native enzyme. Kinetic parameters of the activity are presumably modified by methylation (Hirshfield et al., 1977). Mutants with decreased specific activities (Hirshfield and Zamecnik, 1972) or with altered K_m's (Boy et al., 1976) have been isolated.

The synthesis of the enzyme is repressed by lysine (Reinisch et al., 1975) and induced by fructose, alanine, or leucine dipeptides (Buklad et al., 1973). The physiological meaning of the latter effect remains unclear.

XI. CONCLUSION

Owing to the genetic complexity and the multiplicity of potential effectors, lysine regulation is far from completely understood. Very few regulatory mutants affecting lysine biosynthesis have been described. The *lysX* mutations (Jenkins et al., 1974) identified by their ability to excrete lysine are located at min 60 of the *E. coli* chromosome; they could actually be *rel* mutations, since *relA* influences the activities of several steps in lysine biosynthesis (Patte et al., 1980).

ACKNOWLEDGMENTS

I wish to thank all the members of my laboratory for helpful discussions. The work in my laboratory has been supported by the Centre National de la Recherche Scientifique, the Délégation Générale à la Recherche Scientifique et Technique, and the Fondation pour la Recherche Médicale Française.

REFERENCES

Antia, M., Hoare, D. S., and Work, E. (1957) *Biochem. J. 65,* 448–459.
Artz, S. W., and Broach, J. R. (1975) *Proc. Nat. Acad. Sci. U.S.A. 72,* 3453–3457.
Bachmann, B. J., and Low, K. B. (1980) *Microbiol. Rev. 44,* 1–56.

Bendiak, D. S., and Friesen, J. D. (1981) *Mol. Gen. Genet. 181,* 356–362.
Boy, E., and Patte, J.-C. (1972) *J. Bacteriol. 112,* 84–92.
Boy, E., Reinisch, F., Richaud, C., and Patte, J.-C. (1976) *Biochimie 58,* 213–218.
Boy, E., Borne, F., and Patte, J.-C. (1978) *Mol. Gen. Genet. 159,* 33–38.
Boy, E., Borne, F., and Patte, J.-C. (1979a) *Biochimie 61,* 1151–1160.
Boy, E., Richaud, C., and Patte, J.-C. (1979b) *FEMS Microbiol. Lett. 5,* 287–290.
Bruni, C. B., Colantuoni, V., Sbordone, L., Cortese, R., and Blasi, F. (1977) *J. Bacteriol. 130,* 4–10.
Bukhari, A. I., and Taylor, A. L. (1971a) *J. Bacteriol. 105,* 844–854.
Bukhari, A. I., and Taylor, A. L. (1971b) *J. Bacteriol. 105,* 988–998.
Buklad, N. E., Sanborn, D., and Hirshfield, I. N. (1973) *J. Bacteriol. 116,* 1477–1478.
Butour, J. L., Felenbok, B., and Patte, J.-C. (1974) *Ann. Microbiol. Inst. Pasteur 125B,* 459–462.
Casadaban, M. J. (1976) *J. Mol. Biol. 104,* 541–555.
Cassan, M., Boy, E., Borne, F., and Patte, J.-C. (1975) *J. Bacteriol. 123,* 391–399.
Chenais, J. (1981) "Cloning and regulation of expression of the *lysA* gene of *Escherichia coli,*" Ph.D. thesis, Université Paris-Sud, Paris.
Chenais, J, Richaud, C., Ronceray, J., Cherest, H., Surdin-Kerjan, Y., and Patte, J.-C. (1981) *Mol. Gen. Genet. 182,* 456–461.
Cohen, G. N., Patte, J.-C., and Boezi, J. (1963) *C. R. Acad. Sci. Paris 256,* 2939–2941.
Cohen, G. N., Patte, J.-C., Truffa-Bachi, P., Sawas, C., and Doudoroff, M. (1965) in *Mécanismes de régulations des activités cellulaires chez les micro-organismes* pp. 243–253, Centre National de la Recherche Scientifique, Paris.
Dauce, L. B., Boitel, M., Deschamps, A. M., Lebeault, J. M., Sano, K., Takinami, K., and Patte, J.-C. (1982) Eur. J. Appl. Microb. Biotec. *15,* 227–231.
Davis, B. D. (1952) *Nature 169,* 534–536.
Davis, B. D., Kornberg, H. L., Nagler, A., Miller, P., and Mingioli, E. (1959) *Fed. Proc. 18,* 211.
Dewey, D. L., and Work, E. (1952) *Nature 169,* 533–534.
Farkas, W., and Gilvarg, C. (1965) *J. Biol. Chem. 240,* 4717–4722.
Gilvarg, C. (1959) *J. Biol. Chem. 234,* 2955–2959.
Gilvarg, C. (1961) *J. Biol. Chem. 236,* 1429–1431.
Hirshfield, I. N., and Zamecnik, P. C. (1972) *Biochim. Biophys. Acta 259,* 330–343.
Hirshfield, I. N., Yeh, F. M., and Zamecnik, P. C. (1976) *Biochim. Biophys. Acta 435,* 290–305.
Hirshfield, I. N., Liu, C., and Yeh, F. M. (1977) *J. Bacteriol. 131,* 589–597.
Jacob, F., and Wollman, E. L. (1961) *Sexuality and the Genetics of Bacteria,* Academic Press, New York.
Jenkins, S. J., Sparkes, C. A., and Jones-Mortimer, M. C. (1974) *Heredity 32,* 409–412.
Kaminski, M., Falcoz-Kelly, F., Truffa-Bachi, P., Patte, J.-C., and Cohen, G. N. (1969) *Eur. J. Biochem. 11,* 278–282.
Kaplan, M., and Flavin, M. (1964) *Bact. Proc., 64,* 90.
Kindler, S. H., and Gilvarg, C. (1960) *J. Biol. Chem. 235,* 3532–3535.
Lafuma, C., Gros, C., and Patte, J.-C. (1970) *Eur. J. Biochem. 15,* 111–115.
Lee, F., and Yanofsky, C. (1977) *Proc. Nat. Acad. Sci. U.S.A. 74,* 4365–4369.

Mackie, G. A. (1980) *J. Biol. Chem. 255,* 8928–8935.
Mazat, J. P., and Patte, J.-C. (1976) *Biochemistry 15,* 4053–4058.
Monod, J., Wyman, J., and Changeux, J. P. (1965) *J. Mol. Biol. 12,* 88–118.
Mouhli, H., Zakin, M. M., Richaud, C., and Cohen, G. N. (1980) *Biochemistry Int. 1,* 403–409.
Niles, E. G., and Westhead, E. W. (1970) *Fed. Proc. 29,* 912.
Niles, E. G., and Westhead, E. W. (1973a) *Biochemistry 12,* 1715–1722.
Niles, E. G., and Westhead, E. W. (1973b) *Biochemistry 12,* 1723–1729.
Pacaud, M., and Richaud, C. (1975) *J. Biol. Chem. 250,* 7771–7779.
Patte, J.-C., and Cohen, G. N. (1964) *C. R. Acad. Sci. Paris 259,* 1255–1257.
Patte, J.-C. Loviny, T., and Cohen, G. N. (1962) *Biochem. Biophys. Acta 58,* 359–360.
Patte, J.-C., Loviny, T., and Cohen, G. N. (1965) *Biochem. Biophys. Acta 99,* 523–530.
Patte, J.-C., Morand, P., Boy, E., Richaud, C., and Borne, F. (1980) *Mol. Gen. Genet. 179,* 319–325.
Peterkofsky, B., and Gilvarg, C. (1961) *J. Biol. Chem. 236,* 1432–1438.
Popkin, P. S., and Maas, W. K. (1980) *J. Bacteriol. 141,* 485–492.
Reinisch, F., Boy, E., and Patte, J.-C. (1975) *C. R. Acad. Sci. Paris Ser. D 280,* 2041–2043.
Richaud, C., Mazat, J. P., Gros, C., and Patte, J.-C. (1973) *Eur. J. Biochem. 40,* 619–629.
Richaud, C., Mazat, J. P., Felenbok, B., and Patte, J.-C. (1974) *Eur. J. Biochem. 48,* 147–156.
Richaud, F., Phuc, N. H., Cassan, M., and Patte, J.-C. (1980) *J. Bacteriol. 143,* 513–515.
Richaud, F., Richaud, C., Haziza, C., and Patte, J.-C. (1981) *C. R. Acad. Sci. Paris 293,* 507–512.
Sabo, D. L., Boeker, E. A., Byers, B., Waron, H., and Fischer, E. H. (1974) *Biochemistry 13,* 662–670.
Shedlarski, J. G., and Gilvarg, C. (1970) *J. Biol. Chem. 245,* 1362–1373.
Shimada, K., Weisberg, R. A., and Gottesman, M. E. (1972) *J. Mol. Biol. 63,* 483–503.
Stadtman, E. R., Cohen, G. N., LeBras, G., and de Robichon-Szulmajster, H. (1961) *J. Biol. Chem. 236,* 2033–2038.
Stephens, J. C., Artz, S. W., and Ames, B. N. (1975) *Proc. Nat. Acad. Sci. U.S.A. 72,* 4389–4393.
Stern, R., and Peterkofsky, A. (1969) *Biochemistry 8,* 4346–4354.
Tabor, H., Hafner, E. W., and Tabor, C. W. (1980) *J. Bacteriol. 144,* 952–956.
Tamir, H., and Gilvarg, C. (1974) *J. Biol. Chem. 249,* 3034–3040.
Thèze, J., Margarita, D., Cohen, G. N., Borne, F., and Patte, J.-C. (1974) *J. Bacteriol. 117,* 133–143.
Tosaka, O., and Takinami, K. (1978) *Agr. Biol. Chem. 42,* 95–100.
Truffa-Bachi, P., and Cohen, G. N. (1966) *Biochem. Biophys. Acta 113,* 531–541.
Truffa-Bachi, P., Patte, J.-C., and Cohen, G. N. (1967) *C. R. Acad. Sci. Paris 265,* 928–929.
Umbarger, H. E. (1978) *Ann. Rev. Biochem. 47,* 533–606.
von Dippe, P. J., Abraham, A., Nelson, C. A., and Smith, W. G. (1972) *J. Biol. Chem. 247,* 2433–2438.

Wampler, D. E., and Westhead, E. W. (1968) *Biochemistry 7,* 1661–1671.
White, P. J. (1976) *J. Gen Microbiol. 96,* 51–62.
White, P. J., and Kelly, B. (1965) *Biochem. J. 96,* 75–84.
Work, E. (1962a) in *Methods in Enzymology* Vol. V (Colowick, S. P., and Kaplan, N. O., eds.) pp. 858–864, Academic Press, New York.
Work, E. (1962b) in *Methods in Enzymology* Vol. V (Colowick, S. P., and Kaplan, N. O., eds.) pp. 864–870, Academic Press, New York.
Yugari, Y., and Gilvarg, C. (1962) *Biochem. Biophys. Acta 62,* 612–614.

CHAPTER

13

Lysine Biosynthesis in Eukaryotes

Jnanendra K. Bhattacharjee

I. INTRODUCTION AND DICHOTOMOUS EVOLUTION OF LYSINE PATHWAYS

Lysine is biosynthesized via two mutally exclusive pathways: the diaminopimelate pathway and the α-aminoadipate pathway (Vogel, 1964). Aspartate serves as a precursor for the diaminopimelate pathway, found in bacteria (Chapter 12), some Phycomycetes (Hypochytriales, Saprolegniales, and Leptomytales), and green plants. The α-aminoadipate pathway is used by Euglenids, some Phycomycetes (Chytridiales, Blastocladiales, and Mucorales), yeasts, and higher fungi (Ascomycetes and Basidiomycetes). In animals, α-aminoadipate, α-ketoadipate, and glutarate have long been known as products of lysine catabolism (Borsook et al., 1948; Cavallini and Mondovi, 1952). α-Aminoadipate supports the growth of *Neurospora crassa lys1* but not of *lys3, lys4,* or *lys2* mutants (Mitchell and Houlahan, 1948). The flow of isotope label from α-aminoadipate to lysine in *N. crassa* (Windsor, 1951) and growth of certain Ophiostoma lysine auxotrophs on α-amino and α-ketoadipate (Bergstrom and Rotenberg, 1950) provided further evidence for the role of these compounds as lysine precursors. However, glutarate is not an intermediate of lysine biosynthesis (Bhattacharjee and Tucci, 1969).

Metabolic studies with ^{14}C-acetate suggest that, in *Saccharomyces cere-*

visiae, the carbon skeleton of lysine is derived from acetate (Gilvarg and Bloch, 1951). In their pioneering studies with *Torulopsis utilis,* Strassman and Weinhouse (1952, 1953) postulated a condensation of acetate with α-ketoglutarate to yield the 7-carbon citrate homolog, homocitrate, which would yield homoaconitate, homoisocitrate, oxaloglutarate, α-ketoadipate, and α-aminoadipate through a series of reactions analogous to those of the citric acid cycle. The α-aminoadipate is coverted to lysine via α-aminoadipate semialdehyde and saccharopine (Fig. 13.1). The enzymes and intermediates of the α-aminoadipate pathway have been elucidated (Strassman and Ceci, 1964, 1965, 1966; Baroquist and Trupin, 1966; Maragoudakis and Strassman, 1966; Broquist, 1971). This review focuses on recent work on the genetics and regulation of this pathway. Since the α-aminoadipate pathway has not been observed in prokaryotes, it may have evolved totally after the emergence of eukaryotes. The evolutionary significance of the dichotomous nature of lysine biosynthesis is of considerable interest for biochemical evolution and taxonomy (LeJohn, 1971; Vogel, 1960).

II. BIOCHEMISTRY OF THE α-AMINOADIPATE PATHWAY

A. Enzymes and Intermediates

Lysine biosynthesis has been investigated in detail in *S. cerevisiae,* where seven enzymes have been partially or fully characterized (Strassman and Ceci, 1966; Broquist, 1971) (Table 13.1). Several lysine biosynthetic enzymes have also been characterized from *N. crassa* (Broquist, 1971), *Saccharomycopsis lipolytica* (Heslot et al., 1979), *Rhodotorula glutinis* (Kurtz and Bhattacharjee, 1975), *Candida pelliculosa* (Takenouchi et al., 1981), and *Penicillium chrysogenum* (Demain and Masurekar, 1974). The reactions leading to α-aminoadipate are catalyzed by a synthase, isomerase, dehydrogenase, and transaminase, quite analogous to the synthesis of glutamate, leucine, isoleucine, and valine (Strassman and Ceci, 1963), although the intermediates and the enzymes are different for each pathway.

Two homocitrate synthase isoenzymes are found in *S. cerevisiae*. Enzyme A (isoelectric at pH 4.9) and enzyme B (isoelectric at pH 5.8) are both inhibited by lysine, but only enzyme A is repressed by lysine (Tucci and Ceci, 1972). The presence of two isozymes may explain the failure to isolate yeast mutants blocked in this step. Only one homocitrate synthase enzyme is found in *N. crassa* and *S. lipolytica* (Hogg and Broquist, 1968; Gaillardin et al., 1976b).

The accumulation of homocitrate by *lys7* and *lys8* mutants and the accumulation of homocitrate and homoaconitate by *S. cerevisiae lys4* indicate two steps in the conversion of homocitrate to homoisocitrate. A similar conclusion has been reached in studies with *S. lipolytica* (C. M. Gaillardin, personal communication). The equilibrium between the two acids greatly

TABLE 13.1 Selected Properties of Lysine Biosynthetic Enzymes of *S. cerevisiae*

Enzyme	*Purification*	*Molecular Weight*	*K_m for Substrate*	*Other Properties*	*Reference*
Homocitrate synthase A	Partial	—	—	Inhibited and repressed by lysine	Tucci and Ceci (1972)
Homocitrate synthase B	Partial	—	—	Inhibited by lysine	Tucci and Ceci (1972)
Homoaconitate hydratase	Partial	—	1.4×10^{-4} M for homoaconitate, 2×10^{-4} M for homoisocitrate	—	Strassman and Ceci (1966)
Homoisocitrate dehydrogenase	500-fold	48,000	$<10^{-5}$ M for homoisocitrate, 6.5×10^{-5} M for NADPH	K_m is higher for impure enzyme	Rowley and Tucci (1970)
Aminoadipate aminotransferase (mitochondrial)	Partial	100,000	1.4×10^{-2} M for α-ketoadipate	Inactive in reverse reaction	Matsuda and Ogur (1969)
Aminoadipate aminotransferase (cytosolic)	Partial	140,000	1.4×10^{-2} M for α-ketoadipate, 2×10^{-2} M for α-aminoadipate, 2×10^{-2} M for α-ketoglutarate	Active in forward and reverse reactions	Matsuda and Ogur (1969)
	380-fold	99,000	1.06×10^{-3} M for α-aminoadipate	—	Sinha and Bhattacharjee (1971)
Aminoadipate semialdehyde dehydrogenase					Sagisaka and Shimura (1962)
Saccharopine dehydrogenase (glutamate forming)	129-fold	73,000	0.92×10^{-3} M for saccharopine, 0.02×10^{-3} M for NADP	pH optimum 7.0 for forward reaction and >9.0 for reverse action	Broquist (1971)
Saccharopine dehydrogenase (lysine forming)	708-fold	49,000	1.2×10^{-2} M for lysine, 4.4×10^{-4} M for α-ketoglutarate, 4.6×10^{-5} M for NADH	pH optimum >9 for forward and 7.0 for reverse	Broquist (1971)

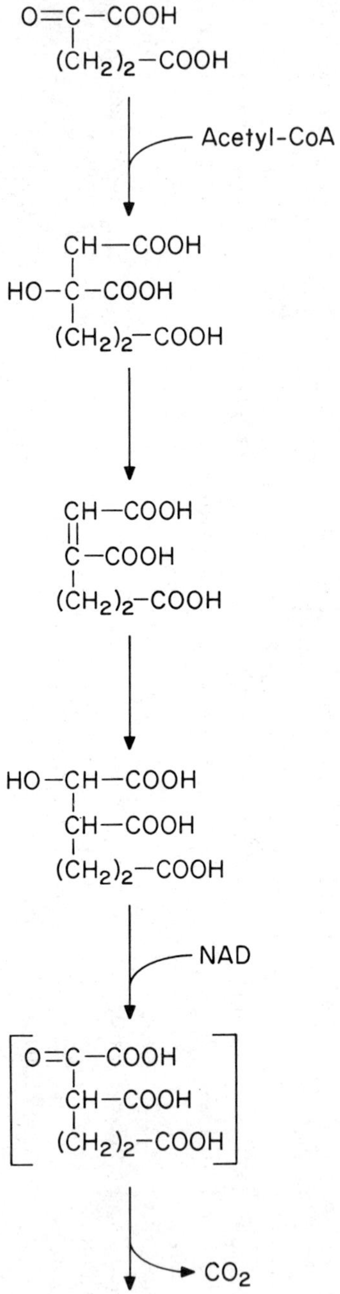

α-Ketoglutarate

N. crassa: *lys5*
S. lipolytica: *lys1, lys11*
S. cerevisiae: —

Homocitrate synthase (EC 4.1.3.21)

Homocitrate

N. crassa: —
S. lipolytica: —
S. cerevisiae: *lys7, lys8*

Homocitrate dehydrase

Homoaconitate

N. crassa: —
S. lipolytica: *lys6, lys7*
S. cerevisiae: *lys4*

Homoaconitate hydratase (EC 4.2.1.36)

Homoisocitrate

N. crassa: —
S. lipolytica: *lys9, lys10*
S. cerevisiae: *lys12*

Homoisocitrate dehydrogenase (EC 1.1.1.87)

Oxaloglutarate

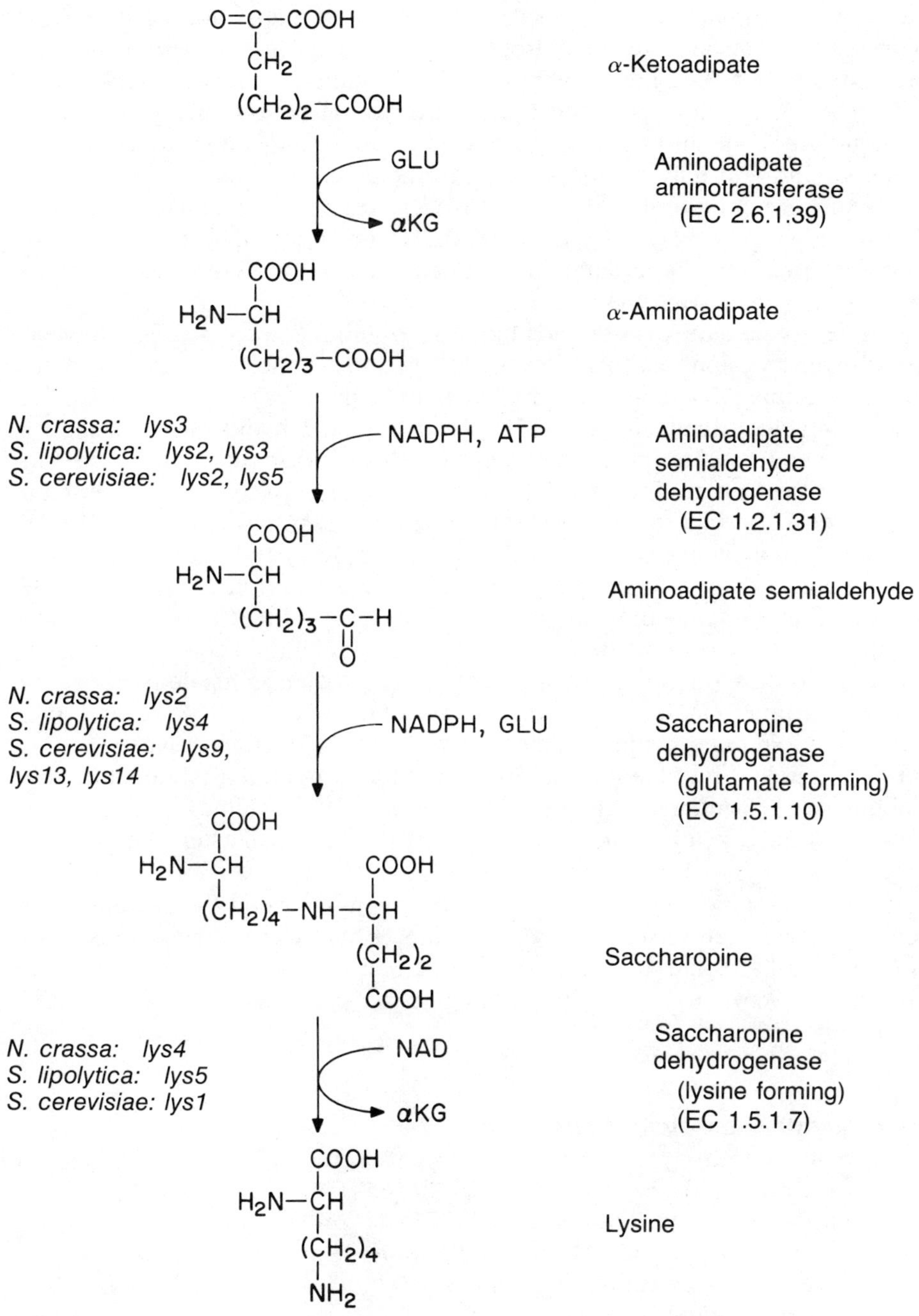

FIGURE 13.1 **α-Aminoadipate pathway for the biosynthesis of lysine.**

favors the accumulation of homocitrate (Bhattacharjee and Strassman, 1967a). Purified homoisocitrate dehydrogenase converts homoisocitrate to α-ketoadipate but does not decarboxylate oxaloglutarate (Rowley and Tucci, 1970). Also, mutants of *S. cerevisiae* blocked after α-ketoadipate do not accumulate oxaloglutarate. Like oxalosuccinate, oxaloglutarate may be enzyme-bound and may not be a free intermediate.

One mitochondrial and one cytosolic aminoadipate aminotransferase are found in *S. cerevisiae* (Matsuda and Ogur, 1969). The cytosolic aminotransferase may be a catabolic enzyme, since it is derepressed in cells grown in a lysine-rich medium.

The conversion of α-aminoadipate to α-aminoadipate semialdehyde is a multistep reaction catalyzed by a single ATP, Mg^{2+}, and NADPH-dependent enzyme, aminoadipate semialdehyde dehydrogenase (Sagisaka and Shimura, 1962; Sinha and Bhattacharjee, 1971). Saccharopine dehydrogenase (glutamate forming) and saccharopine dehydrogenase (lysine forming) catalyze the formation of saccharopine and lysine, respectively. Both enzymes exhibit different pH optima for the forward and reverse reactions (Broquist, 1971). A pH of 10.0 favors the forward reaction catalyzed by saccharopine dehydrogenase (lysine forming), while the reverse reaction is optimal at pH 7.0. *Lys1* mutants accumulate saccharopine. *Lys1* mutant extracts catalyze neither the forward nor the reverse reaction, excluding separate enzymes. Kinetic studies on the forward reaction of saccharopine dehydrogenase (lysine forming) indicate an ordered mechanism where NAD^+ and saccharopine bind and lysine, α-ketoglutarate, and NADH are released in this order (Fujioka and Nakatani, 1970). The reverse reaction is competitively inhibited by ornithine, leucine, norleucine, and norvaline; oxaloacetate, pyruvate, and α-ketobutyrate inhibit the reverse reaction allosterically (Fujioka and Nakatani, 1972).

Mutant studies with *S. cerevisiae* have demonstrated the accumulation of all intermediates except oxaloglutarate (Bhattacharjee and Strassman, 1967a, b; Bhattacharjee and Sinha, 1972; Mattoon et al., 1961; Jones and Broquist, 1965), suggesting that oxaloglutarate is not a free intermediate.

B. Biosynthetic Role of Pipecolate

Pipecolate can satisfy the lysine requirement of *Aspergillus nidulans* (Aspen and Meister, 1962), *Euglena gracilis* (Rothstein and Saffran, 1963), and *R. glutinis* (Kurtz and Bhattacharjee, 1975; Kinzel and Bhattacharjee, 1979) but not of *S. cerevisiae*. These observations remained unexplained until the conversion of pipecolate to lysine was elucidated in *R. glutinis*. In this organism, pipecolate oxidase catalyzes the conversion of pipecolate to piperideine 6-carboxylate, which equilibrates spontaneously with α-aminoadipate semialdehyde (Fig. 13.2) (Kinzel and Bhattacharjee, 1982). The role of

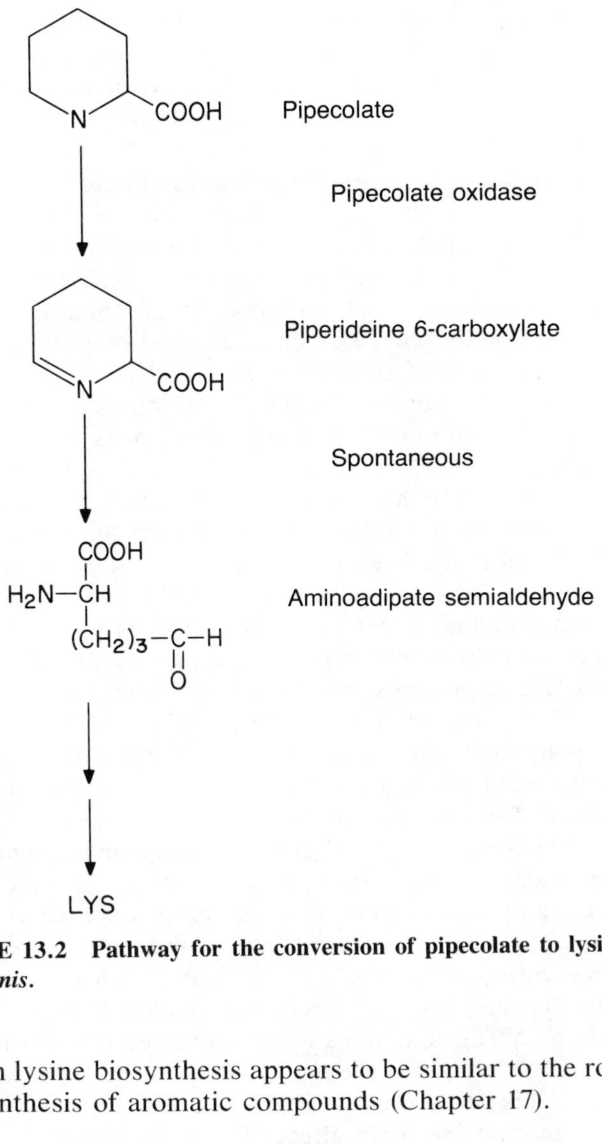

FIGURE 13.2 Pathway for the conversion of pipecolate to lysine in *R. glutinis*.

pipecolate in lysine biosynthesis appears to be similar to the role of quinate in the biosynthesis of aromatic compounds (Chapter 17).

C. Subcellular Localization of Enzymes and Lysine

In *S. cerevisiae*, the enzymes of α-aminoadipate biosynthesis, namely, homocitrate synthase, homoaconitate hydratase, homoisocitrate dehydrogenase, and aminoadipate aminotransferase, are localized in the mitochondria. These enzymes are encoded by nuclear genes and are synthesized in the cytoplasm. A second aminoadipate aminotransferase and the enzymes catalyzing the last three steps of lysine biosynthesis are nonmitochondrial (Betterton et al., 1968); Matsuda and Ogur, 1969). Intracellular free lysine, like arginine, is sequestered in vacuoles of *S. cerevisiae* and *S. lipolytica* (Wiemken and Duerr, 1974; Chapter 5; H. Heslot, personal communication).

D. Toxicity and Growth Inhibition by α-Aminoadipate

In bacteria, growth inhibition by specific amino acids is attributable to antagonistic effects of related amino acids on the regulation of biosynthetic pathways (De Felice et al., 1979). Basic amino acids, including lysine, inhibit the growth of *S. cerevisiae* on nitrogen-poor media (Sumrada and Cooper, 1978). In eukaryotic cells, the study of amino acid cross-inhibition is complicated by the existence of separate amino acid pools in cytoplasm and organelles such as mitochondria, nucleus, and vacuoles. α-Aminoadipate satisfies the growth requirement of lysine auxotrophs blocked before the α-aminoadipate step. Not only are wild-type cells unable to use α-aminoadipate as a sole nitrogen source (Chattoo et al., 1979a), but their viability is also lost after 72 h in a medium containing α-aminoadipate as the sole nitrogen source (M. K. Winston and J. K. Bhattacharjee, unpublished observation). α-Aminoadipate also inhibits the growth of wild-type *S. cerevisiae* in a minimal medium containing ammonia. Mutant cells grown with elevated levels of α-aminoadipate exhibit impaired cell division and altered cell morphologies, including an increase in size, multiple buds, pseudohyphae, and germ tubes. Both the toxicity and the growth inhibition by α-aminoadipate are reversed by arginine, aspartate, glutamine, homoserine, methionine, or serine. Mutants *lys2, lys5,* and *lys14* grow poorly in a medium containing α-aminoadipate as the nitrogen source. Growth is improved by the single amino acid supplements mentioned above. During such growth of mutant and wild-type strains, α-ketoadipate accumulates in the culture medium, indicating that α-aminoadipate serves as a nitrogen source via reversal of the glutamate-α-ketoadipate transamination. Aminoadipate aminotransferase of *lys2* mutants and α-aminoadipate semialdehyde dehydrogenase of wild-type cells grown in the presence of selected supplemental amino acids and α-aminoadipate as nitrogen sources are derepressed twofold to fivefold in comparison to enzyme levels of ammonia-grown cells (Winston and Bhattacharjee, 1982). This derepression may explain the ability of *lys2* and wild-type cells to overcome the toxic effect of α-aminoadipate and the use of it as nitrogen source. The role of the above-mentioned amino acids in the detoxification and reversal of growth inhibition is not known.

E. Catabolism of Lysine

There is considerable diversity in the catabolism of lysine in yeasts and fungi. Some yeasts, for example, *S. cerevisiae,* can use lysine as a carbon source, whereas others, for example, *R. glutinis,* can use it only as a nitrogen source (Schultz and Pomper, 1948; Walters et al., 1953). Lysine is catabolized via the pipecolate pathway in plants and molds (Scheet, et al., 1954; Gupta and Spenser, 1969) and via acetylated intermediates in *Hansenula saturnus* and *S. lipolytica* (Rothstein, 1965; Gaillardin et al., 1976a). Both the pipecolate and the acetylated pathways lead to the formation of glutarate, which serves as a carbon source. However, not all of the individual steps for either pathway have been elucidated.

In *R. glutinis,* lysine is catabolized to α-aminoadipate semialdehyde and glutamate through a lysine α-ketoglutarate aminotransferase. Mutant strains that lack saccharopine dehydrogenase (lysine forming) can use lysine as the sole nitrogen source and accumulate α-aminoadipate semialdehyde in the culture medium. These catabolic processes are not direct reversals of the biosynthetic pathways (J. J. Kinzel and J. K. Bhattacherjee, unpublished observation). In contrast, animals catabolize lysine through direct reversal of the biosynthetic pathway (Borsook et al., 1948; Cavallini and Mondovi, 1952; Higashino et al., 1965).

III. GENETICS OF THE α-AMINOADIPATE PATHWAY

A. Gene–enzyme Relationships (Fig. 13.1)

The gene–enzyme relationships for the biosynthesis of lysine have been investigated in detail in *S. cerevisiae* and *S. lipolytica* and to a lesser extent in *N. crassa.* For this purpose, lysine auxotrophs were analyzed (a) nutritionally for their response to selected intermediates such as α-aminoadipate, (b) genetically by complementation and recombination, and (c) biochemically for the accumulation of intermediates and lack of specific enzymes. The lysine auxotrophs of *S. cerevisiae* are classified into 15 complementation groups, *lys1* through *lys15*. *Lys3* is considered to be the same as *lys4* (Hwang et al., 1966; Biswas and Bhattacharjee, 1974). Mutants belonging to the *lys4, lys6, lys7, lys8, lys10, lys11, lys12,* and *lys15* groups are blocked before the α-aminoadipate step, *lys7* and *lys8* lack homocitrate dehydrase, *lys4* lacks homoaconitate hydratase, and *lys12* lacks homoisocitrate dehydrogenase. To date, no homocitrate synthase or aminoadipate aminotransferase mutants have been isolated from *S. cerevisiae*. Mutants *lys6* and *lys8* are respiratory-deficient and aconitate hydratase-less and exhibit a glutamate requirement, *glu2* (Sherman, 1963; Ogur et al., 1965). The pleiotropic properties of *glu2 lys6* and *glu2 lys8* mutants cannot be separated by back-mutation and recombination. The biochemical lesions of *lys6, lys10, lys11,* and *lys15* are currently being investigated (J. E. Rogers, P. A. Bricmont, and J. K. Bhattacharjee, unpublished results). Mutants *lys1, lys2, lys5, lys9, lys13,* and *lys14* are blocked between α-aminoadipate and lysine; *lys2* and *lys5* lack aminoadipate semialdehyde dehydrogenase; *lys9, lys13,* and *lys14* lack saccharopine dehydrogenase (glutamate forming), and *lys1* lacks saccharopine dehydrogenase (lysing forming) (Bhattacharjee and Sinha, 1972).

In *S. lipolytica,* lysine auxotrophs have been classified into 11 complementation groups (Heslot et al., 1979). *Lys2* and *lys3* are linked; the other lysine genes are scattered throughout the genome. Mutants *lys1* and *lys11* lack homocitrate synthase, *lys6* and *lys7* lack homoaconitate hydratase, *lys9* and *lys10* lack homoisocitrate dehydrogenase, *lys2* and *lys3* lack aminoadipate semialdehyde dehydrogenase, *lys4* lacks saccharopine dehydrogenase (glutamate forming), and *lys5* lacks saccharopine dehydrogenase (lysine forming). No mutant has been identified lacking aminoadipate aminotransferase. *Lys1* is believed to be the structural gene for homocitrate synthase (C. M. Gaillardin, personal communication).

The number of genes exceeds the number of enzymes required for lysine biosynthesis in both *S. cerevisiae* and *S. lipolytica*. Because two or more genes are responsible for a biosynthetic step, some enzymes may be heterologous oligomers. Alternatively, one of the genes may encode a positive regulatory element required for transcription of a structural gene.

Only five *lys* loci have been identified in *N. crassa*. The mutants *lys1* and STL-7 (*lys5*) are blocked before α-aminoadipate production; *lys5* lacks homocitrate synthase (Hogg and Broquist, 1968), indicating that *N. crassa* contains only a single form of this enzyme. *Lys3, lys2,* and *lys4* are blocked after α-aminoadipate production and lack aminoadipate semialdehyde dehydrogenase and the two saccharopine dehydrogenases, respectively (Jones and Broquist, 1965). A summary of the gene–enzyme relationships for all three organisms is given in Fig. 13.1.

B. Gene Mapping, Mutation Specificity, and Suppression

Five *N. crassa lys* genes reside on three different linkage groups (Perkins and Bjorkman, 1982). However, the genetics of lysine biosynthesis via α-aminoadipate have been investigated in greater detail in *S. cerevisiae* than in any other yeast or mold. The genes are not linked to each other; seven of the genes are located on six different chromosomes: *lys1* and *lys11* on chromosome IX, *lys2* on chromosome II, *lys4* on chromosome IV, *lys5* on chromosome VII, *lys7* on chromosome XIII, and *lys9* on chromosome XIV (Mortimer and Schild, 1980).

Tetrad analyses indicate that *lys10, lys11,* and *lys12* are "leaky" mutations at a single locus; *lys9* and *lys13* also represent a single locus; and *lys14*, also leaky, is centromere-linked (J. K. Bhattacharjee et al., unpublished observation).

More than 50% of *S. cerevisiae* mutagen-induced lysine auxotrophs are *lys4* alleles. *Lys1, lys2,* and *lys4* mutants together account for more than 85% of all the induced auxotrophs, while mutations at some loci are relatively rare (Biswas and Bhattacharjee, 1974). A novel procedure for the selection of *lys2* and *lys5* mutants exploits the fact that wild-type cells do not grow in media containing α-aminoadipate as the sole nitrogen source (Chattoo et al., 1979a). Cells that grow on α-aminoadipate as the sole nitrogen source are *lys2* or *lys5* mutants. The underlying basis for this selection is not fully understood.

Allele-specific, locus-independent supersuppressor mutations reverse the phenotypes of a wide variety of *lys* mutations (Hawthorne and Leupold, 1974; Gilmore et al., 1971). Mutants *lys2-1, lys9-1,* and *lys1-1* of *S. cerevisiae* exhibit the wild-type phenotype in the presence of supersuppressor mutation S1 or S2 (Fjellstedt and Ogur, 1970). Significant amounts of aminoadipate semialdehyde dehydrogenase and the two saccharopine dehydrogenases are present in the supersuppressed strains but are undetectable in the unsuppressed mutants. Significantly higher K_m values of the enzymes from the suppressed strains compared to those of wild-type strains suggest that altered polypeptides are synthesized. A study of the suppressibility of 358 *lys2* mutants revealed that 68% of the mutants were not suppressible by any of the known suppressors, 24% were suppressible by UAA suppressors, 4.5% by UAG suppressors, and only one by a UGA suppressor (Chattoo et al., 1979b).

C. Cloning of *lys* Genes

Two *lys* genes, *LYS1* (J. Marmur, personal communication) and *LYS2* (Eibel and Philippsen, 1982), of *S. cerevisiae* have been cloned. The coding region of *LYS2* is located within a 4.2-kb *EcoRI-PstI* restriction fragment. An examination of 200 *lys2* mutants by genome blotting using the cloned gene as a probe revealed that 198 mutants had restriction patterns identical with wild-type DNA, and two of the mutants had altered restriction patterns. The altered restriction pattern was apparently due to insertion of the TY element in the promoter region of *LYS2*. These two mutants exhibited residual enzyme activity, as evidenced by their slow growth in a medium containing α-aminoadipate as the sole nitrogen source. None of the mutants had any DNA rearrangement in the 4.2-kb fragment.

IV. REGULATION OF THE α-AMINOADIPATE PATHWAY

A. Feedback Inhibition and High Lysine-Producing-Mutants

Lysine inhibits the first enzyme of the pathway, homocitrate synthase, of *S. cerevisiae* (Maragoudakis et al., 1967; Tucci, 1969; Sinha et al., 1971), *N. crassa* (Hogg and Broquist, 1968), *S. lipolytica* (Gaillardin et al., 1976a), *C. pelliculosa* (Takenouchi et al., 1981), *R. glutinis* (Glass and Bhattacharjee, 1971), and *P. chrysogenum* (Masurekar and Demain, 1972; Demain and Masurekar, 1974). In *S. lipolytica,* the lysine inhibition of homocitrate synthase is reversed by α-ketoglutarate; lysine inhibition causes the saturation isotherm for α-ketoglutarate to become cooperative (Gaillardin et al., 1976b).

The lysine analogs S-(2-aminoethyl)-L-cysteine (thialysine) and δ-hydroxylysine inhibit the growth of wild-type *S. cerevisiae* (Sinha et al., 1971; Haidaris and Bhattacharjee, 1977). Mutants resistant to thialysine and δ-hydroxylysine overproduce and excrete lysine. Lysine excretion, enhanced by the addition of urea to the culture medium (Haidaris and Bhattacharjee, 1978), reaches 100 μg/ml in some thialysine-resistant mutants. Both dominant and recessive thialysine-resistant mutants fall into three complementation groups and exhibit pleiotropic effects including impaired lysyl-tRNA synthetase activity and thermosensitivity (Zwolshen and Bhattacharjee, 1981). Homocitrate synthase of analog-resistant mutants is insensitive to lysine, thialysine, or δ-hydroxylysine (Gray and Bhattacharjee, 1976). A thialysine-resistant mutant of *C. pelliculosa* produces 3 mg of lysine per milliliter of culture medium when grown in a medium supplemented with α-aminoadipate (Takenouchi et al., 1977, 1981). The UV-induced, feedback-insensitive *lex* mutants of *S. lipolytica* also excrete lysine (Gaillardin et al., 1975).

Lysine is an essential amino acid. Most cereal grains are deficient in this amino acid (Sure, 1948; Bhattacharjee, 1970). Although the amount of lysine produced by yeast mutants is low in comparison to *Brevibacterium flavum* (Sano and Shiio, 1970), the capability of food and feed yeasts to overproduce lysine or other essential amino acids could significantly improve their nutritional value and marketability.

B. Regulation of Penicillin Biosynthesis by Lysine

Lysine inhibits penicillin production; this inhibition is reversed by α-aminoadipate (Demain, 1966; Masurekar and Demain, 1972), a common precursor of lysine and penicillin biosynthesis. Feedback inhibition of homocitrate synthase in *P. chrysogenum* (Demain and Masurekar, 1974; Friedrich and Demain, 1977) would limit the α-aminoadipate level and thereby impair penicillin production. It should be possible to select lysine analog–resistant, feedback-insensitive mutants of *P. chrysogenum* and *Cephalosporium* for increased production of penicillin and cephalosporin, respectively.

C. Repression and General Control of Lysine Biosynthetic Enzymes

The lysine biosynthetic enzymes, homocitrate synthase, homoaconitate hydratase, homoisocitrate dehydrogenase, aminoadipate semialdehyde dehydrogenase, and the two saccharopine dehydrogenases are repressed twofold to fourfold when *S. cerevisiae* is grown in a lysine-supplemented medium (Sinha et al., 1971; L. A. Urrestarazu and J. K. Bhattacharjee, unpublished observation). Homoisocitrate dehydrogenase and saccharopine dehydrogenase (glutamate forming) are more sensitive to repression than the other enzymes. Homocitrate synthase of *N. crassa* and saccharopine dehydrogenase (lysine forming) of *S. lipolytica* are also repressed by lysine (Hogg and Broquist, 1968; Gaillardin et al., 1979).

Two of the lysine biosynthetic enzymes, aminoadipate semialdehyde dehydrogenase and saccharopine dehydrogenase (lysine forming), respond to general amino acid regulation (Wolfner et al., 1975; Messenguy, 1979), which is unique to yeast and molds (Carsiotis and Jones, 1974; Schürch et al., 1974). This regulation involves the simultaneous derepression of one or more enzymes of several unrelated amino acid biosynthetic pathways in response to starvation for one of the member amino acids. Not all biosynthetic enzymes of a given pathway respond to general regulation. Homocitrate synthase, homoaconitate hydratase, and homoisocitrate dehydrogenase are repressed in a lysine bradytroph grown in a minimal medium and in a minimal medium supplemented with lysine. On the other hand, aminoadipate semialdehyde dehydrogenase and saccharopine dehydrogenase (lysine forming) are derepressed when the mutant is grown in a minimal medium. Evidently, the enzymes before the α-aminoadipate step are not subject to general regulation.

Expression of the *HIS4* and *TRP5* genes of *S. cerevisiae* is also under general amino acid control. While the molecular mechanism of this regulatory phenomenon is unknown, the presence of a highly conserved DNA sequence at nucleotides −355 to −185 and the consensus sequence TGACTC preceding each structural gene offers a potential recognition site for the transcriptional control of general regulation (Hauge and Greer, 1982). Such a consensus sequence is not found upstream of *LEU2* and *TRP1*, genes that are not controlled by general amino acid regulation.

ACKNOWLEDGMENTS

I thank Dr. C. C. Griffin for reading the manuscript, my past and present graduate students and research associates for their contributions to the biosynthesis of lysine in yeast, and Mrs. Darlene Davidson for typing the manuscript. The author's research has been supported by the National Science Foundation and the Faculty Research Committee, Miami University, Oxford, Ohio.

REFERENCES

Aspen, A. J., and Meister, A. (1962) *Biochemistry 1,* 606–612.

Bergstrom, S., and Rottenberg, M. (1950) *Acta Chem. Scand. 4,* 553.

Betterton, H., Fjellstedt, T., Matsuda, M., Ogur, M., and Tate, R. (1968) *Biochim. Biophys. Acta 170,* 459–461.

Bhattacharjee, J. K. (1970) *Adv. Appl. Microbiol. 13,* 139–161.

Bhattacharjee, J. K., and Sinha, A. K. (1972) *Mol. Gen. Genet. 115,* 26–30.

Bhattacharjee, J. K., and Strassman, M. (1967a) *J. Biol. Chem. 242,* 2542–2546.

Bhattacharjee, J. K., and Strassman, M. (1967b) *J. Albert Einstein Med. Center 15,* 44–49.

Bhattacharjee, J. K., and Tucci, A. F. (1969) *J. Biol. Chem. 244,* 1417–1423.

Biswas, G. D., and Bhattacharjee, J. K. (1974) *Antonie Van Leeuwenhoek J. Microbiol. Serol. 40,* 221–231.

Borsook, H., Deasy, C. L., Haagen-Smit, A. J., Keighley, G., and Lowy, P. H. (1948) *J. Biol. Chem. 176,* 1395–1400.
Broquist, H. P. (1971) in *Methods in Enzymology* vol. 17B, (Colowick, S. P., and Kaplan, N. O., eds.) pp. 112–129, Academic Press, New York.
Broquist, H. P., and Trupin, J. S. (1966) *Ann. Rev. Biochem. 35,* 231–274.
Carsiotis, M., and Jones, R. F. (1974) *J. Bacteriol. 119,* 889–892.
Cavallini, D., and Mondovi, B. (1952) *Arch. Sci. Biol. 36,* 468–473.
Chattoo, B. B., Sherman, F., Azubalis, D. A., Fjellstedt, T. A., Mehnert, D., and Ogur, M. (1979a) *Genetics 93,* 51–65.
Chattoo, B. B., Palmer, E., Ono, B., and Sherman, F. (1979b) *Genetics 93,* 67–79.
DeFelice, M., Levinthal, M., Iccarino, M., and Guardiola, J. (1979) *Microbiol. Rev. 43,* 42–58.
Demain, A. L. (1966) *Adv. Appl. Microbiol. 8,* 1–27.
Demain, A. L., and Masurekar, P. S. (1974) *J. Gen. Microbiol. 82,* 143–151.
Eibel, H., and Philippsen, P. (1982) abstract from the Eleventh International Conference on Yeast and Genetic Molecular Biology, Montpellier, France.
Fjellstedt, T. A., and Ogur, M. (1970) *J. Bacteriol. 101,* 108–117.
Friedrich, C. G., and Demain, A. L. (1977) *J. Antibiot. 30,* 760–761.
Fujioka, M., and Nakatani, Y. (1970) *Eur. J. Biochem. 16,* 180–186.
Fujioka, M., and Nakatani, Y. (1972) *Eur. J. Biochem. 25,* 301–307.
Gaillardin, C. M., Sylvestre, G., and Heslot, H. (1975) *Arch. Microbiol. 104,* 89–94.
Gaillardin, C. M., Fournier, P., Sylvestre, G., and Heslot, H. (1976a) *J. Bacteriol. 125,* 48–57.
Gaillardin, C. M., Poirier, L., and Heslot, H. (1976b) *Biochim. Biophys. Acta 422,* 390–406.
Gaillardin, C. M., Poirier, L., Ribet, A. M., and Heslot, H. (1979) *Biochimie 61,* 473–482.
Gilmore, R. A., Stewart, J. W., and Sherman, F. (1971) *J. Mol. Biol. 61,* 157–173.
Gilvarg, C., and Bloch, K. (1951) *J. Biol. Chem. 193,* 339–346.
Glass, J., and Bhattacharjee, J. K. (1971) *Genetics 67,* 365–376.
Gray, G. S., and Bhattacharjee, J. K. (1976) *J. Gen. Microbiol. 97,* 117–120.
Gupta, R. N., and Spenser, I. D. (1969) *J. Biol. Chem. 244,* 88–94.
Haidaris, C. G., and Bhattacharjee, J. K. (1977) *J. Ferment. Technol. 55,* 189–192.
Haidaris, C. G., and Bhattacharjee, J. K. (1978) *J. Ferment. Technol. 56,* 189–192.
Hauge, B., and Greer, H. (1982) abstract from the Eleventh International Conference on Yeast and Genetic Molecular Biology, Montpellier, France.
Hawthorne, D. C., and Leupold, U. (1974) *Curr. Topics Microbiol. Immunol. 64,* 1–47.
Heslot, H., Gaillardin, C. M., Beckerich, J. M., and Fournier, P. (1979) in *Genetics of Industrial Microorganisms,* (Sebek, O. K., and Laskin, A. I., eds.) pp. 54–60. American Society for Microbiology, Washington, D.C.
Higashiuno, K., Tsukada, K., and Lieberman, I. (1965) *Biochem. Biophys. Res. Comm. 20,* 285–290.
Hogg, R. W., and Broquist, H. P. (1968) *J. Biol. Chem. 243,* 1839–1845.
Hwang, Y. L., Lindegren, G., and Lindegren, C. C. (1966) *Can. J. Genet. Cytol. 8,* 471–480.
Jones, E. E., and Broquist, H. P. (1965) *J. Biol. Chem. 240,* 2531–2536.

Kinzel, J. J., and Bhattacharjee, J. K. (1979) *J. Bacteriol. 138,* 410–417.
Kinzel, J. J., and Bhattacharjee, J. K. (1982) *J. Bacteriol. 151,* 1073–1077.
Kurtz, M., and Bhattacharjee, J. K. (1975) *J. Gen. Microbiol. 86,* 103–110.
LeJohn, H. B. (1971) *Nature 231,* 164–168.
Maragoudakis, M. E., and Strassman, M. (1966) *J. Biol. Chem. 241,* 695–699.
Maragoudakis, M. E., Holmes, H., and Strassman, M. (1967) *J. Bacteriol. 93,* 1677–1680.
Masurekar, P. S., and Demain, A. L. (1972) *Can. J. Microbiol. 18,* 1045–1048.
Matsuda, M., and Ogur, M. (1969) *J. Biol. Chem. 244,* 5153–5158.
Mattoon, J. R., Moshier, T. A., and Kreiser, T. H. (1961) *Biochim. Biophys. Acta 51,* 615–617.
Messenguy, F. (1979) *Mol. Gen. Genet. 169,* 85–95.
Mitchell, H. K., and Houlahan, M. B. (1948) *J. Biol. Chem. 174,* 883–887.
Mortimer, R. K., and Schild, D. (1980) *Microbiol. Rev. 44,* 519–571.
Ogur, M., Roshanmanesh, A., and Ogur, S. (1965) *Science 147,* 1590.
Perkins, D. D., and Bjorkman, M. (1982) *Neurospora News Lett. 29,* 31–44.
Rothstein, M. (1965) *Arch. Biochem. Biophys. 111,* 467–476.
Rothstein, M., and Saffran, E. M. (1963) *Arch. Biochem. Biophys. 101,* 373–377.
Rowley, B., and Tucci, A. F. (1970) *Arch. Biochem. Biophys. 141,* 499–510.
Sagisaka, S., and Shimura, K. (1962) *J. Biochem.* (Tokyo) *52,* 155–161.
Sano, K., and Shiio, I. (1970) *J. Gen. Appl. Microbiol. 16,* 373-391.
Schultz, A. S., and Pomper, S. (1948) *Arch. Biochem. 19,* 184–192.
Schürch, A., Miozzari, J., and Hutter, R. (1974) *J. Bacteriol. 117,* 1131–1140.
Schweet, R. S., Holden, J. T., and Lowy, P. H. (1954) *J. Biol. Chem. 211,* 517–529.
Sherman, F. (1963) *Genetics 48,* 375–385.
Sinha, A. K., and Bhattacharjee, J. K. (1971) *Biochem. J. 125,* 743–749.
Sinha, A. K., Kurtz, M., and Bhattacharjee, J. K. (1971) *J. Bacteriol. 108,* 715–719.
Strassman, M., and Ceci, L. N. (1963) *J. Biol. Chem. 238,* 2445–2452.
Strassman, M., and Ceci, L. N. (1964) *Biochem. Biophys. Res. Comm. 14,* 262–267.
Strassman, M., and Ceci, L. N. (1965) *J. Biol. Chem. 240,* 4357–4361.
Strassman, M., and Ceci, L. N. (1966) *J. Biol. Chem. 241,* 5401–5407.
Strassman, M., and Weinhouse, S. (1952) *J. Amer. Chem. Soc. 74,* 3457–3458.
Strassman, M., and Weinhouse, S. (1953) *J. Amer. Chem. Soc. 75,* 1680–1684.
Sumrada, R., and Cooper, T. G. (1978) *J. Gen. Microbiol. 108,* 45–56.
Sure, B. (1948) *J. Nutr. 36,* 59–63.
Takenouchi, E., Nikolova, D. K., Awano, K., Soda, K., and Tanaka, H. (1977) *Agr. Biol. Chem. 41,* 615–616.
Takenouchi, E., Tanaka, H., and Soda, K. (1981) *J. Ferment. Technol. 59,* 429–433.
Tucci, A. F. (1969) *J. Bacteriol. 99,* 624–625.
Tucci, A. F., and Ceci, L. N. (1972) *Arch. Biochem. Biophys. 153,* 742–750.
Vogel, H. J. (1960) *Biochim. Biophys. Acta 41,* 172–173.
Vogel, H. J. (1964) *Amer. Natur. 98,* 435–446.
Walters, L. S., Ci, A. R. A., and Thisselton, M. R. (1953) *J. Inst. Brew. 59,* 401–404.
Wiemken, A., and Duerr, M. (1974) *Arch. Microbiol. 101,* 45–57.

Windsor, E. (1951) *J. Biol. Chem. 192,* 607–609.

Winston, M. K., and Bhattacharjee, J. K. (1982) *J. Bacteriol. 152,* 874–879.

Wolfner, M., Yep, D., Messenguy, F., and Fink, G. R. (1975) *J. Mol. Biol. 96,* 273–290.

Zwolshen, J. H., and Bhattacharjee, J. K. (1981) *J. Gen. Microbiol. 122,* 281–287.

CHAPTER 14

The Biosynthesis of Isoleucine and Valine and its Regulation

H. Edwin Umbarger

I. INTRODUCTION

The biosynthesis of isoleucine and valine in nearly all bacteria, fungi, and plants that have been studied occurs by the pathways shown in Fig. 14.1. The pathways are parallel, because the enzymes catalyzing the last four specific steps in isoleucine biosynthesis also catalyze the four steps converting pyruvate to valine. Valine biosynthesis starts with the transfer of an active acetaldehyde derived from pyruvate to a second molecule of pyruvate. The corresponding step in isoleucine biosynthesis consists of the transfer of active acetylaldehyde to α-ketobutyrate, which is not an intermediate of central metabolism. Therefore an additional enzyme, usually threonine deaminase, is needed for isoleucine biosynthesis.

A few organisms that cannot convert threonine to α-ketobutyrate employ more complex pathways. One such pathway involves the lengthening of the carbon chain of an α-keto acid by reactions analogous to those by which α-ketoisovalerate is converted to α-ketoisocaproate in leucine biosynthesis (Chapters 15 and 16). This other route may involve the same enzymes that the cell uses to form α-ketoisocaproate (Charon et al., 1974). A threonine deaminase deficiency is indeed bypassed in cells with derepressed levels

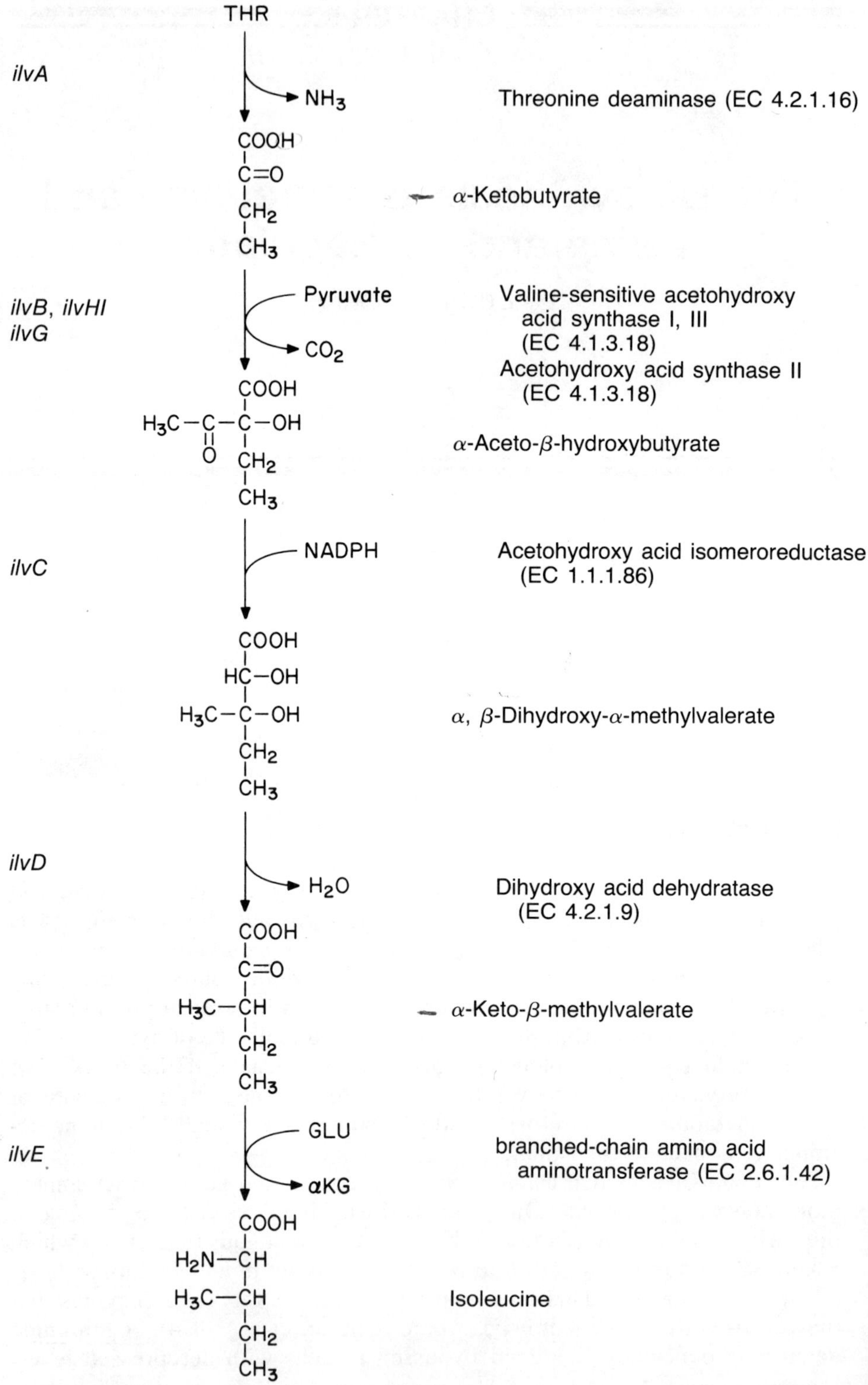
THR
ilvA
NH3
Threonine deaminase (EC 4.2.1.16)
α-Ketobutyrate
ilvB, ilvHI
ilvG
Pyruvate
CO2
Valine-sensitive acetohydroxy acid synthase I, III (EC 4.1.3.18)
Acetohydroxy acid synthase II (EC 4.1.3.18)
α-Aceto-β-hydroxybutyrate
ilvC
NADPH
Acetohydroxy acid isomeroreductase (EC 1.1.1.86)
α, β-Dihydroxy-α-methylvalerate
ilvD
H2O
Dihydroxy acid dehydratase (EC 4.2.1.9)
α-Keto-β-methylvalerate
ilvE
GLU
αKG
branched-chain amino acid aminotransferase (EC 2.6.1.42)
Isoleucine

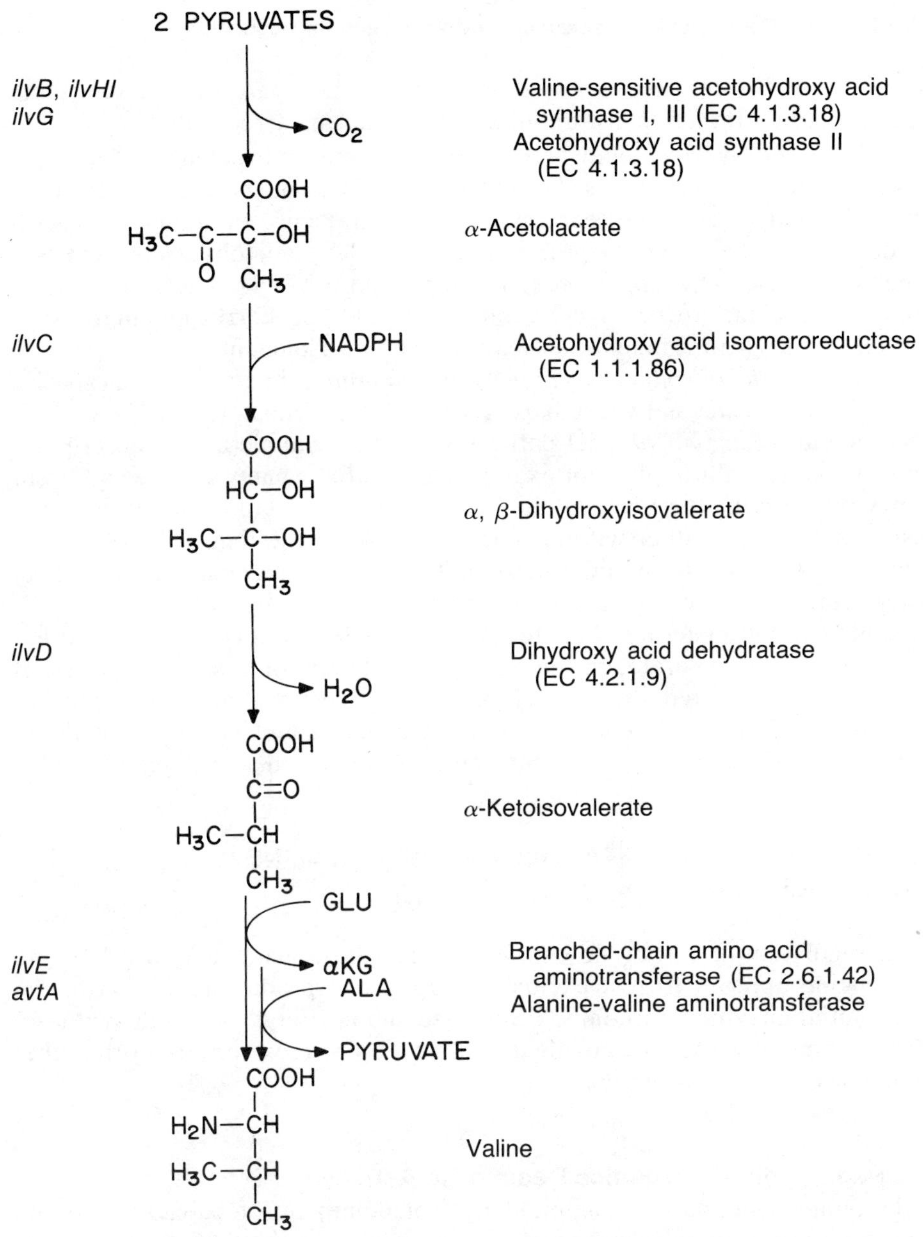

FIGURE 14.1 The biosynthetic pathways to isoleucine and valine.

of leucine biosynthetic enzymes, including a feedback-resistant isopropylmalate synthase (Kisumi et al., 1976). In another organism, glutamate is converted to β-methylaspartate, which, upon transamination to the unstable β-methyloxaloacetate, is decarboxylated to α-ketobutyrate (Phillips et al., 1972). Finally, threonine deaminase deficiency in *Bacillus subtilis* can be bypassed by derepressed levels of threonine synthase because a secondary activity of this enzyme converts homoserine phosphate to α-ketobutyrate and ammonia rather than threonine (Skarstedt and Greer, 1973).

The acetohydroxy acid precursors of valine and isoleucine have the same number of carbons as the two amino acids, but the carbon skeletons must be rearranged. The rearrangement which occurs simultaneously with reduction of the carbonyl groups is catalyzed by acetohydroxy acid isomeroreductase. The enzyme requires Mg^{2+} and NADPH. The two resulting dihydroxy acids are then dehydrated to yield the corresponding α-keto acids. Finally, transamination yields valine and isoleucine.

In *Escherichia coli* and *Salmonella typhimurium,* the two bacterial species in which isoleucine and valine biosynthesis has been most thoroughly studied, the operation of two parallel pathways is not as simple as the description given above might imply. For example, there are as many as three different enzymes forming the two acetohydroxy acids. The relationship of the three isozymes to each other will become clearer when the regulation of carbon flow is considered in the next section. There is also a departure at the final step from the pattern of each enzyme being shared by both pathways. A second aminotransferase (Chapter 2) can aminate α-ketoisovalerate, but not α-keto-β-methylvalerate, at the expense of alanine or α-aminobutyrate. It might be anticipated that other minor variations will be found in other organisms. This chapter mainly covers isoleucine and valine biosynthesis in *E. coli* and *S. typhimurium* but refers to some other systems as well.

II. REGULATION OF METABOLITE FLOW TO VALINE AND ISOLEUCINE

The regulation of carbon flow into and through the pathways to valine and isoleucine appears to rest exclusively upon end product inhibition of the two initial enzymes, threonine deaminase and acetohydroxy acid synthase. There is no evidence for covalent modification of these or any of the other enzymes of the two pathways.

A. Regulation of Threonine Deaminase Activity

Threonine deaminase is inhibited by isoleucine; this feedback inhibition provides an ideal explanation of how carbon flows into a specific branch of a complex biosynthetic system only when the end product is absent (Umbarger, 1956). Simple but elegant isotope incorporation studies show convincingly that the control of carbon flow into the isoleucine pathway is efficient (Roberts et al., 1955). Essentially, *E. coli* strain B is grown in a mineral salts medium for one doubling with uniformly labeled glucose of known specific activity as the carbon and energy source. The cell protein is hydrolyzed, the amino acids are separated by paper chromatography, and their quantities are determined directly by scanning the paper with a Geiger counter. There is virtually no radioactivity in the isoleucine spot of the chromatogram when an ample supply of unlabeled isoleucine is present in the growth medium. There is also no labeled isoleucine excreted into the medium, a result that would indicate both a preferential utilization of exogenous isoleucine and a quenching of endogenous isoleucine formation, rather than a rapid exchange and swamping of the endogenously formed pool.

These results can be explained if any of the five steps to isoleucine are susceptible to inhibition by exogenous isoleucine. However, only the inhibition of threonine deaminase explains why isoleucine exerts a strong sparing effect on the threonine requirement of threonine auxotrophs (Umbarger and Brown, 1953).

The inhibitor and substrate binding sites of *E. coli* threonine deaminase are distinct, and the interactions between the sites can be abolished by mercuric ions or by heating the enzyme (Changeux, 1961). Valine is an activator of the enzyme (Changeux, 1961). Detailed kinetic studies of the cooperative binding of inhibitor and substrate (when both are present) have been interpreted (Changeux, 1963) in terms of the allosteric transition model (Monod et al., 1965). The model proposes an equilibrium between the inactive and active forms of the enzyme and a preferential binding of substrate and inhibitor to the active and inactive form, respectively. According to the model, valine maintains the enzyme in the active state by binding at a separate activator site. The cooperative effects are attributed to a concerted transition of all the subunits of the multimeric enzyme to the inactive or active state. Thus binding at a single site favors binding to the other subunits.

The kinetic behavior of the nearly identical *S. typhimurium* threonine deaminase has been interpreted differently. The *Salmonella* enzyme is a tetramer of apparently identical subunits, each with a molecular weight of about 48,500 (Zarlengo et al., 1968). Surprisingly, only two of the subunits are able to bind the coenzyme pyridoxal phosphate (Hatfield and Burns, 1970). Binding of pyridoxal phosphate to only half of the potential sites may be an example of a strong negative cooperativity, as invoked by Stallcup and Koshland (1973) to account for half-of-the-sites reactivity of yeast glyceraldehyde-3-phosphate dehydrogenase. Alternatively, the subunit arrangement in the disulfide-linked dimer may not allow access to both potential pyridoxal phosphate sites. In the tetrameric form of the enzyme, a similar event may occur, since there is but a single activator binding site for valine.

The interactions between effectors and threonine deaminase are complex (Decedue et al., 1975; Hofler and Burns, 1978). Binding of isoleucine to the two inhibitor binding sites of the enzyme in the absence or presence of valine is a noncooperative process. However, to prevent the binding of valine at the single activator site or to inhibit threonine deamination, both isoleucine sites must be occupied, and the interaction is therefore a function of $[\text{isoleucine}]^2$. In the absence of the inhibitor, isoleucine, the saturation of threonine deaminase with threonine is noncooperative; that is, binding to each catalytic site is independent of binding to the other. In the presence of isoleucine, the threonine saturation curve is sigmoid because threonine must displace isoleucine by binding at the activator site before it can occupy the substrate binding site. Valine can displace isoleucine, whereupon threonine binding to the active site is noncooperative. Binding to the activator site by either valine or threonine is necessary only when isoleucine is present.

It is still unclear whether the activator and inhibitor sites are physically distinct. It is possible that valine binds at one of the isoleucine sites and that, by a pronounced negative cooperativity, a second molecule of valine as well as isoleucine is excluded from the second site.

Two other threonine deaminases, both concerned with isoleucine biosynthesis, should be mentioned. The *Pseudomonas putida* threonine deaminase does not contain pyridoxal phosphate (Cohn and Phillips, 1974). However, the enzyme has several kinetic properties akin to those of the pyridoxal phosphate–dependent enzymes of *S. typhimurium* and *E. coli*. Isoleucine inhibits the enzyme and induces cooperativity with respect to threonine binding. The *Pseudomonas* enzyme is about two orders of magnitude less efficient than the pyridoxal phosphate–dependent threonine deaminases of *E. coli* or *S. typhimurium*. *Saccharomyces cerevisiae* threonine deaminase differs from the bacterial enzyme in that it shows cooperative binding of threonine even in the absence of isoleucine (de Robichon-Szulmajster and Magee, 1968).

B. Regulation of Acetohydroxy Acid Synthase Activity

The activity that converts two pyruvate molecules to acetolactate, first demonstrated in *E. coli* K12, is inhibited by valine, indicating that the reaction is indeed the initial specific step in valine biosynthesis (Umbarger and Brown, 1953). This seemingly rational conclusion is at variance with isotope incorporation studies of *E. coli* B (Roberts et al., 1955). The same kind of experiment that indicates a quenching of isoleucine biosynthesis by isoleucine reveals that the incorporation of exogenous valine into protein is explainable by a swamping of the endogenously formed valine pool. Radioactive valine formed from glucose is found in the medium in amounts that suggest very little quenching of valine biosynthesis.

However, the difference between the K12 and B strains of *E. coli* is due to a difference in valine sensitivities of the respective acetohydroxy acid synthases. Of all enterics tested, only *E. coli* K12 is growth-inhibited by valine, as can be expected if the enzyme that forms acetolactate is also required for acetohydroxybutyrate formation and if this enzyme is significantly inhibited by valine. It was later demonstrated that *E. coli* W as well as *S. typhimurium* contains two acetohydroxy acid synthase activities, one sensitive to inhibition by valine and the other insensitive (Blatt et al., 1972; O'Neill and Freundlich, 1972). The two activities can be separated by centrifugation in sucrose density gradients or by hydroxylapatite chromatography. The two activities are further differentiated by genetic criteria (see p. 252). In contrast, the K12 strain contains only a valine-sensitive activity.

The K12 strain does produce a valine-insensitive acetohydroxy acid synthase activity following a mutation in *ilvO* (Brown et al., 1975). Historically, this locus has been designated "*ilvO*" because, long before it was recognized that *ilvO* mutations lead to expression of the *ilvG* gene, it was known that *ilvO* mutations result in elevated expression of the *ilvE, ilvD,* and *ilvA* genes. Therefore *ilvO* was thought to be an operator locus (Ramakrishnan and Adelberg, 1965).

The presence of two different acetohydroxy acid synthases in the valine-insensitive strains such as *S. typhimurium* appears to account for the fact that no mutants lacking acetohydroxy acid synthase have ever been isolated. Indeed, through prior selection of valine-sensitive strains, that is, *ilvG* mutants, acetohydroxy acid synthase–negative strains can be isolated that are *ilvB-ilvG* double mutants and that require isoleucine and valine for growth (J. Pledger, unpublished results). However, *E. coli* K12, with only a valine-sensitive acetohydroxy acid synthase, also fails to yield single-step mutants that are completely blocked in acetolactate and acetohydroxybutyrate formation. A clearer picture of the acetohydroxy acid synthase isozymes emerges through the analysis of a second kind of mutation in *E. coli* K12 giving rise to valine resistance (DeFelice et al., 1974). The mutation lies in *ilvH,* a locus linked to the *leu* operon. Unlike *ilvO, ilvH* mutations have no effect on *ilvE, ilvD,* and *ilvA* expression. In order to identify the structural gene for this valine-resistant acetohydroxy acid synthase, a mutant lacking the activity, that is, one that has become valine-sensitive again, has been isolated. This mutant does not arise by reversion of the original *ilvH* mutation, since the valine-resistant activity is restored by an amber-suppressing phage. The locus of the amber mutation, designated *ilvI,* is closely linked to *ilvH*. The idea that *ilvH* specifies a subunit rendering the *ilvI*-encoded acetohydroxy acid synthase valine-sensitive has since been supported by nucleotide sequence analysis indicating two separate cistrons (J. Calvo, personal communication).

In parallel experiments with *ilvO* mutants of *E. coli* K12, valine-sensitive derivatives with lesions in *ilvG* have been selected (Guardiola et al., 1974). An *ilvI* mutation, transduced into an *ilvO-ilvG* double mutant, yields transductants that remain valine-sensitive but that are still prototrophic. Upon mutagenesis, these transductants gave rise to *ilvB* mutants that are devoid of acetohydroxy acid synthase activity and exhibit an absolute requirement for isoleucine and valine. The *ilvB* gene is not linked to the *ilv* gene cluster as has been long assumed, but rather to *uhp* at min 81.5 on the *E. coli* chromosome (Fig. 14.2) (Newman and Levinthal, 1980). With the availability of mutants lacking all three acetohydroxy acid synthases, strains carrying only one of the three isozymes can be prepared. Recently, valine-resistant *ilvB* mutants of *E. coli* K12 have been described (Sutton et al., 1981).

Whether all three acetohydroxy acid synthase isozymes are formed in many of the valine-sensitive bacterial species is as yet uncertain. If the *ilvHI* operon is expressed in *S. typhimurium,* expression is probably at a very low level. When all three isozymes can be formed, as in *ilvO* derivatives

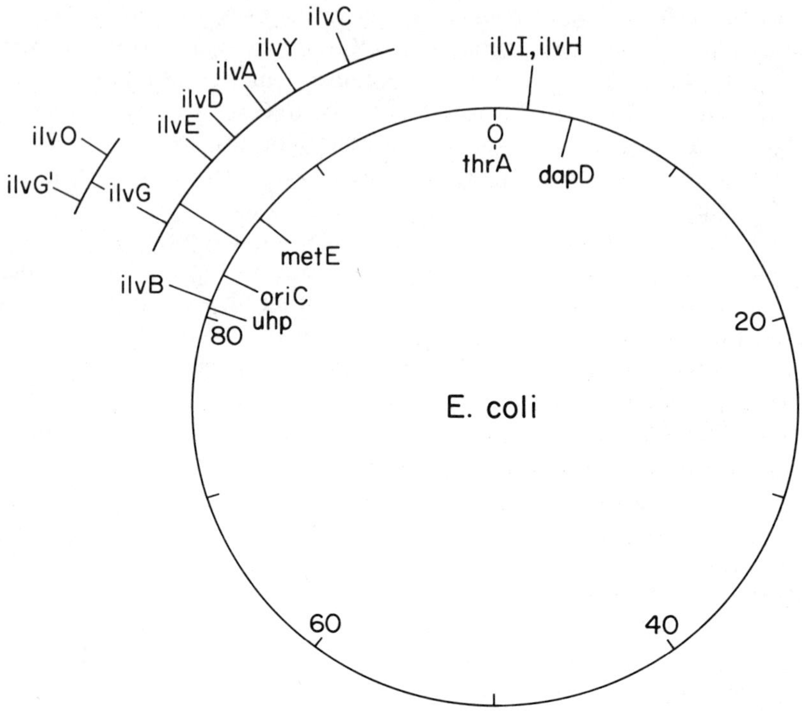

FIGURE 14.2 Chromosomal localization of the genes of isoleucine and valine biosynthesis in *E. coli.* In the K12 strain, *ilvG* is not expressed unless the reading frame in the latter part of *ilvG* is altered by a frameshift mutation in the *ilvO* portion of the gene (see p. 254).

of the K12 strain, it is not yet certain to what extent each isoenzyme mediates acetolactate and acetohydroxybutyrate synthesis. To a limited extent, one or the other isozyme may be relatively more important for valine or isoleucine formation. However, the prototrophy of strains bearing only a single isozyme proves that such a commitment is not an absolute one.

Additional genes in the *E. coli* chromosome may be involved in acetohydroxy acid synthesis. For example, pyruvate dehydrogenase (decarboxylases) catalyzes the synthesis of acetolactate when confronted with a high concentration of pyruvate (Juni and Heym, 1956). Even though the product formed is racemic, the natural isomer could probably be converted to valine if the pyruvate dehydrogenase of a valine-forming cell were to so function. Similarly, acetohydroxybutyrate may also be formed if α-ketobutyrate is present at a high concentration. Such a reaction, which occurs at a lower pH (6.0) than is optimal for the usual acetohydroxy acid biosynthesis (pH 7.5–8.0), has been demonstrated in *E. coli* (Radhakrishnan and Snell, 1960).

Furthermore, valine resistance of *E. coli* K12 has been brought about by mutations at several previously unrecognized loci specifying additional acetohydroxy acid synthases (Davis et al., 1977). One such mutation is linked to *metE,* and another lies between the *ara* and *leu* operons, which distinguish it from the closely linked *ilvH* mutation (Jackson et al., 1981). Studies demonstrating a unique protein product of these genes have not been reported.

Perhaps related to these observations are results obtained when plasmids containing the *ilvB* gene are isolated (Smith, 1981). The DNA used in these studies is from a strain with a deletion of *ilvHI* and a deletion extending over parts of the *ilvG* and *ilvE* genes. Yet among the isolated plasmids, some specify a valine-insensitive acetohydroxy acid synthase and are able to transform *ilvHI-ilvB* double mutants to prototrophy. The plasmids or their encoded products have not been further examined; the plasmids may contain either some of the genes identified by Davis et al. (1977) or the *aceE* gene, which specifies the carboxylase component of pyruvate dehydrogenase. This protein could perhaps catalyze both acetolactate and acetohydroxybutyrate synthesis in amounts sufficient for growth when its gene is carried on a high-copy-number plasmid.

III. REGULATION OF THE FORMATION OF THE ISOLEUCINE AND VALINE BIOSYNTHETIC ENZYMES

The regulation of the amount of isoleucine and valine biosynthetic enzymes, studied extensively in *E. coli* and *S. typhimurium,* is controlled exclusively at the level of transcription. Both the initiation of transcription and the extent of transcription can be subject to control. In both organisms, there appear to be at least five separate units of transcription, each with its own specific pattern of control.

A. Regulation of the *ilvGEDA* Operon

The *ilvGEDA* operon of *E. coli* K12 once appeared to be more complex than it is. For example, mutations enhance the level of expression of the *ilvE, ilvD,* and *ilvA* genes and simultaneously activate a cryptic *ilvG* gene, which is silent in *ilvO*$^{+}$ strains. The *ilvO* region is now known to be a ten-base sequence preceding a TGA codon near the middle of the *ilvG* gene in the wild-type K12 strain (F. Nargang and E. Harms, unpublished results; Lawther et al., 1981). The *ilvO* mutations are either a one-base deletion or a two-base insertion (Fig. 14.3). Plasmids carrying an *ilvG* gene with an *ilvO* mutation specify the formation in minicells of the approximately 60,000-dalton *ilvG* product instead of a fragment about half that size specified by the *ilvG*$^{+}$ plasmid (Fig. 14.4). The *ilvO* mutations also abolish a polar effect on the downstream genes of the operon, *ilvE, ilvD,* and *ilvA* (Gayda et al., 1980).

Frame	
	Sequence in *ilvO2096* DNA:
	....GGCACATGTGGCATTACAAGGTGATTTAAATGCTCTGTTACCAGCATTACAGCAGCCGTTAAATCAATATGACTGGCAGCAA....
0	...AlaHisGlyAlaLeuGlnGlyAspLeuAsnAlaLeuLeuProAlaLeuGlnGlnProLeuAsnGlnTyrAspTrpGlnGln
	↑ Insertion of AT (or TA)
	Sequence in *ilvO*$^{+}$ DNA:
	....GGCACATGTGGCATTACAAGGTGATTTAAATGCTCTGTTACCAGCATTACAGCAGCCGTTAAATCAATGACTGGCAGCAA....
0	...AlaHisGlyAlaLeuGlnGlyAspLeuAsnAlaLeuLeuProAlaLeuGlnGlnProLeuAsnGlnStop
+1	....HisMetTrpHisTyrLysGlyIleStop
-1	..GlyThrCysGlyIleThrThrStop
	Deletion of C ↓
	Sequence in *ilvO2200* DNA:
	..GGCACATGTGGCATTACAAGGTGATTTAAATGCTCTGTTACCAGCATTACAGCAGCCGTTAAATAATGACTGGCAGCAA....
0	...AlaHisGlyAlaLeuGlnGlyAspLeuAsnAlaLeuLeuProAlaLeuGlnGlnProLeuAsnIleAspTrpGlnGln....

FIGURE 14.3 Two *ilvO* mutations conferring the Val^{r} phenotype to *E. coli* K12. Nucleotide sequences in the *ilvO* region: top, following a two-base insertion (Lawther et al., 1981); center, wild-type K12 sequence; bottom, following a one-base deletion (F. Nargang and E. Harms, unpublished results).

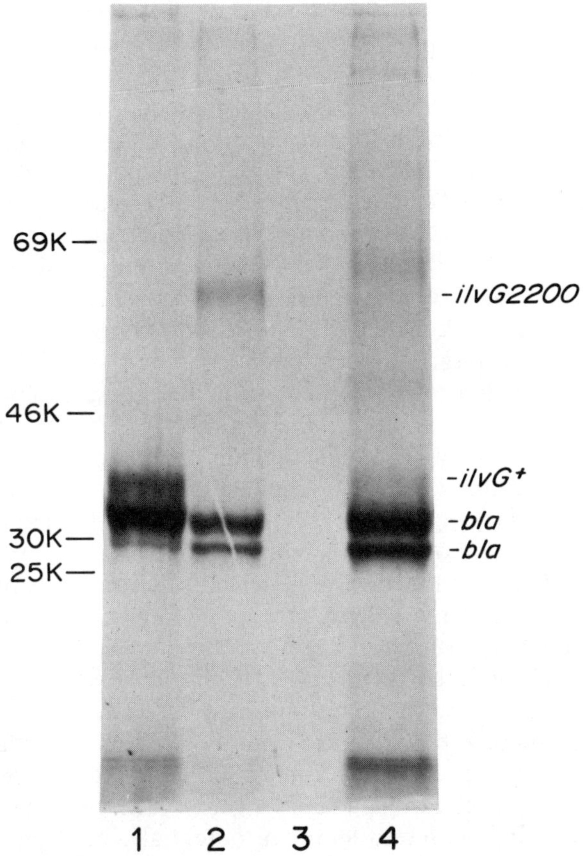

FIGURE 14.4 **Formation of the *ilvG*$^+$ and *ilvG2200* products in minicells. A minicell strain of *E. coli* K12 is infected with plasmids and used for protein synthesis in the presence of ^{35}S-methionine in a minimal salts medium supplemented with a mixture of 17 amino acids; valine, isoleucine, and carrier methionine are not added. The labeled proteins are detected by fluorography following discontinuous electrophoresis in 10% polyacrylamide gels. Lane 1: Minicells infected with pBR322 carrying the 4.6-kb *HindIII* fragment encoding *ilvG*$^+$. Lane 2: Minicells infected with pBR322 carrying the homologous *ilvG2200* fragment. Lane 3: Uninfected minicells. Lane 4: Minicells infected with pBR322. Numbers to left are molecular weights of protein standards in kilodaltons. The gene products recognized are indicated at the right (E. Garrison, unpublished experiments).**

Another apparent complexity relates to multivalent control of *ilvGEDA* expression. Full repression requires all three branched-chain amino acids, leucine, valine, and isoleucine, to be present in excess. Derepression results if any one of the three is limiting. Furthermore, if charging of the cognate tRNA's is restricted, for example, in a mutant containing a low-affinity aminoacyl-tRNA synthetase, derepression occurs even in the presence of excess amino acids. Derepression also results from several mutations that probably involve modification and maturation of tRNA isoacceptors. In no case, however, has any evidence been obtained for the existence of a repressor protein comparable to that encoded by the *trpR* gene.

Regulation of the *ilvGEDA* operon is by attenuation of transcription. In this respect, the *ilvGEDA* operon resembles other amino acid biosynthetic operons that respond to tRNA charging levels: tryptophan (Lee and Yanofsky, 1977), phenylalanine (Zurawski et al., 1978), histidine (DiNocera et al, 1978; Barnes, 1978), threonine (Gardner, 1979), and leucine (Gemmill et al., 1979). In the specific case of the bivalently controlled *thr* operon (Gardner, 1979) (Chapter 10), transcripts escape attenuation when translation of the *thr* leader is stalled at either the threonine or the isoleucine codon. By analogy, one expects a leader transcript for the *ilvGEDA* operon to encode a short polypeptide rich in the three branched-chain amino acids. This expectation has been confirmed through DNA sequence analysis of a region upstream of the *ilvG* gene (Nargang et al., 1980; Lawther and Hatfield, 1980).

The DNA sequence of the noncoding strand is shown in Fig. 14.5. The eight thymidine residues that would appear as eight uridine residues in the transcript are preceded by a DNA stretch encoding a potential stem and loop RNA secondary structure. Together they provide an *ilv*-independent termination signal for RNA polymerase. Evidence for the efficiency of this terminator is a 183-nucleotide-long transcript derived from DNA templates carrying the polymerase-binding site. Upstream of this potential termination site, the transcript has one reading frame encoding a 32-residue polypeptide, 15 of which are branched-chain amino acids. The upstream arm of the proposed terminating stem and loop can potentially base pair instead with bases farther upstream, and these in turn can also pair with others still farther upstream. Some of these competing structures should be prevented by translating ribosomes trailing close by the transcribing polymerase. A model incorporating these structural features accounts for early termination unless the formation of the terminating stem and loop is prevented. In the model (Fig. 14.6), the competing RNA secondary structures are referred to as terminator, preemptor, and protector, in accord with the terminology of Keller and Calvo (1979).

The model for attenuation of transcription initiated at the *ilvGEDA* promoter accounts for all the known facts. The significance of the RNA secondary structures will become more certain when attenuator sequences of mutants with either increased or decreased expression of the *ilvGEDA* operon are examined. Such studies are in progress. The attenuator structure found in *E. coli* K12 appears to be functionally important because it is completely conserved in *S. typhimurium* in the region specifying the 32-residue polypeptide and the predicted RNA stems and loops (Taillon et al., 1981; C. S. Subrahmanyam, unpublished observation). In contrast, there are a number of base differences outside this region and even third-base (synonymous) codon changes in the early part of the *ilvG* structural gene.

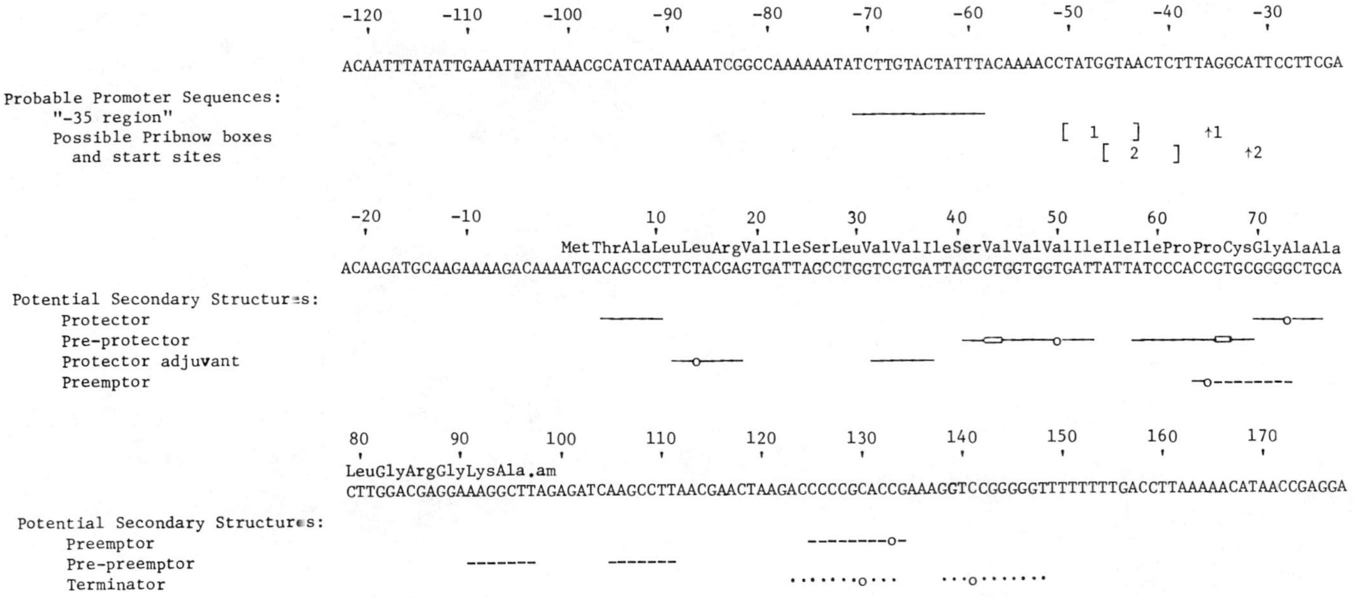

FIGURE 14.5 **DNA sequence of noncoding strand of *ilv* leader showing corresponding amino acid sequence of *ilv* leader peptide. Two possible Pribnow boxes are indicated by brackets; two possible transcription initiation sites are indicated by arrows. The regions of secondary structure expected in the corresponding *ilv* leader transcript are shown as solid, dashed, or dotted lines under the appropriate sequence. Circle: single-base bulge in the stem; open box: regions of G·U base pairing. (Reprinted from *Proceedings of the National Academy of Sciences U.S.A.* with permission.)**

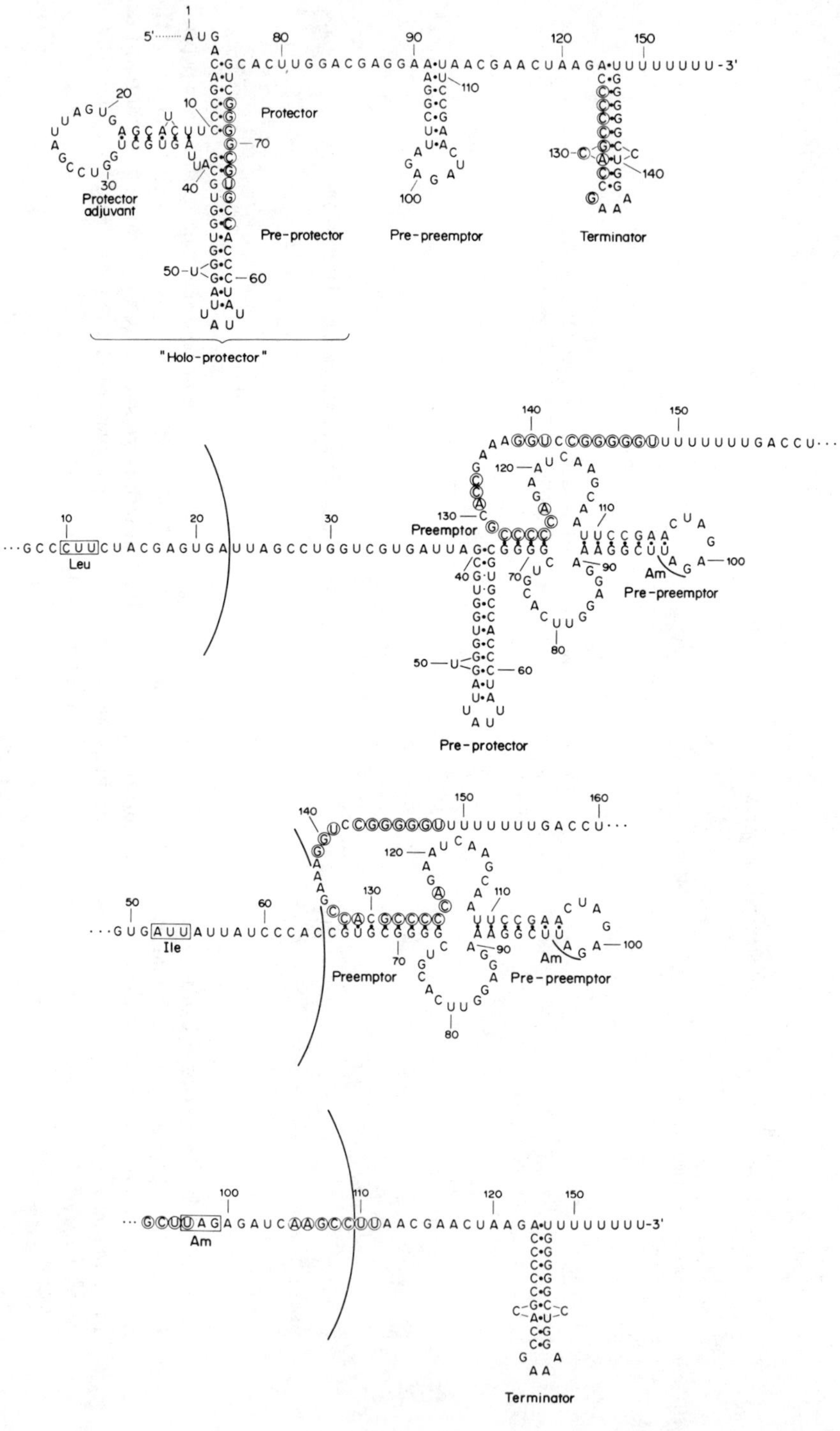
Protector
Protector adjuvant
Pre-protector
Pre-preemptor
Terminator
"Holo-protector"
Leu
Preemptor
Pre-preemptor
Am
Pre-protector
Ile
Preemptor
Pre-preemptor
Am
Am
Terminator

While attenuation is the only *ilv*-specific control of the *ilvGEDA* operon in *E. coli,* expression of the operon is strongly repressed in cells grown in a rich medium. This reduced expression can be regarded as a nonspecific control. In rapidly growing cells, more of the synthetic activities may be primarily committed to protein and nucleic acid–forming enzymes. RNA polymerase as well as ribosomes may in turn be disproportionately devoted to these functions, so that the operons concerned with small molecule formation become superrepressed.

Perhaps related to rich medium control is the in vitro guanosine tetraphosphate–mediated stimulation of coupled translation–transcription initiated at the *ilvGEDA* promoter (Noti and Umbarger, 1980). One cannot be certain that guanosine tetraphosphate stimulates transcription at the *ilvGEDA* promoter because the relative extent of *ilv*-specific transcription and readthrough transcription have not been determined. Thus whether the rich medium effect is related to the guanosine tetraphosphate effect is not known. However, it would be of interest to examine how the growth rate affects *ilv* operon expression in the complete absence of attenuation control, that is, when the operon is constitutive.

The possible role of threonine deaminase or the *ilvA* gene itself regulating the expression of the *ilv* genes has been reviewed (Calhoun and Hatfield, 1975). Only one observation cannot be accommodated directly by the attenuation model: a mutation in the *ilvA* gene (*ilvA538*) that reduces expression of the *ilvGEDA* operon (Levinthal et al., 1973). Although the mutation is nitrosoguanidine-induced, the fact that small deletions encompassing the *ilvA538* mutation restore operon function favors the view that the *ilvA538* mutation is responsible for the reduced expression. The biochemical basis of the effect remains unknown. However, the branched-chain aminoacyl-tRNA synthetases are unstable in a strain containing a deletion of the *ilvA* gene, suggesting that threonine deaminase affects attenuation indirectly via stabilization of the synthetases (Coleman et al., 1975).

Finally, in *Neurospora crassa,* a highly pleiotropic locus, *leu3,* apparently allows derepression of all the enzymes of branched-chain amino acid biosynthesis in response to a limitation of one or another of the branched-chain amino acids (Olshan and Gross, 1974). As in other eukaryotic genetic regulatory circuits, the mechanism by which the *leu3* gene functions is not understood.

◄ **FIGURE 14.6 Proposed secondary structure of mRNA derived from the leader region of the *ilvGEDA* operon. Top: Attenuated transcript obtained when transcript occurs in the absence of translating ribosomes. Second from top: Nonattenuated transcript resulting from ribosome stalled at a leucine codon; preemptor formed rather than terminator. Second from bottom: Same except that ribosome has stalled at an isoleucine codon. Bottom: Attenuated transcript resulting from complete translation of the leader peptide.**

B. Regulation of the *ilvHI* Operon

The product of the *E. coli* K12 *ilvHI* operon, acetohydroxy acid synthase III, has not yet been purified. Expression of *ilvHI* is controlled by leucine alone, and it seems reasonable to infer a regulatory function for leucyl-tRNA.

More definitive are the results of nucleotide sequence determination of the entire *ilvHI* operon. The genetic complementation studies mentioned above favor the idea of separate *ilvH* and *ilvI* genes. DNA sequence analysis reveals two separate genes that appear to be *ilvH* and *ilvI* specifying acetohydroxy acid synthase III (J. Calvo, personal communication).

C. Regulation of *ilvB* Expression

The *ilvB* gene is bivalently repressed by leucine and valine. Fully functional valyl-tRNA synthetase is necessary for valine to participate in the repression signal (Eidlic and Neidhardt, 1965). These experiments antedate the recognition of acetohydroxy acid synthase III and were performed in the presence of excess leucine, which represses *ilvHI* expression. Thus whether leucyl-tRNA synthetase function is required for *ilvB* expression was not addressed in the early work.

In an *ilvHI* mutant, *ilvB* expression does indeed respond to limitations in the charging of both $tRNA^{Val}$ and $tRNA^{Leu}$ (Smith, 1981). The *ilvB* gene from *E. coli* has been cloned (Smith, 1981; Neuman et al., 1982). The cloned fragment contains an RNA polymerase binding site from which an attenuated transcript has been obtained. Nucleotide sequence of this region, reported by Friden et al. (1982) and by Hauser and Hatfield (1983), has revealed that the attenuated transcript specifies a 32-amino acid peptide containing nine valine and three leucine residues. The two groups have proposed slightly different systems of competing stem and loop structures that might account for the control of the *ilvB* gene. This region also contains a site reminiscent of a cAMP-dependent binding site for catabolite repressor protein. In vitro synthesis of the leader transcript is strongly stimulated by catabolite repressor protein and cAMP (Friden et al., 1982). This observation is in accord with the earlier observations on regulation of acetohydroxy acid synthase activity in *E. coli* by catabolite repression (Coukell and Polglase, 1969; Sutton and Freundlich, 1980).

Thus it appears that the bivalent repression of the *ilvB* gene is achieved by attenuation of transcription like that found for other genes that are affected by the charging level of a tRNA.

D. Regulation of *ilvC* Expression

The *ilvC* gene specifies acetohydroxy acid isomeroreductase. Unlike the repressible genes specifying the other isoleucine and valine biosynthetic enzymes, *ilvC* expression is inducible. The inducers are either of the two acetohydroxy acid isomeroreductase substrates (Ratzkin et al., 1972). Induction does not occur as does that of the *lac* operon by inducer-dependent inactivation of a repressor but, rather, as does that of the *mal* operons, by an inducer-dependent activation of a positive control element. The positive control element is specified by the *ilvY* gene.

The role of the *ilvY* gene is revealed in in vitro coupled transcription–translation experiments with templates containing *lacZ* fused to the *ilvC* promoter (Watson et al., 1979). Any *E. coli* S-30 extract serves to form acetohydroxybutyrate–induced β-galactosidase except extracts prepared from strains bearing deletions in the *ilvA-C* region (Fig. 14.2). Such strains yield active extracts if transformed with plasmids containing the DNA between the *ilvA* and *ilvC* genes, which is the locus of *ilvY*.

The sequential in vitro synthesis of the *ilvY* gene product, upsilon, and either *ilvC*-controlled β-galactosidase or the *ilvC* product itself has been demonstrated (Biel, 1981). Inducer-independent mutations in *ilvY* allow *ilvC* expression in the absence of inducer (Biel and Umbarger, 1981). The constitutive mutations are *trans*-dominant to the wild-type alleles, so that it appears unlikely that upsilon plays a negative control role in the absence of inducer. These results support a model for *ilvY* function in which upsilon, in the presence of inducer, binds to a site near the *ilvC* promoter to facilitate transcription of the *ilvC* gene. However, this simple model must be questioned in view of the fact that neither the wild-type nor the constitutive allele of the *ilvY* gene of *E. coli* is active when placed in *trans* to an *ilvY* gene bearing an insertion of Mu or Tn*10* (J. Falk, unpublished observation). Thus, it may be that, while transcription into the *ilvC* gene is dependent upon an active form of upsilon, that transcription must be initiated either within the *ilvY* gene or at the *ilvY* promoter. It is possible that upsilon is an antiterminator preventing termination of transcription at a site between the *ilvY* and *ilvC* genes.

A general control over *ilvC* expression is indicated by a strong in vitro stimulation by guanosine tetraphosphate (Wild et al., 1977). Together with the other *ilv* genes, expression of *ilvC* is strongly depressed during rapid growth in a rich medium. Although *ilvY* expression is also depressed in rich medium-grown cells, it does not seem to be affected by any specific control mediated by the branched-chain amino acids (J. Falk, unpublished observations). The *ilvY* gene is not autogenously controlled; high levels of the product are made by cells carrying high copy number *ilvY* plasmids (Biel, 1981).

IV. CLONING OF THE *ILV* GENES AND THE CORRELATION BETWEEN THE GENETIC AND PHYSICAL MAPS

Defective lambda *ilv* transducing phages (Favre et al., 1976) have provided the initial material for cloning of the genes in the *ilv* cluster of *E. coli* K12. A variety of fragments of the *ilv* cluster inserted into pBR313 and pBR322 have been subjected to restriction analysis; together with electron microscopic examination of heteroduplexes between several *ilv* phages as well as *ilv*-

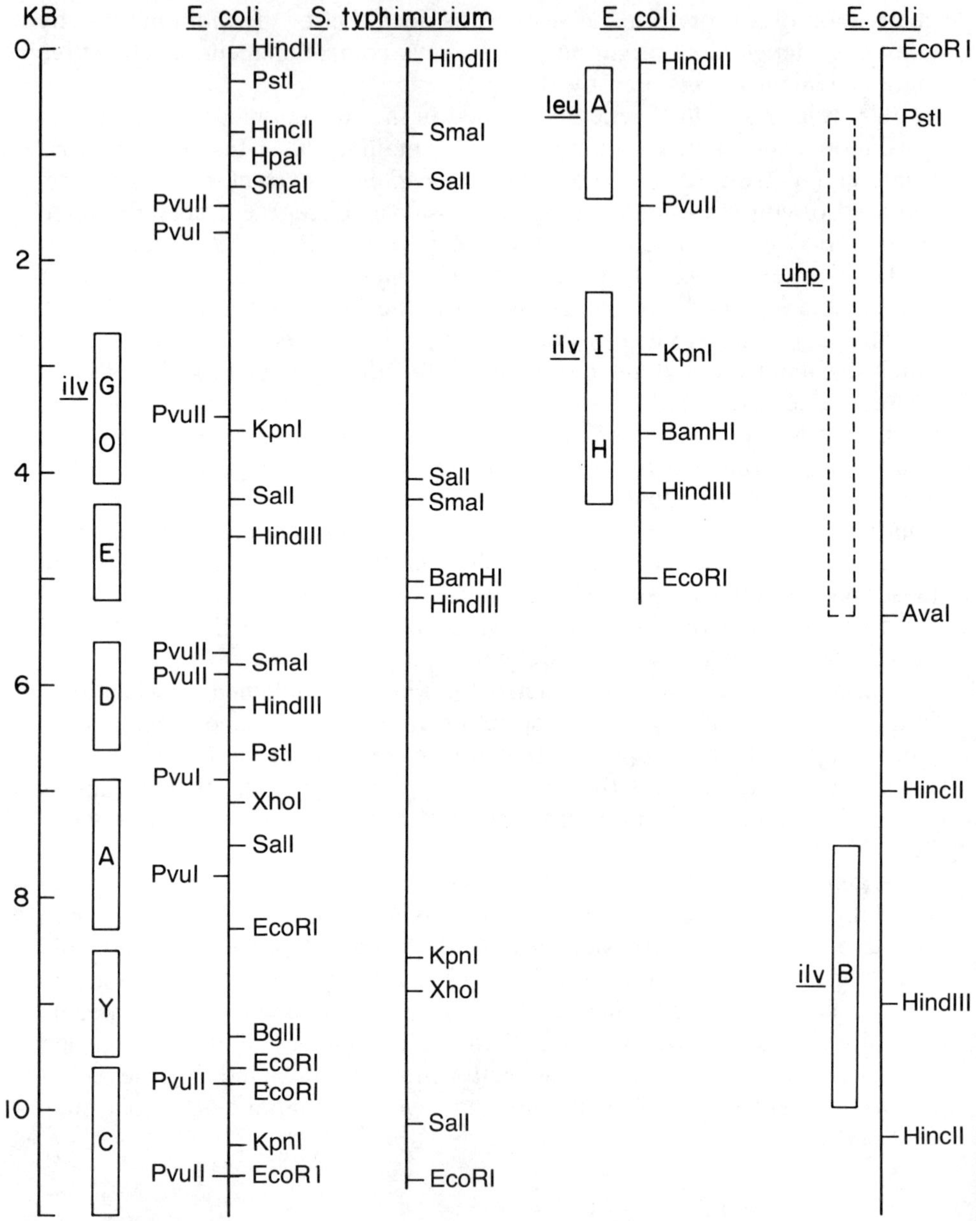

FIGURE 14.7 Correlation of the physical and genetic map of *ilv* cluster DNA from *E. coli* and *S. typhimurium* and partial restriction maps of *E. coli ilvHI* and *ilvB* genes. Transcription of the respective *ilv* genes is from top to bottom. The DNA fragments are drawn to scale; coding sequences are indicated by boxes.

bearing plasmids, the results permit the construction of a composite map that correlates physical and genetic data for the *ilv* cluster (Fig. 14.7) (McCorkle et al., 1978).

Also shown in Fig. 14.7 is a physical and genetic map of the corresponding region in *S. typhimurium* adapted from the work of Blazey et al. (1981), who initially cloned *S. typhimurium* chromosomal DNA partially digested with *EcoRI* into pVH2124 and subsequently subcloned fragments into pBR322. The initial *S. typhimurium ilv* clone carried *ilv* DNA extending only into but not through the *ilvC* gene.

The *ilvHI* genes of *E. coli* K12 have been subcloned from a lambda *leu* transducing phage that also contains the closely linked *ilvHI* gene (Squires et al., 1981). A partial physical map of that region is also shown in Fig. 14.7.

As indicated above, DNA in the region of the *ilvB* gene has been cloned in three different laboratories from the chromosomal DNA of the K12 strain, and there is agreement that the entire *ilvB* gene lies on a 3.4 kb region between two *Hinc*II sites. The attenuator region that accounts for the bivalent control of *ilvB* by valine and leucine lies close to the *Hinc*II site at about 10 kb on the map in Fig. 14.7. The precise location of the *ilvB* gene is still uncertain. Friden et al. (1982) have observed two peptide products arising from the 3.4 kb plasmid insert. One is 60,000 daltons as expected of the *ilvB* gene (Grimminger and Umbarger, 1979), and the other, the function of which is unknown, is 9,000 daltons. One possibility is that the *ilvB* gene is part of an operon in which the *ilvB* gene might be proximal or distal to the leader region.

The physical maps for the various *ilv* regions provide hints to other investigators who may wish to clone specific fragments of the *ilv* genes. High-copy-number plasmids carrying the *ilv* gene cluster with an *ilvO* mutation allowing *ilvG* expression and a stable mutation inactivating the *ilvA* gene, would be expected to produce large quantities of valine. Such a plasmid should be effective because the valine pathway draws its initial substrate, pyruvate, directly from central metabolism. In contrast, the corresponding DNA fragment carrying a feedback resistance marker in the *ilvA* gene might not be so effective in producing isoleucine, unless it was placed in a background carrying mutations that would result in a high internal pool of endogenously formed threonine. Mutations resulting in elevated threonine production might be a derepressed *thr* operon and a *thrA* gene with a mutation resulting in resistance of aspartate kinase I–homoserine dehydrogenase I to threonine.

ACKNOWLEDGMENT

The work on isoleucine and valine biosynthesis conducted in the author's laboratory has been supported for the past thirty-one years exclusively by grants from the National Institutes of Health and since 1964 by Public Health Service Grant GM12522.

REFERENCES

Barnes, W. M. (1978) *Proc. Nat. Acad. Sci. U.S.A. 75,* 4281–4285.

Biel, A. J. (1981) "Regulation of the Expression of the *Escherichia coli ilvC* Gene," Ph.D. thesis, Purdue University, West Lafayette, Ind.

Biel, A. J., and Umbarger, H. E. (1981) *J. Bacteriol. 146,* 718–724.

Blatt, J. M., Pledger, W. J., and Umbarger, H. E. (1972) *Biochem. Biophys. Res. Comm. 48,* 444–450.

Blazey, D. L., and Burns, R. O. (1979) *Mol. Gen. Genet. 177,* 1–11.

Blazey, D. L., Kim, R., and Burns, R. O. (1981) *J. Bacteriol. 147,* 452–462.

Brown, C. S., Kline, E. L., and Umbarger, H. E. (1975) in *Isozymes II, Physiological Function* (Markert, C. L., ed.) pp. 249–273, Academic Press, New York.

Calhoun, D. H., and Hatfield, G. W. (1975) *Ann. Rev. Microbiol. 29,* 275–299.

Changeux, J.-P. (1961) *Cold Spring Harbor Symp. Quant. Biol. 26,* 313–318.

Changeux, J.-P. (1963) *Cold Spring Harbor Symp. Quant. Biol. 28,* 497–504.

Charon, N. W., Johnson, R. C., and Peterson, D. (1974) *J. Bacteriol. 117,* 203–211.

Cohn, M. S., and Phillips, A. T. (1974) *Biochemistry 13,* 1208–1214.

Coleman, W., Jr., Kline, E. L., Brown, C. S., and Williams, L. S. (1975) *J. Bacteriol. 121,* 785–793.

Davis, E. J., Blatt, J. M., Henderson, E. K., Whittaker, J. J., and Jackson, J. H. (1977) *Mol. Gen. Genet. 156,* 239–249.

Decedue, C. J., Hofler, J. G., and Burns, R. O. (1975) *J. Biol. Chem. 250,* 1563–1570.

DeFelice, M., Guardiola, J., Esposito, B., and Iaccarino, M. (1974) *J. Bacteriol. 120,* 1068–1077.

Cowkell, M. B., and Polglase, W. T. (1969) *Biochem. J. 111,* 273–278.

DiNocera, P. P., Blasi, F., DiLauro, R., Frunzio, R., and Bruni, C. B. (1978) *Proc. Nat. Acad. Sci. U.S.A. 75,* 4276–4280.

Eidlic, L., and Neidhardt, F. C. (1965) *Proc. Nat. Acad. Sci. U.S.A. 53,* 539–543.

Favre, R., Wiater, A., Puppo, S., Iaccarino, M., Noelle, R., and Freundlich, M. (1976) *Mol. Gen. Genet. 143,* 243–252.

Friden, P., Newman, T., and Freundlich, M. (1982) *Proc. Nat. Acad. Sci. U.S.A. 79,* 6156–6160.

Gardner, J. F. (1979) *Proc. Nat. Acad. Sci. U.S.A. 76,* 1706–1710.

Gayda, D. J., Leathers, T. D., Noti, J. D., Smith, F. J., Smith, J. M., Subrahmanyam, C. S., and Umbarger, H. E. (1980) *J. Bacteriol. 142,* 556–567.

Gemmill, R. M., Wessler, S. R., Keller, E. B., and Calvo, J. M. (1979) *Proc. Nat. Acad. Sci. U.S.A. 76,* 4941–4945.

Guardiola, J., DeFelice, M., and Iaccarino, M. (1974) *J. Bacteriol. 120,* 536–538.

Hatfield, G. W., and Burns, R. O. (1970) *J. Biol. Chem. 245,* 787–791.

Hawser, C. A., and Hatfield, G. W. (1983) *Nucl. Acid Res. 11,* 127–139.

Hofler, J. G., and Burns, R. O. (1978) *J. Biol. Chem. 253,* 1245–1251.

Jackson, J. H., Davis, E. J., Madu, A. C., and Braxter, S. E. (1981) *Mol. Gen. Genet. 181,* 417–419.

Juni, E., and Heym, G. A. (1956) *J. Biol. Chem. 218,* 365–378.

Keller, E. B., and Calvo, J. M. (1979) *Proc. Nat. Acad. Sci. U.S.A. 76,* 6186–6190.

Kisumi, M., Sugiura, M., and Chibata, I. (1976) *J. Biochem.* (Tokyo) *80,* 333–339.
Lawther, R. P., and Hatfield, G. W. (1980) *Proc. Nat. Acad. Sci. U.S.A. 77,* 1862–1866.
Lawther, R. P., Calhoun, D. H., Adams, C. W., Hauser, C. A., Gray, J., and Hatfield, G. W. (1981) *Proc. Nat. Acad. Sci. U.S.A. 78,* 922–925.
Lee, F., and Yanofsky, C. (1977) *Proc. Nat. Acad. Sci. U.S.A. 74,* 4365–4369.
Levinthal, M., Williams, L. S., Levinthal, M., and Umbarger, H. E. (1973) *Nature New Biol. 246,* 65–68.
McCorkle, G. M., Leathers, T. D., and Umbarger, H. E. (1978) *Proc. Nat. Acad. Sci. U.S.A. 75,* 89–93.
McKenney, K., Shimtake, H., Court, D., Schmeissner, U., Brody, C., and Rosenberg, M. (1981) in *Gene Amplification and Analysis* Vol. II, *Analysis of Nucleic Acids by Enzymatic Methods* (Chirikjian, J. C., and Papas, T. S., eds.) pp. 383–415, Elsevier/North Holland, Amsterdam.
Monod, J., Wyman, J., and Changeux, J.-P. (1965) *J. Mol. Biol. 12,* 88–118.
Nargang, F. E., Subrahmanyam, C. S., and Umbarger, H. E. (1980) *Proc. Nat. Acad. Sci. U.S.A. 77,* 1823–1827.
Newman, T. C., and Levinthal, M. (1980) *Genetics 96,* 59–77.
Newman, T., Friden, P., Sutton, A., and Freundlich, M. (1982) *Molec. Gen. Genet. 186,* 378–384.
Noti, J. D., and Umbarger, H. E. (1980) *J. Bacteriol. 144,* 291–299.
Olshan, A. R., and Gross, S. R. (1974) *J. Bacteriol. 118,* 374–384.
O'Neill, J. P., and Freundlich, M. (1972) *Biochem. Biophys. Res. Comm. 48,* 437–443.
Phillips, A. T., Nuss, J. I., Moosic, J., and Foshay, C. (1972) *J. Bacteriol. 109,* 714–719.
Radhakrishnan, A. N., and Snell, E. E. (1960) *J. Biol. Chem. 235,* 2316–2321.
Ramakrishnan, T., and Adelberg, E. A. (1965) *J. Bacteriol. 89,* 654–660.
Ratzkin, B., Arfin, S., and Umbarger, H. E. (1972) *J. Bacteriol. 112,* 131–141.
Roberts, R. B., Cowie, D. B., Abelson, P. H., Bolton, E. T., and Britten, R. J. (1955) in *Studies of Biosynthesis in Escherichia coli* Chap. 19, pp. 406–417, Publication 607, Carnegie Institution, Washington, D.C.
de Robichon-Szulmajster, H., and Magee, P. T. (1968) *Eur. J. Biochem. 3,* 492–501.
Skarstedt, M. T., and Greer, S. B. (1973) *J. Biol. Chem. 248,* 1032–1044.
Smith, F. J. (1981) "Genetic and physical analysis of the *ilvGEDA* operon and the *ilvB* gene of *Escherichia coli* K12," Ph.D. thesis, Purdue University, West Lafayette, Ind.
Squires, C. H., DeFelice, M., Wessler, S. R., and Calvo, J. M. (1981) *J. Bacteriol. 147,* 797–804.
Stallcup, W. B., and Koshland, D. E., Jr. (1973) *J. Mol. Biol. 80,* 41–62.
Sutton, A., and Freundlich, M. (1980) Molec. Gen. Genet. 178, 179–183.
Sutton, A., Newman, T., Francis, M., and Freundlich, M. (1981) *J. Bacteriol. 148,* 998–1001.
Taillon, M. P., Gotto, D. A., and Lawther, R. P. (1981) *Nucl. Acids Res. 9,* 3419–3432.
Umbarger, H. E. (1956) *Science 123,* 848.
Umbarger, H. E., and Brown, B. (1953) *J. Bacteriol. 65,* 203–209.

Watson, M. D., Wild, J., and Umbarger, H. E. (1979) *J. Bacteriol. 139,* 1014–1020.
Wild, J., Smith, J. M., and Umbarger, H. E. (1977) *J. Bacteriol. 132,* 876–883.
Zarlengo, M. H., Robinson, G. W., and Burns, R. O. (1968) *J. Biol. Chem. 243,* 186–191.
Zurawski, G., Brown, K., Killingly, D., and Yanofsky, C. (1978) *Proc. Nat. Acad. Sci. U.S.A. 75,* 4271–4275.

CHAPTER
15

Leucine Biosynthesis in Prokaryotes

Joseph M. Calvo

I. INTRODUCTION

The major pathway to leucine, found in almost all bacteria and fungi studied to date, is termed the isopropylmalate pathway (Fig. 15.1). This pathway has been investigated most extensively in *Salmonella typhimurium* and *Escherichia coli* among bacteria and in *Neurospora crassa* and *Saccharomyces cerevisiae* among fungi. Less extensive studies suggest that this pathway functions in more than 15 genera of bacteria (Stieglitz and Calvo, 1974; Martin et al., 1962) and in maize, sorghum, and flax (Oaks, 1965; Butler and Shen, 1963). While the isopropylmalate pathway is certainly the major pathway, it is not the only pathway to leucine. In certain rumen microorganisms, including *Bacteroides ruminocola* and *Peptostreptococcus elsdenii*, leucine is synthesized by a pathway that involves reductive carboxylation of isovalerate (Allison et al., 1966). This latter pathway is well suited to rumen prokaryotes, since their environment is rich in short-chain fatty acids produced during the decomposition of cellulose.

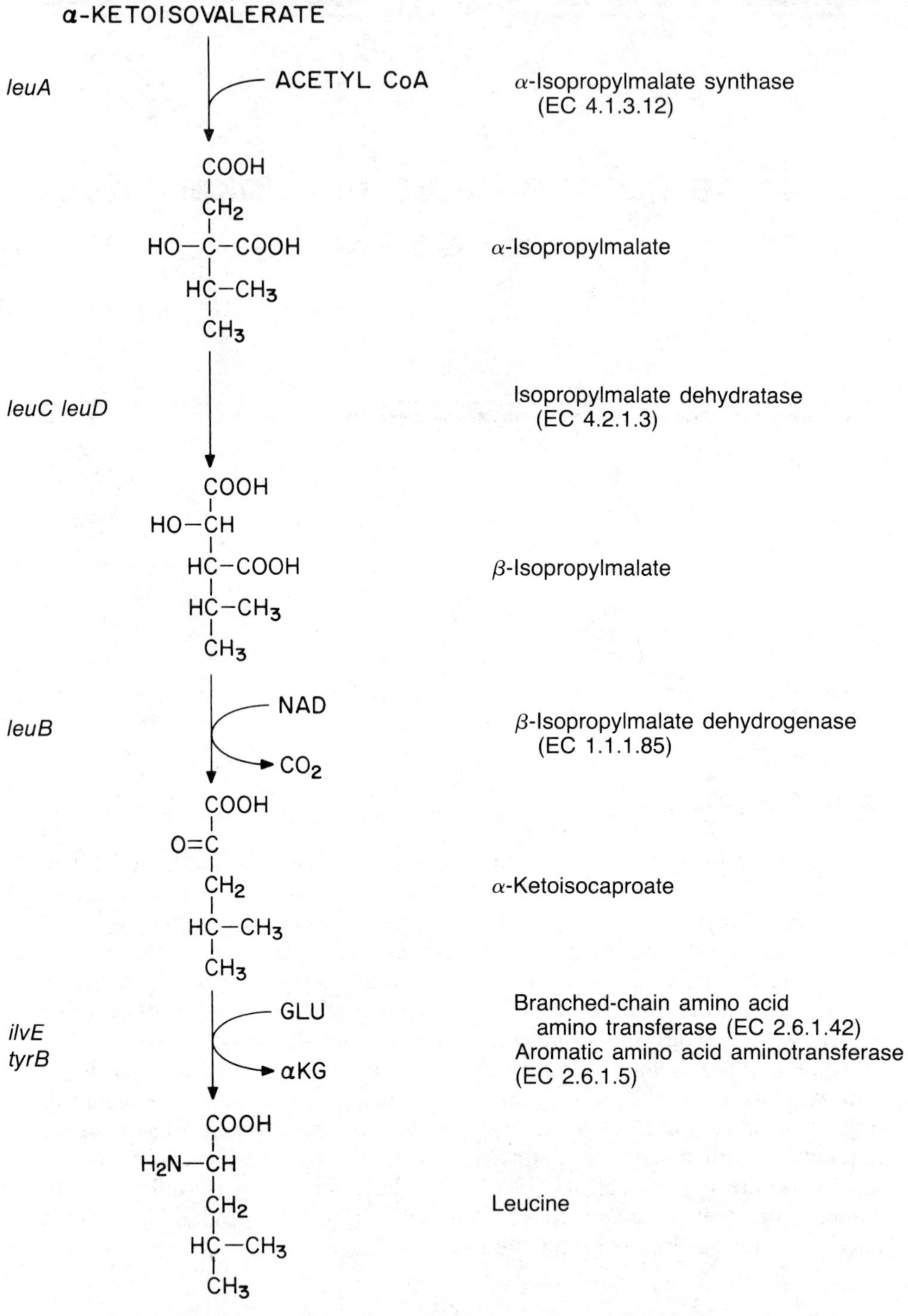

FIGURE 15.1 The pathway of leucine biosynthesis in *E. coli* and *S. typhimurium*.

II. THE PATHWAY

In *S. typhimurium* and *E. coli*, there are three enzymes unique to leucine biosynthesis: α-isopropylmalate synthase, β-isopropylmalate dehydrogenase, and isopropylmalate dehydratase (Fig. 15.1). Transamination of α-ketoisocaproate can be catalyzed by either the branched-chain amino acid or the aromatic amino acid aminotransferase (Umbarger, 1978). α-Isopropylmalate synthase from *S. typhimurium* consists of up to four identical polypeptides, each having a molecular weight of 50,000 (Bartholomew and Calvo, 1971a; Kohlhaw and Boatman, 1971). The molecular weight of the native enzyme varies, depending upon the concentration of this protein and upon the presence of certain ligands (Leary and Kohlhaw, 1970; Kohlhaw and Boatman, 1971). No cofactors or metal ions are required for its activity. β-Isopropylmalate dehydrogenase from *S. typhimurium* is composed of two identical subunits having a molecular weight of 35,000 (Parsons and Burns, 1969). The absolute configuration of the β-isopropylmalate stereoisomer recognized by this enzyme is *erythro*-D_s (Calvo et al., 1964). Isopropylmalate dehydratase from enteric bacteria (Gross et al., 1963) is relatively unstable, and its purification has not been reported. Purification of α-isopropylmalate synthase and β-isopropylmalate dehydrogenase has been considerably simplified because of the availability of plasmids encoding *leu* enzymes. Under suitable conditions of growth, more than 20% of the total soluble protein of appropriate plasmid-bearing strains is α-isopropylmalate synthase, and another 20% is β-isopropylmalate dehydrogenase (Gemmill et al., 1983).

In both *E. coli* and *S. typhimurium*, four contiguous structural genes encode the three *leu*-unique enzymes (Margolin, 1963; Yang and Kessler, 1974; Somers et al., 1973). These four genes, located at min 4 on the chromosome, are transcribed as a unit and thus comprise an operon (Burns et al., 1966). A detailed genetic map of the *leu* operon from *S. typhimurium* results from a genetic analysis of more than 50 deletions and 600 point mutations (Calvo and Warden, 1970).

III. REGULATION OF ENZYME ACTIVITY

α-Isopropylmalate synthase, the first enzyme unique to leucine biosynthesis, is inhibited by low concentrations of leucine (Calvo et al., 1969b). Detailed studies of this enzyme and of the mechanism by which it is inhibited show that leucine inhibition is strongly pH-dependent, decreasing some 30-fold between pH 6.5 and 8.5 (Kohlhaw et al., 1969). The kinetics of inhibition are noncompetitive with respect to α-ketoisovalerate but partially competitive with respect to acetyl CoA. The native enzyme exists in an association–dissociation equilibrium that fits an ideal monomer–tetramer equilibrium with an association constant of $k_4 = [A_4]/[A]^4$ of $9 \times 10^{16}\ M^{-3}$ (Leary and Kohlhaw, 1972). Leucine shifts the equilibrium in favor of dissociation.

Neither substrate alone affects the equilibrium, but together they favor association. By protecting the enzyme with substrates during reaction with dimethyl suberimidate, it is possible to isolate a cross-linked, tetrameric, enzymatically active protein. The activity of this cross-linked enzyme is not significantly inhibited by leucine.

Equilibrium dialysis measurements indicate that leucine binding (up to 0.5 leucine per monomer) is positively cooperative and that the affinity for leucine decreases with increasing enzyme concentration, with increasing pH, and with the presence of acetyl CoA (Leary and Kohlhaw, 1973). The binding constant for leucine is similar to the concentration of leucine required for 50% inhibition (about 20 μM).

A thorough comparison was made between the wild-type enzyme and a mutant enzyme that is relatively insensitive to inhibition (K_i 22 μM versus 20 mM, respectively) (Leary and Kohlhaw, 1973; Soper et al., 1976). Leucine does not bind to the mutant enzyme with high affinity, nor does it promote the dissociation of the enzyme to dimers and monomers. These altered properties are the result of the absence from the mutant enzyme of several C-terminal amino acids (Bartholomew and Calvo, 1971b).

Leucine promotes the dissociation of the enzyme into monomers and dimers. Leucine also causes reduced reactivity of SH groups (Kohlhaw, 1970) and changes the ultraviolet absorption spectrum, results suggesting that leucine causes a conformational change in the enzyme (Leary and Kohlhaw, 1973). Leucine may inhibit the enzyme by causing or stabilizing conformational changes or by promoting dissociation of the enzyme into dimers and monomers. Resolution of this question will depend upon determining the relative enzymatic activities of monomers, dimers, and tetramers.

IV. REGULATION OF *LEU* OPERON EXPRESSION

The *leu* enzyme levels in enteric bacteria vary over a several thousand–fold range, depending upon the strain and upon the conditions of growth. The following relative values define convenient points of reference: wild-type strain grown in a minimal medium containing excess leucine, 1; the same strain grown in a minimal medium, 2–3; a leucine auxotroph grown in a minimal medium in a chemostat under conditions of leucine limitation, 20–50; wild-type strain grown in a medium containing yeast extract and tryptone broth, 0.02–0.1 (Davis and Calvo, 1977b).

The remainder of this discussion relates to cells grown in a minimal medium at 37°C. For these growth conditions, there is good evidence that attenuation of transcription is the major (and probably only) control mechanism by which *leucine* regulates expression of the *leu* operon. Repression in rich media is probably independent of attenuation in the *leu* leader region, since both wild-type and an attenuator mutant are repressed equally in rich media (Searles et al., 1983).

A. The *leu* Promoter and Control Region

An *EcoRI* fragment of the *leu* operon of *S. typhimurium* containing *leuA*, *leuB*, *leuC*, and part of *leuD* has been cloned in plasmids pSC101 and pMB9 (Hertzberg et al., 1980). A restriction map of the *EcoRI* fragment was correlated with a genetic map by Southern blotting techniques and by analyzing cloned DNA from deletion-containing strains (Gemmill et al., 1983). A deletion, with an end point between *leuP* and the beginning of *leuA*, was particularly useful in identifying a fragment of DNA containing the *leu* promoter and control region. The nucleotide sequence of this region is shown in Fig. 15.2 (Gemmill et al., 1979). The *leu* promoter, defined by mutant analysis and in vitro transcription studies, extends from nucleotides −7 to −37. The coding sequence of *leuA*, deduced from the amino acid sequence of the N-terminal portion of α-isopropylmalate synthase, is at position 202 (Gemmill et al., 1983). Between *leuP* and *leuA* is a strong rho-independent transcription termination signal, the *leu* attenuator, spanning positions 132–160. In vitro transcription initiated at the *leu* promoter terminates with greater than 98% efficiency at the attenuator site, resulting in the production of a 160-nucleotide leader RNA (Gemmill et al., 1979). The *leu* leader RNA can potentially code for a 28 amino acid peptide (ribosome-binding site, nucleotides 15–18; start codon AUG, 28; stop codon UAA, 112) containing four adjacent leucine residues (codons CUA CUA CUA CUC, 54–65). The formation of this peptide in vivo has not been demonstrated. The *leu* operon control region has important features in common with the control regions of the *trp*, *phe*, *thr*, *his*, and *ilv* operons (Yanofsky, 1981).

A similar analysis of the *leu* operon of *E. coli* K12 was carried out. Specialized transducing phages carrying portions of the *leu* operon were isolated from a lysogen having phage λ inserted in *leuA* (Shimada et al., 1973), and an intact *leu* operon was generated from two of them by in vivo recombination (Davis and Calvo, 1977a). Identification of a DNA fragment containing the *leu* promoter was effected by cross-hybridizing *leu* operon DNA from *S. typhimurium* and *E. coli* and by determining which DNA fragments hybridized with *leu* mRNA (Wessler and Calvo, 1982). The *E. coli leu* operon control region (Fig. 15.2b), shares common structural features with the analogous region of the *Salmonella* genome: promoter, −7 to −37; beginning of *leuA*, 208; attenuator, 135–160; leader ribosome-binding site, 16–19; leader peptide start codon AUG, 29; leader peptide stop codon UAA, 114; four consecutive leucine codons, CUA CUA CUA CUA, 55–66. In vitro transcription of the *E. coli leu* operon leads almost exclusively to a 165-nucleotide leader RNA (Wessler and Calvo, 1982). RNA sequence studies establish the transcription start point (Fig. 15.2, position 1) and the position for transcription termination (*S. typhimurium*, 160, 161, 162; *E. coli*, 163, 164, 165) (Gemmill et al., 1983).

The *leu* promoters of *S. typhimurium* and *E. coli* contain a -35 region almost identical to the consensus sequence of Rosenberg and Court (1979)

```
        -30         -20             -10          1          10          20          30
GTT GAC ATT  AAA CGG CAT ATC CAG TAC CAC TAA AAG CAT AAC GCA TTC GCT GGA GCT GAA TTA ATG TCA CAT ATC
                                                                                    Met Ser His Ile

40              50              60              70              80              90              100             110
GTT CGT TTC ACT GGG CTA CTA CTA CTC AAC GCA TTT ATT GTG CGC GGT AGA CCG GTG GGC GGC ATT CAA CAT TAA
Val Arg Phe Thr Gly Leu Leu Leu Leu Asn Ala Phe Ile Val Arg Gly Arg Pro Val Gly Gly Ile Gln His

        120         130             140             150             160             170             180
GTC AGC TCG AAG TCA AAC AAA ACC CGC GCC GTT GCG CGG GTT TTT TTA TGC CTG ACG CAA GGC GCC CCT GGA GAC

190         200             210             220             230
AAG GAC CAC ATC ATG AGC CAG CAA GTC ATT ATT TTT GAT ACG
                Met Ser Gln Gln Val Ile Ile Phe Asp Thr
```

(A)

```
        -30         -20         -10           1     10          20          30
 G TTG ACA TCC GTT TTT GTA TCC AGT AAC TCT AAA AGC ATA TCG CAT TCA TCT GGA GCT CAT TTA ATG ACT CAC
                                                                                       Met Thr His

 40          50          60           70          80          90           100          110
ATC GTT CGC TTT ATC GGT CTA CTA CTA CTA AAC GCA TCT TCT TTG CGC GGT AGA CGA GTG AGC GGC ATC CAG CAT
Ile Val Arg Phe Ile Gly Leu Leu Leu Leu Asn Ala Ser Ser Leu Arg Gly Arg Arg Val Ser Gly Ile Gln His

        120         130         140         150          160         170         180
TAA GCC AGC ACG CAG TCA AAC AAA AAA CCC GCG CCA TTG CGC GGG TTT TTT TAT GCC CGA AGC GAG GCG CTC TAA

 190         200            210         220         230
AAG AGA CAA GGA CCC AAA CG  ATG AGC CAG CAA GTC ATT ATT TTC GAT ACC
                            Met Ser Gln Gln Val Ile Ile Phe Asp Thr
```

(B)

FIGURE 15.2 **The control region of the *leu* operon: (a) *S. typhimurium* and (b) *E. coli*.**

```
gTTGACA ttt              atttgtTATAATg     CAT        Consensus sequence
            6–9 bp                    4–7 bp

  -35 Region                    -10 Region

   -30        -20       -10        1
GTTGACATCCGTTTTTGTATCCAGTAACTCTAAAAGCA       E. coli
GTTGACATTAAACGGCATATCCAGTACCACTAAAAGCA       S. typhimurium
```

FIGURE 15.3 The promoter of the *leu* operon.

(Fig. 15.3). This is not so for the -10 region. The *leu*-500 mutation, studied extensively by Margolin and his co-workers (Dubnau et al., 1973), is an A → G transition at position -12 (Gemmill, 1981). A second site mutation that restores low level *leu* operon expression and that brings such expression under control of catabolite repression (Friedman and Margolin, 1968) is a G → A transition at position -34 (Gemmill, 1981).

A schematic comparison of the *leu* leader regions of *S. typhimurium* and *E. coli* is shown in Fig. 15.4. The genomes of these two organisms

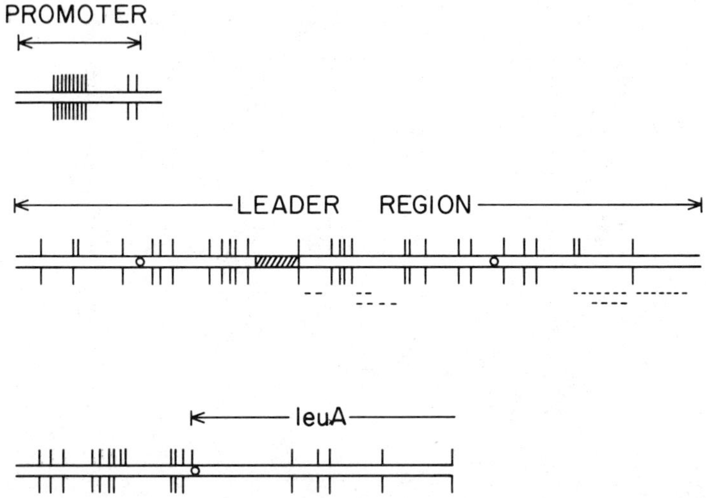

FIGURE 15.4 Conserved sequences in the *leu* operon control region of (top line) *E. coli* and (bottom line) *S. typhimurium*. Two vertical lines opposite one another indicate that the sequences differ in that position; an unopposed vertical line indicates an extra nucleotide at that position; open circles indicate translation start and stop codons; hatching indicates the position of the four leucine codons in the leader peptide; dashed lines indicate regions encoding RNA that is presumably involved in stem and loop structure formation.

have diverged extensively within the *leu* attenuator. Close inspection reveals, however, that sequences within critical regions (ribosome-binding sites, translational start and stop signals, and the four adjacent leucine codons) have been highly conserved. Other sequence similarities related to the mechanism of attenuation will be described later.

A comparison of the hypothetical leader peptides of *S. typhimurium* and *E. coli* is shown in Fig. 15.5. The number (7/28) and nature of the differences (for example, Ser/Phe, Arg/Pro) suggest that the peptides do not play a direct role in regulation of the operons.

A region of about 500 base pairs immediately upstream from the *leu* promoter differs markedly between *S. typhimurium* and *E. coli* in sequence but is rich in AT base pairs in both (about 75% AT) (G. Haughn, S. Wessler, and R. Gemmill, unpublished data). Presumably, AT richness in this region has some important function that has been preserved during evolution. Removal of this AT-rich region does not affect expression of the *leu* operon of *S. typhimurium* (G. Haughn, unpublished data).

B. Evidence for Attenuation Control

Transcription initiated at the *leu* promoter terminates with nearly complete efficiency before the first structural gene. Evidence that transcription is prematurely terminated in vivo comes from two types of experiments. The *leu* operons cloned from three strains that constitutively produce high levels of *leu* enzymes (Calvo et al., 1969a) contain in each case a mutation in the GC-rich region of the attenuator, so as to reduce the stability of the terminator stem (Searles et al., 1983). The mutations, formerly classified as operator-constitutive mutations, are attenuator mutations. A *leu* attenuator mutation has been introduced into a strain auxotrophic for leucine and tyrosine. Both strains are grown in a chemostat under conditions of leucine limitation and, at the same growth rate by limiting for tyrosine, under conditions of leucine excess. For the strain with the attenuator mutation, limitation for leucine does not substantially increase *leu* enzyme levels over those of leucine excess (Searles et al., 1983). If the cells had two separate leucine-specific control mechanisms, for example, an operator–repressor mechanism and an attenuator mechanism, then limitation for leucine should have caused further derepression of an attenuator-defective *leu* operon. Attenuation is thus considered to be the major (and perhaps only) mechanism by which leucine regulates expression of the *leu* operon.

Studies of *leu* mRNA synthesis lead to the same conclusion. Pulse-labeled RNA isolated from a *leu–tyr* double auxotroph grown in a chemostat under conditions of leucine limitation or leucine excess is hybridized to two different DNA probes, specific for either leader RNA or readthrough RNA. The fraction of the total counts hybridizing to a leader probe is the same for RNA samples from leucine repressed and derepressed cells. How-

E. coli

```
Met Thr His Ile Val Arg Phe Ile Gly Leu Leu Leu Leu Asn Ala Ser Ser Leu Arg Gly Arg Arg Val Ser Gly Ile Gln His

    A T C           C  T   TC   T               A          C  TC  T                GA      A         C   G
    T A T           T  C   CT   G               C          T  AT  G                CG      G         T   A

    Ser                        Thr                            Phe Ile Val             Pro     Gly
```

S. typhimurium

FIGURE 15.5 **The amino acid sequences of the putative *leu* leader peptides of (top) *E. coli* and (bottom) *S. typhimurium*. For the *S. typhimurium* sequence, only the differences are shown. The two middle lines show the nucleotide differences in the corresponding DNA (noncoding strands).**

ever, the rate of synthesis of readthrough RNA is 12-fold higher in derepressed cells than in repressed cells. The rates of degradation of leader and readthrough RNAs are similar under both growth conditions (Searles et al., 1983). Thus leucine limitation does not alter the frequency of transcription initiation at the *leu* promoter but does affect the rate of synthesis of readthrough RNA. Since transcription initiates only at a single promoter, the conclusion is inescapable that derepression results from increased readthrough of previously initiated transcripts.

C. Involvement of $tRNA^{Leu}$

In a number of mutants of *S. typhimurium* and *E. coli* resistant to trifluoroleucine (Rennert and Anker, 1963) or azaleucine (Steiglitz and Calvo, 1971), regulation of *leu* operon expression is altered (Calvo et al., 1969b; Mikulka et al., 1972; Kline, 1972). Other regulatory mutations have been isolated by selecting for arabinose utilization in an *E. coli* B/r strain containing a *leu–ara* fusion in which the *ara* operon is under *leu* operon control (Brown et al., 1978). The mutations fall within one of six distinct loci: *leuA* (feedback-insensitive α-isopropylmalate synthase; Bartholomew and Calvo, 1971b); *leuL* (attenuator region, formerly *leuO*; Calvo et al., 1969a); *leuS* (leucyl tRNA synthetase; Alexander et al., 1971; Mikulka et al., 1972); *flrA* (Kline, 1972); *flrB* (Alexander and Calvo, 1969; Friedberg et al., 1974); *leuK* (Brown et al., 1978). Mutations in *leuS, flrA,* and *flrB* lead to elevated, constitutive expression of both the *leu* and *ilvGEDA* operons. Mutations in *leuK* affect not only these two operons but others as well. In this respect, *leuK* mutations resemble *hisT* mutations, known to pleiotropically affect several operons including the *leu* operon (Cortese et al., 1974; Rizzino et al., 1974).

Leucyl-tRNA synthetases of several *leuS* mutants have higher K_m values for leucine than the parent enzymes. Since these mutants have elevated levels of *leu* and *ilv* biosynthetic enzymes, in spite of high intracellular levels of leucine, charged $tRNA^{Leu}$ rather than leucine appears to be more directly involved in repression (Alexander et al., 1971). This finding is consistent with current views on the mechanism of attenuation control of amino acid biosynthetic operons that postulate prominent roles for particular charged tRNAs (Yanofsky, 1981).

Mutations in *hisT* that prevent the formation of pseudouridine near the anticodon of several tRNAs have a distinct effect on expression of the *leu* operon (Cortese et al., 1974; Rizzino et al., 1974). The products specified by *flrB* and *leuK* are apparently also involved in modification of $tRNA^{Leu}$ species. The $tRNA_5^{Leu}$ from the *flrB* mutant has an altered chromatographic mobility (Searles, 1981). Parent and *flrB* mutant $tRNA_5^{Leu}$ differ at position 33 in the anticodon stem. In *leuK* strains, both $tRNA_5^{Leu}$ and $tRNA_4^{Leu}$ show altered chromatographic mobilities. Both mutant tRNAs lack the wild-type modification of U at position 40. Isoleucine and valine tRNAs do not show altered chromatographic mobilities in *leuK* strains. While these results add to the evidence that $tRNA^{Leu}$ participates in regulating the *leu* and *ilvGEDA* operons, they leave a number of important questions unresolved. Both $tRNA_5^{Leu}$ and $tRNA_4^{Leu}$ are known to decode codons UUA and UUG, whereas the four adjacent codons in the *leu* leader RNA are $(CUA)_4$ for *E. coli* or $(CUA)_3CUC$ for *S. typhimurium*. Of five distinct $tRNA^{Leu}$ species resolved on two-dimensional gels and identified by sequencing (Searles, 1981), none decodes CUA when standard base pairing is applied. A $tRNA^{Leu}$ species that recognizes CUA has recently been identified (Nakajima et al., 1981). It remains to be determined whether mutations in *leuK* or *flrB* affect the structure of this tRNA.

The unusual properties of *flrA* mutants of *E. coli* B/r deserve special comment. Whereas *leu* enzyme levels in haploid *flrA* mutants and in *flrA*/*flrA* merodiploids are only severalfold higher than in wild-type strains, the enzyme levels are elevated almost 20-fold in *flrA*/*flrA*$^+$ merodiploids (Kline, 1972). The $tRNA_2^{Leu}$ and $tRNA_5^{Leu}$ from these merodiploids have altered chromatographic mobilities, a finding suggesting that the unusual phenotype associated with the heterozygote may ultimately be explained in terms of alterations in tRNA modification (J. Jones and J. M. Calvo, unpublished data).

D. Mechanism of Transcriptional Control at the *leu* Attenuator

A plausible model of attenuation control, proposed for the *trp* operon (Lee and Yanofsky, 1977), can readily be adapted to the *leu* operon. A prominent feature of this model are two stem and loop structures within the leader RNA, termed the terminator (Fig. 15.2a, nucleotides 132–142 pair with 147–157) and the preemptor (Fig. 15.2a, nucleotides 80–85 pair with 137–142) (Keller and Calvo, 1979). The terminator, required for termination, and the preemptor, required for readthrough, are mutually exclusive, that is, formation of one secondary structure precludes formation of the other. Which of these two structures is formed during transcription is supposedly determined by events occurring during translation of the *leu* leader. A reduced rate of translation of the four adjacent leucine codons imposed by a limited supply of leucine, or by an altered $tRNA^{Leu}$ in the case of certain mutants, favors formation of the preemptor stem and loop structure, whereas unimpeded translation favors formation of the termination stem and loop (Gemmill et al., 1979). A third region of secondary structure, termed the protector (Fig. 15.2a, nucleotides 68–71 pair with 79–82), is inferred following a comparison of five attenuator regions (Keller and Calvo, 1979). With the inclusion of this third element, transcription from the *leu* promoter is expected to terminate at the attenuator under almost all circumstances: transcription in the absence of translation, coupled transcription and translation in the presence of excess amino acids, and translation limited by some amino acid other than leucine. The only condition under which substantial readthrough is expected is translation limited by the availability of leucine.

All the results obtained to date for the *leu* operon are consistent with this model. Its credibility rests mainly upon similarities with the more extensively studied *trp* operon (Yanofsky, 1981) and with the *phe, thr, his,* and *ilv* operons (Zurawski et al., 1978; Gardner, 1979; DiNocera et al., 1978; Barnes, 1978; Lawther and Hatfield, 1980; Nargang et al., 1980; Keller and Calvo, 1979). In addition, the comparison of the *S. typhimurium* and *E. coli leu* leader regions (Fig. 15.4) offers some further support for this model. In the face of so many differences in the sequences of these two regions, it seems significant that the protector, preemptor, and terminator subregions are, with one exception, free of differences. Indeed, the exception is a difference in the attenuator region, located at a position that should not affect the function of the terminator.

V. OUTLOOK

Several questions concerning attenuation, not yet adequately considered, pertain to the number, location, and identity of control codons within leader regions. Why, for example, do the *trp, leu,* and *his* operons have two, four, and seven control codons, respectively?

The number of control codons probably does not determine the magnitude of the derepression response, since the *leu* and *his* operons have about the same ratio of minimum versus maximum response. A more reasonable explanation is that the number of control codons determines the sensitivity of the operon response to amino acid limitation. One experiment (Fig. 15.6) serves to illustrate this point and to define some terms. The specific activity of β-isopropylmalate dehydrogenase is a measure of *leu* operon expression (output of the system). The rate of synthesis of *leu* mRNA would have served to define this function equally well. The number of doublings per unit time for a chemostat-grown leucine auxotroph under conditions of leucine limitation is a measure of the extent of starvation for leucine (the magnitude of the signal). Unit time (48 min) is the growth rate in minimal medium containing excess leucine. The chemostat growth rate should be directly related to the intracellular concentration of charged $tRNA^{Leu}$. The operon response pattern is defined as the shape of the curve when output is plotted as a function of the signal magnitude (Fig. 15.6). The response pattern of the *leu* operon has several noteworthy features: first, the output of the operon does not begin to rise until the degree of starvation is considerable (0.6 generation per unit time); second, the output rises to a maximum over a narrow range of generation times; and third, under conditions of severe leucine limitation, the operon maintains maximum output. It is easy to imagine that the steep slope of the curve in Fig. 15.6 has some adaptive significance to *S. typhimurium,* but it is more difficult to rationalize that the operon responds only to a relatively strong signal for leucine limitation.

How do number, location, and identity of control codons affect operon response patterns? To my knowledge, the only published data on an operon

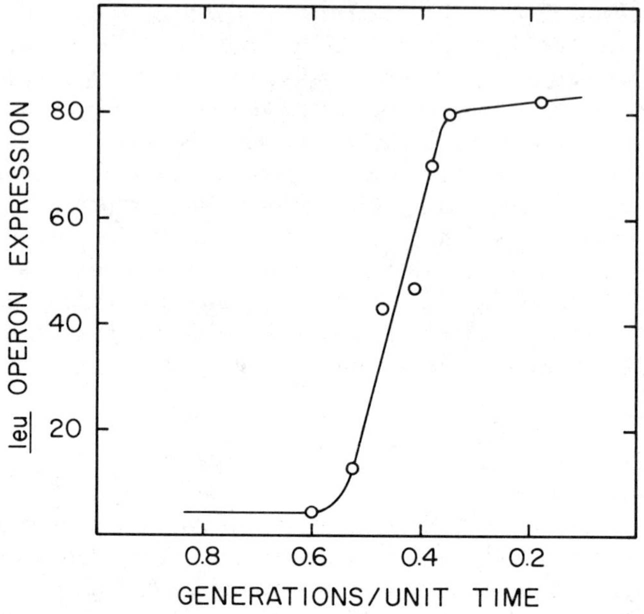

FIGURE 15.6 The *leu* operon response pattern. The specific activity for β-isopropylmalate dehydrogenase (*leu* operon expression) is plotted as a function of the generations per unit time for chemostat-grown cells limited by leucine. Unit time (48 min) is the doubling time of *S. typhimurium* grown in a minimal medium with excess leucine.

response pattern are those shown in Fig. 15.6; therefore the discussion below is intended only to outline some basic theoretical considerations and to make some predictions.

A priori, one would expect the number of control codons to have a sizable effect upon the operon response pattern. As a case in point, consider the *trp* operon, which has two adjacent tryptophan codons. The current model for attenuation control postulates that, under conditions of tryptophan limitation, a ribosome will be at or very near these tryptophan codons when RNA polymerase is at the attenuator. The critical pause time is defined as the time required for the polymerase to transcribe DNA from a point just downstream of the control codons (that is, the expected position of the polymerase when the ribosome first begins to pause) to the attenuator. If the degree of tryptophan starvation is such that the sum of pauses at two tryptophan codons equals the critical pause time, then derepression is expected. For a given critical pause time, it is expected that the number of control codons will determine the degree of starvation required to initiate derepression. Thus increasing the number of tryptophan codons should shift the operon response curve to the left; that is, a smaller degree of tryptophan starvation should elicit the derepression response.

In the case of the *his* operon, which has seven control codons, it might be expected that derepression would be triggered by a very small degree of histidine limitation. This prediction has not yet been tested experimentally. However, with such a large number of control codons, another factor may come into play, namely, that with a ribosome stalled at one of the first histidine codons owing to a very severe limitation of histidine, the signal for readthrough might not occur. Thus operon expression in this system might be expected to decrease under conditions of severe histidine limitation. Indeed, the results of some preliminary experiments indicate that *his* operon expression decreases dramatically below 0.2 doubling per unit time (J. Jones and J. Calvo, unpublished data).

Another factor determining the operon response pattern is the magnitude of the critical pause time. RNA polymerase is known to slow down at regions of high GC content (Platt, 1981); therefore both the distance the polymerase travels between the control codons and the attenuator and the base composition of this region are important parameters.

The location of control codons relative to the protector pairing region has been discussed (Keller and Calvo, 1979); however, questions relating to the identity of control codons and to the clustering of control codons have not been considered. In the *trp, leu,* and *his* leader regions, the control codons are adjacent to one another, and it seems likely that all or almost all of them are involved in the regulatory mechanism. On the other hand, for the *phe, thr,* and *ilv* leader regions, the relevant codons, although generally organized in groups of adjacent codons, are more spread out. For these cases, it is not clear that each related codon is, in fact, a control codon. For example, of the four leucine codons within the *ilv* leader region (which contains 32 codons in all), two are near the beginning (positions 4 and 5) and one is near the end (position 27). It seems unlikely that the most distal leucine codon plays any role in the regulatory response. On the other hand, the two leucine codons near the beginning might play a significant role by contributing to the critical pause time. According to this view, it is only necessary for one pair of control codons to be appropriately positioned relative to the protector, while others may be at any proximal position. It would be interesting to know the response pattern of the *ilvGEDA* operon, separately, to limitations of leucine, valine, and isoleucine. Such data could help to evaluate the contribution of leucine to the multivalent repression phenomenon (Freundlich et al., 1962) and perhaps give a clue as to the number of leucine codons involved in the regulatory response.

Another factor that may affect the operon response pattern is the identity of individual control codons. This consideration is particularly relevant in the case of the *leu* operon, since there are six different leucine codons. Among leucine codons within the *leu* leaders of *S. typhimurium* and *E. coli,* three out of four and four out of four, respectively, are CUA codons. This is striking because CUA is very seldom found as a codon in structural genes that are expressed at high levels in enteric bacteria (Grantham et al.,

1980). It is possible that the ribosome step time for CUA codons is longer than that for other leucine codons and that, as a consequence, CUA has been selected during evolution for a special role in regulation of the *leu* operon. It is also conceivable that adjacent CUA codons exhibit a context effect (Bossi and Roth, 1980) that results in lengthened step times. It will be of some interest to determine the effect upon operon expression of replacing CUA codons by other *leu* codons.

There are other factors that may affect the response pattern of an operon, for example, incorporation of incorrect amino acids by ribosomes stalled at control codons (Parker et al., 1978). I have chosen to limit this discussion to certain questions related to the number, location, and identity of control codons. While not exhaustive, these comments may focus attention on aspects of regulation that heretofore have not been considered. With few exceptions, only maximum and minimum levels of expression have been determined until now for amino acid biosynthetic operons. A deeper understanding of the physiological significance of the regulatory process, and of the regulatory mechanism itself, will require more detailed determination of the response patterns for several operons.

ACKNOWLEDGMENTS

The research in the author's laboratory was supported by grants from the National Science Foundation (BMS74-21987) and the National Institutes of Health (AI14340).

REFERENCES

Alexander, R. R., and Calvo, J. M. (1969) *Genetics 61,* 539–556.

Alexander, R. R., Calvo, J. M., and Freundlich, M. (1971) *J. Bacteriol. 106,* 213–220.

Allison, M. J., Bucklin, J. A., and Robinson, I. M. (1966) *Appl. Microbiol. 14,* 807–814.

Barnes, W. M. (1978) *Proc. Nat. Acad. Sci. U.S.A. 75,* 4281–4285.

Bartholomew, J. C., and Calvo, J. M. (1971a) *Biochim. Biophys. Acta 250,* 568–576.

Bartholomew, J. C., and Calvo, J. M. (1971b) *Biochim. Biophys. Acta 250,* 577–587.

Bossi, L., and Roth, J. R. (1980) *Nature 286,* 123–127.

Brown, C. S., West, R., Hilderman, R. H., Bayliss, F. T., and Kline, E. L. (1978) *J. Bacteriol. 135,* 542–550.

Burns, R. O., Calvo, J. M., Margolin, P., and Umbarger, H. E. (1966) *J. Bacteriol. 91,* 1570–1576.

Butler, G. W., and Shen, L. (1963) *Biochim. Biophys. Acta 71, 456–458.*

Calvo, J. M., and Worden, H. E. (1970) Genetics 64, 199–214.

Calvo, J. M., Stevens, C. M., Kalyanpur, M. G., and Umbarger, H. E. (1964) *Biochemistry 3,* 2024–2027.

Calvo, J. M., Margolin, P., and Umbarger, H. E. (1969a) *Genetics 61,* 777–787.
Calvo, J. M., Freundlich, M., and Umbarger, H. E. (1969b) *J. Bacteriol. 97,* 1272–1282.
Cortese, R., Landsberg, R., VonderHaar, R. A., Umbarger, H. E., and Ames, B. N. (1974) *Proc. Nat. Acad. Sci. U.S.A. 71,* 1857–1861.
Davis, M. G., and Calvo, J. M. (1977a) *J. Bacteriol. 129,* 1078–1090.
Davis, M. G., and Calvo, J. M. (1977b) *J. Bacteriol. 131,* 997–1007.
DiNocera, P. P., Blasi, F., DiLauro, R., Frunzio, R., and Bruni, C. (1978) *Proc. Nat. Acad. Sci. U.S.A. 75,* 4276–4280.
Dubnau, E., Lenny, A. B., and Margolin, P. (1973) *Mol. Gen. Genet. 126,* 191–200.
Freundlich, M., Burns, R. O., and Umbarger, H. E. (1962) *Proc. Nat. Acad. Sci. U.S.A. 48,* 1804–1808.
Friedberg, D., Mikulka, T. W., Jones, J., and Calvo, J. M. (1974) *J. Bacteriol. 118,* 942–951.
Friedman, S. B., and Margolin, P. (1968) *J. Bacteriol. 95,* 2263–2269.
Gardner, J. F. (1979) *Proc. Nat. Acad. Sci. U.S.A. 76,* 1706–1710.
Gemmill, R. M. (1981) "Attenuation Control: The Leader and the Promoter of the Salmonella *leu* Operon," Ph.D. thesis, Cornell University, Ithaca, N.Y.
Gemmill, R. M., Wessler, S. R., Keller, E. B., and Calvo, J. M. (1979) *Proc. Nat. Acad. Sci. U.S.A. 76,* 4941–4945.
Gemmill, R. M., Jones, J. W., Haughn, G. W., and Calvo, J. M. (1983) *J. Mol. Biol.* (in press)
Grantham, R., Gaütier, C., and Gouy, M. (1980) *Nucl. Acids Res. 8,* 1893–1912.
Gross, S. R., Burns, R. O., and Umbarger, H. E. (1963) *Biochemistry 2,* 1046–1052.
Hertzberg, K. M., Gemmill, R., Jones, J., and Calvo, J. M. (1980) *Gene 8,* 135–152.
Keller, E. B., and Calvo, J. M. (1979) *Proc. Nat. Acad. Sci. U.S.A. 76,* 6186–6190.
Kline, E. L. (1972) *J. Bacteriol. 110,* 1127–1134.
Kohlhaw, G. (1970) *Biochim. Biophys. Acta. 212,* 58–64.
Kohlhaw, G., and Boatman, G. (1971) *Biochem. Biophys. Res. Comm. 43,* 741–746.
Kohlhaw, G., Leary, T. R., and Umbarger, H. E. (1969) *J. Biol. Chem. 244,* 2218–2225.
Lawther, R. P., and Hatfield, G. W. (1980) *Proc. Nat. Acad. Sci. U.S.A. 77,* 1862–1866.
Leary, T. R., and Kohlhaw, G. (1970) *Biochem. Biophys. Res. Comm. 39,* 494–501.
Leary, T. R., and Kohlhaw, G. B. (1972) *J. Biol. Chem. 247,* 1089–1095.
Leary, E. T., and Kohlhaw, G. B. (1973) *Biochemistry 12,* 2980–2986.
Lee, F., and Yanofsky, C. (1977) *Proc. Nat. Acad. Sci. U.S.A. 74,* 4365–4369.
Margolin, P. (1963) *Genetics 48,* 441–457.
Martin, W. R., Coleman, W. H., Wideburg, N. E., Cantrell, R., Jackson, M., and Denison, F. W., Jr. (1962) *Biochim. Biophys. Acta 62,* 165–167.
Mikulka, T. W., Stieglitz, B., and Calvo, J. M. (1972) *J. Bacteriol. 109,* 584–593.
Nakajima, N., Ozeki, H., and Shimura, Y. (1981) *Cell 23,* 239–249.

Nargang, F. E., Subrahmanyam, C. S., and Umbarger, H. E. (1980) *Proc. Nat. Acad. Sci. U.S.A. 77,* 1823–1827.

Oaks, A. (1965) *Biochim. Biophys. Acta 111,* 79–89.

Parker, J., Pollard, J. W., Friesen, J. D., and Stanners, C. P. (1978) *Proc. Nat. Acad. Sci. U.S.A. 75,* 1091–1095.

Parsons, S. J., and Burns, R. O. (1969) *J. Biol. Chem. 244,* 996–1003.

Platt, T. (1981) *Cell 24,* 10–23.

Rennert, O. M., and Anker, H. S. (1963) *Biochemistry 2,* 471–476.

Rizzino, A. A., Bresalier, R. S., and Freundlich, M. (1974) *J. Bacteriol. 117,* 449–455.

Rosenberg, M., and Court, D. (1979) *Ann. Rev. Genet. 13,* 319–353.

Searles, L. (1981) "The *leu* Operon of *E. coli* and *S. typhimurium:* Biochemical Analysis of Regulatory Mutations, Expression of the *leu* Operon in the Zubay System," Ph.D. thesis, Cornell University, Ithaca, N.Y.

Searles, L. L., Wessler, S. R., and Calvo, J. M. (1983) *J. Mol. Biol. 163,* 377–394.

Shimada, K., Weisberg, R. A., and Gottesman, M. E. (1973) *J. Mol. Biol. 80,* 297–314.

Somers, J. M., Amzallag, A., and Middleton, R. B. (1973) *J. Bacteriol. 113,* 1268–1272.

Soper, T. S., Doellgast, G. J., and Kohlhaw, G. B. (1976) *Arch. Biochem. Biophys. 173,* 362–374.

Stieglitz, B. I., and Calvo, J. M. (1971) *J. Bacteriol. 108,* 95–104.

Stieglitz, B. I., and Calvo, J. M. (1974) *J. Bacteriol. 118,* 935–941.

Umbarger, H. E. (1978) *Ann. Rev. Biochem. 47,* 533–606.

Wessler, S. R. (1980) "*In vitro* and *in vivo* Expression of the *leu* Operon of *Escherichia coli* and *Salmonella typhimurium,*" Ph.D. thesis, Cornell University, Ithaca, N.Y.

Wessler, S. R., and Calvo, J. M. (1982) *J. Mol. Biol., 149,* 579–597.

Yang, H. L., and Kessler, D. P. (1974) *J. Bacteriol. 117,* 63–72.

Yanofsky, C. (1981) *Nature 289,* 751–758.

Zurawski, G., Brown, K., Killingly, D., and Yanofsky, C. (1978) *Proc. Nat. Acad. Sci. U.S.A. 75,* 4271–4275.

CHAPTER
16

Regulation of Leucine Biosynthesis in Lower Eukaryotes

Gunter B. Kohlhaw

I. INTRODUCTION

Among eukaryotes, *Neurospora crassa* and *Saccharomyces cerevisiae* have been the organisms of choice for the study of leucine biosynthesis and its regulation. In both organisms, those structural genes of the leucine pathway that have been localized lie within different linkage groups (Fig. 16.1). Regulation of the pathway is brought about through intricate combinations of the modulation of enzyme activity, repression, and induction, with α-isopropylmalate synthase playing a principal role.

In *N. crassa,* α-isopropylmalate synthase is inhibited, and its formation is repressed, by leucine. These controls affect not only the flux of carbon through the pathway but also the formation of isopropylmalate dehydratase and β-isopropylmalate dehydrogenase. Apparently, α-isopropylmalate binds to a regulatory protein, encoded in *leu-3,* and the ensuing complex induces the expression of *leu-2* and *leu-1*. In the absence of α-isopropylmalate, very little, if any, dehydratase or dehydrogenase is made. The *leu-3* gene product–α-isopropylmalate complex also has other effects described in detail below.

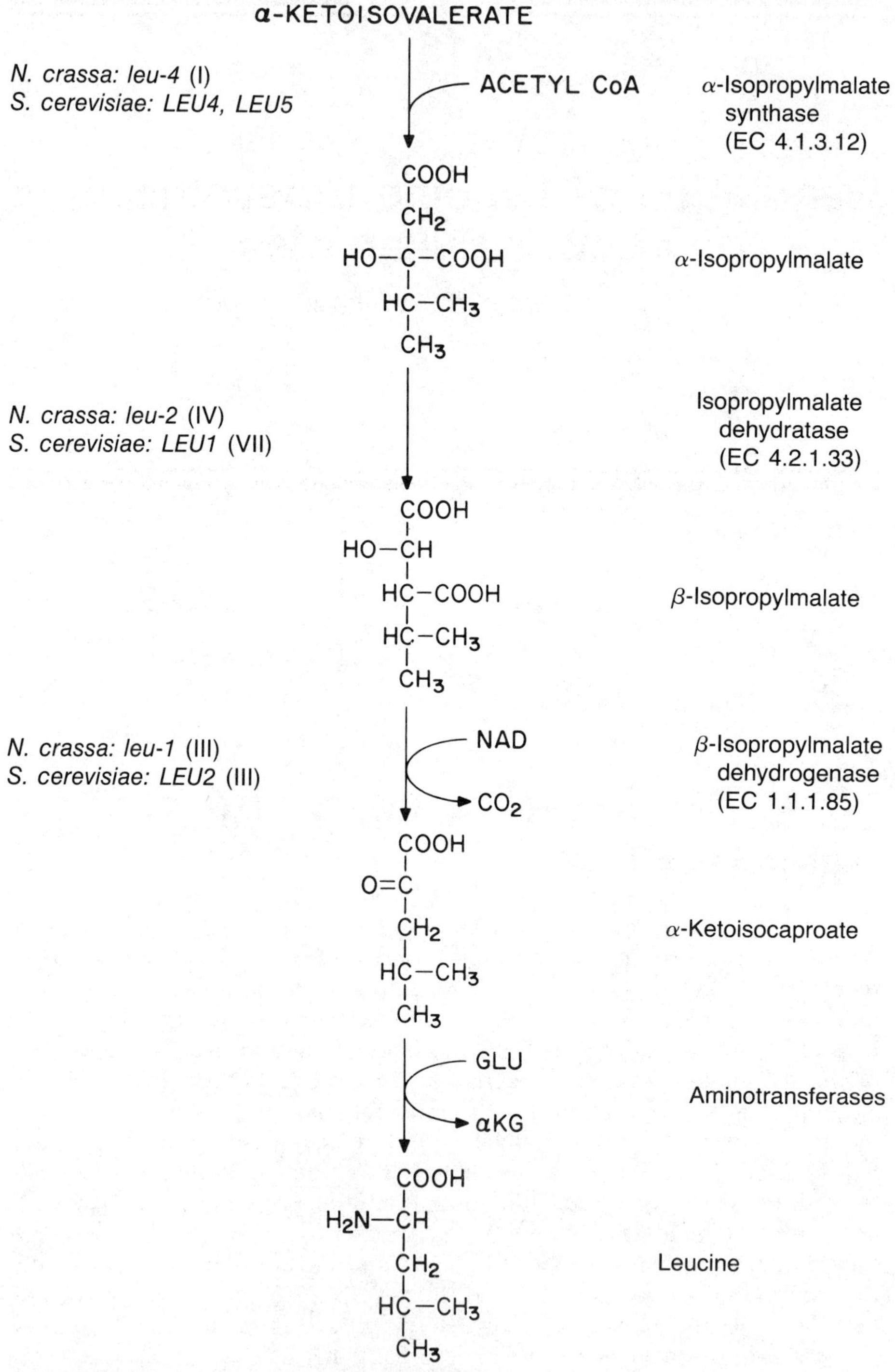

FIGURE 16.1 The pathway of leucine biosynthesis in *N. crassa* and *S. cerevisiae*. The bracketed roman numerals refer to linkage group for *N. crassa* or chromosome for *S. cerevisiae*.

The salient regulatory features of leucine biosynthesis in *S. cerevisiae* are similar to those of *N. crassa*. α-Isopropylmalate synthase is feedback-inhibited by leucine, and the dehydratase and dehydrogenase appear to be induced by α-isopropylmalate. There are nonetheless important differences, especially with respect to α-isopropylmalate synthase. In yeast, much of the α-isopropylmalate synthase is mitochondrial. The enzyme not only is inhibited by leucine but also inactivated by CoA, a product of the reaction. Furthermore, the enzyme is under cross-pathway control, that is, its level is derepressed whenever cells are starved for any one of a number of amino acids. Recently obtained evidence suggests that, in yeast, two genes encode α-isopropylmalate synthase.

II. SUBCELLULAR LOCALIZATION OF THE ENZYMES OF LEUCINE, ISOLEUCINE, AND VALINE BIOSYNTHESIS

In *N. crassa,* all three leucine-specific enzymes are cytosolic (S. Gross, personal communication), in contrast to yeast, in which the three enzymes are distributed between two compartments (Ryan et al., 1973). The dehydratase and dehydrogenase are cytosolic, while much of the synthase is mitochondrial. In intact mitochondria, the synthase is resistant to proteolysis and inaccessible to its substrates unless the mitochondria are rendered permeable (Hampsey et al., 1983). The enzyme is located in the mitochondrial matrix (Hampsey et al., 1983; Gasser et al., 1982). In vitro translation of total yeast RNA generates two polypeptides that cross-react with anti-yeast α-isopropylmalate synthase antibodies (Hampsey et al., 1983). The larger one is identical in molecular weight to the subunit of highly purified α-isopropylmalate synthase (i.e., 65,000) and is incorporated in vitro into intact mitochondria, suggesting that its translocation across the mitochondrial membranes occurs posttranslationally and that it is not made as a larger-molecular weight precursor. The smaller polypeptide (apparent molecular weight 63,000–64,000) is not taken up by isolated mitochondria. The relationship between the two polypeptides, if any, is not clear, nor is it known whether they represent the products of the two genes encoding α-isopropylmalate synthase.

Branched-chain amino acid aminotransferase activity is present in the cytosol as well as in particulate cell fractions, both in yeast and in *N. crassa* (Ryan et al., 1973; Collins and Wagner, 1973). In both organisms, high proportions of three enzymes involved in the synthesis of isoleucine and valine, that is, acetohydroxy acid synthase, isomeroreductase, and dihydroxy acid dehydratase, are found associated with the mitochondria (Ryan and Kohlhaw, 1974; Cassady et al., 1973). Threonine deaminase, the first enzyme specific to isoleucine biosynthesis, appears to be mitochondrial in yeast (Ryan and Kohlhaw, 1974) but cytosolic in *N. crassa* (Olshan and Gross, 1974).

III. REGULATION OF ENZYME ACTIVITY

A. End Product Inhibition of α-*N. crassa*

Inhibition by leucine of purified *N. crassa* α-isopropylmalate synthase is very specific. Only trifluoroleucine is comparable to leucine as an inhibitor (Webster and Gross, 1965). As is true for the synthase from various sources, leucine inhibition of the *N. crassa* enzyme is pH-dependent. The Hill coefficient, obtained from inhibition data at three different pH values, is 1.51, indicative of positive cooperativity between at least two inhibitor binding sites. Leucine probably alters the conformation of the synthase, since it protects the enzyme against thermal inactivation. Protection by leucine is not found for a leucine-insensitive mutant form of the enzyme.

Available data on the quaternary structure of the synthase are consistent with either a trimer or a tetramer (Gross, 1970). The subunit molecular weight of the *N. crassa* enzyme (about 43,000) (Webster et al., 1965; Gross, 1970) is significantly lower than that of the yeast enzyme (about 65,000) (Tracy and Kohlhaw, 1977). It will be of interest to determine whether the *N. crassa* enzyme differs from the yeast enzyme in other respects, especially in its sensitivity to CoA.

B. *S. cerevisiae*

(i) End Product Inhibition of α-Isopropylmalate Synthase. α-Isopropylmalate synthase of yeast is allosterically inhibited by leucine (Satyanarayana et al., 1968b; Ulm et al., 1972). The enzyme that has been purified (probably representing one of the two postulated isoenzymes) consists of two apparently identical subunits with a molecular weight of about 65,000 (Tracy and Kohlhaw, 1977). Moderate positive cooperativity between the leucine sites is indicated by a Hill coefficient of 1.4 (Ulm et al., 1972). Leucine probably causes conformational changes, since the binding of the enzyme to hydroxylapatite and the elution profile in gel permeation chromatography are altered by leucine. The leucine pools in mutants with feedback-resistant synthase are approximately 15 times larger than in wild-type (Hampsey, 1982). The mutants excrete leucine and/or α-ketoisocaproate into the medium to an extent that roughly parallels the degree of feedback resistance (Fig. 16.2). End product inhibition of yeast α-isopropylmalate synthase thus constitutes a physiologically important feedback loop.

(ii) Inactivation of α-Isopropylmalate Synthase by CoA. Yeast α-isopropylmalate synthase is not only subject to feedback inhibition by leucine, but in the presence of Zn^{2+} also undergoes a very specific, time-dependent, reversible inactivation by CoA (Ulm et al., 1972; Tracy and Kohlhaw, 1975,

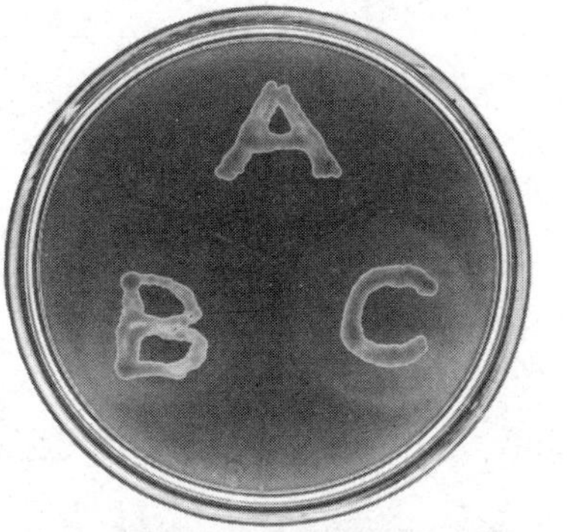

FIGURE 16.2 Excretion of leucine and/or α-ketoisocaproate by yeast mutants containing feedback-resistant α-isopropylmalate synthase. Three yeast strains were streaked on top of agar to which had been added a suspension of α-isopropylmalate dehydrogenase-deficient *Escherichia coli* cells. A: wild-type; B: yeast containing α-isopropylmalate synthase that is about 2 orders of magnitude less sensitive to leucine than enzyme from wild-type cells; C: yeast containing α-isopropylmalate synthase resistant to as much as 50 mM leucine (from Hampsey, 1982).

1977; Hampsey and Kohlhaw, 1981). Less than 20% of the initial activity remains after a 10-min incubation with 10 μM each of CoA and Zn^{2+} in vitro. With permeabilized cells, the CoA–Zn concentration must be increased about fivefold to achieve the same effect.

The CoA effect is not a simple product inhibition: first, CoA inactivation can be neither prevented nor reversed by acetyl CoA; second, CoA inactivation requires the addition of divalent metal ions, preferably Zn^{2+}, while no such requirement exists for catalytic activity, even though the enzyme is a zinc metalloenzyme (Roeder and Kohlhaw, 1980); and third, each monomer of the synthase contains two binding sites for CoA, one of which becomes available only in the presence of Zn^{2+} (Tracy and Kohlhaw, 1977).

A schematic representation of the various sites likely to exist on the surface of each monomer is given in Fig. 16.3. Binding of CoA at the product site requires an intact nucleotide moiety, but not the sulfhydryl group of CoA. In contrast, the regulatory CoA site is highly specific for CoA, since related compounds have essentially no inactivating effect. However, it is possible that ATP binds at or near the regulatory site, since ATP efficiently protects the synthase against CoA inactivation in an apparently competitive manner (Hampsey and Kohlhaw, 1981; Hampsey, 1982). The regulatory CoA site is apparently also distinct from the leucine binding site because the synthases of two feedback-resistant mutants are still CoA-sensitive (Hampsey, 1982).

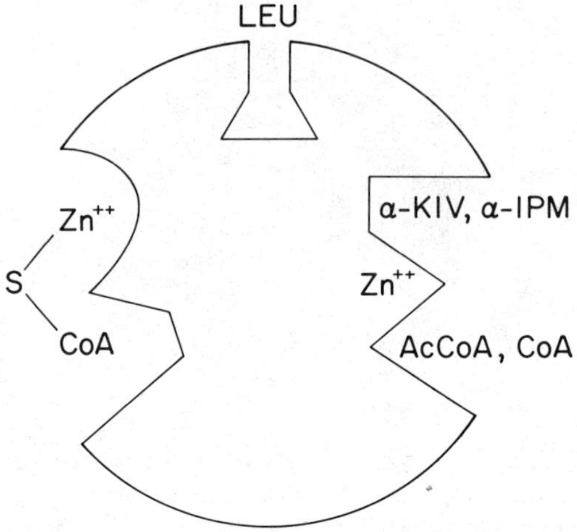

FIGURE 16.3 Proposed binding sites on yeast α-isopropylmalate synthase; α-KIV, α-ketoisovalerate; α-IPM, α-isopropylmalate; AcCoA, acetyl CoA.

The inactivation of α-isopropylmalate synthase by CoA is thought to be part of a regulatory system that distributes acetyl CoA among anabolic and catabolic pathways. At least two other enzymes are susceptible to CoA inactivation: homocitrate synthase (Tracy and Kohlhaw, 1975), which catalyzes the first committed step in lysine biosynthesis (Chapter 13), and β-hydroxy-β-methylglutaryl-CoA reductase, a key enzyme in sterol biosynthesis (Tan-Wilson and Kohlhaw, 1978). The actual inactivating principle of β-hydroxy-β-methylglutaryl-CoA reductase is probably CoA disulfide (Gilbert and Stewart, 1981). The CoA-sensitive enzymes are all mitochondrial and have regulatory functions in their respective pathways. Together, they could deplete the cell's acetyl CoA supply. Since citrate synthase is resistant to CoA inactivation, the emerging pattern is consistent with a control mechanism that functions to channel acetyl CoA to catabolic reaction sequences under conditions of energy depletion (Fig. 16.4). N-Acetylglutamate synthase, the first enzyme in ornithine and arginine biosynthesis, is also located in the mitochondria. The enzyme catalyzes an acetyl CoA utilizing reaction. However, the acetyl group is recycled (Chapter 5). Thus the in vivo consumption of acetyl-CoA by this enzyme is probably minute. N-Acetylglutamate synthase is not inactivated by CoA (Wipf and Leisinger, 1979).

IV. REGULATION OF ENZYME SYNTHESIS

A. *N. crassa*

Synthesis of the leucine biosynthetic enzymes in *N. crassa* is regulated by interrelated repression–induction: leucine-mediated repression of α-

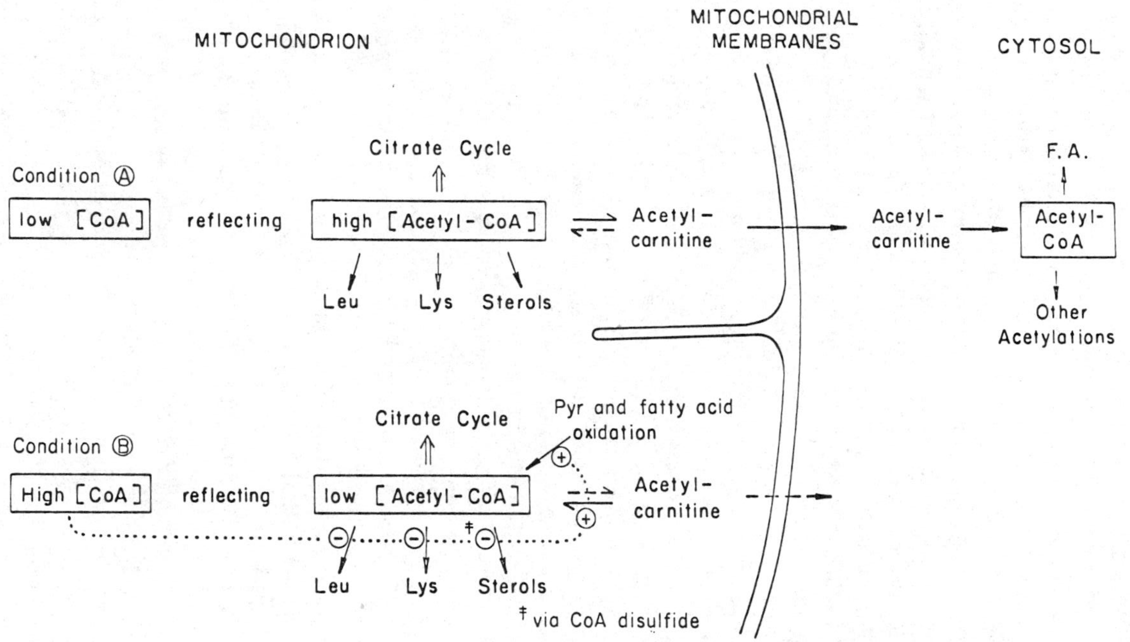

FIGURE 16.4 Proposed control of acetyl CoA utilization in yeast mitochondria (from Hampsey and Kohlhaw, 1981). A and B signify two extreme values of the acetyl charge, defined as [acetyl-CoA]/([CoA] + [acetyl-CoA]). Minus signs refer to inactivation by CoA or CoA disulfide of α-isopropylmalate synthase, homocitrate synthase, and α-hydroxy-α-methylglutaryl CoA reductase, respectively, while plus signs indicate stimulation by increased concentrations of CoA. Dashed arrows indicate slow processes. The assumption that acetyl-carnitine serves as an intermediate in the transfer of acetyl groups out of the mitochondria is supported by the relatively high specific activity of carnitine acetyltransferase and its comparatively high affinity for acetyl CoA (Kohlhaw and Tan-Wilson, 1977).

isopropylmalate synthase, encoded by *leu-4,* and α-isopropylmalate-mediated induction of the dehydratase, encoded by *leu-2,* and the dehydrogenase, encoded by *leu-1* (Gross, 1969). When grown on limiting leucine, *leu-2* and *leu-1* mutants display elevated levels of the synthase and the dehydrogenase or the dehydratase, respectively. With varying amounts of leucine, the levels of dehydratase and dehydrogenase change coordinately, but the production of synthase does not exactly parallel that of dehydratase or dehydrogenase (Gross, 1965). *Leu-4* mutants that produce only a trace amount of synthase produce little, if any, dehydratase and dehydrogenase. Likewise, the production of the dehydratase in *leu-1, leu-4* and of the dehydrogenase in a *leu-2, leu-4* double mutant, grown with limiting leucine, is severely reduced. Mutants with a feedback-resistant α-isopropylmalate synthase, which produce either as much or less of the synthase when compared with wild-type, exhibit a significant increase in the production of dehydratase and dehydrogenase (Gross, 1965; Kashmiri and Gross, 1970), even though they overproduce α-ketoisocaproate and leucine. *Leu-1* and *leu-2* mutants with a genetic background of a feedback-resistant synthase display elevated dehydratase and dehydrogenase levels that are essentially independent of the amount of leucine present in the medium (Gross, 1965). Thus, the production of dehydratase and dehydrogenase is apparently related to the intracellular level of α-isopropylmalate, which is practically nil in *leu-4* auxotrophs, reaches intermediate levels in feedback-resistant mutants, and is very high in *leu-1* or *leu-2* mutants grown with limiting leucine and in *leu-1* and *leu-2* double mutants with a feedback-resistant synthase.

The production of the synthase is predominantly under leucine control. The wild-type enzyme can be repressed by excess leucine (Kashmiri and Gross, 1970), although this effect is not always observed and may depend on the growth phase (Polacco and Gross, 1973). In *leu-1* and *leu-2* mutants with or without a genetic background of a feedback-resistant synthase, the synthase levels vary inversely with the leucine concentration in the medium. Ways were devised to separate *leu-4* gene expression from that of *leu-1* and *leu-2*. For example, 4-azaleucine increases the level of α-isopropylmalate synthase without affecting the other two enzymes (Gross, 1965). Mutants were also isolated that display elevated levels of synthase but normal levels of dehydratase, or vice versa (Kashmiri and Gross, 1970).

These results are consistent with a model implicating at least two regulatory elements in the synthesis of the three leucine specific enzymes. One element binds leucine or a leucine derivative, and the complex functions as a repressor for *leu-4* expression. The other element interacts with α-isopropylmalate and functions as an inducer for *leu-1* and *leu-2* expression. Obviously, the production of α-isopropylmalate can be controlled either through repression of the synthase by leucine or through feedback inhibition, thus allowing for fast and slow inductive responses.

Analysis of *leu-3* mutants supports the model. *Leu-3* mutants are identified on the basis of their very low content of dehydratase and dehydrogenase.

Leu-3 mutants contain synthase, but the enzyme cannot be fully derepressed by leucine starvation. In addition, in *leu-3* mutants, the levels of four *ilv* enzymes cannot be derepressed by limiting end product(s) (Olshan and Gross, 1974). Thus, a *leu-3* gene product-α-isopropylmalate complex serves apparently as positive regulatory element for transcription of seven or eight cistrons (Polacco and Gross, 1973; Olshan and Gross, 1974). The idea that the *leu-3* cistron specifies a positive regulatory element is buttressed by the isolation of partial revertants of *leu-3* mutants, designed *leu-3*^{cc} (Polacco and Gross, 1973). In these revertants, the production of dehydratase and dehydrogenase is conditionally constitutive, that is, independent of α-isopropylmalate but at the same time dependent on the availability of either leucine or isoleucine plus valine for realization of its full induction potential.

More direct evidence for the induction of the dehydratase by α-isopropylmalate comes from experiments with *N. crassa* strains capable of utilizing exogenous α-isopropylmalate (Reichenbecher et al., 1978). Mutations in two genes, identified as *ipm-1* and *ipm-2,* mediate maximal growth of α-isopropylmalate. In a quadruple mutant with mutations in *ipm-1, ipm-2, leu-1* and *leu-4* the response of *leu-2* expression to exogenously supplied α-isopropylmalate was studied. A more than 100-fold increase in the dehydratase activity occurs shortly after α-isopropylmalate is added to cells grown in a repression medium, that is, a medium containing excess leucine. As little as 0.1 mM α-isopropylmalate produces maximal induction. The experiments also suggest that, with respect to regulatory consequences there is no direct competition between leucine and α-isopropylmalate. The effects of translational and transcriptional inhibitors suggest that the dehydratase synthesis is controlled at the transcriptional level (Reichenbecher et al., 1978). Detectable levels of dehydratase mRNA, identified as enzyme-forming capacity, appear about 12 min after induction; the mRNA half-life is 4–8 min. By contrast, the dehydratase appears to be stable in vivo. The exact mechanism of induction is unknown.

The *leu-3* gene product–α-isopropylmalate complex may play an even larger role than has heretofore been assumed, since it apparently mediates transient repression of rRNA and total protein synthesis (*Neurospora Newsletter* No. 27, June 1980, p. 10).

B. *S. cerevisiae*

The regulation of *leu* enzyme synthesis in yeast is not as well understood as that of *N. crassa*. Available data may be summarized as follows: Wild-type cells (strain 60615 of the Lindegren collection) in a minimal medium respond to the addition of up to 2 mM leucine by only a slight drop in the levels of the dehydratase and dehydrogenase, while actually showing an increase in the synthase activity; a sharp decline (fivefold to 20-fold) of the dehydratase and dehydrogenase levels occurs when leucine and threonine are added together or when tryptone is added to a minimal medium

(Satyanarayana et al., 1968b; Brown et al., 1975). Under these conditions the synthase level drops to 35–60% of the level observed in a minimal medium. Experiments with a different wild-type (S288c) have shown that 1 mM leucine alone is less effective than leucine and threonine together (1 mM each) in lowering the dehydrogenase levels (Kohlhaw et al., 1980; Hsu and Kohlhaw, 1982). However, with this strain, 2 mM leucine alone has a strong repressing effect (V. Baichwal and G. B. Kohlhaw, unpublished results). Kinetic studies of repression and derepression of the dehydrogenase show that the enzyme remains stable for at least 7 hours after repressing conditions are established and is only diluted out by cell division (Kohlhaw et al., 1980). Derepression is prevented by cycloheximide or the RNA polymerase inhibitor lomofungin; derepression is reduced in an organism with a temperature-sensitive *rna1* gene product when cells are transferred to a nonpermissive temperature (Y.-P. Hsu and G. B. Kohlhaw, unpublished results). Thus, leucine and threonine probably exert their effect at a pretranslational level. The underlying molecular events are not known.

However, mutant analysis is shedding some light on the system. Dehydratase-deficient or dehydrogenase-deficient mutants exhibit elevated levels of the dehydrogenase and the dehydratase, respectively, when leucine is limiting. In the presence of excess leucine, the enzyme levels are low but well above those of repressed wild type cells. The synthase level in these mutants is elevated twofold to threefold under both conditions, that is, with limiting or excess leucine in the medium (V. Baichwal and G. B. Kohlhaw, unpublished results). Synthase-deficient yeast mutants, like their *N. crassa* counterparts, contain negligible amounts of dehydratase and dehydrogenase, even after severe leucine starvation (Satyanarayana et al., 1968b; V. Baichwal, Y.-P. Hsu, and G. B. Kohlhaw, unpublished results). Feedback-resistant synthase mutants show moderately increased levels of synthase, dehydratase, and dehydrogenase; more importantly, there is no repression of the latter two enzymes by excess leucine in these mutants. Feedback-resistant *leu1* and feedback-resistant *leu2* double mutants also show twofold and threefold elevated levels of the synthase regardless of whether they are grown with limiting or excess leucine. In these mutants, the dehydrogenase and isomerase levels are 13-fold to 18-fold and 9-fold to 11-fold, respectively, above wild-type levels and are again refractory to leucine repression (V. Baichwal and G. B. Kohlhaw, unpublished results).

As in *N. crassa,* derepression of the yeast dehydratase and dehydrogenase is thus apparently a function of the amount of α-isopropylmalate produced in vivo. This conclusion is reinforced by two experimental results: first, the concentration of α-isopropylmalate excreted into the culture medium at the time of harvest correlates well with the degree of derepression of dehydratase and dehydrogenase (V. Baichwal and G. B. Kohlhaw, unpublished results); second, when synthaseless, α-isopropylmalate permeable mutants are exposed to α-isopropylmalate, the level of dehydratase and dehydrogenase increases at least fivefold within 30 min. The increase is not affected by leucine and threonine but is prevented by proflavine, 8-hydroxy-quinoline, or cycloheximide (Y.-P. Hsu and G. B. Kohlhaw, unpublished results). While these results are consistent with an induction mechanism as described for *N. crassa,* they must be regarded as preliminary, since the experimental strains have not yet been fully characterized genetically.

The similarity to *N. crassa* becomes even more striking when *leu3*-type mutants are considered. Such mutants were first isolated in G. R. Fink's laboratory. Of 500 independently isolated, leucine-requiring yeast mutants, 498 were either *leu1* or *leu2;* the remaining two mutants fell into a different complementation group designated *leu3* (G. R. Fink, personal communication). The *leu3* mutants are leaky, contain elevated levels of the synthase when grown with excess or limiting leucine, and contain very low levels of dehydratase and dehydrogenase even when grown under leucine limitation; they excrete sizable amounts of α-isopropylmalate (V. Baichwal, Y.-P. Hsu, and G. B. Kohlhaw, unpublished results). The leucine-leaky mutant RH657 (Niederberger et al., 1981) belongs to the same complementation group and has the same phenotype as Fink's *leu3* mutants (V. Baichwal, T. Cunningham, and G. B. Kohlhaw, unpublished results). A *leu2–leu3* double mutant contains elevated levels of the synthase and very low levels of the dehydratase, irrespective of whether the cells are grown on limiting or excess leucine.

A significant difference between yeast and *N. crassa* synthase is observed with respect to the regulation of α-isopropylmalate synthase. In yeast, the synthase is under general amino acid control (Hsu et al., 1982). This control system constitutes a metabolic interlock that, in yeast, encompasses biosynthetic pathways leading to aromatic and basic amino acids, most notably tryptophan, histidine, arginine, and lysine, as well as the aspartate family of amino acids, leucine, and possibly others (Schürch et al., 1974; Delforge et al., 1975; Wolfner et al., 1975; Niederberger et al., 1981; Hsu et al., 1982). Limitation for any one of the amino acids involved causes derepression of enzymes in many or all of the concerned pathways. Mutants have been isolated that contain either constitutively derepressed levels (*cdr; tra*) or nonderepressible levels (*ndr; aas*) of those enzymes (Schürch et al., 1974; Wolfner et al., 1975; Miozzari et al., 1978). Of the three leucine specific enzymes, only the synthase is significantly derepressed in a *cdr* strain; the specific activity of this enzyme is also elevated in lysine- and arginine-leaky strains grown under limiting conditions and in wild-type cells limited for tryptophan by the addition of 5-methyltryptophan to the growth medium; finally, the level of the synthase is low in an *ndr* strain grown under the same conditions (Hsu et al., 1982). The increase in synthase activity in the *cdr* mutant is paralleled by an increase in immunoreactive material. The *cdr* mutant does not display elevated levels of dehydratase or dehydrogenase, and does not excrete α-isopropylmalate. Apparently, the synthase, even in an elevated state, remains inhibited by leucine and hence does not generate additional induction signal.

The fact that yeast α-isopropylmalate synthase is under general control makes it more difficult to determine whether the enzyme is also under specific repression control by leucine. Appropriate mutants will have to be constructed and characterized to attack this question.

Yeast *leu* enzyme levels also depend upon the carbon source. Thus, acetate grown cells produce 2 to 3 times more synthase than glucose-grown cells and about half the dehydratase and dehydrogenase (Brown et al., 1975). The apparent contradiction to the induction hypothesis may be explained by a discrepancy in the synthase activity measured in cell-free extracts and that which is operative in vivo, although the existence of other mechanisms of control under these conditions cannot be ruled out. When cells are shifted from an acetate medium to glucose medium, the synthase level drops by about 60% within 15 min. Differences in the properties of enzyme from acetate-grown versus glucose-grown cells have been interpreted to mean that there are two α-isopropylmalate synthases (Brown et al., 1975). Two genes encoding α-isopropylmalate synthase in yeast have recently been identified by tetrad analysis (V. Baichwal, T. Cunningham, P. Gatzek, and G. B. Kohlhaw, unpublished observations).

V. GENE CLONING AND SEQUENCING

The only leucine gene of yeast cloned and sequenced is *LEU2*. This gene was one of the first eukaryotic genes to be cloned and expressed in *E. coli* (Ratzkin and Carbon, 1977). The specific activity of β-isopropylmalate dehydrogenase in *E. coli* K12 *leuB* cells containing the *LEU2* gene is 10–15% of that found in wild-type *S. cerevisiae* cells. There is no repression by leucine or leucine plus threonine (Kohlhaw et al., 1980). Yeast β-isopropylmalate dehydrogenase synthesized in *E. coli* is cold-labile, even though the enzyme is stable in frozen yeast cells.

The *LEU2* gene, fused to plasmid ColEl, has served as the prototype gene to develop the yeast transformation system (Hinnen et al., 1978). The gene has since been used as a convenient genetic marker in a number of transformation experiments, including the construction of hybrid plasmids that allow higher frequencies of transformation (Beggs, 1978; Broach et al., 1979), and the expression of *Drosophila* and rabbit genes in yeast (Beggs et al., 1980; Henikoff et al., 1981).

The *LEU2* gene has been transposed from its normal location on chromosome III to a new location on chromosome XII. This was achieved by transforming a *leu2* recipient with a recombinant plasmid carrying both the *LEU2* gene and pieces of yeast rDNA. Since the ca. 100 copies of the rDNA gene are located in a cluster on chromosome XII (Petes, 1979), ample homology exists to allow stable integration of the plasmid into this chromosome (Szostak and Wu, 1979; Petes, 1980). The transposed *LEU2* gene is regulated in normal fashion by leucine and leucine plus threonine (Kohlhaw et al., 1980).

Normal regulation is also observed with a transformant that harbors a hybrid plasmid consisting of the bacterial pBR325 sequence, the 2-μm yeast plasmid, and a relatively short yeast DNA fragment carrying the *LEU2* gene and flanking regions (Cohen et al., 1980; Hsu and Kohlhaw, 1982). The 2-μm plasmid replicates autonomously at levels of 50–100 copies per cell (Beggs, 1978). The transformed strain apparently contains multiple copies of the plasmid and overproduces the *LEU2* gene product, α-isopropylmalate dehydrogenase, to such an extent that it constitutes approximately 2% of the extractable protein, allowing a great simplification of a previously developed, rather cumbersome purification procedure for this enzyme (Hsu and Kohlhaw, 1980; 1982). The *LEU2* gene and flanking regions have recently been sequenced (A. Andreadis, Y.-P. Hsu, G. B. Kohlhaw, M. Hermodson, and P. R. Schimmel, in preparation). The 5′ region preceding *LEU2* codes for a gene for $tRNA_3^{Leu}$, the major $tRNA^{Leu}$ isoacceptor in yeast; another part of this region contains potential stem and loop structures that could be involved in regulating *LEU2* gene expression, either at the transcriptional or at the translational level (Andreadis et al., 1982). Further scrutiny of this region of *LEU2,* together with similar information now being sought for *LEU1,* is expected to help explain the coordinate control of these two unlinked genes.

ACKNOWLEDGMENT

Work in the author's laboratory has been supported by a grant from the National Institute of General Medical Sciences.

REFERENCES

Andreadis, A., Hsu, Y.-P., Kohlhaw, G. B., and Schimmel, P. (1982) *Cell 31,* 319–325.

Beggs, J. D. (1978) *Nature 275,* 104–109.

Beggs, J. D., van den Berg, J., van Ooyen, A., and Weissmann, C. (1980) *Nature 283,* 835–840.

Bigelis, R., and Umbarger, H. E. (1976) *J. Biol. Chem. 251,* 3545–3552.

Broach, J. R., Strathern, J. N., and Hicks, J. B. (1979) *Gene 8,* 121–133.

Brown, H. D., Satyanarayana, T., and Umbarger, H. E. (1975) *J. Bacteriol. 121,* 959–969.

Burns, R. O., Umbarger, H. E., and Gross, S. R. (1963) *Biochemistry 2,* 1053–1058.

Cassady, W. E., Leither, E. H., Bergquist, A., and Wagner, R. P. (1972) *J. Cell Biol. 53,* 66–72.

Cohen, J. D., Eccleshall, T. R., Needleman, R. B., Federoff, H., Buchferer, B. A., and Marmur, J. (1980) *Proc. Nat. Acad. Sci. U.S.A. 77,* 1078–1082.

Collins, M., and Wagner, R. P. (1973) *Arch. Biochem. Biophys. 155,* 184–193.

Delforge, J., Messenguy, F., and Wiame, J. M. (1975) *Eur. J. Biochem. 57,* 231–239.

Gasser, S. M., Daum, G., and Schatz, G. (1982) *J. Biol. Chem. 257,* 13034–13041.
Gilbert, H. F., and Stewart, M. D. (1981) *J. Biol. Chem. 256,* 1782–1785.
Gross, S. R. (1965) *Proc. Nat. Acad. Sci. U.S.A. 54,* 1538–1546.
Gross, S. R. (1969) *Ann. Rev. Genet. 3,* 395–424.
Gross, S. R. (1970) *Methods Enzymol. 17A,* 777–781.
Hampsey, D. M. (1982) "Yeast α-Isopropylmalate synthase: CoA inactivation, leucine inhibition and subcellular location," Ph.D. thesis, Purdue University, West Lafayette, Ind.
Hampsey, D. M., and Kohlhaw, G. B. (1981) *J. Biol. Chem. 256,* 3791–3796.
Hampsey, D. M., Lewin, A. S., and Kohlhaw, G. B. (1983) *Proc. Natl. Acad. Sci. U.S.A. 80.* (in press)
Henikoff, S., Tatchell, K., Hall, B. D., and Nasmyth, K. A. (1981) *Nature 289,* 33–37.
Hinnen, A., Hicks, J. B., and Fink, G. R. (1978) *Proc. Nat. Acad. Sci. U.S.A. 75,* 1929–1933.
Hsu, Y.-P., and Kohlhaw, G. B. (1980) *J. Biol. Chem. 255,* 7255–7260.
Hsu, Y.-P., and Kohlhaw, G. B. (1982) *J. Biol. Chem. 257,* 39–41.
Hsu, Y.-P., Kohlhaw, G. B., and Niederberger, P. (1982) *J. Bacteriol. 150,* 969–972.
Kashmiri, S. V. S., and Gross, S. R. (1970) *Genetics 64,* 423–440.
Kohlhaw, G. B., and Tan-Wilson, A. (1977) *J. Bacteriol. 129,* 1159–1161.
Kohlhaw, G. B., Hsu, Y.-P., Lemmon, R. D., and Petes, T. D. (1980) *J. Bacteriol. 144,* 852–855.
Messenguy, F., and Cooper, T. G. (1977) *J. Bacteriol. 130,* 1253–1261.
Miozzari, G., Niederberger, P., and Hütter, R. (1978) *J. Bacteriol. 134,* 48–59.
Mortimer, R. K., and Schild, D. (1980) *Microbiol. Rev. 44,* 519–571.
Niederberger, P., Miozzari, G., and Hütter, R. (1981) *Mol. Cell. Biol. 1,* 584–593.
Olshan, A. R., and Gross, S. R. (1974) *J. Bacteriol. 118,* 374–384.
Petes, T. D. (1979) *Proc. Nat. Acad. Sci. U.S.A. 76,* 410–414.
Petes, T. D. (1980) *Cell 19,* 765–774.
Polacco, J. C., and Gross, S. R. (1973) *Genetics 74,* 443–459.
Ratzkin, B., and Carbon, J. (1977) *Proc. Nat. Acad. Sci. U.S.A. 74,* 487–491.
Reichenbecher, V. E., Fischer, M., and Gross, S. R. (1978) *J. Bacteriol. 133,* 794–801.
Reichenbecher, V. E., and Gross, S. R. (1978) *J. Bacteriol. 133,* 802–810.
Roeder, P. R., and Kohlhaw, G. B. (1980) *Biochim. Biophys. Acta 613,* 482–487.
Ryan, E. D., and Kohlhaw, G. B. (1974) *J. Bacteriol. 120,* 631–637.
Ryan, E. D., Tracy, J. W., and Kohlhaw, G. B. (1973) *J. Bacteriol. 116,* 222–225.
Satyanarayana, T., Umbarger, H. E., and Lindegren, G. (1968a) *J. Bacteriol. 96,* 2012–2017.
Satyanarayana, T., Umbarger, H. E., and Lindegren, G. (1968b) *J. Bacteriol. 96,* 2018–2024.
Schürch, A., Miozzari, G., and Hütter, R. (1974) *J. Bacteriol. 117,* 1131–1140.
Szostak, J. W., and Wu, R. (1979) *Plasmid 2,* 536–554.
Tan-Wilson, A., and Kohlhaw, G. B. (1978) *Biochem. Biophys. Res. Commun. 85,* 70–76.
Tracy, J. W., and Kohlhaw, G. B. (1975) *Proc. Nat. Acad. Sci. U.S.A. 72,* 1802–1806.
Tracy, J. W., and Kohlhaw, G. B. (1977) *J. Biol. Chem. 252,* 4085–4091.

Ulm, E. H., Böhme, R., and Kohlhaw, G. B. (1972) *J. Bacteriol. 110,* 1118–1126.
Webster, R. E., and Gross, S. R. (1965) *Biochemistry 4,* 2309–2318.
Webster, R. E., Nelson, C. A., and Gross, S. R. (1965) *Biochemistry 4,* 2319–2327.
Wipf, B., and Leisinger, T. (1979) *J. Bacteriol, 140,* 874–880.
Wolfner, M., Yep, D., Messenguy, F., and Fink, G. R. (1975) *J. Mol. Biol. 96,* 273–290.

CHAPTER
17

The Common Aromatic Biosynthetic Pathway

Klaus M. Herrmann

I. INTRODUCTION

The aromatic amino acids phenylalanine and tryptophan are synthesized by bacteria and plants only. As "essential" amino acids, they are required ingredients in the diet of animals, who generally synthesize the third aromatic amino acid, tyrosine, by hydroxylation from phenylalanine. In bioenergetic terms, the aromatic amino acids represent the most expensive protein building blocks; for example, the synthesis of each mole of tryptophan requires 78 moles of ATP (Atkinson, 1977).

Accordingly, the biosynthesis of tyrosine, phenylalanine, and tryptophan is complex and best considered in two parts: the synthesis of chorismate, the last common precursor, and the conversion of chorismate to the three aromatic amino acids. Chapters 18, 19, and 20 of this volume deal with the individual terminal pathways from chorismate to phenylalanine, tyrosine, and tryptophan, respectively. This chapter describes the synthesis of chorismate with major emphasis on the results of work with enterobacteriaceae.

The reaction sequence outlined in Fig. 17.1 is called either the shikimate pathway (Haslam, 1974), after the metabolic intermediate identified first, or the common aromatic biosynthetic pathway because chorismate is a

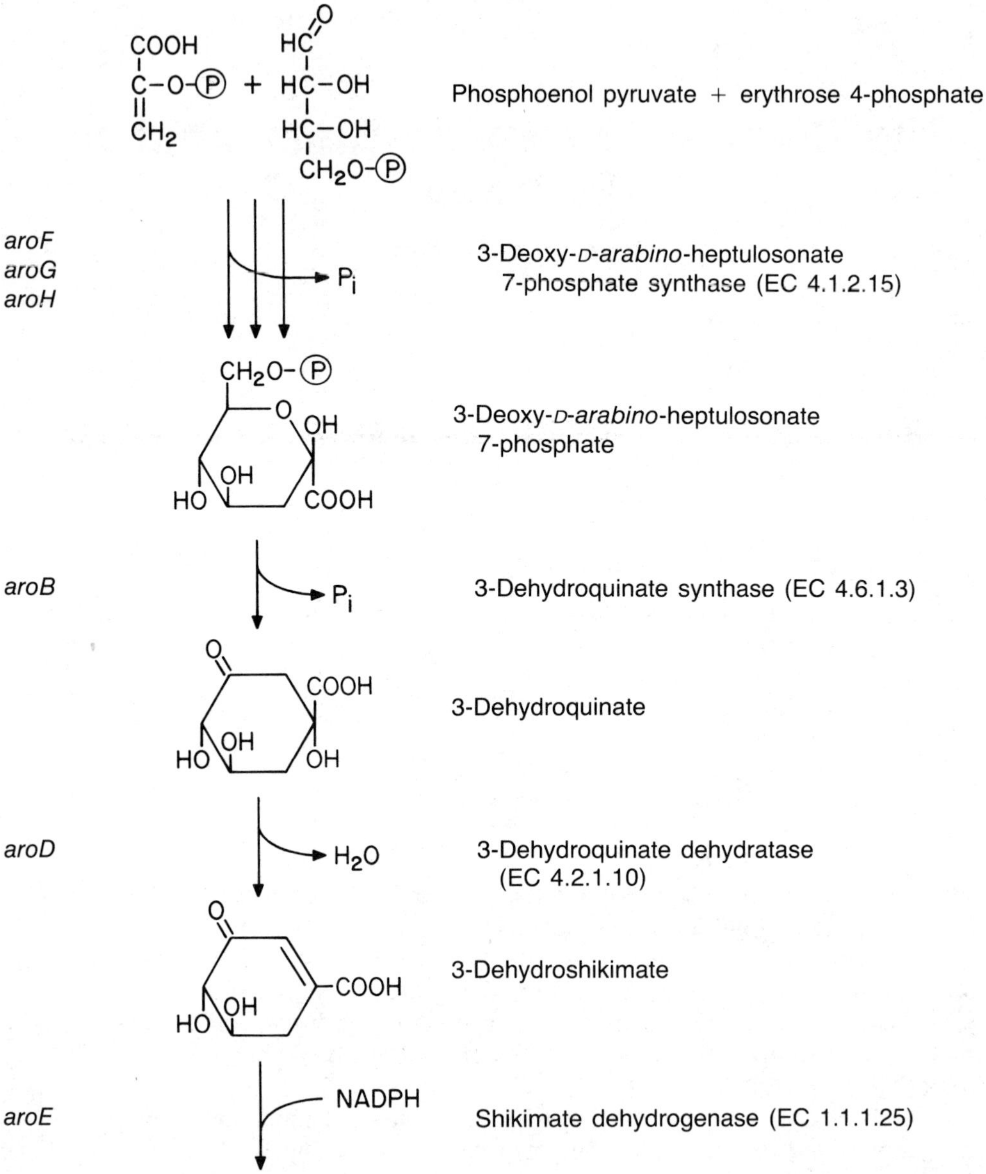
Phosphoenol pyruvate + erythrose 4-phosphate
aroF
aroG
aroH
Pi
3-Deoxy-D-arabino-heptulosonate 7-phosphate synthase (EC 4.1.2.15)
3-Deoxy-D-arabino-heptulosonate 7-phosphate
aroB
Pi
3-Dehydroquinate synthase (EC 4.6.1.3)
3-Dehydroquinate
aroD
H2O
3-Dehydroquinate dehydratase (EC 4.2.1.10)
3-Dehydroshikimate
aroE
NADPH
Shikimate dehydrogenase (EC 1.1.1.25)

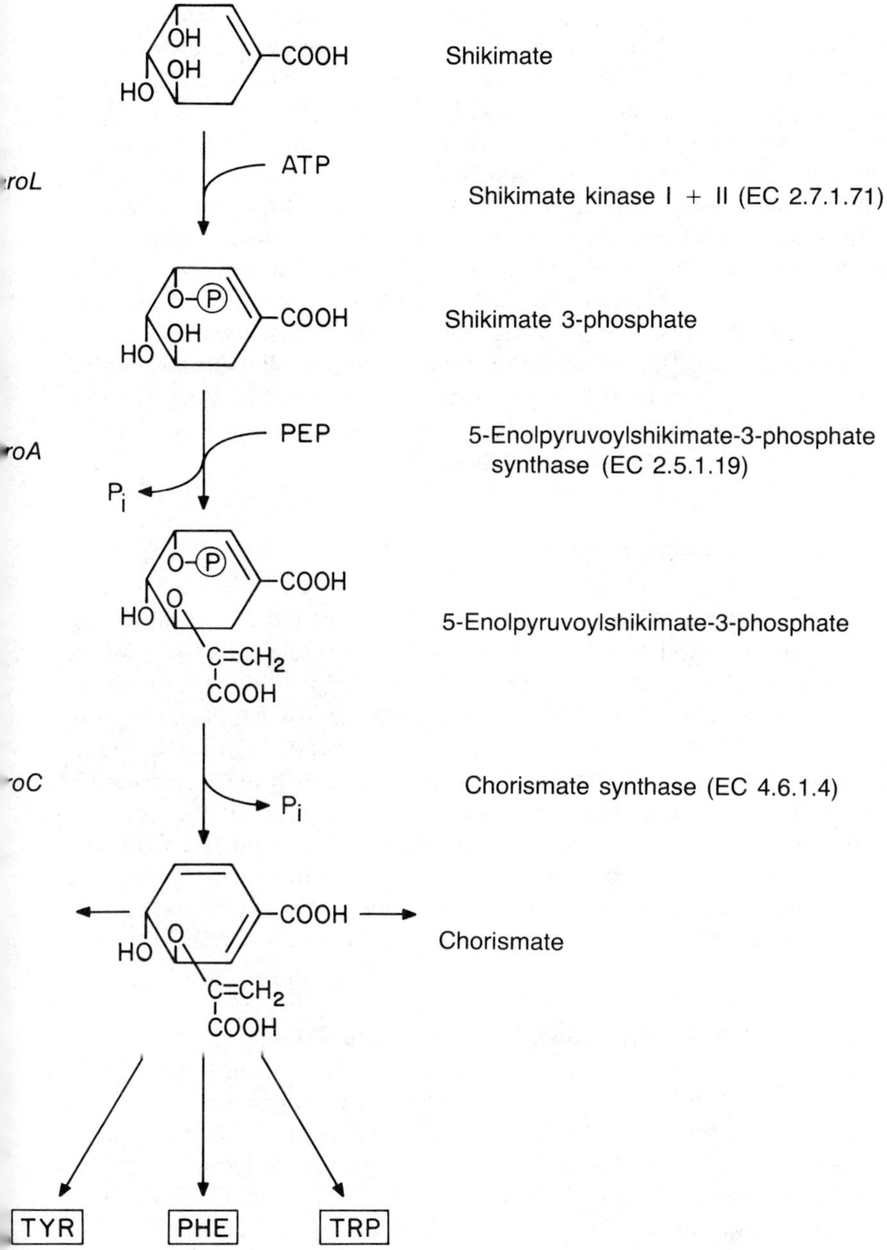

FIGURE 17.1 The shikimate pathway.

precursor not only for the aromatic amino acids but also for many other aromatic compounds, such as tetrahydrofolate (an essential cofactor in one-carbon metabolism) and enterochelin (an iron chelator). The seven enzyme-catalyzed reactions were elucidated mainly through the classical work of Davis (1955) and Sprinson (1961) and their co-workers. Historically, the second half of the pathway was described first. The early steps in this metabolic sequence proved to be much more elusive. Clues were obtained after the hexose monophosphate shunt was discovered and—in particular—after the chemical synthesis of erythrose 4-phosphate was accomplished (Ballou et al., 1955). Through two fairly complex reactions, phosphoenolpyruvate and erythrose 4-phosphate are converted into a six-membered carbon skeleton destined to become the benzene ring in all aromatic amino acids. Full aromatization of the ring system via the introduction of three double bonds requires the remaining five steps of chorismate biosynthesis plus additional steps in the terminal pathways.

II. CONTROL OF CARBON FLOW

Carbon flow through the shikimate pathway is controlled at the first step by modulation of the activity of 3-deoxy-D-*arabino*-heptulosonate 7-phosphate (DAHP) synthase. The enzyme is subject to feedback inhibition, specific degradation, and control at the level of transcription. In *Escherichia coli,* phenylalanine, tyrosine, and tryptophan, the end products of the major terminal pathways, serve as sensors to control carbon flow. Chorismate, the end product of the common pathway, is not involved in the control. Shikimate kinase represents a second potential point of control in the shikimate pathway. The remaining five constitutive enzymes (Tribe et al., 1976) have a remarkable catalytic capacity, since they readily accommodate carbon flow changes of several orders of magnitude (Ogino et al., 1982).

A. 3-Deoxy-D-*arabino*-heptulosonate 7-phosphate (DAHP)

The first committed step in the pathway is the condensation of phosphoenolpyruvate and erythrose 4-phosphate to form the seven-carbon compound DAHP and inorganic phosphate (Srinivasan and Sprinson, 1959). Srinivasan et al. (1955) were the first to suggest DAHP as a precursor for all aromatic amino acids and demonstrated (Srinivasan et al., 1956) that C-1 to C-3 of DAHP originate from glucose C-3 to C-1 or C-4 to C-6 and that C-4 to C-7 of DAHP originate from glucose C-3 to C-6 with some scrambling due to cycling in the hexose monophosphate shunt. These studies also refuted earlier claims that acetate was a precursor for phenylalanine, tyrosine, and tryptophan (Ehrensvard, 1955). The principal structure of DAHP was confirmed by chemical synthesis (Sprinson et al., 1963; Herrmann and Poling, 1975). Fine-structure analysis by ^{1}H and ^{13}C NMR identified DAHP as

1-carboxy-2-deoxy-α-D-glucose-6-phosphate; the oxygen-containing six-membered ring results from hemiacetal formation between C-2 and the hydroxyl on C-6 of the primary condensation product (Garner and Herrmann, 1983).

The biosynthesis of DAHP can be considered a special aldol condensation between an intermediate carbanion of C-3 of phosphoenolpyruvate and the carbonyl C-1 of erythrose-4-phosphate (Floss et al., 1972). The majority of the phosphoenolpyruvate molecules react stereospecifically with respect to the configuration on C-3, excluding the formation of a freely rotating intermediate methyl group (DeLeo and Sprinson, 1968). However, there is some incorporation of tritium into DAHP when the reaction is performed in ^3H_2O (DeLeo and Sprinson, 1968; Floss et al., 1972), and the inorganic phosphate is formed through a C-0 rather than a P-O cleavage (DeLeo and Sprinson, 1968; Nagano and Zalkin, 1970), which would favor an intermediate tetrahedral C-2 of phosphoenolpyruvate and consequently a CH_3 configuration for C-3. The paradox that exists between the two sets of data is best resolved by assuming a reversible protonation of the carbanion as a side reaction permitting the tritium incorporation from the medium (Floss et al., 1972). A critical evaluation of all available data led to the proposal of a mechanism for DAHP synthesis involving an enzyme-bound sulfur-containing enolpyruvoyl intermediate (Ganem, 1978).

B. DAHP Synthase from *E. coli*

1. Structure of the Enzyme. The condensation of phosphoenolpyruvate and erythrose-4-phosphate is catalyzed by DAHP synthase (Srinivasan and Sprinson, 1959). Wild-type *E. coli* produces three DAHP synthase isoenzymes: phenylalanine-sensitive DAHP synthase (Phe), tyrosine-sensitive DAHP synthase (Tyr), and tryptophan-sensitive DAHP synthase (Trp) (Doy and Brown, 1965; Pittard and Gibson, 1970). The three isoenzymes have been purified to homogeneity (Schoner and Herrmann, 1976; McCandliss et al., 1978; Poling et al., 1981). The tetrameric DAHP synthase (Phe) has a subunit molecular weight of 35,000 (McCandliss et al., 1978), and the dimeric DAHP synthase (Tyr) and DAHP synthase (Trp) have subunit molecular weights approaching 40,000 (Schoner and Herrmann, 1976; Poling et al., 1981). The native forms of the enzymes are probably protein–phosphoenolpyruvate adducts.

Antibodies raised against homogeneous DAHP synthase (Phe) form antigen–antibody complexes that maintain full enzyme activity and are still inhibited by phenylalanine, indicating that neither the active site nor the feedback inhibitor binding site is mechanistically connected to amino acid sequences that are antigenic determinants (McCandliss and Herrmann, 1979). While DAHP synthase (Phe) is quantitatively removed from solution by immunoprecipitation with soluble or immobilized antibodies, neither DAHP synthase (Tyr) nor DAHP synthase (Trp) forms any detectable complexes with such antibodies, a result suggesting structural differences between the isoenzymes. Similar results are obtained for antibodies raised against homogeneous preparations of DAHP synthase (Tyr) and DAHP synthase (Trp) (Suzich and Herrmann, 1983).

The structural differences of the isoenzymes inferred from immunochemical studies are confirmed in part through a comparison of the amino-terminal sequences of the three isoenzymes (Shultz et al., 1981). Only five of the first 40 residues are identical. From such data, one might infer that the proteins are only distantly related. The complete amino acid sequences of all three isoenzymes have been obtained through a combination of protein and DNA sequence efforts (Davies and Davidson, 1982; O'Connell et al., 1983; Herrmann et al., 1983; Synenki et al., 1983). A comparison of the three amino acid sequences shows that regions of very strong homology alternate with stretches of no apparent similarity. This arrangement may suggest that the three structural genes encoding DAHP synthases evolved not by gene duplication but rather through evolutionary reshuffling of functionally defined sequences encoding protein domains, for example, binding sites for phosphoenolpyruvate or erythrose 4-phosphate.

2. The Metal Content of the Enzyme. DAHP synthase (Phe) and DAHP synthase (Tyr) are inhibited by chelating agents, and partially purified preparations of the enzymes are activated by Co (Staub and Denes, 1969a; Dusha and Denes, 1976). Purified preparations of both isoenzymes are still inhibited by metal chelators, but no longer activated by Co (Schoner and Herrmann, 1976; McCandliss et al., 1978). Since the second enzyme in the pathway, dehydroquinate synthase, requires Co for activity, it is tempting to speculate that DAHP synthase and dehydroquinate synthase form a loose complex in vivo and that an active dehydroquinate synthase either stabilizes or activates DAHP synthase, even though there is no direct in vitro evidence for such a complex. However, both DAHP synthase and dehydroquinate synthase of crude cell extracts are inhibited by the herbicide N-[phosphonomethyl]glycine, and the addition of Co reverses this inhibition (Roisch and Lingens, 1980). Mung beans contain a DAHP synthase isoenzyme that requires Co for activity and is sensitive to the herbicide (Rubin et al., 1982). However, the primary point of attack for this herbicide may be the penultimate step of the shikimate pathway, since it is inhibited by 10^3-fold lower concentrations (Steinruecken and Amrhein, 1980).

DAHP synthase (Phe) and DAHP synthase (Tyr) purified from extracts of wild-type *E. coli* cells have a unique absorption with a maximum at 350 nm, and they contain about one mole of Fe per mole of native enzyme, but no Co (McCandliss and Herrmann, 1978). The amino acid sequence of residues 10–18 of DAHP synthase (Tyr), His-Ile-Thr-Asp-Glu-Gln-Val-Leu-Met, is highly homologous to the sequence of residues 54–62 of hemerythrin from the sea worm *Phascolopsis gouldii,* His-Phe-Leu-Asn-Glu-Gln-Val-Leu-Met (Herrmann et al., 1980). Since His-54 and Glu-58 of hemerythrin are Fe ligands (Stenkamp et al., 1976; Loehr et al., 1978; Gormley et al., 1978), residues 10–18 of the *E. coli* DAHP synthase (Tyr) may represent part of the Fe-binding fold in this protein, and His-10 and Glu-14 could be Fe ligands (Herrmann et al., 1980). DAHP synthase (Phe) does not contain a sequence homologous to residues 10–18 of DAHP synthase (Tyr).

In some *E. coli* cells bearing plasmids encoding DAHP synthase (Tyr), this enzyme may constitute 10–15% of the total soluble protein of the organism. DAHP synthase (Tyr) isolated from extracts of such cells contains Mn, Cu, Zn, and Fe in addition to traces of other metals, to the extent of about one mole of metal per mole of enzyme (Dean and Herrmann, 1983). While the enzyme requires a metal for activity, it appears that several different metals can satisfy this requirement. When the "physiological" metal is scarce, the polypeptide *in statu nascendi* may act as a metal scavenger and incorporate whatever metal is available.

3. Feedback Inhibition. DAHP synthase (Phe) is 50% inhibited by 13 μM phenylalanine. The inhibition is fairly specific; D-phenylalanine, tyrosine, and tryptophan do not inhibit this activity (McCandliss et al., 1978), but β-thienylalanine is a potent inhibitor (Ezekiel, 1965). Mutants resistant to β-thienylalanine contain a feedback-insensitive DAHP synthase (Phe) (Ezekiel, 1965).

DAHP synthase (Tyr) is 50% inhibited by 20 μM tyrosine (Dusha and Denes, 1976). Phenylalanine does not inhibit the pure enzyme. Tyrosine is a competitive inhibitor with respect to phosphoenolpyruvate and a non-competitive inhibitor with respect to erythrose 4-phosphate (Schoner and Herrmann, 1976). At concentrations above 10 μM, tyrosine acts as a classical dead-end inhibitor; the kinetic mechanism of the enzyme-catalyzed reaction is ordered sequential (Schoner and Herrmann, 1976), as was shown previously for the enzyme from *Salmonella typhimurium* (DeLeo et al., 1973), and not ping-pong, as was suggested in earlier studies on partially purified preparations (Staub and Denes, 1969b; Nagano and Zalkin, 1970). In addition, tyrosine at very low concentrations alters some undefined kinetic properties of the enzyme without preventing the subsequent binding of the substrates and the ensuing reaction (Schoner and Herrmann, 1976). Among the tyrosine analogs inhibiting DAHP synthase (Tyr), 3-fluorotyrosine is superior for the selection of mutants with a feedback-insensitive DAHP synthase (Tyr) (K. M. Herrmann et al., unpublished results). Phosphono and homophosphono analogs of erythrose 4-phosphate are substrates and inhibit DAHP synthase (Tyr) (LeMarechal et al., 1980).

DAHP synthase (Trp) is only partially inhibited by tryptophan (Pittard et al., 1969). Whereas phenylalanine and tyrosine inhibit DAHP synthase (Phe) and DAHP synthase (Tyr), respectively, by more than 95%, tryptophan inhibition of the pure DAHP synthase (Trp) does not exceed 40% (Fig. 17.2) (K. M. Herrmann et al., in preparation). The inability of tryptophan to totally inhibit DAHP synthase (Trp) ensures a sufficient supply of chorismate for the biosynthesis of aromatic compounds other than amino acids, when phenylalanine, tyrosine, and tryptophan are present in excess. Mutants

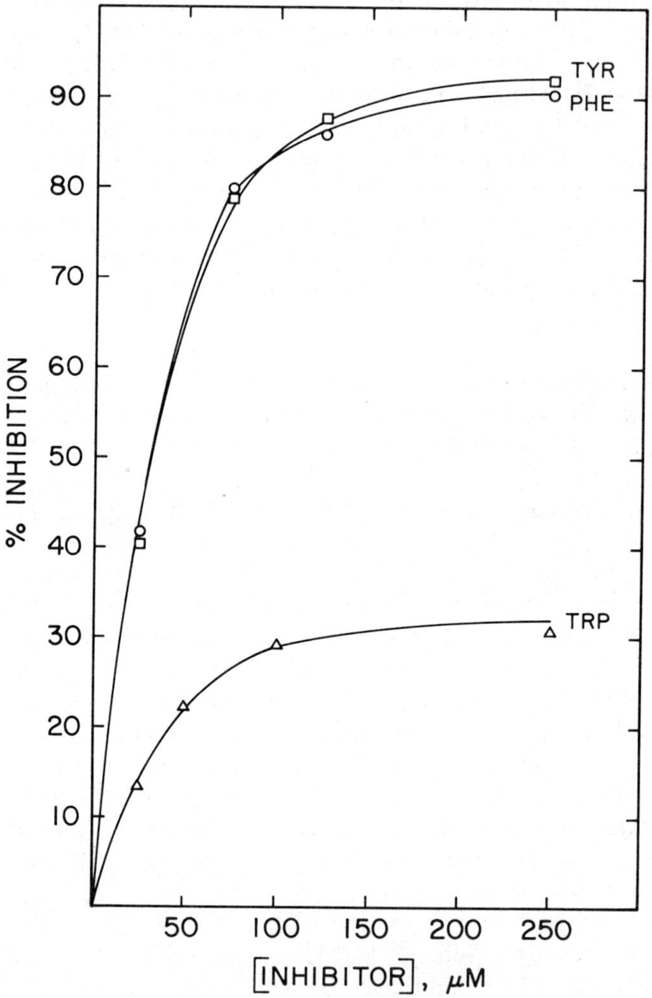

FIGURE 17.2 Feedback inhibition of DAHP synthase from *E. coli*. Homogeneous preparations of the three isoenzymes are assayed in the presence of variable amounts of their specific feedback inhibitors. Plotted is the percent inhibition of the enzyme activity as a function of the inhibitor concentration; squares are for DAHP synthase (Tyr), circles are for DAHP synthase (Phe), and triangles are for DAHP synthase (Trp).

containing a feedback-insensitive DAHP synthase (Trp) have been obtained (Pittard et al., 1969).

C. Modulation of Gene Expression

The structural genes for DAHP synthase (Tyr), DAHP synthase (Phe), and DAHP synthase (Trp) are *aroF, aroG,* and *aroH,* respectively, with gene locations at 56, 17, and 37 min on the *E. coli* linkage map (Bachmann and Low, 1980). In wild-type *E. coli,* 80% of the total DAHP synthase activity is contributed by the phenylalanine-sensitive isoenzyme (Jensen and Nasser, 1968), and 20% is contributed by the tyrosine-sensitive isoenzyme. There are only traces of the DAHP synthase (Trp). Since the turnover numbers of the pure isoenzymes are about the same, the isoenzyme distribution reflects differential expression of the respective genes, which in turn mirrors somewhat the total aromatic amino acid output of the *E. coli* cell: phenylalanine, tyrosine, and tryptophan make up 3.3, 2.2, and 1%, respectively, of total cellular protein. Thus the three DAHP synthases may be the remnants of three individual pathways that evolved into today's common pathway feeding from a single DAHP pool.

Expression of *aroG* is controlled by phenylalanine and tryptophan mediated through *tyrR* (Brown and Somerville, 1971; Im et al., 1971; Camakaris and Pittard, 1973) rather than *pheR* (Gollub et al., 1973; Gowrishankar and Pittard, 1982). However, maximal variation of *aroG* expression is less than two-fold. The monocistronic *aroG* operon is not controlled by an attenuation mechanism (Davies and Davidson, 1982). The *aroG* gene has been cloned into a multicopy plasmid and sequenced independently from a wild-type (Davies and Davidson, 1982) and a feedback-resistant mutant (Synenki et al., 1983). The two sequences are very similar; their translated amino acid sequences differ only in one residue. However, the codon usage in the two strains is not identical.

Genes *aroF* and *tyrA,* the structural gene for chorismate mutase–prephenate dehydrogenase, form the *tyr* operon that is under the control of tyrosine, and possibly phenylalanine, mediated by *tyrR* (Wallace and Pittard, 1969; Brown and Somerville, 1971; Im et al., 1971; Mattern and Pittard, 1971; Im and Pittard, 1973). The enzyme levels of DAHP synthase (Tyr) under different growth conditions vary more than 100-fold. Furthermore, in *aroG* mutants grown on a minimal salts medium, the *tyr* operon is sufficiently derepressed to supply enough chorismate for maximal growth (Pittard and Gibson, 1970). In addition to derepression, elevation of *aroF* expression in *tyrR, trpS* double mutants suggests attenuation control for the *tyr* operon (Camakaris et al., 1980) by a mechanism that resembles *trp* operon attenuation (Yanofsky, 1981). One can also interpret the specific derepression of the *tyr* operon in response to Fe starvation (McCray and Herrmann, 1976) by an attenuation mechanism, since one of the enzymes involved in full modification of several tRNA species requires Fe (Rosenberg and Gefter, 1969). Direct evidence for attenuation control of the *tyr* operon should soon come from DNA sequence data, since the *aroF* gene has been cloned into a multicopy plasmid (Zurawski et al., 1978) and the coding region of the gene has already been sequenced (O'Connell et al., 1983).

Both modes of transcriptional control on the *tyr* operon discussed above do not readily explain the discoordinate expression of *aroF* and *tyrA* observed when growth resumes after carbon starvation (DeLucia and Herrmann, 1983). During starvation or when cells reach the stationary phase of growth, DAHP synthase (Tyr) is specifically inactivated (DeLucia et al., 1979; Tribe and Pittard, 1979) by proteolytic degradation that does not affect the *tyrA* product (DeLucia and Herrmann, 1983).

Expression of *aroH* is controlled by tryptophan (Brown, 1968) mediated through *trpR* (Pittard et al., 1969). The *aroH* gene has been cloned and sequenced (Zurawski et al., 1981; Herrmann et al., 1983). On the basis of DNA sequence data, attenuation has been ruled out as a potential transcriptional control for *aroH* expression (Zurawski et al., 1981). The DNA region preceding the coding sequence of *aroH* is homologous to the operator of the *trp* operon (Chapter 20), a result suggesting that repression is the sole specific regulatory mechanism for *aroH* expression.

D. Control of DAHP Synthase Activity

The control mechanism for carbon flow through the shikimate pathway is complex and not yet fully understood. The structural genes encoding the seven enzymes of the pathway are scattered over the *E. coli* linkage map (Pittard and Wallace, 1966; Wallace and Pittard, 1967), and most of them form single-gene operons. Three regulatory genes, *trpR, tyrR,* and *pheR,* control the expression of some of these operons as well as those encoding the terminal pathways to the aromatic amino acids. Since repression mediated by *trpR* and *tyrR* affects more than one operon, two regulons are found in the biosynthesis of aromatic amino acids. While the terminal pathways to phenylalanine and tryptophan are partly regulated by attenuation, there is only indirect evidence of such a mechanism for one of the ten structural genes encoding enzymes of the shikimate pathway.

The most striking control feature in the shikimate pathway is the existence of three DAHP synthase isoenzymes that are controlled by feedback inhibition. The concept of feedback inhibition, discovered in isoleucine (Umbarger, 1956) and pyrimidine biosynthesis (Yates and Pardee, 1956), was shown to apply to the shikimate pathway 20 years ago (Smith et al., 1962). Genetic analysis demonstrates the high degree of efficiency of feedback control in the biosynthesis of the aromatic amino acids (Wallace and Pittard, 1967; Tribe and Pittard, 1979). Feedback inhibition appears to be the quantitatively major regulatory mechanism for chorismate biosynthesis in vivo (Ogino et al., 1982), supporting the importance of the allosteric concept of which

Jaques Monod once remarked, ''J'ai découvert le deuxième secret de la vie.''

E. DAHP Synthases From Other Organisms

The control pattern for carbon flow through the shikimate pathway of *Salmonella* (Hu and Sprinson, 1977), *Klebsiella, Serratia,* or *Shigella* is similar, but not identical, to the *E. coli* one; substantial variations, found in other bacteria (Jensen et al., 1967), can serve to classify taxa (Whitaker et al., 1981a, 1981b).

Bacillus subtilis contains only a single DAHP synthase. The enzyme has been purified to homogeneity from strain 168 (Nakatsukasa and Nester, 1972), in which DAHP synthase and chorismate mutase form a bifunctional enzyme that has four identical subunits of molecular weight 38,500 (Huang et al., 1974a). The bifunctional enzyme is part of a trifunctional complex that includes shikimate kinase. The DAHP synthase of the complex is insensitive to aromatic amino acids but inhibited by chorismate and prephenate (Huang et al., 1974b). However, the shikimate kinase of the complex appears to be the major control point of the shikimate pathway in *B. subtilis* (Huang et al., 1975). *B. subtilis* strain 168 is a derivative of the wild-type Marburg strain, which also contains only a single DAHP synthase. However, the wild-type DAHP synthase is not associated with chorismate mutase. There is good evidence that the mutant bifunctional DAHP synthase–chorismate mutase has artificially evolved from a monofunctional DAHP synthase since 1967 (Llewellyn et al., 1980). Transcriptional control of *B. subtilis* DAHP synthase is mediated by tyrosine (Nester et al., 1969), and biosynthesis of aromatic amino acids and histidine may share common regulatory elements (Chapman and Nester, 1968).

Brevibacterium flavum DAHP synthase purified to homogeneity is a tetramer having a subunit molecular weight of 55,000 (Sugimoto and Shiio, 1980a). The enzyme is synergistically inhibited by phenylalanine and tyrosine (Shiio et al., 1974), and 3-fluorophenylalanine resistant mutants have a feedback-insensitive DAHP synthase (Shiio and Sugimoto, 1979). As in *B. subtilis,* DAHP synthase and chorismate mutase form a bifunctional enzyme. However, unlike *B. subtilis,* a second polypeptide, component B, is required for chorismate mutase activity and stimulates DAHP synthase activity. Component B purified to homogeneity is a dimer of subunit molecular weight 13,500 (Sugimoto and Shiio, 1980b). An $\alpha_4\beta_8$ quarternary structure has been proposed for the native complex in which chorismate mutase is inhibited by phenylalanine and tyrosine; tryptophan activates chorismate mutase but not DAHP synthase through stabilization of the $\alpha_4\beta_8$ complex (Shiio and Sugimoto, 1981). The synthesis of both polypeptides is subject to regulation by tyrosine (Shiio and Sugimoto, 1979).

DAHP synthase activities have been demonstrated in various higher plants (Minamikawa, 1967; Huisman and Kosuge, 1974; Rothe et al., 1976;

Saijo and Takeo, 1979; Graziana and Boudet, 1980; Rubin et al., 1982). One isoenzyme has recently been obtained in near-homogeneous form after a more than 10,000-fold purification from *Daucus carota;* the dimeric hysteretic enzyme is insensitive to phenylalanine but activated by tryptophan and tyrosine (Suzich and Herrmann, 1983).

III. THE REMAINING ENZYMES OF THE PATHWAY

A. Dehydroquinate Synthase

The second reaction of the shikimate pathway is an intramolecular exchange of the DAHP ring oxygen with C-7, accompanied by an oxidation at C-6 and a reduction at C-2. Cleavage of the phosphoester provides the driving force to form dehydroquinate, first isolated by Weiss et al. (1953).

The reaction is catalyzed by dehydroquinate synthase. The pure enzyme from *E. coli* is a single polypeptide chain having a molecular weight of 57,000 with a K_m for DAHP of 0.33 mM (Maitra and Sprinson, 1978). The enzyme requires Co and NAD for activity, the latter in catalytic amounts (K_m, 1 μM) (Srinivasan et al., 1963). 5-Dehydro-DAHP is an intermediate in the enzyme-catalyzed reaction (Rotenberg and Sprinson, 1978), a result confirming earlier data that suggested C-5 of DAHP as the site of intermediate oxidation–reduction (LeMarechal and Azerad, 1976a). A small kinetic isotope effect on C-6 of DAHP reflects the removal of a proton during the phosphate elimination (Rotenberg and Sprinson, 1978). NAD is apparently recycled from enzyme-bound NADH, since there is no tritium incorporation into the product when the reaction is performed in ^3H_2O. The formation of dehydroquinate from DAHP is stereospecific and occurs with inversion of the conformation on C-7 of DAHP without exchange of hydrogen with the medium (Rotenberg and Sprinson, 1970, 1978; Turner et al., 1975). NADP cannot substitute for NAD, and NADH inhibits the reaction at concentrations exceeding 1 μM.

Dehydroquinate synthase is competitively inhibited by C-3 (LeMarechal and Azerad, 1976b) and phosphono analogs of DAHP (LeMarechal et al., 1981). The latter also protect the enzyme against thermal denaturation. The *E. coli* enzyme is also inhibited by N-[phosphonomethyl]glycine (Roisch and Lingens, 1980). In *E. coli* mutants with unregulated DAHP synthesis, dehydroquinate synthase becomes rate-limiting for carbon flow through the shikimate pathway (Ogino et al., 1982). The *B. subtilis* enzyme, purified to homogeneity, has a molecular weight of 17,000 and forms a trifunctional complex with chorismate synthase and a flavin reductase (Hasan and Nester, 1978a). Association with chorismate synthase is required for enzyme activity of the dehydroquinate synthase.

B. Dehydroquinate Dehydratase

A stereospecific *syn*-dehydration of dehydroquinate (Hanson and Rose, 1963; Turner et al., 1975) introduces the first double bond of the aromatic ring system to yield dehydroshikimate (Salamon and Davis, 1953; Srinivasan et al., 1956). The reaction is catalyzed by dehydroquinate dehydratase (Mitsuhashi and Davis, 1954). Schiff base formation between enzyme and substrate (Butler et al., 1974) causes a conformational change in the substrate (twisted boat) that leads to the stereospecific course of the reaction (Vaz et al., 1975).

There are conflicting data about the structure of the enzyme from *E. coli*. The homogeneous enzyme appears to be a tetramer with a subunit molecular weight of 40,000 (Vaz, 1980). However, the structural gene encoding dehydroquinate dehydratase, *aroD,* cloned into a multicopy plasmid, directs the synthesis of a dimeric protein with a subunit molecular weight of 31,500 (Kinghorn et al., 1981). The *N. crassa* structural gene encoding dehydroquinate dehydratase, cloned into an *E. coli* vector, complements *aroD* mutants (Vapnek et al., 1977). The expression of the fungal gene in *E. coli* is increased 100-fold in *pnp* mutants devoid of polynucleotide phosphorylase, reflecting inherent structural differences between prokaryotic and eukaryotic mRNA's (Hautala et al., 1979).

Dehydroquinate dehydratase has also been purified to homogeneity from mung beans (Koshiba, 1978a) and *Physcomitrella pateus* (Polley, 1978). In both organisms, a single polypeptide contains dehydroquinate dehydratase and shikimate dehydrogenase, the enzyme that catalyzes the next step in the metabolic sequence. The bifunctional enzyme from *P. pateus* has a molecular weight of 49,500. Furthermore, mung beans (Koshiba, 1978a) and corn (Graziana et al., 1980) contain isoenzymic forms of the bifunctional dehydroquinate dehydratase–shikimate dehydrogenase.

C. Shikimate Dehydrogenase

Shikimate, the central metabolic intermediate of the pathway, was first isolated from the Japanese tree shikimi-no-ki (Eijkman, 1885). The absolute configuration of the molecule (Fischer and Dangschat, 1937) is confirmed through conformational analysis by NMR (Hall, 1964) and numerous total syntheses (Haslam, 1974; Ganem, 1978).

Shikimate biosynthesis from 3-dehydroshikimate is catalyzed by shikimate dehydrogenase (Yaniv and Gilvarg, 1955). The NADP-specific enzyme from *E. coli* (Dansette and Azerad, 1974) and pea seedlings (D. D. Davies et al., 1972) facilitates the hydrogen transfer from the A-side of NADPH. Product inhibition and isotope exchange studies on the 200-fold purified enzyme from pea seedlings suggest an ordered kinetic mechanism for the enzyme-catalyzed reaction (Balinski et al., 1971). *B. subtilis* mutants lacking shikimate dehydrogenase require not only aromatic amino acids and *p*-aminobenzoate but also shikimate to restore wild-type growth rates (Farrand and Taber, 1973).

Partially purified preparations of the enzyme have been studied from mung bean (Nandy and Ganguli, 1961), tea plant (Sanderson, 1966; Saijo and Takeo, 1979), bamboo (Higuchi and Shimada, 1967), tobacco cell cultures

(Kuboi and Yamada, 1978), and corn (Byng et al., 1981). Shikimate dehydrogenase activity changes during plant development (Blume and McClure, 1979) and is inhibited by herbicides (Cole et al., 1980; Blume and McClure, 1980). The enzyme activity appears in isoenzymic forms located in different cell organelles but encoded by nuclear DNA (Feierabend and Brassel, 1977; Weeden and Gottlieb, 1980; Linhart et al., 1981).

D. Shikimate Kinase

Shikimate is phosphorylated to shikimate 3-phosphate (Weiss and Mingioli, 1956) by shikimate kinase. *S. typhimurium* (Morell and Sprinson, 1968) and *E. coli* (Berlyn and Giles, 1969) contain two isoenzymes, shikimate kinase I and II. While shikimate kinase of *S. typhimurium* appears not to be regulated (Gollub et al., 1967) and the *E. coli* enzyme I is constitutive, *E. coli* enzyme II (molecular weight, 20,000) encoded by *aroL* is under the control of *tyrR* (Ely and Pittard, 1979). In wild-type *E. coli, aroL* specifies about 50%, but in *tyrR* mutants, it specifies 90% of the total kinase activity. Enzyme II is repressed by tyrosine or tryptophan, but *trpR* mutants do not show elevated levels of the enzyme. The existence of shikimate kinase isoenzymes could indicate shikimate as a branch point intermediate to hitherto unrecognized biosynthetic pathways (Weiss and Edwards, 1980).

The *B. subtilis* strain 168 shikimate kinase is a polypeptide of 10,000 daltons that is complexed with the bifunctional DAHP synthase–chorismate mutase (see Section II.E). The kinase, only active in the complex, has been purified to homogeneity (Huang et al., 1975). Since the enzyme is inhibited by chorismate, prephenate, ADP, and 5-enolpyruvoylshikimate 3-phosphate (Huang et al., 1975) and derepressed by growth on limiting tyrosine (Nasser et al., 1969), shikimate kinase may represent the key allosteric control point of the pathway in *B. subtilis*.

Shikimate kinase has been partially purified from mung beans (Koshiba, 1979) and sorghum (Bowen and Kosuge, 1979).

E. 5-Enolpyruvoylshikimate 3-Phosphate Synthase

Shikimate 3-phosphate reacts with phosphoenolpyruvate to form 5-enolpyruvoylshikimate 3-phosphate and inorganic phosphate (Levin and Sprinson, 1964). The reversible enzyme-catalyzed reaction is a rare transfer of an unchanged enolpyruvoyl moiety of phosphoenolpyruvate. Protonation of C-3 of phosphoenolpyruvate combined with a nucleophilic attack of the 5-hydroxyl of shikimate leads to a presumed intermediate from which 5-envolpyruvoylshikimate 3-phosphate is obtained in a 1,2-elimination of orthophosphate. The mechanism is supported by NMR and isotope exchange studies (Bondinell et al.,1971).

The reaction is catalyzed by 5-enolpyruvoylshikimate 3-phosphate synthase that has been partially purified from *E. coli* (Levin and Sprinson,

1964) and *S. typhimurium* (Bondinell et al., 1971). The enzyme appears to be the major site of inhibition for N-[phosphonomethyl]glycine, because 5- to 7-μM solutions of this herbicide inhibit 50% of the enzyme activity in *K. aerogenes, E. coli,* and mung beans (Steinruecken and Amrhein, 1980). Mung bean seedlings are the only higher plant material shown to contain this enzyme (Koshiba, 1978b).

F. Chorismate Synthase

The second double bond in the aromatic ring system is introduced through an unusual *trans*-1,4-elimination of orthophosphate from 5-enolpyruvoylshikimate 3-phosphate (Hill and Newkome, 1969; Onderka and Floss, 1969; Floss et al., 1972) to yield chorismate (F. Gibson and Jackman, 1963; M. I. Gibson and Gibson, 1964). The absolute stereochemical structure of chorismate (Edwards and Jackman 1965) has recently been confirmed by total synthesis (McGowan and Berchtold, 1982; Ganem et al., 1982).

The reaction is catalyzed by chorismate synthase. The partially purified enzyme from *E. coli* (Morell et al., 1967) requires an $FADH_2$ or NADH regenerating system in a N_2 atmosphere. The enzyme is sensitive to O_2, presumably because Fe^{2+} is an integral part of the protein.

The *B. subtilis* chorismate synthase, purified to homogeneity, has a subunit molecular weight of 24,000 (Hasan and Nester, 1978b) and is part of a trimeric complex with an NADPH-dependent flavin reductase (Hasan and Nester, 1978c) and dehydroquinate synthase (Hasan and Nester, 1978a). The flavin reductase, also purified to homogeneity, has a subunit molecular weight of 13,000. While the *E. coli* reductase uses NADH and can be replaced by pig heart dihydrolipoate dehydrogenase (Morell et al., 1967), the *B. subtilis* enzyme is specific for NADPH. In *N. crassa,* the chorismate synthase and the flavin reductase reside on the same polypeptide (Welch et al., 1974; see also Section IV). The role of the reductase in the catalytic process remains to be elucidated. Chorismate synthase has not yet been studied in higher plants.

IV. THE SHIKIMATE PATHWAY OF LOWER EUKARYOTES

Like enterobacteriaceae, *N. crassa* contains three DAHP synthase isoenzymes that are inhibited by tyrosine, phenylalanine, and tryptophan, respectively (Jensen and Nasser, 1968). The tyrosine-sensitive isoenzyme from *N. crassa* was the first DAHP synthase obtained in near-homogeneous form from any organism (Hoffmann et al., 1972). The homogeneous tryptophan-sensitive DAHP synthase of *N. crassa* is a tetramer with a subunit molecular weight of 52,000 (Nimmo and Coggins, 1981a). The reaction mechanism has been classified as ordered sequential (Nimmo and Coggins, 1981b) and is therefore similar to that found for bacterial enzymes (DeLeo et al., 1973; Schoner and Herrmann, 1976). Yeasts contain two DAHP synthases, one sensitive to phenylalanine and one sensitive to tyrosine (Lingens et al., 1966); the latter has been purified to homogeneity from *Schizosaccharomyces pombe* (Bracher and Schweingruber, 1977). Carbon flow in the shikimate pathway of *Saccharomyces cerevisiae* (Lingens et al., 1967) and *Hansenula polymorpha* (Denenu and Demain, 1981) is regulated by feedback inhibition rather than by transcriptional control of DAHP synthase.

In contrast to the bacteria, the five central enzymes of the shikimate pathway in *N. crassa* (Giles et al., 1967), *Euglena* (Berlyn et al., 1970; Kempner et al., 1980), *S. pombe* (Strauss, 1979), and *S. cerevisiae* (Larimer et al., 1982) are encoded by the *arom* gene cluster, which specifies a pentafunctional enzyme on a single polypeptide chain (Gaertner and Cole, 1977). The *S. cerevisiae arom* cluster cloned into a multicopy plasmid is expressed in yeast and *E. coli* (Larimer et al., 1982). The pentafunctional enzyme from *N. crassa,* purified to homogeneity (Burgoyne et al., 1969; Gaertner, 1972), is a dimer with a subunit molecular weight of 165,000 (Lumsden and Coggins, 1977). This enzyme catalyzes the conversion of DAHP into 5-enolpyruvoylshikimate 3-phosphate by processes in which the intermediates remain enzyme-bound during catalysis (Welch and Gaertner, 1975). The pentafunctional enzyme is protected against proteolytic attack (Gaertner and Cole, 1976), particularly noted during the late log phase of growth (Gaertner and Cole, 1977), by a specific protease inhibitor (Lampkin et al., 1976) as well as by DAHP (Vitto and Gaertner, 1978). DAHP may induce a physiologically significant conformational change in the complex. Four of the five activities of the pentafunctional enzyme are coordinately activated by DAHP; at elevated DAHP concentrations, shikimate kinase becomes rate-limiting for carbon flow in the shikimate pathway (Welch and Gaertner, 1976). Through ADP inhibition of the kinase, energy charge becomes a major regulatory signal under conditions of rising DAHP supply. The pentafunctional enzyme is not ubiquitous among the lower eukaryotes; *Anabaena* and *Chlamydomonas* (Berlyn et al., 1970), *Hansenula, Candida,* and others (Bode and Birnbaum, 1981) do not contain *arom* clusters.

N. crassa and other fungi contain two dehydroquinate dehydratases and two shikimate dehydrogenases (Cain, 1972). In addition to the biosynthetic enzymes encoded in the *arom* cluster, degradative isoenzymes of the two activities are induced by growth on quinate. Mutants lacking the degradative enzymes can still grow on quinate, and the degradative enzymes can also take over the function of the biosynthetic enzymes.

N. crassa chorismate synthase purified to homogeneity is a dimer with a subunit molecular weight of 55,000 (Welch et al., 1974). Like other chorismate synthases, the *N. crassa* activity is dependent upon a flavin reductase using NADPH and FMN. Pure chorismate synthase has diaphorase activity; thus the flavin reductase and the chorismate synthase form a bicephalic enzyme.

V. OUTLOOK

The structures of all the metabolic intermediates of the shikimate pathway are firmly established, even though some of them have not yet been obtained by total organic synthesis. Several of the bacterial enzymes in the pathway remain to be purified, and amino acid sequence data are only available for DAHP synthase from *E. coli*. A comparison of the total amino acid sequences of all seven enzymes from a single organism will most likely be accomplished through cloning and sequencing of the respective structural genes combined with some confirmatory peptide sequence efforts. The cloning of the structural genes will also provide access to DNA sequences of regions preceding the coding sequences and thus contribute some knowledge about promoter differences between constitutive and regulated genes. The elucidation of the overall control of carbon flow along the pathway remains a challenging problem. The feedback inhibitor binding sites of the *E. coli* DAHP synthases will soon be characterized. However, the integration of transcriptional control of *aroF*, *aroG*, and *aroH* may require new models to explain how the expression of unlinked yet functionally related genes are coordinated. The cloning of genes encoding plant DAHP synthases will help to understand mechanisms of gene expression in higher eukaryotes, which process in excess of 25% of total carbon flow through the shikimate pathway.

ACKNOWLEDGMENTS

The author thanks all his co-workers and colleagues who made the work on the biosynthesis of aromatic compounds an exciting and rewarding experience. Supported by research grant GM-17678 from the United States Public Health Service, this is paper no. 9260 of the Purdue University Agricultural Experiment Station.

REFERENCES

Atkinson, D. E. (1977) *Cellular Energy Metabolism and Its Regulation*. Academic Press, New York.

Bachmann, B. J., and Low, K. B. (1980) *Microbiol. Rev. 44,* 1–56.

Balinski, D., Dennis, A. W., and Cleland, W. W. (1971) *Biochemistry 10,* 1947–1952.

Ballou, C. E., Fischer, H. O. L., and McDonald, D. L. (1955) *J. Amer. Chem. Soc. 77,* 5967–5970.

Berlyn, M. B., and Giles, N. H. (1969) *J. Bacteriol. 99,* 222–230.

Berlyn, M. B., Ahmed, S. I., and Giles, N. H. (1970) *J. Bacteriol. 104,* 768–774.

Blume, D. E., and McClure, J. W. (1979) *Z. Pflanzenphysiol. 95,* 121–128.

Blume, D. E., and McClure, J. W. (1980) *Plant Physiol. 65,* 238–244.

Bode, R., and Birnbaum, D. (1981) *Z. Allg. Mikrobiol. 21,* 417–422.

Bondinell, W. E., Vnek, J., Knowles, P. F., Sprecher, M., and Sprinson, D. B. (1971) *J. Biol. Chem. 246,* 6191–6196.
Bowen, J. R., and Kosuge, T. (1979) *Plant Physiol. 64,* 382–386.
Bracher, M., and Schweingruber, M. E. (1977) *Biochim. Biophys. Acta 485,* 446–451.
Brown, K. D. (1968) *Genetics 60,* 31–48.
Brown, K. D., and Somerville, R. L. (1971) *J. Bacteriol. 108,* 386–399.
Burgoyne, L., Case, M. E., and Giles, N. H. (1969) *Biochim. Biophys. Acta 191,* 452–462.
Butler, J. R., Alworth, W. L., and Nugent, M. J. (1974) *J. Amer. Chem. Soc. 96,* 1617–1618.
Byng, G., Whitaker, R., Flick, C., and Jensen, R. A. (1981) *Phytochemistry 20,* 1289–1292.
Cain, R. B. (1972) *Biochem. J. 127,* 15p–16p.
Camakaris, H., and Pittard, J. (1973) *J. Bacteriol. 115,* 1135–1144.
Camakaris, H., Camakaris, J., and Pittard, J. (1980) *J. Bacteriol. 143,* 613–620.
Chapman, L. F., and Nester, E. W. (1968) *J. Bacteriol. 96,* 1658–1663.
Cole, D. J., Dodge, A. D., and Caseley, J. C. (1980) *J. Exp. Bot. 31,* 1665–1674.
Dansette, P., and Azerad, R. (1974) *Biochimie 56,* 751–755.
Davies, D. D., Teixeira, A., and Kenworthy, P. (1972) *Biochem. J. 127,* 335–343.
Davies, W. D., and Davidson, B. E. (1982) *Nucl. Acid Res. 10,* 4045–4058.
Davis, B. D. (1955) *Adv. Enzymol. 16,* 247–312.
Dean, J., and Herrmann, K. M. (1983) *Bioinorg. Chem.* (in preparation)
DeLeo, A. B., and Sprinson, D. B. (1968) *Biochem. Biophys. Res. Comm. 32,* 873–877.
DeLeo, A. B., Dayan, J., and Sprinson, D. B. (1973) *J. Biol. Chem. 248,* 2344–2353.
DeLucia, A., and Herrmann, K. M. (1983) *J. Bacteriol.* (in preparation)
DeLucia, A., Schoner, R., and Herrmann, K. M. (1979) *Int. Congr. Biochem. XI,* 252.
Denenu, E. O., and Demain, A. L. (1981) *Appl. Environ. Microbiol. 42,* 497–501.
Doy, C. H., and Brown, K. D. (1965) *Biochim. Biophys. Acta 104,* 377–389.
Dusha, I., and Denes, G. (1976) *Biochim. Biophys. Acta 438,* 563–573.
Edwards, J. M., and Jackman, L. M. (1965) *Austral. J. Chem. 18,* 1227–1239.
Ehrensvard, G. (1955) *Ann. Rev. Biochem. 24,* 275–310.
Eijkman, J. F. (1885) *Rec. Trav. Chim. Pays-Bas 4,* 32–54.
Ely, B., and Pittard, J. (1979) *J. Bacteriol. 138,* 933–943.
Ezekiel, D. H. (1965) *Biochim. Biophys. Acta 95,* 54–62.
Farrand, S. K., and Taber, H. W. (1973) *J. Bacteriol. 115,* 1021–1034.
Feierabend, J., and Brassel, D. (1977) *Z. Pflanzenphysiol. 82,* 334–346.
Fischer, H. O. L., and Dangschat, G. (1937) *Helv. Chim. Acta 20,* 705–716.
Floss, H. G., Onderka, D. K., and Carroll, M. (1972) *J. Biol. Chem. 247,* 736–744.
Gaertner, F. H. (1972) *Arch. Biochem. Biophys. 151,* 277–284.
Gaertner, F. H., and Cole, K. W. (1976) *Arch. Biochem. Biophys. 177,* 566–573.
Gaertner, F. H., and Cole, K. W. (1977) *Biochem. Biophys. Res. Comm. 75,* 259–264.
Ganem, B. (1978) *Tetrahedron 34,* 3353–3383.
Ganem, B., Ikota, N., Muralidharan, V. B., Wade, W. S., Young, S. D., and Yukimoto, Y. (1982) *J. Am. Chem. Soc. 104,* 6787–6788.

Garner, C., and Herrmann, K. M. (1983) *Carbohydrate Res.* (submitted)
Gibson, F., and Jackman, L. M. (1963) *Nature 198,* 388–389.
Gibson, M. I., and Gibson, F. (1964) *Biochem. J. 90,* 248–256.
Giles, N. H., Case, M. E., Partridge, C. W. H., and Ahmed, S. I. (1967) *Proc. Nat. Acad. Sci. U.S.A. 58,* 1453–1460.
Gollub, E., Zalkin, H., and Sprinson, D. B. (1967) *J. Biol. Chem. 242,* 5323–5328.
Gollub, E. G., Liu, K. P., and Sprinson, D. B. (1973) *J. Bacteriol. 115,* 121–128.
Gormley, P. M., Loehr, J. S., Brimhall, B., and Hermodson, M. A. (1978) *Biochem. Biophys. Res. Comm. 85,* 1360–1366.
Gowrishankar, J., and Pittard, J. (1982) *J. Bacteriol. 150,* 1130–1137.
Graziana, A., and Boudet, A. M. (1980) *Plant Cell. Physiol. 21,* 793–802.
Graziana, A., Boudet, A., and Boudet, A. M. (1980) *Plant Cell Physiol. 21,* 1163–1174.
Hall, L. D. (1964) *J. Org. Chem. 29,* 297–299.
Hanson, K. R., and Rose, I. A. (1963) *Proc. Nat. Acad. Sci. U.S.A. 50,* 981–988.
Hasan, N., and Nester, E. W. (1978a) *J. Biol. Chem. 253,* 4999–5004.
Hasan, N., and Nester, E. W. (1978b) *J. Biol. Chem. 253,* 4993–4998.
Hasan, N., and Nester, E. W. (1978c) *J. Biol. Chem. 253,* 4987–4992.
Haslam, E. (1974) *The Shikimate Pathway,* John Wiley, New York.
Hautala, J. A., Bassett, C. L., Giles, N. H., and Kushner, S. R. (1979) *Proc. Nat. Acad. Sci. U.S.A. 76,* 5774–5778.
Herrmann, K. M., and Poling, M. D. (1975) *J. Biol. Chem. 250,* 6817–6821.
Herrmann, K. M., Shultz, J., and Hermodson, M. A. (1980) *J. Biol. Chem. 255,* 7079–7081.
Herrmann, K. M., Poling, M. D., Hermodson, M. A., and Yanofsky, C. (1983) *J. Biol. Chem.* (in preparation)
Higuchi, T., and Shimada, M. (1967) *Plant Cell Physiol, 8,* 61–69.
Hill, R. K., and Newkome, G. R. (1969) *J. Amer. Chem. Soc. 91,* 5893–5894.
Hoffmann, P. J., Doy, C. H., and Catcheside, D. E. A. (1972) *Biochim. Biophys. Acta 268,* 550–561.
Hu, C.-Y., and Sprinson, D. B. (1977) *J. Bacteriol. 129,* 177–183.
Huang, L., Nakatsukasa, W. M., and Nester, E. W. (1974a) *J. Biol. Chem. 249,* 4467–4472.
Huang, L., Montoya, A. L., and Nester, E. W. (1974b) *J. Biol. Chem. 249,* 4473–4479.
Huang, L., Montoya, A. L., and Nester, E. W. (1975) *J. Biol. Chem. 250,* 7675–7681.
Huisman, O. C., and Kosuge, T. (1974) *J. Biol. Chem. 249,* 6842–6848.
Im, S. W. K., and Pittard, J. (1973) *J. Bacteriol. 115,* 1145–1152.
Im, S. W. K., Davidson, H., and Pittard, J. (1971) *J. Bacteriol. 108,* 400–409.
Jensen, R. A., and Nasser, D. S. (1968) *J. Bacteriol. 95,* 188–196.
Jensen, R. A., Nasser, D. S., and Nester, E. W. (1967) *J. Bacteriol. 94,*1582–1593.
Kempner, E. S., Miller, J. H., Schlegel, W., and Hearon, J. Z. (1980) *J. Biol. Chem. 255,* 6826–6831.
Kinghorn, J. R., Schweizer, M., Giles, N. H., and Kushner, S. R. (1981) *Gene 14,* 73–80.
Koshiba, T. (1978a) *Biochim. Biophys. Acta 522,* 10–18.
Koshiba, T. (1978b) *Z. Pflanzenphysiol. 88,* 353–355.
Koshiba, T. (1979) *Plant Cell Physiol. 20,* 803–809.

Kuboi, T., and Yamada, Y. (1978) *Biochim. Biophys. Acta 542,* 181–190.
Lampkin, S. L., Cole, K. W., Vitto, A., and Gaertner, F. H. (1976) *Arch. Biochem. Biophys.177,* 561–565.
Larimer, F. W., Beck, A. K., Morse, C., Cole, K. W., and Gaertner, F. H. (1982) *Proc. Nat. Acad. Sci. U.S.A.* (in press)
LeMarechal, P., and Azerad, R. (1976a) *Biochimie 58,* 1123–1128.
LeMarechal, P., and Azerad, R. (1976b) *Biochimie 58,* 1145–1148.
LeMarechal, P., Froussios, C., Level, M., and Azerad, R. (1980) *Biochem. Biophys. Res. Comm. 92,* 1097–1103.
LeMarechal, P., Froussios, C., Level, M., and Azerad, R. (1981) *Carbohydr. Res. 94,* 1–10.
Levin, J. G., and Sprinson, D. B. (1964) *J. Biol. Chem. 239,* 1142–1150.
Lingens, F., Goebel, W., and Uesseler, H. (1966) *Biochem. Z. 346,* 357–367.
Lingens, F., Goebel, W., and Uesseler, H. (1967) *Eur. J. Biochem. 1,* 363–374.
Linhart, Y. B., Davis, M. L., and Mitton, J. B. (1981) *Biochem. Genet. 19,* 641–646.
Llewellyn, D. J., Daday, A., and Smith, G. D. (1980) *J. Biol. Chem. 255,* 2077–2084.
Loehr, J. S., Lammers, P. J., Brimhall, B., and Hermodson, M. A. (1978) *J. Biol. Chem. 253,* 5726–5731.
Lumsden, J., and Coggins, J. R. (1977) *Biochem. J. 161,* 599–607.
Maitra, U. S., and Sprinson, D. B. (1978) *J. Biol. Chem. 253,* 5426–5430.
Mattern, I. E., and Pittard, J. (1971) *J. Bacteriol. 107,* 8–15.
McCandliss, R. J., and Herrmann, K. M. (1978) *Proc. Nat. Acad. Sci. U.S.A. 75,* 4810–4813.
McCandliss, R. J., and Herrmann, K. M. (1979) *J. Biol. Chem. 254,* 3761–3764.
McCandliss, R. J., Poling, M. D., and Herrmann, K. M. (1978) *J. Biol. Chem. 253,* 4259–4265.
McCray, J. W., Jr., and Herrmann, K. M. (1976) *J. Bacteriol. 125,* 608–615.
McGowan, D. A., and Berchtold, G. A. (1982) *J. Amer. Chem. Soc. 104,* 1153–1154.
Minamikawa, T. (1967) *Plant Cell Physiol. 8,* 695–707.
Mitsuhashi, S., and Davis, B. D. (1954) *Biochim. Biophys. Acta 15,* 54–61.
Morell, H., and Sprinson, D. B. (1968) *J. Biol. Chem. 243,* 676–677.
Morell, H., Clark, M. J., Knowles, P. F., and Sprinson, D. B. (1967) *J. Biol. Chem. 242,* 82–90.
Nagano, H., and Zalkin, H. (1970) *Arch. Biochem. Biophys. 138,* 58–65.
Nakatsukasa, W. M., and Nester, E. W. (1972) *J. Biol. Chem. 247,* 5972–5979.
Nandy, M., and Gaguli, N. C. (1961) *Arch. Biochem. Biophys. 92,* 399–408.
Nasser, D., Henderson, G., and Nester, E. W. (1969) *J. Bacteriol. 98,* 44–50.
Nester, E. W., Jensen, R. A., and Nasser, D. S. (1969) *J. Bacteriol. 97,* 83–90.
Nimmo, G. A., and Coggins, J. R. (1981a) *Biochem. J. 197,* 427–436.
Nimmo, G. A., and Coggins, J. R. (1981b) *Biochem. J. 199,* 657–665.
O'Connell, A. M., Killingly, D. J., Brown, K. D., Shultz, J., Hermodson, M. A., and Herrmann, K. M. (1983) *J. Biol. Chem.* (in press)
Ogino, T., Garner, C., Markley, J. L., and Herrmann, K. M. (1982) *Proc. Nat. Acad. Sci. U.S.A. 79,* 5828–5832.
Onderka, D. K., and Floss, H. G. (1969) *J. Amer. Chem. Soc. 91,* 5894–5896.
Pittard, J., and Gibson, F. (1970) in *Current Topics in Cellular Regulation* (Horecker,

B. L., and Stadtman, E. R., eds.) Vol. II, pp. 29–63, Academic Press, New York.
Pittard, J., Wallace, B. J. (1966) *J. Bacteriol. 91,* 1494–1508.
Pittard, J., Camakaris, J., and Wallace, B. J. (1969) *J. Bacteriol. 97,* 1242–1247.
Poling, M. D., Suzich, J., Shultz, J., and Herrmann, K. M. (1981) *Fed. Proc. 40,* 1581.
Polley, L. D. (1978) *Biochim. Biophys. Acta 526,* 259–266.
Roisch, U., and Lingens, F. (1980) *Hoppe-Seyler's Z. Physiol. Chem. 361,* 1049–1058.
Rosenberg, A. H., and Gefter, M. L. (1969) *J. Mol. Biol. 46,* 581–584.
Rotenberg, S. L., and Sprinson, D. B. (1970) *Proc. Nat. Acad. Sci. U.S.A. 67,* 1669–1672.
Rotenberg, S. L., and Sprinson, D. B. (1978) *J. Biol. Chem. 253,* 2210–2215.
Rothe, G. M., Maurer, W., and Mielke, C. (1976) *Ber. Deutsch. Bot. Ges. 89,* 163–173.
Rubin, J. L., Gaines, C. G., and Jensen, R. A. (1982) *Plant. Physiol. 70,* 833–839.
Saijo, R., and Takeo, T. (1979) *Agr. Biol. Chem. 43,* 1427–1432.
Salamon, I. I., and Davis, B. D. (1953) *J. Amer. Chem. Soc. 75,* 5567–5571.
Sanderson, G. W. (1966) *Biochem. J. 98,* 248–252.
Schoner, R., and Herrmann, K. M. (1976) *J. Biol. Chem. 251,* 5440–5447.
Shiio, I., and Sugimoto, S. (1979) *J. Biochem. 86,* 17–25.
Shiio, I., and Sugimoto, S. (1981) *J. Biochem. 89,* 1483–1492.
Shiio, I., Sugimoto, S., and Miyajima, R. (1974) *J. Biochem. 75,* 987–997.
Shultz, J., Hermodson, M. A., and Herrmann, K. M. (1981) *FEBS Lett. 131,* 108–110.
Smith, L. C., Ravel, J. M., Lax, S. R., and Shive, W. (1962) *J. Biol. Chem. 237,* 3566–3570.
Sprinson, D. B. (1961) *Adv. Carbohydrate Chem. 15,* 235–270.
Sprinson, D. B., Rothschild, J., and Sprecher, M. (1963) *J. Biol. Chem. 238,* 3170–3175.
Srinivasan, P. R., and Sprinson, D. B. (1959) *J. Biol. Chem. 234,* 716–722.
Srinivasan, P. R., Katagiri, M., Sprinson, D. B. (1955) *J. Amer. Chem. Soc. 77,* 4943–4944.
Srinivasan, P. R., Shigeura, H. T., Sprecher, M., Sprinson, D. B., and Davis, B. D. (1956) *J. Biol. Chem. 220,* 477–497.
Srinivasan, P. R., Rothschild, J., and Sprinson, D. B. (1963) *J. Biol. Chem. 238,* 3176–3182.
Staub, M., and Denes, G. (1969a) *Biochim. Biophys. Acta 178,* 588–598.
Staub, M., and Denes, G. (1969b) *Biochim. Biophys. Acta 178,* 599–608.
Steinruecken, H. C., and Amrhein, N. (1980) *Biochem. Biophys. Res. Comm. 94,* 1207–1212.
Stenkamp, R. E., Sieker, L. C., and Jensen, L. H. (1976) *Proc. Nat. Acad. Sci. U.S.A. 73,* 349–351.
Strauss, A. (1979) *Mol. Gen. Genet. 172,* 233–241.
Sugimoto, S., and Shiio, I. (1980b) *J. Biochem. 88,* 167–176.
Sugimoto, S., and Shiio, I. (1980a) *J. Biochem. 87,* 881–890.
Suzich, J. A., and Herrmann, K. M. (1983) *J. Biol. Chem.* (submitted)
Synenki, R., Bond, W., and Herrmann, K. M. (1983) *J. Biol. Chem.* (in preparation)
Tribe, D. E., and Pittard, J. (1979) *Appl. Environ. Microbiol. 38,* 181–190.

Tribe, D. E., Camakaris, H., and Pittard, J. (1976) *J. Bacteriol. 127,* 1085–1097.
Turner, M. J., Smith, B. W., and Haslam, E. (1975) *J. Chem. Soc. Perkin I,* 52–55.
Umbarger, H. E. (1956) *Science 123,* 847.
Vapnek, D., Hautala, J. A., Jacobson, J. W., Giles, N. H., and Kushner, S. R. (1977) *Proc. Nat. Acad. Sci. U.S.A. 74,* 3508–3512.
Vaz, A. D. N. (1980) *Diss. Abstr. Int. B. 41,* 941–942.
Vaz, A. D. N., Butler, J. R., and Nugent, M. J. (1975) *J. Amer. Chem. Soc. 97,* 5914–5915.
Vitto, A., and Gaertner, F. H. (1978) *Biochem. Biophys. Res. Comm. 82,* 977–981.
Wallace, B. J., and Pittard, J. (1967) *J. Bacteriol. 93,* 237–244.
Wallace, B. J., and Pittard, J. (1969) *J. Bacteriol. 97,* 1234–1241.
Weeden, N. F., and Gottlieb, L. D. (1980) *J. Hered. 71,* 392–396.
Weiss, U., and Edwards, J. M. (1980) *The Biosynthesis of Aromatic Compounds,* John Wiley, New York.
Weiss, U., and Mingioli, E. S. (1956) *J. Amer. Chem. Soc. 78,* 2894–2898.
Weiss, U., Davis, B. D., and Mingioli, E. (1953) *J. Amer. Chem. Soc. 75,* 5572–5576.
Welch, G. R., and Gaertner, F. H. (1975) *Proc. Nat. Acad. Sci. U.S.A. 72,* 4218–4222.
Welch, G. R., and Gaertner, F. H. (1976) *Arch. Biochem. Biophys. 172,* 476–489.
Welch, G. R., Cole, K. W., and Gaertner, F. H. (1974) *Arch. Biochim. Biophys. 165,* 505–518.
Whitaker, R. J., Byng, G. S., Gherna, R. L., and Jensen, R. A. (1981a) *J. Bacteriol. 145,* 752–759.
Whitaker, R. J., Byng, G. S., Gherna, R. L., and Jensen, R. A. (1981b) *J. Bacteriol. 147,* 526–534.
Yaniv, H., and Gilvarg, C. (1955) *J. Biol. Chem. 213,* 787–795.
Yanofsky, C. (1981) *Nature 289,* 751–758.
Yates, R. A., and Pardee, A. B. (1956) *J. Biol. Chem. 221,* 757– 770.
Zurawski, G., Brown, K., Killingly, D., and Yanofsky, C. (1978) *Proc. Nat. Acad. Sci. U.S.A. 75,* 4271–4275.
Zurawski, G., Gunsalus, R. P., Brown, K. D., and Yanofsky, C. (1981) *J. Mol. Biol. 145,* 47–73.

CHAPTER
18

Biosynthesis of Phenylalanine

Craig Garner
Klaus M. Herrmann

I. INTRODUCTION

Chorismate, the end product of the shikimate pathway, is converted to phenylalanine in three steps via either phenylpyruvate (Fig. 18.1) or arogenate (Fig. 18.2), depending upon the organism (Jensen and Pierson, 1975). The phenylpyruvate pathway, operative in *Escherichia, Salmonella,* and *Klebsiella,* was discovered first (Weiss et al., 1954; Cotton and Gibson, 1965). The enterics contain a bifunctional chorismate mutase–prephenate dehydratase encoded within a single polypeptide chain. For tyrosine biosynthesis, these organisms utilize a bifunctional chorismate mutase–prephenate dehydrogenase. The two mutases have thus been termed channeling isoenzymes (Jensen and Byng, 1981), although strict channeling of chorismate does not actually prevail (Duggleby et al., 1978). Chorismate mutase–prephenate dehydratase is also a bifunctional enzyme in *Alcaligenes eutrophus,* an organism with a monofunctional prephenate dehydrogenase. Such an arrangement requires for tyrosine biosynthesis the dissociation of prephenate from the bifunctional enzyme (B. Friedrich et al., 1976; C. G. Friedrich et al., 1976).

Bacillus, Brevibacterium, and the lower eukaryotes *Neurospora* and *Saccharomyces* also synthesize phenylalanine by transamination of phenylpyruvate; however, their chorismate mutase and prephenate dehydratase

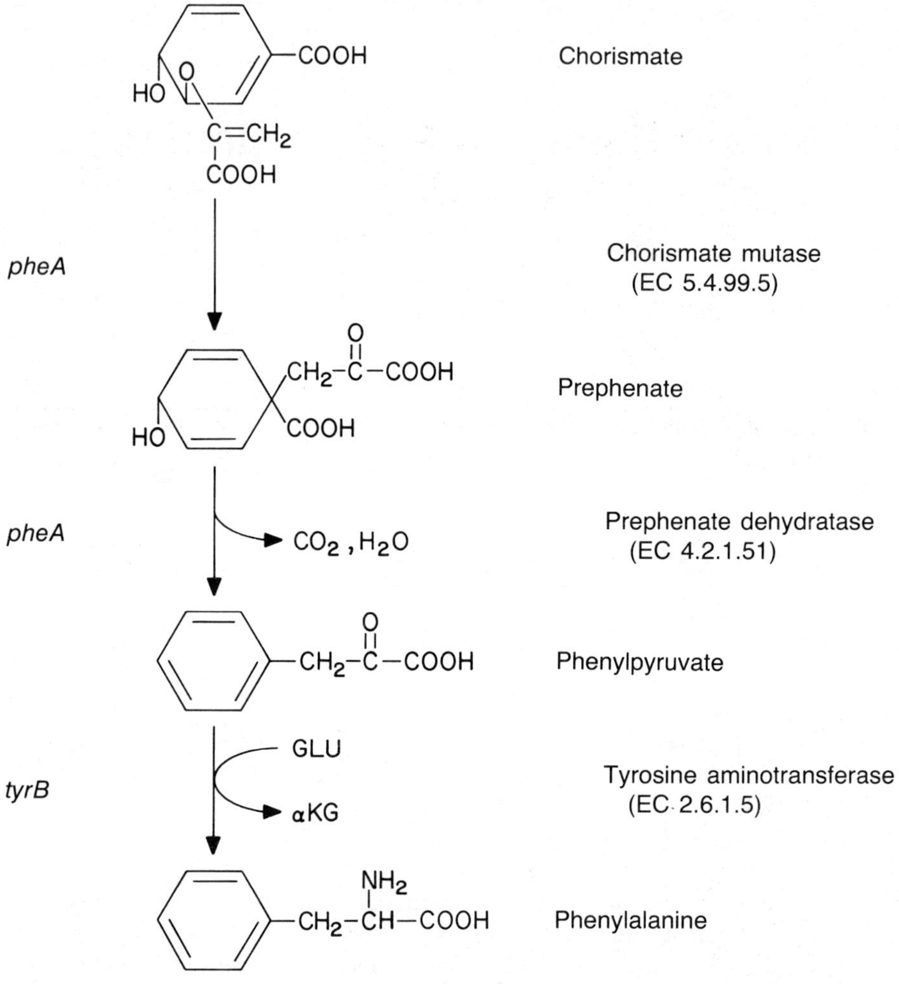

FIGURE 18.1 The phenylpyruvate pathway to phenylalanine.

activities reside on separate polypeptide chains. In these organisms, chorismate mutase is truly the last activity of the common pathway to phenylalanine and tyrosine, both of which draw upon a single prephenate pool. Even though the wild-type Marburg strain of *Bacillus subtilis* contains a monofunctional chorismate mutase, the activity has been studied the most from *B. subtilis* strain 168, which, like *Brevibacterium flavum*, contains a bifunctional 3-deoxy-D-*arabino*-heptulosonate 7-phosphate (DAHP synthase–chorismate mutase). There is good evidence that the bifunctional *Bacillus* enzyme evolved artificially (Llewellyn et al., 1980).

The arogenate pathways of tyrosine (Stenmark et al., 1974) and phenyl-

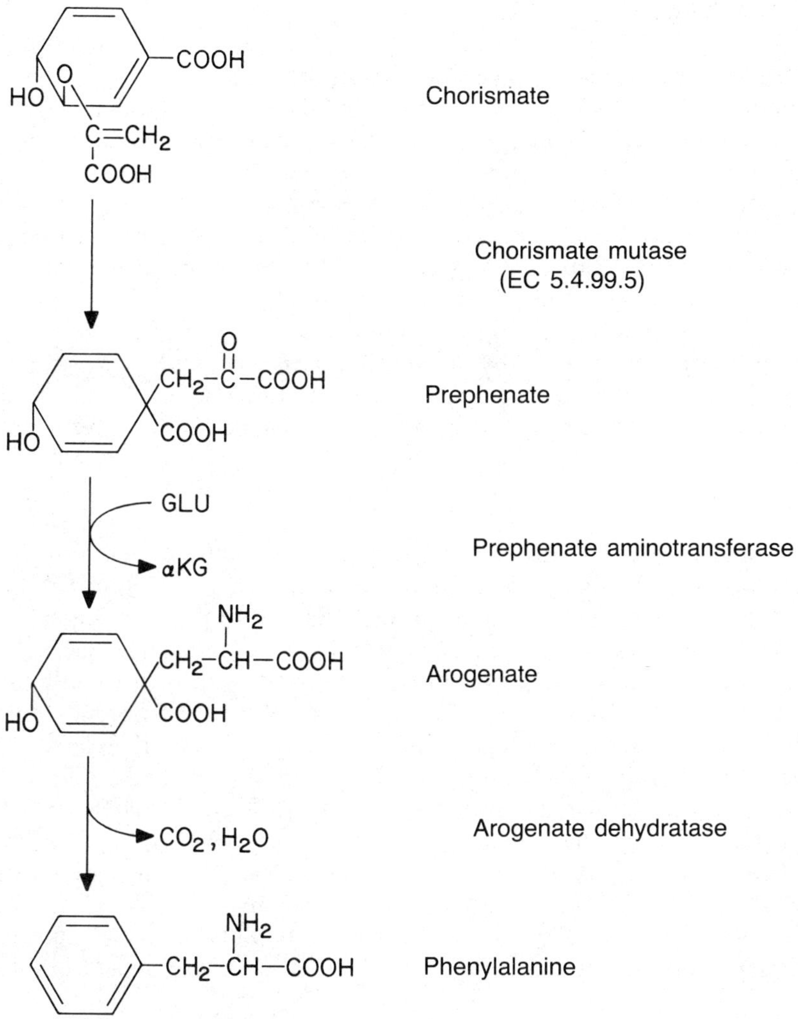

FIGURE 18.2 The arogenate pathway to phenylalanine.

alanine (Patel et al., 1977) biosynthesis were discovered in blue-green bacteria and *Pseudomonas aeruginosa,* respectively. Here transamination precedes the dehydration–decarboxylation step; thus prephenate yields arogenate. Full aromatization of the benzene ring is the very last step of the sequence. *P. aeruginosa* contains both the phenylpyruvate and the arogenate pathway. Arogenate is a dead-end product in *Neurospora crassa,* since this organism lacks both arogenate dehydratase and arogenate dehydrogenase (Jensen et al., 1977).

II. ENTEROBACTERIACEAE

A. Chorismate mutase–prephenate dehydratase

The spontaneous (Andrews et al., 1973) and the enzyme-catalyzed (Edwards and Jackman, 1965; Andrews et al., 1977) conversion of chorismate to prephenate are Claisen rearrangements that are rarely found as enzyme-catalyzed reactions (Ife et al., 1976). Prephenate was first isolated by Weiss et al. (1954), who suggested a structure for this relatively unstable compound that was confirmed through two independent total syntheses (Danishefsky et al., 1979; Gramlich and Plieninger, 1979). In acidic solutions, prephenate decomposes nonenzymatically to phenylpyruvate (Katagiri and Sato, 1953; Davis, 1953; Young et al., 1969). In enterobacteriaceae, both the rearrangement of chorismate to prephenate and the dehydration–decarboxylation of prephenate are catalyzed by a bifunctional chorismate mutase–prephenate dehydratase (Cotton and Gibson, 1965).

1. The structure of the bifunctional enzyme. The pure enzyme from *Escherichia coli* is a dimer with a subunit molecular weight of 40,000 (Davidson et al., 1972). Amino acid sequence analysis of the polypeptide, now 75% complete, suggests that the chorismate mutase–prephenate dehydratase of *E. coli* evolved through a gene fusion/duplication process (Baldwin and Davidson, 1979). Each catalytic center is located in a different domain of a single polypeptide chain as shown through differential inactivation by chemical modification (Gething and Davidson, 1977a, b), equilibration of prephenate with the medium (Duggleby et al., 1978), and isolation of mutants that lack either chorismate mutase or prephenate dehydratase activity (Baldwin and Davidson, 1981). These studies confirm earlier observations on the partially purified enzyme from *Salmonella typhimurium* (Schmit and Zalkin, 1969, 1971), which clearly demonstrated functionally distinct catalytic sites for chorismate mutase and prephenate dehydratase (Schmit et al., 1970).

2. Feedback inhibition by phenylalanine. Both reactions of chorismate mutase–prephenate dehydratase from *E. coli* (Dopheide et al., 1972) or *S. typhimurium* (Schmit and Zalkin, 1971) are competitively inhibited by phenylalanine; this ligand simultaneously alters the aggregation state of the protein (Schmit and Zalkin, 1971). With increasing phenylalanine concentrations, the active dimeric enzyme from *E. coli* is converted into inactive tetramers and octamers (Baldwin et al., 1981). Maximal inhibition is obtained when each subunit binds one molecule of phenylalanine. The inhibition is fairly specific; D-phenylalanine, tyrosine, and tryptophan do not inhibit. Phenylalanine derivatives with monosubstitutions in the benzene ring or at the amino group inhibit to various degrees, while carboxy derivatives do not inhibit at all (Dopheide et al., 1972; Gething and Davidson, 1978b).

While changes in the aggregation state of the enzyme do not induce gross secondary structure changes (Gething and Davidson, 1978a), aggregation causes cooperativity between the active sites (Gething et al., 1976; Baldwin et al., 1981). Circular dichroism studies reveal the movement of a tryptophanyl residue into a more hydrophobic microenvironment upon exposure of the enzyme to phenylalanine (Gething and Davidson, 1978a). This tryptophanyl residue may be part of the phenylalanine binding site, since modification of the two tryptophanyl residues of the enzyme with dimethyl[2-hydroxy-5-nitrobenzyl sulphomium bromide] results in a partially active enzyme that is feedback-resistant (Gething and Davidson, 1977b). Desensitization of the enzyme is also caused by modification of cysteinyl residues with 5,5-dithiobis[nitro-benzoate] (Gething and Davidson, 1978b).

3. Control of gene expression. In *E. coli,* the structural gene for chorismate mutase–prephenate dehydratase, *pheA,* is a monocistronic operon controlled by *pheR* (Gowrishankar and Pittard, 1982) and mediated through phenylalanine (Brown, 1968). The *pheR* regulatory locus, first identified in *S. typhimurium* (Gollub et al., 1973), specifies a classical Jacob-Monod repressor that binds at a presumed *pheA* operator site (Im and Pittard, 1971). The maximal derepression of *pheA* is about tenfold. The expression of *aroG,* a gene also regulated by phenylalanine, is not controlled by *pheR*. Thus in contrast to *tyrR* and *trpR,* there is no evidence for a *pheR*-defined regulon.

The *E. coli pheA* gene has been cloned into a multicopy plasmid, and the regulatory region preceding the coding region has been sequenced (Zurawski et al., 1978). The DNA sequence suggests an attenuation mechanism similar to those identified for other biosynthetic operons (Artz and Broach, 1975; Yanofsky, 1981). In temperature-sensitive *pheS* mutants, the level of chorismate mutase–prephenate dehydratase is unaltered when cells are grown at the restrictive temperature (Pittard and Gibson, 1970). However, *pheA* is derepressed in *miaA* mutants (Gowrishankar and Pittard, 1982). The *miaA* gene encodes a tRNA-modifying activity involved in thiomethylation of adenosine in several tRNA species; lack of this enzyme causes a relief of transcription termination at the attenuator site of the *trp* operon (Yanofsky and Soll, 1977). Also, tRNA modification mediated by the *miaA* gene product requires Fe (Rosenberg and Gefter, 1969), and in *E. coli* cells grown under Fe limitation, chorismate mutase–prephenate dehydratase is elevated fivefold (McCray and Herrmann, 1976). The enzyme activity is also elevated in *rpoB* mutants (Gowrishankar and Pittard, 1982). Certain *rpoB* mutants affect transcription termination at the *trp* attenuator (Yanofsky and Horn, 1981).

Since the *pheR* and *rpoB* effects are additive, *pheA* expression must be controlled by at least two distinct regulatory mechanisms involving initiation and termination of transcription. Thus carbon flow from chorismate to phenylalanine is controlled by the end product through repression and attenuation of *pheA* and through feedback inhibition of the *pheA* product.

Ultimately, the rate of phenylalanine biosynthesis is dependent upon chorismate supply, which is also controlled by phenylalanine (Ogino et al., 1982).

B. Transamination of phenylpyruvate

Phenylpyruvate is converted to phenylalanine by the aromatic transaminase that also catalyzes the last step in tyrosine biosynthesis. The enzyme is encoded by *tyrB* (Gelfand and Steinberg, 1977). The expression of *tyrB* is controlled by *tyrR* (Wallace and Pittard, 1969). Phenylalanine biosynthesis is not absolutely dependent upon the *tyrB* product, since other transaminases can replace this activity (Chapter 2). However, in wild-type cells, the aromatic transaminase appears to be the major enzyme responsible for phenylalanine and tyrosine biosynthesis.

The aromatic transaminase forms a noncovalent complex with either chorismate mutase–prephenate dehydratase or chorismate mutase–prephenate dehydrogenase (Powell and Morrison, 1979). The transaminase–dehydrogenase complex dissociates when exposed to tyrosine, whereas the transaminase-dehydratase complex is stabilized by phenylpyruvate, reflecting the cellular needs for channeling of chorismate under different growth conditions.

III. *A. EUTROPHUS*

The bifunctional chorismate mutase–prephenate dehydratase from *A. eutrophus* has been purified to homogeneity; the enzyme is a tetramer with a subunit molecular weight of 47,000 (B. Friedrich et al., 1976). In addition to its role in phenylalanine biosynthesis, the bifunctional enzyme also supplies prephenate for tyrosine biosynthesis that requires only two specific enzymes: a monofunctional prephenate dehydrogenase and a transaminase. *A. eutrophus* can also synthesize tyrosine by hydroxylation of phenylalanine (B. Friedrich and Schlegel, 1972). The chorismate mutase activity of the bifunctional enzyme is inhibited by phenylalanine and prephenate. A more relevant point of control is the dehydratase activity that is 50% inhibited by 2.6 μM phenylalanine or 23 μM tryptophan but strongly activated by tyrosine. Tyrosine activation involves a decrease in the binding constant for prephenate (C. G. Friedrich et al., 1976).

IV. ORGANISMS WITH CHORISMATE MUTASE AND PREPHENATE DEHYDRATASE ACTIVITIES ON SEPARATE POLYPEPTIDE CHAINS

A. *B. subtilis*

1. Chorismate mutase. Chorismate mutase of *B. subtilis* appears in three forms (Lorence and Nester, 1967) specified by two genetic loci (Nasser and Nester, 1967). Chorismate mutase-1 and -2 found in the wild-type Marburg strain are different aggregation states of the same polypeptide (Lorence and Nester, 1967) having a molecular weight of 40,000 (Llewellyn et al., 1980). The chorismate mutase-1 and -2 activities are inhibited by prephenate but unaffected by tyrosine, phenylalanine, or tryptophan (Kane et al., 1971). The synthesis of this enzyme is repressed by phenylalanine, tyrosine, and tryptophan (Lorence and Nester, 1967).

B. subtilis strain 23, a derivative of the wild-type Marburg strain, contains, in addition to chorismate mutase-1 and -2, chorismate mutase-3, which has only one tenth of the specific activity of enzyme-1 or -2 (Lorence and Nester, 1967). In *B. subtilis* strain 168, derived from strain 23, chorismate mutase-3 is the only representative of this activity (Kane et al., 1971). Chorismate mutase-3 from strain 168 has been purified to homogeneity; the enzyme is a bifunctional tetramer with a subunit molecular weight of 38,500 (Huang et al., 1974a). DAHP synthase is the other reaction catalyzed (Nester et al., 1967). Although there are separate domains for the DAHP synthase and the chorismate mutase activities, the mutase active site appears to be the feedback inhibitor site of the DAHP synthase (Huang et al., 1974b). The bifunctional enzyme presumably evolved from wild-type DAHP synthase through a change in the prephenate (feedback inhibitor) binding site into a chorismate mutase active site (Llewellyn et al., 1980). Wild-type *B. subtilis* DAHP synthase is monofunctional and inhibited by prephenate only (Llewellyn et al., 1980); DAHP synthase from strain 168 is a bifunctional enzyme and inhibited by prephenate and chorismate (Huang et al., 1974b).

The bifunctional DAHP synthase–chorismate mutase-3 of strain 168 forms a trifunctional complex with shikimate kinase (Nakatsukasa and Nester, 1972). Complex formation is absolutely required for kinase activity. When the ratio of kinase to bifunctional enzyme in an in vitro reconstituted complex is between 1 and 2, the bifunctional activities are 30% activated; when the ratio becomes larger than 2, chorismate mutase activity starts to decrease, while DAHP synthase activity is unaffected (Huang et al., 1975). Synthesis of the bifunctional enzyme is repressed by tyrosine and phenylalanine (Nester et al., 1969).

2. Prephenate dehydratase. Pure prephenate dehydratase from *B. subtilis* has a subunit molecular weight of 28,000 (Riepl and Glover, 1978). Its catalytic activity depends upon reversible, effector-mediated aggregation to dimeric and octameric forms. The inactive monomer predominates in the absence of hydrophobic amino acids (Riepl and Glover, 1979). Low concentrations of phenylalanine and higher concentrations of methionine and other amino acids favor the formation of active dimers. In the presence of methionine, leucine, and prephenate, dimers aggregate to equally active octamers (Pierson and Jensen, 1974). Enzyme activity of the octamer is not affected by any amino acid (Riepl and Glover, 1979), but phenylalanine and tryptophan cause dissociation of the octamer to the dimer (Pierson and Jensen, 1974). The dimer is competitively inhibited by phenylalanine and tryptophan (Rebello and Jensen, 1970). Tryptophan inhibition is reversed by tyrosine. Phenylalanine-mediated octamer dissociation is strongly favored over association caused by methionine, leucine, and prephenate. However, tryptophan-mediated octamer dissociation is effectively antagonized by methionine, leucine, and prephenate. Thus activation by leucine or methionine, which is noncompetitive with respect to prephenate, is only significant at low phenylalanine concentrations. The activation is independent of phenylalanine inhibition of the dimers; the activators increase the activity of the remaining free enzyme (Jensen, 1969).

A model to explain association and dissociation of *B. subtilis* prephenate dehydratase proposes two effector binding sites, the association activator site and the feedback inhibitor site (Riepl and Glover, 1979). Binding of phenylalanine, methionine, or other hydrophobic amino acids at the association activator site causes oligomer formation; phenylalanine appears to be the preferred effector. The feedback inhibitor site is accessible to phenylalanine only in the dimer; octamer formation makes this site inaccessible. Even though phenylalanine does not bind to the octamer, higher concentrations of this amino acid shift the dimer–octamer equilibrium to the inhibitable state of lower aggregation.

Of all the effectors that change the aggregation state of prephenate dehydratase and thereby the activity of this enzyme, only phenylalanine and prephenate are metabolites of the pathway. The involvement of metabolites from other pathways, termed metabolic interlock, may be a mechanism for general control of amino acid pool levels (Jensen, 1969). Since the synthesis of *B. subtilis* prephenate dehydratase is not affected by phenylalanine, the specific control of carbon flow through the phenylpyruvate pathway of this organism appears to be by transcriptional control of chorismate mutase and feedback inhibition of prephenate dehydratase.

B. *B. flavum*

Phenylalanine biosynthesis in *B. flavum* proceeds via the phenylpyruvate pathway. Chorismate mutase and prephenate dehydratase reside on separate polypeptide chains. A single chorismate mutase provides for both phenylalanine and tyrosine biosynthesis. The wild-type *B. flavum* mutase activity is part of a bifunctional DAHP synthase–chorismate mutase (Shiio and Sugimoto, 1979). This enzyme, described in some detail in the preceding chapter, is composed of two dissimilar subunits, both of which are required for chorismate mutase activity (Sugimoto and Shiio, 1980a, b). While the active sites for DAHP synthase and chorismate mutase are distinct, the feedback inhibitor sites for both activities are the same (Sugimoto and Shiio, 1980a). The chorismate mutase activity is controlled by the aggregation state of the subunits, which depends in turn upon the aromatic amino acids (Shiio and Sugimoto, 1981). Either phenylalanine or tyrosine competitively inhibits the chorismate mutase by causing dissociation of the subunits, while tryptophan activates the mutase through stabilization of the complex (Shiio and Sugimoto, 1976). When present together, phenylalanine and tyrosine cause cumulative inhibition (Sugimoto and Shiio, 1980b). The synthesis of both subunits of this enzyme is discoordinately controlled by tyrosine (Shiio and Sugimoto, 1979). Steady state kinetic studies on the bifunctional DAHP synthase-chorismate mutase show that phenylalanine and tyrosine biosyntheses are controlled by regulation of DAHP synthase rather than by chorismate mutase (Shiio and Sugimoto, 1981).

The partially purified prephenate dehydratase from *B. flavum* has a native molecular weight of 220,000 (Sugimoto and Shiio, 1974). The enzyme is 50% inhibited by 2.5 μM phenylalanine or 25 μM tryptophan but activated by very low tyrosine concentrations (Sugimoto et al., 1973). This regulatory pattern is also found in other coryneforms, *B. subtilis, A. eutrophus,* and cyanobacteria (Hall and Jensen, 1980). However, unlike *B. subtilis,* the effectors in *B. flavum* do not cause changes in the aggregation state of the enzyme (Fazel and Jensen, 1980). Mutants of *B. flavum* with a feedback-insensitive prephenate dehydratase excrete phenylalanine (Sugimoto et al., 1973), a result suggesting that prephenate dehydratase, but not DAHP synthase, is the major control point of phenylalanine biosynthesis.

In *B. flavum,* carbon flow to both phenylalanine and tyrosine is controlled by aromatic amino acid–mediated modulation of prephenate dehydratase activity and by phenylalanine-mediated repression of prephenate dehydratase synthesis (Fazel and Jensen, 1980). Tyrosine is synthesized via the arogenate pathway (Fig. 18.2) (Fazel et al., 1980). However, regulation of tyrosine synthesis involves prephenate dehydratase of the phenylpyruvate pathway. Excess phenylalanine inhibits the dehydratase, leaving prephenate for tyrosine biosynthesis; excess tyrosine activates the dehydratase, diverting more prephenate into phenylalanine biosynthesis (Fazel and Jensen, 1980).

The *Corynebacterium glutamicum* prephenate dehydratase is very similar to the enzyme from *B. flavum* (Hagino and Nakayama, 1974). In *C. glutamicum,* tyrosine is also synthesized by the arogenate pathway (Fazel and Jensen, 1979). The prephenate dehydratase of *C. glutamicum* copurifies with an aromatic aminotransferase that acts upon phenylpyruvate, 4-hydroxyphenylpyruvate, and prephenate (Fazel and Jensen, 1980). This would suggest some channeling of chorismate if formation of a complex between the transaminase and the chorismate-utilizing enzymes were to be demonstrated in vivo.

C. *Streptomyces aureofaciens*

S. aureofaciens chorismate mutase, purified to homogeneity, is a trimer with a subunit molecular weight of 14,500; the activity of the enzyme is not regulated (Goerisch and Lingens, 1973, 1974). Phenylalanine biosynthesis in *Streptomyces* is apparently regulated by modulation of prephenate dehydratase. In a chloramphenicol-producing strain, this enzyme is inhibited by phenylalanine (Lowe and Westlake, 1973).

V. THE AROGENATE PATHWAY

Euglena gracilis synthesizes phenylalanine exclusively by the arogenate pathway (Fig. 18.2) (Byng et al., 1981a). Phenylalanine and tyrosine biosynthesis from chorismate via prephenate and arogenate leaves only a single step for the terminal "pathways" to these two amino acids. In acidic solutions, arogenate converts nonenzymatically to phenylalanine (Jensen et al., 1977) in a manner reminiscent of the prephenate-to-phenylpyruvate decomposition. Arogenate has been prepared in pure form from *Neurospora* (Jensen et al., 1977) and from *Synechocystis* (Zamir et al., 1982). The structure of arogenate, established through various spectroscopic techniques (Zamir et al., 1980), has been confirmed through total synthesis (Danishefsky et al., 1981).

A. *E. gracilis*

The single chorismate mutase of *E. gracilis,* purified to near homogeneity, has a native molecular weight of 160,000 (Weber and Böck, 1970). The enzyme is competitively inhibited by phenylalanine and tyrosine; tryptophan antagonizes the inhibition even at high phenylalanine and tyrosine concentrations. Arogenate does not affect the enzyme activity.

The resulting prephenate is not converted to phenylpyruvate or 4-hydroxyphenylpyruvate but is instead transaminated to arogenate. *E. gracilis* contains at least four different prephenate–arogenate aminotransferases (Byng et al., 1981a). One relatively specific glutamate-dependent aminotransferase appears to be the major arogenate-synthesizing enzyme. The arogenate dehydratase of *E. gracilis* has been identified but not yet studied in detail (Byng et al., 1981a; Shapiro et al., 1981).

B. *Pseudomonas*

P. aeruginosa presumably synthesizes phenylalanine either by the arogenate or the phenylpyruvate pathway, since the enzymes of both metabolic sequences are present in this organism (Patel et al., 1977). The phenylpyruvate pathway, resembling phenylalanine biosynthesis in the enterobacteriaceae, consists of a bifunctional chorismate mutase–prephenate dehydratase (Calhoun et al., 1973; Ahmed and Campbell, 1973) and a glutamate-dependent aminotransferase (Patel et al., 1977, 1978). The chorismate mutase–prephenate dehydratase has a native molecular weight of 134,000; both activities of the enzyme are competitively inhibited by phenylalanine, and in addition, the mutase is competitively inhibited by prephenate (Calhoun et al., 1973).

The dehydratase is activated by tyrosine (Cerutti and Guroff, 1965; Waltho, 1973). The synthesis of the bifunctional enzyme from *P. aureofaciens* appears to be constitutive (Blumenstock et al., 1980) like most *Pseudomonas* enzymes (Jensen and Pierson, 1975; Clarke and Ornston, 1975). The glutamate-dependent aromatic aminotransferase has a native molecular weight of 70,000; the enzyme is specific for phenylpyruvate and 4-hydroxyphenylpyruvate (Patel et al., 1977).

P. aeruginosa can also synthesize both tyrosine and phenylalanine by the arogenate pathway. The first enzyme of this pathway, chorismate mutase II, is distinct from the bifunctional chorismate mutase–prephenate dehydratase of the phenylpyruvate pathway (Patel et al., 1977). Prephenate synthesized by mutase II can also be converted to tyrosine via the 4-hydroxyphenylpyruvate route. Chorismate mutase II, having a native molecular weight of 27,000, is apparently a constitutive enzyme (Patel et al., 1977). Prephenate aminotransferase, the second enzyme of the arogenate pathway, is likewise common to phenylalanine and tyrosine biosynthesis. The enzyme catalyzes the reversible transamination of prephenate and leucine to arogenate and α-ketoisocaproate. The enzyme, having a native molecular weight of 200,000, is repressed by phenylalanine and tyrosine (Patel et al., 1978).

The final step in phenylalanine biosynthesis via the arogenate pathway is catalyzed by arogenate dehydratase, an enzyme initially designated prephenate dehydratase II (Calhoun et al., 1973) but now termed arogenate dehydratase (Patel et al., 1977). Arogenate dehydratase has a native molecular weight of 80,000. Whether or not the enzyme activity can be regulated remains to be seen.

Thus *P. aeruginosa* has two quite distinct routes to phenylalanine (and tyrosine). While the regulation of the arogenate pathway seems less refined (Jensen and Pierson, 1975), elucidation of the specific physiological function of each pathway and the evolutionary relationship between the two pathways remains a challenging problem. Finally, like *A. eutrophus* and *Flavobacterium devorans* (Krauss et al., 1980), *Pseudomonas* can directly hydroxylate phenylalanine to tyrosine (Guroff and Rhoads, 1967).

VI. PHENYLALANINE BIOSYNTHESIS IN EUKARYOTES

A. Fungi and yeast

Some enzymes of phenylalanine biosynthesis have been found in fungi and yeast. Chorismate mutase and prephenate dehydratase activities have been demonstrated in *Claviceps* (Lingens et al., 1967; Schmauder and Groeger, 1973), *Neurospora* (Baker, 1968; El-Eryani, 1969), *Saccharomyces* (Sprossler et al., 1970; Kradolfer et al., 1977), *Penicillium* (Woodin and Nishioka, 1973b), and *Hansenula* (Bode et al., 1980). *Hansenula* synthesizes phenylalanine by the phenylpyruvate route but forms tyrosine via the arogenate pathway (Bode and Birnbaum, 1979).

B. Plants

All three enzymes of the phenylpyruvate pathway to phenylalanine have been identified in *Phaseolus* (Gamborg and Simpson, 1964; Gamborg, 1965; Gilchrist et al., 1972). In this organism, chorismate mutase appears in two isoenzymic forms, an aromatic amino acid–sensitive chorismate mutase-1 (Gilchrist and Kosuge, 1974) and an insensitive chorismate mutase-2 (Gilchrist and Kosuge, 1975). Chorismate mutase-1, purified 1200-fold, is inhibited by phenylalanine and tyrosine, but this inhibition is reversed by equimolar concentrations of tryptophan. A similar regulatory pattern has been found for the enzyme from *Pisum sativum* (Cotton and Gibson, 1968), *Quercus rubra* (Carasco et al., 1976), and a chorismate mutase extracted from cell suspension cultures of *Nicotiana tabacum* (Widholm, 1974). In *Phaseolus,* chorismate mutase-2 represents about one half of the total mutase activity. The partially purified enzyme has a molecular weight of 36,000. Its specific function is unknown. In a variety of other higher plants, either two or three chorismate mutase isoenzymes are differentially affected by various primary and secondary aromatic metabolites (Woodin and Nishioka, 1973a; Woodin et al., 1978).

Phaseolus can synthesize tyrosine via the arogenate pathway (Rubin and Jensen, 1979), and this is the exclusive route to the amino acid in *Zea mays* (Byng et al., 1981b). Thus in higher plants, as in bacteria, different patterns for phenylalanine biosynthesis prevail in different species.

VII. OUTLOOK

This brief summary of the current status of phenylalanine biosynthesis draws attention to several variations, particularly with respect to regulatory patterns, for various organisms. While phenylalanine is always synthesized from prephenate, this metabolite is converted to phenylalanine either via phenylpyruvate or via arogenate. Some organisms contain enzymes of both pathways. A more detailed structural analysis of these proteins followed by comparative biochemistry will undoubtedly generate some new insight into the evolution of these metabolic sequences.

Our understanding of the regulation of phenylalanine biosynthesis in organisms with a bifunctional chorismate mutase–prephenate dehydratase is well advanced. In organisms with a monofunctional chorismate mutase, prephenate dehydratase frequently acts as the controlling enzyme, with phenylalanine as a negative effector. However, in some bacteria, prephenate dehydratase is activated by phenylalanine (Krauss et al., 1980). Detailed studies on the prephenate dehydratase of higher plants are conspicuously absent from the literature.

ACKNOWLEDGMENTS

C. G. is supported by United States Public Health Service Training Grant GM-07211. This is paper No. 9285 of the Purdue University Agricultural Experiment Station.

REFERENCES

Ahmed, S. I., and Campbell, J. J. R. (1973) *J. Bacteriol. 115,* 205–212.

Andrews, P. R., Smith, G. D., and Young, I. G. (1973) *Biochemistry 12,* 3492–3498.

Andrews, P. R., Cain, E. N., Rizzardo, E., and Smith, G. D. (1977) *Biochemistry 16,* 4848–4852.

Artz, S. W., and Broach, J. R. (1975) *Proc. Nat. Acad. Sci. U.S.A. 72,* 3453–3457.

Baker, T. I. (1968) *Genetics 58,* 351–359.

Baldwin, G. S., and Davidson, B. E. (1979) *Biochim. Biophys. Acta 579,* 483–486.

Baldwin, G. S., and Davidson, B. E. (1981) *Arch. Biochem. Biophys. 211,* 66–75.

Baldwin, G. S., McKenzie, G. H., and Davidson, B. E. (1981) *Arch. Biochem. Biophys. 211,* 76–85.

Blumenstock, E., Salcher, O., and Lingens, F. (1980) *J. Gen. Microbiol. 117,* 81–87.

Bode, R., and Birnbaum, D. (1979) *Z. Allg. Mikrobiol. 19,* 83–88.

Bode, R., Boettcher, F., and Birnbaum, D. (1980) *Cell. Mol. Biol. 26,* 615–620.

Brown, K. D. (1968) *Genetics 60,* 31–48.

Byng, G. S., Whitaker, R. J., Shapiro, C. L., and Jensen, R. A. (1981a) *Mol. Cell. Biol. 1,* 426–438.

Byng, G., Whitaker, R., Flick, C., and Jensen, R. A. (1981b) *Phytochemistry 20,* 1289–1292.

Calhoun, D. H., Pierson, D. L., and Jensen, R. A. (1973) *J. Bacteriol. 113,* 241–251.

Carasco, A., Boudet, A., and Gadal, P. (1976) *Marcellia 39,* 119–123.

Cerutti, P., and Guroff, G. (1965) *J. Biol. Chem. 240,* 3034–3038.

Clarke, P. H., and Ornston, L. N. (1975) in *Genetics and Biochemistry of Pseudomonas* (Clarke, P. H., and Richmond, M. H., eds.) pp. 263–340, John Wiley, New York.

Cotton, R. G. H., and Gibson, F. (1965) *Biochim. Biophys. Acta 100,* 76–88.

Cotton, R. G. H., and Gibson, F. (1968) *Biochim. Biophys. Acta 156,* 187–189.

Danishefsky, S., Hirama, M., Fritsch, N., and Clardy, J. (1979) *J. Amer. Chem. Soc. 101,* 7013–7018.

Danishefsky, S., Morris, J., and Clizbe, L. A. (1981) *J. Amer. Chem. Soc. 103,* 1602–1604.

Davidson, B. E., Blackburn, E. H., and Dopheide, T. A. A. (1972) *J. Biol. Chem. 247,* 4441–4446.

Davis, B. D. (1953) *Science 118,* 251–252.

Dopheide, T. A. A., Crewther, P., and Davidson, B. E. (1972) *J. Biol. Chem. 247,* 4447–4452.

Duggleby, R. G., Sneddon, M. K., and Morrison, J. F. (1978) *Biochemistry 17,* 1548–1554.

Edwards, J. M., and Jackman, L. M. (1965) *Austral. J. Chem. 18,* 1227–1239.
El-Eryani, A. A. (1969) *Genetics 62,* 711–723.
Fazel, A. M., and Jensen, R. A. (1979) *J. Bacteriol. 138,* 805–815.
Fazel, A. M., and Jensen, R. A. (1980) *Arch. Biochem. Biophys. 200,* 165–176.
Fazel, A. M., Bowen, J. R., and Jensen, R. A. (1980) *Proc. Nat. Acad. Sci. U.S.A. 77,* 1270–1273.
Friedrich, B., and Schlegel, H. G. (1972) *Arch. Mikrobiol. 83,* 17–31.
Friedrich, B., Friedrich, C. G., and Schlegel, H. G. (1976) *J. Bacteriol. 126,* 712–722.
Friedrich, C. G., Friedrich, B., and Schlegel, H. G. (1976) *J. Bacteriol. 126,* 723–732.
Gamborg, O. L. (1965) *Can. J. Biochem. 43,* 723–730.
Gamborg, O. L., and Simpson, F. J. (1964) *Can. J. Biochem. 42,* 583–591.
Gelfand, D. H., and Steinberg, R. A. (1977) *J. Bacteriol 130,* 429–440.
Gething, M.-J. H., and Davidson, B. E. (1977a) *Eur. J. Biochem. 78,* 103–110.
Gething M.-J. H., and Davidson, B. E. (1977b) *Eur. J. Biochem. 78,* 111–117.
Gething, M.-J. H., and Davidson, B. E. (1978a) *Eur. J. Biochem. 86,* 159–164.
Gething, M.-J. H., and Davidson, B. E. (1978b) *Eur. J. Biochem. 86,* 165–174.
Gething, M.-J. H., Davidson, B. E., and Dopheide, T. A. A. (1976) *Eur. J. Biochem. 71,* 317–325.
Gilchrist, D. G., and Kosuge, T. (1974) *Arch. Biochem. Biophys. 164,* 95–105.
Gilchrist, D. G., and Kosuge, T. (1975) *Arch. Biochem. Biophys. 171,* 36–42.
Gilchrist, D. G., Woodin, T. S., Johnson, M. L., and Kosuge, T. (1972) *Plant Physiol. 49,* 52–57.
Goerisch, H., and Lingens, F. (1973) *J. Bacteriol. 114,* 645–651.
Goerisch, H., and Lingens, F. (1974) *Biochemistry 13,* 3790–3794.
Gollub, E. G., Liu, K. P., and Sprinson, D. B. (1973) *J. Bacteriol. 115,* 121–128.
Gowrishankar, J., and Pittard, J. (1982) *J. Bacteriol. 150,* 1130–1137.
Gramlich, W., and Plieninger, H. (1979) *Ber. 112,* 1571–1584.
Guroff, G., and Rhoads, C. A. (1967) *J. Biol. Chem. 242,* 3641–3645.
Hagino, H., and Nakayama, K. (1974) *Agr. Biol. Chem. 38,* 2367–2376.
Hall, G. C., and Jensen, R. A. (1980) *J. Bacteriol. 144,* 1034–1042.
Huang, L., Nakatsukasa, W. M., and Nester, E. W. (1974a) *J. Biol. Chem. 249,* 4467–4472.
Huang, L., Montoya, A. L., and Nester, E. W. (1974b) *J. Biol. Chem. 249,* 4473–4479.
Huang, L., Montoya, A. L., and Nester, E. W. (1975) *J. Biol. Chem. 250,* 7675–7681.
Ife, R. J., Ball, L. F., Lowe, P., and Haslam, E. (1976) *J. Chem. Soc. Perkin Trans. I,* 1776–1783.
Im, S. W. K., and Pittard, J. (1971) *J. Bacteriol. 106,* 784–790.
Jensen, R. A. (1969) *J. Biol. Chem. 244,* 2816–2823.
Jensen, R. A., and Byng, G. S. (1981) *Isoenzymes: Current Topics in Biological and Medical Research 5,* 143–174.
Jensen, R. A., and Pierson, D. L. (1975) *Nature 254,* 667–671.
Jensen, R. A., Zamir, L., St. Pierre, M., Patel, N., and Pierson, D. L. (1977) *J. Bacteriol. 132,* 896–903.
Kane, J. F., Stenmark, S. L., Calhoun, D. H., and Jensen, R. A. (1971) *J. Biol. Chem. 246,* 4308–4316.

Katagiri, M., and Sato, R. (1953) *Science 118,* 250–251.
Kradolfer, P., Zeyer, J., Miozzari, G., and Huetter, R. (1977) *FEMS Microbiol. Lett. 2,* 211–216.
Krauss, G., Suessmuth, R., and Lingens, F. (1980) *Hoppe-Seyler's Z. Physiol. Chem. 361,* 809–818.
Lingens, F., Goebel, W., and Uesseler, H. (1967) *Eur. J. Biochem. 2,* 442–447.
Llewellyn, D. J., Daday, A., and Smith, G. D. (1980) *J. Biol. Chem. 255,* 2077–2084.
Lorence, J. H., and Nester, E. W. (1967) *Biochemistry 6,* 1541–1552.
Lowe, D. A., and Westlake, D. W. S. (1973). *Biochem. Soc. Trans. 1,* 219–222.
McCray, J. W., Jr., and Herrmann, K. M. (1976) *J. Bacteriol. 125,* 608–615.
Nakatsukasa, W. M., and Nester, E. W. (1972) *J. Biol. Chem. 247,* 5972–5979.
Nasser, D., and Nester, E. W. (1967) *J. Bacteriol. 94,* 1706–1714.
Nester, E. W., Lorence, J. H., and Nasser, D. S. (1967) *Biochemistry 6,* 1553–1563.
Nester, E. W., Jensen, R. A., and Nasser, D. S. (1969) *J. Bacteriol. 97,* 83–90.
Ogino, T., Garner, C., Markley, J. L., and Herrmann, K. M. (1982) *Proc. Nat. Acad. Sci. U.S.A. 79,* 5828–5832.
Patel, N., Pierson, D. L., and Jensen, R. A. (1977) *J. Biol. Chem. 252,* 5839–5846.
Patel, N., Stenmark-Cox, S. L., and Jensen, R. A. (1978) *J. Biol. Chem. 253,* 2972–2978.
Pierson, D. L., and Jensen, R. A. (1974) *J. Mol. Biol. 90,* 563–579.
Pittard, J., and Gibson, F. (1970) in *Current Topics in Cellular Regulation* (Horecker, B. L., and Stadtman, E. R., eds.) Vol. II, 29–63, Academic Press, New York.
Powell, J. T., and Morrison, J. F. (1979) *Biochim. Biophys. Acta 568,* 467–474.
Rebello, J. L., and Jensen, R. A. (1970) *J. Biol. Chem. 245,* 3738–3744.
Riepl, R. G., and Glover, G. I. (1978) *Arch. Biochem. Biophys. 191,* 192–197.
Riepl, R. G., and Glover, G. I. (1979) *J. Biol. Chem. 254,* 10321–10328.
Rosenberg, A. H., and Gefter, M. L. (1969) *J. Mol. Biol. 46,* 581–584.
Rubin, J. L., and Jensen, R. A. (1979) *Plant Physiol. 64,* 727–734.
Schmauder, H. P., and Groeger, D. (1973) *Biochem. Physiol. Pflanz. 164,* 41–57.
Schmit, J. C., and Zalkin, H. (1969) *Biochemistry 8,* 174–181.
Schmit, J. C., and Zalkin, H. (1971) *J. Biol. Chem. 246,* 6002–6010.
Schmit, J. C., Artz, S. W., and Zalkin, H. (1970) *J. Biol. Chem. 245,* 4019–4027.
Shapiro, C. L., Jensen, R. A., Wilson, K. A., and Bowen, J. R. (1981) *Anal. Biochem. 110,* 27–30.
Shiio, I., and Sugimoto, S. (1976) *J. Biochem. 79,* 173–183.
Shiio, I., and Sugimoto, S. (1979) *J. Biochem. 86,* 17–25.
Shiio, I., and Sugimoto, S. (1981) *J. Biochem. 89,* 1483–1492.
Sprossler, B., Lenssen, U., and Lingens, F. (1970) *Hoppe-Seyler's Z. Physiol. Chem. 351,* 1178–1182.
Stenmark, S. L., Pierson, D. L., Jensen, R. A., and Glover, G. I. (1974) *Nature 247,* 290–292.
Sugimoto, S., and Shiio, I. (1974) *J. Biochem. 76,* 1103–1111.
Sugimoto, S., and Shiio, I. (1980a) *J. Biochem. 87,* 881–890.
Sugimoto, S., and Shiio, I. (1980b) *J. Biochem. 88,* 167–176.
Sugimoto, S., Nakagawa, M., Tsuchida, T., and Shiio, I. (1973) *Agr. Biol. Chem. 37,* 2327–2336.
Wallace, B. J., and Pittard, J. (1969) *J. Bacteriol. 97,* 1234–1241.

Waltho, J. A. (1973) *Biochim. Biophys. Acta 320,* 232–241.
Weber, H. L., and Böck, A. (1970) *Eur. J. Biochem. 16,* 244–251.
Weiss, U., Gilvarg, C., Mingioli, E. S., and Davis, B. D. (1954) *Science 119,* 774–775.
Widholm, J. M. (1974) *Physiol. Plant. 30,* 13–18.
Woodin, T. S., and Nishioka, L. (1973a) *Biochim. Biophys. Acta 309,* 211–223.
Woodin, T. S., and Nishioka, L. (1973b) *Biochim. Biophys. Acta 309,* 224–231.
Woodin, T. S., Nishioka, L., and Hsu, A. (1978) *Plant Physiol. 61,* 949–952.
Yanofsky, C. (1981) *Nature 289,* 751–758.
Yanofsky, C., and Horn, V. (1981) *J. Bacteriol. 145,* 1334–1341.
Yanofsky, C., and Soll, L. (1977) *J. Mol. Biol. 113,* 663–677.
Young, I. G., Gibson, F., and MacDonald, C. G. (1969) *Biochim. Biophys. Acta 192,* 62–72.
Zamir, L. O., Jensen, R. A., Arison, B. H., Douglas, A. W., Albers-Schönberg, G., and Bowen, J. R. (1980) *J. Amer. Chem. Soc. 102,* 4499–4504.
Zamir, L. O., Jung, E. D., and Tiberio, R. D. (1982) *Bioorg. Chem. 11,* 32–42.
Zurawski, G., Brown, K., Killingly, D., and Yanofsky, C. (1978) *Proc. Nat. Acad. Sci. U.S.A. 75,* 4271–4275.

CHAPTER 19

Tyrosine Biosynthesis

Helen Camakaris
James Pittard

I. THE PATHWAY

Chorismate, the branch point compound of the common pathway of aromatic biosynthesis, is converted to tyrosine by either of two alternative pathways (Fig. 19.1), referred to as the 4-hydroxyphenylpyruvate and the pretyrosine pathways, respectively (Jensen and Pierson, 1975). Both pathways begin with the conversion of chorismate to prephenate, but subsequent steps of dehydrogenation, decarboxylation, and transamination take place in different orders. In the 4-hydroxyphenylpyruvate pathway, found in *Escherichia coli, Klebsiella aerogenes, Salmonella typhimurium, Bacillus subtilis, Pseudomonas, Neurospora crassa,* and *Claviceps paspali* (Cotton and Gibson, 1965; Zalkin, 1967; Nester and Jensen, 1966; Calhoun et al., 1973; Baker, 1968; Lingens et al., 1967), prephenate is first converted to 4-hydroxyphenylpyruvate, which undergoes transamination to tyrosine. In the pretyrosine pathway, found in prokaryotes such as the blue-green alga *Agmenellum quadruplicatum, Pseudomonas,* coryneforms, and the eukaryote *N. crassa,* albeit in an incomplete form, prephenate is transaminated to pretyrosine (arogenate), which is subsequently oxidized and decarboxylated to yield tyrosine (Stenmark et al., 1974; Jensen and Stenmark, 1975; Fazel and Jensen, 1979; Jensen and Pierson, 1975). In *N. crassa* a single enzyme

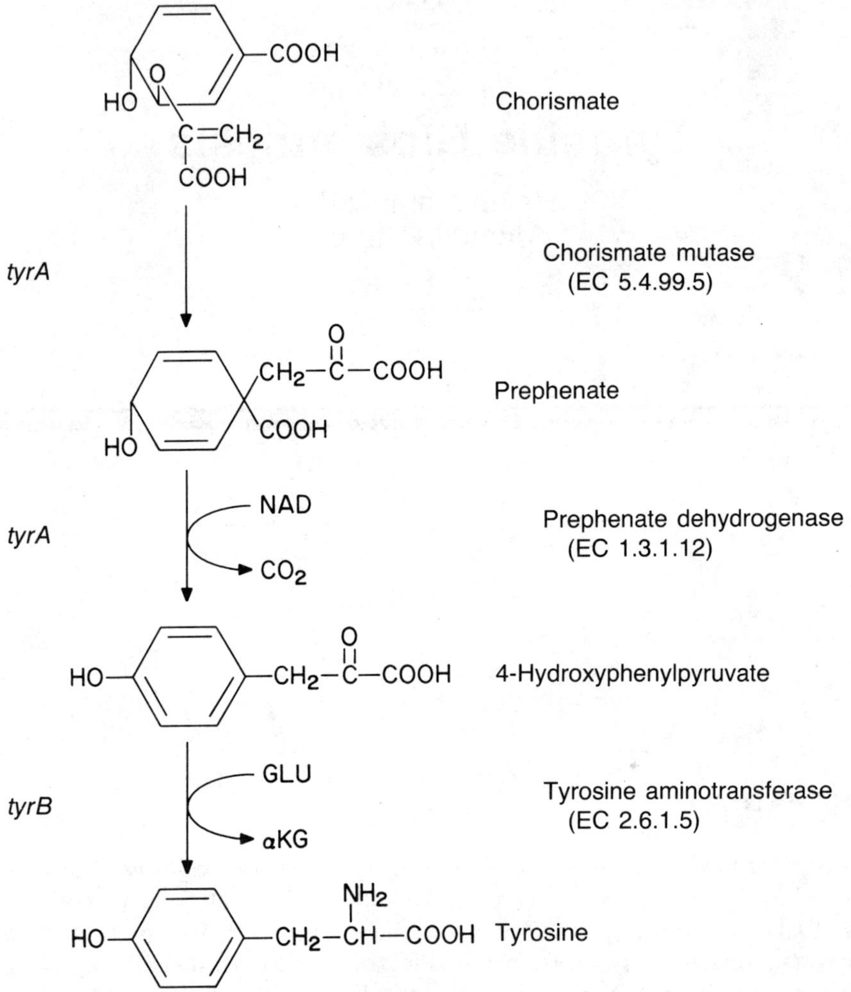

FIGURE 19.1(a) The 4-hydroxyphenylpyruvate pathway to tyrosine.

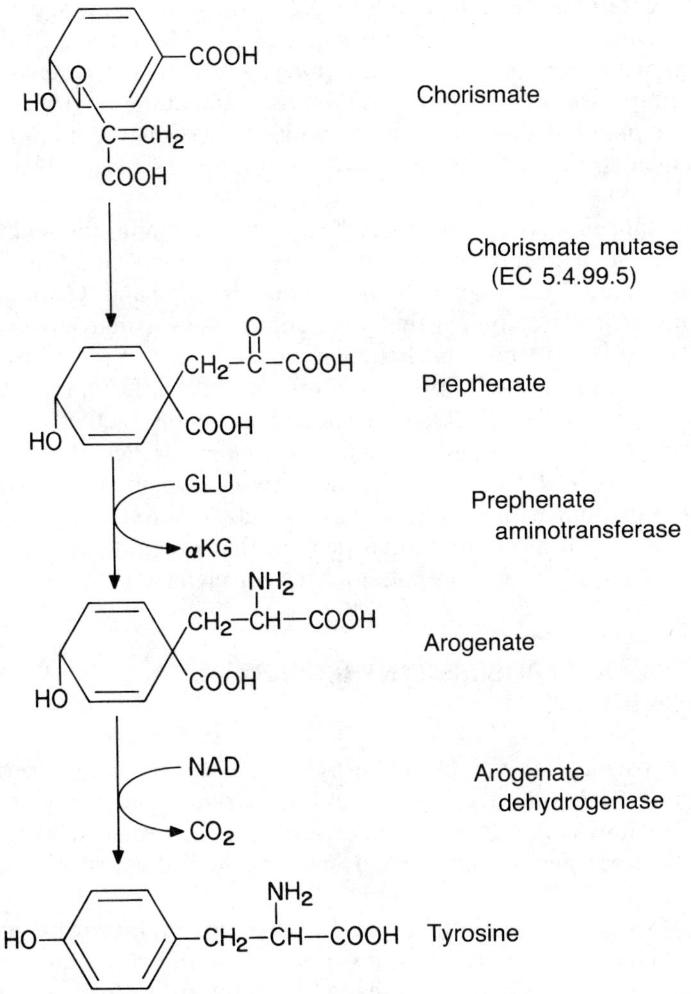

FIGURE 19.1(b) **The arogenate pathway to tyrosine.**

appears to catalyze the transamination of both prephenate and 4-hydroxyphenylpyruvate. Activities for both pathways are found in *P. aeruginosa*. In this organism, both prephenate and pretyrosine are substrates for a single dehydrogenase. Some marine *Pseudomonads* and some microfungi directly hydroxylate phenylalanine to yield tyrosine. However, the quantitative role of the phenylalanine hydroxylase in tyrosine biosynthesis is not clear (Chandra and Vining, 1968).

Prephenate is also an intermediate in phenylalanine biosynthesis, and the mechanism of prephenate synthesis is species-specific. In *E. coli, S. typhimurium,* and *K. aerogenes,* both the phenylalanine and tyrosine pathways involve bifunctional enzymes that first convert chorismate to prephenate. Enzyme-bound prephenate is then converted to either phenylpyruvate or 4-hydroxyphenylpyruvate (Cotton and Gibson, 1967; Dayan and Sprinson, 1971; Koch et al., 1971). In *Bacillus,* chorismate mutase activity is associated with neither prephenate dehydrogenase nor prephenate dehydratase (Lorence and Nester, 1967). *Pseudomonads* have two chorismate mutases, one a bifunctional protein with prephenate dehydratase activity, the other (chorismate mutase II) a monofunctional protein that presumably furnishes prephenate as a substrate for prephenate dehydrogenase (Patel et al., 1977).

II. CONTROL OF TYROSINE SYNTHESIS BY FEEDBACK INHIBITION

Since the tyrosine pathway starts with chorismate, a common precursor of tyrosine, phenylalanine, tryptophan and the aromatic vitamins, partitioning of metabolite flow is of particular importance. Distinct regulatory mechanisms occur in different genera, particularly where the pathways of synthesis are different.

Those bacteria which utilize the 4-hydroxyphenylpyruvate pathway of tyrosine biosynthesis have tyrosine-sensitive prephenate dehydrogenases, e.g., in *E. coli* 1 mM tyrosine causes 90% inhibition of this activity (Cotton and Gibson, 1965; Nester and Jensen, 1966; Lingens et al., 1967; Catcheside, 1969; Stenmark-Cox and Jensen, 1975). Tyrosine also inhibits the chorismate mutase activity in the presence of NAD (G. Hudson and B. Davidson, personal communication). The first enzymes of the phenylalanine and tryptophan branches are similarly subject to feedback inhibition by their respective end products (Egan and Gibson, 1966; Schmit et al., 1970; Dopheide et al., 1972).

In *B. subtilis* a single chorismate mutase is a bifunctional enzyme with DAHP synthase activity (Huang et al., 1975). In this organism, prephenate is therefore a branch point intermediate for tyrosine and phenylalanine biosynthesis. Prephenate inhibits the *B. subtilis* chorismate mutase, while prephenate dehydrogenase is inhibited by tyrosine (Nester and Jensen, 1966;

Champney and Jensen, 1970). Prephenate dehydratase, the first enzyme in the phenylalanine pathway in this organism, is inhibited not only by phenylalanine, but also by tryptophan. Tryptophan inhibition is antagonized by tyrosine (Rebello and Jensen, 1970).

In bacteria that synthesize tyrosine by the pretyrosine pathway, control of tyrosine biosynthesis involves the phenylalanine pathway. In coryneforms, the blue-green algae, and *P. aeruginosa,* prephenate dehydratase is stimulated by tyrosine and inhibited by phenylalanine (Sugimoto et al., 1973; Stenmark et al., 1974; Patel et al., 1977; Fazel and Jensen, 1979). Under conditions of tyrosine excess the stimulation of the blue-green algae prephenate dehydratase channels substrates along the phenylalanine pathway, and when phenylalanine is in excess, the inhibition of prephenate dehydratase instead directs prephenate down the tyrosine pathway (Stenmark et al., 1974).

Certain other regulatory mechanisms have been observed for organisms with the pretyrosine pathway. For example, in the coryneform *Brevibacterium flavum,* chorismate mutase is inhibited by tyrosine or phenylalanine (Shiio and Sugimoto, 1976). This presumably explains why tyrosine inhibits pretyrosine accumulation in a mutant lacking pretyrosine dehydrogenase (Fazel et al., 1980). In blue-green algae, phenylalanine is an excellent amino donor for prephenate transaminase, the enzyme that synthesizes pretyrosine (Stenmark et al., 1974). In *P. aeruginosa,* pretyrosine dehydrogenase is inhibited by tyrosine (Patel et al., 1977). Therefore this organism controls both tyrosine biosynthetic pathways by feedback inhibition and controls prephenate dehydratase by tyrosine activation and phenylalanine inhibition.

III. REGULATION AT THE TRANSCRIPTIONAL LEVEL

A. Repression of tyrosine biosynthetic enzymes

Tyrosine mediates repression of one or more of the enzymes of the tyrosine pathway in *Bacillus* and in the enterics (Nester et al., 1969; Cotton and Gibson, 1965; Gollub and Sprinson, 1969; Zalkin, 1967; Wallace and Pittard, 1969). Such repression is not found in the blue-green algae, *Pseudomonads, N. crassa,* or the coryneforms (Stenmark et al., 1974; Patel et al., 1977; Fazel and Jensen, 1979). We shall now consider in some detail, the mechanism of tyrosine-mediated repression, which has been most extensively investigated in *E. coli* K12. We include some information from studies with *S. typhimurium.*

In *E. coli, tyrA,* the structural gene for chorismate mutase-prephenate dehydrogenase, is adjacent to *aroF,* the structural gene for 3-deoxy-D-*arabino*-heptulosonate 7-phosphate (DAHP) synthase (Tyr) (Pittard and Wallace, 1966). Both genes are part of a single operon (Mattern and Pittard, 1971), situated at min 56 on the *E. coli* genetic linkage map (Fig. 19.2). The structural gene for the tyrosine-repressible aminotransferase, *tyrB,* is unlinked to the *aroF tyrA* operon at min 91 (Gelfand and Rudo, 1977).

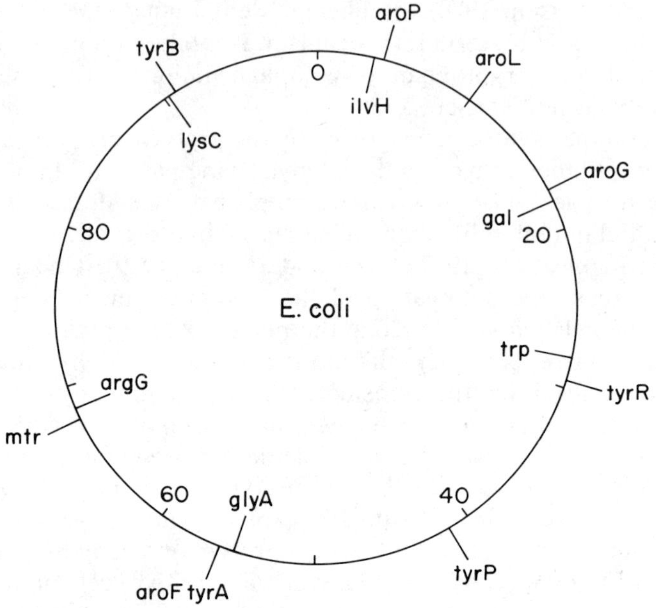

FIGURE 19.2 Chromosomal locations of the genes of the *tyr* regulon in *E. coli*.

The isolation of mutants constitutive for the three enzymes led to the identification of *tyrR*, a regulatory gene situated at min 29 on the chromosome (Wallace and Pittard, 1969). The *tyrR* product controls a number of key enzymes of aromatic biosynthesis and certain components of the aromatic amino acid transport system. The *tyrR* gene and its product are discussed in more detail below.

Other mutants display partially derepressed levels of DAHP synthase (Tyr) and chorismate mutase-prephenate dehydrogenase but unaltered tyrosine-repressible aminotransferase levels. The mutations responsible for the derepression are very tightly linked in transduction to *aroF*. In diploids, expression of only the genes in the *cis* configuration is affected, as is expected for operator mutants (Mattern and Pittard, 1971). One operator mutant with DAHP synthase (Tyr) and chorismate mutase-prephenate dehydrogenase levels characteristic of *tyrR* mutants has since been isolated (H. Camakaris, unpublished observations). Thus the operator mutations presumably interrupt the same regulatory circuit as *tyrR* mutations.

Bacteriophage Mu insertions in *tyrA* abolish chorismate mutase-prephenate dehydrogenase but not DAHP synthase (Tyr) activity. The insertion of Mu within *aroF* causes the loss of both DAHP synthase (Tyr) and chorismate mutase-prephenate dehydrogenase, a finding suggesting the gene order promoter-*aroF*-*tyrA* (S. Morrison and J. Pittard, unpublished observations), which has been confirmed by cloning (Zurawski et al., 1978).

Strains with Mu insertions in *aroF* grow very slowly in the absence of tyrosine, a finding suggesting a weak internal promoter between *aroF* and *tyrA* (S. Morrison and J. Pittard, unpublished observations).

A double mutant defective in the aminotransferases encoded by *ilvE* and *aspC* is derepressed for the aminotransferase encoded by *tyrB*. Thus the regulation of *tyrB* is apparently complex and seems to involve control circuits independent of *tyrR*, for example, a leucine-mediated control (Powell and Morrison, 1978).

Studies of aromatic biosynthesis in *S. typhimurium* have also generated *aroF tyrA* operator mutations. In these mutants, DAHP synthase (Tyr) and prephenate dehydrogenase activity is expressed constitutively, whereas chorismate mutase is not significantly affected (Gollub and Sprinson, 1969; Sprinson et al., 1976). The lack of coordinated expression of chorismate mutase and prephenate dehydrogenase is unexpected.

B. The *tyrR* gene

Mutations in *tyrR* of *E. coli* result in the derepression of DAHP synthase (Tyr), chorismate mutase-prephenate dehydrogenase, and tyrosine aminotransferase (Wallace and Pittard, 1969). Similar observations are made for *S. typhimurium* (Gollub et al., 1973).

The isolation of *tyrR* amber mutants, Mu insertions in *tyrR*, and temperature-sensitive *tyrR* mutants provides evidence that the *tyrR* product is a protein nonessential for cell viability (H. Camakaris and Pittard, 1973). This protein apparently functions directly in repression control, since raising the temperature in a strain carrying a temperature-sensitive *tyrR* allele results in almost immediate derepression (H. Camakaris and Pittard, 1973). Complementation and dominance tests in diploids show that *tyrR* is recessive and consists of one cistron (Im and Pittard, 1973). Thus the *tyrR* gene encodes a protein involved in the regulation of various operons that together constitute the *tyr* regulon.

In addition to *aroF tyrA* and *tyrB*, *tyrR* also controls the expression of *aroG*, the structural gene for DAHP synthase (Phe); *aroL*, the structural gene for shikimate kinase II; *aroP*, which encodes a component of the common aromatic transport system; *tyrP*, which encodes a component of tyrosine-specific transport; *mtr*, a gene whose product is involved in tryptophan-specific transport (Brown and Somerville, 1971; Im et al., 1971; Ely and Pittard, 1979; Whipp and Pittard, 1977); and *tyrR* itself (Camakaris and Pittard, 1982). The map locations of all the known genes of the *tyr* regulon are shown in Fig. 19.2. The effector molecules involved with the *tyrR* product to control some of the genes of the *tyr* regulon have been identified by chemostat and batch culture experiments (Table 19.1) (Brown and Somerville, 1971; Ely and Pittard, 1979; Whipp and Pittard, 1977).

In *E. coli* K12, aromatic amino acid transport is mediated by a high capacity system and by three specific systems (Brown, 1968). The common

TABLE 19.1 The *tyr* Regulon: Genes Controlled by *tyrR*

Operon	*Gene Product*	*Effector*	*Active Repressor (Inducer)*
aroF tyrA	DAHP synthase (Tyr) Chorismate mutase prephenate dehydrogenase	Tyr (Phe at high concentration)	R-Tyr, R-Phe
aroG	DAHP synthase (Phe)	Phe + Trp	R_2-Phe-Trp, *or* R-Phe + R-Trp
aroL	Shikimate kinase II	Tyr/Trp	R-Tyr/R-Trp
tyrB	Tyrosine aminotransferase	Tyr	R-Tyr
aroP	Component of common aromatic transport	Tyr/Phe/Trp	R-Tyr/R-Phe/R-Trp
tyrP	Component of tyrosine-specific transport	Tyr Phe	R-Tyr R-Phe (inducer)
mtr	Component of tryptophan-specific transport	Trp Phe	R-Trp R-Phe (inducer)
tyrR	aporepressor	none, one, or more of Tyr, Phe, Trp	R and one or more of R-Tyr, R-Phe, R-Trp

aromatic transport system is repressed by tyrosine or phenylalanine and, to a lesser extent, by tryptophan. In *tyrR* mutants grown in the presence of aromatic amino acids, uptake remains at the level observed for wild-type cells grown in a minimal medium (Whipp and Pittard, 1977). Apparently, tyrosine, phenylalanine, and tryptophan all combine with the *tyrR*-encoded aporepressor to repress some component of the common aromatic transport system.

The tyrosine-specific transport system is repressed almost completely by tyrosine and is enhanced markedly by phenylalanine. Since cells grown in the presence of all three aromatic amino acids have low tyrosine-specific transport, repression is dominant over induction. Both induction and repression appear to be mediated by the *tyrR* product, since a *tyrR* (amber) mutant grown in excess tyrosine maintains tyrosine-specific transport at a level of wild-type strains grown in a minimal medium (Whipp and Pittard, 1977). In *tyrP–lacZ* fusion strains isolated by the method of Casadaban and Cohen (1979), β-galactosidase is repressed by tyrosine and enhanced by phenylalanine (P. Kasian, unpublished results). Also, the *tyrR* gene product in the presence of phenylalanine appears to enhance tryptophan-specific transport; the enhancement is antagonized by tryptophan (Whipp and Pittard, 1977).

Expression of *tyrR* is autogenously regulated. In *tyrR–lacZ* fusions, where β-galactosidase synthesis is under the control of the *tyrR* promoter, the specific activity of β-galactosidase falls in proportion to the $tyrR^+$ gene copy number (Camakaris and Pittard, 1982).

C. Additional control of the *aroF tyrA* operon

Experiments with *tyrR* mutants have indicated additional control circuits modulating *aroF tyrA* expression. This raises the possibility that expression is regulated by attenuation, but as yet no definitive evidence has been presented.

In *S. typhimurium* a mutation in *tyrS*, the structural gene for tyrosyl-tRNA synthetase, causes a temperature-dependent increase in DAHP synthase (Tyr) above the level characteristic of strains with mutations in *tyrR* alone (Heinonen et al., 1972). Attenuation would be one possible explanation for this result.

In *E. coli* K12 a mutation in *trpS*, the structural gene for tryptophanyl-tRNA synthetase, introduced into a *tyrR* mutant, affects expression of *aroF* (H. Camakaris et al., 1980). The particular *trpS* allele causes increased synthesis of anthranilate synthase, a tryptophan biosynthetic enzyme, in a *trpR* strain (J. Camakaris and Pittard, 1971). This increase may be attributed to reversal of attenuation of the *trp* operon, due to a reduction in the intracellular level of charged tryptophanyl-tRNA (Morse and Morse, 1976; Yanofsky and Soll, 1977). Introduction of the *trpS* allele into the *tyrR* strain results in a substantial increase in the specific activity of DAHP synthase (Tyr). This increase is not due simply to a change in growth rate, nor is it a manifestation of the stringent response. If transcription of *aroF tyrA* is modulated by attenuation, a reduction in the level of tryptophanyl-tRNA could relieve attenuation (H. Camakaris et al., 1980).

Attenuation might also explain the increased resistance of *S. typhimurium hisT* mutants to 3-aminotyrosine (Turnbough et al., 1979). Studies of the DNA sequence of the *aroF tyrA* operon control region may establish the nature of this second regulatory system.

VI. CLONING OF STRUCTURAL GENES

A lambda specialized transducing phage carrying the *aroF tyrA* operon and *pheA* (Zurawski and Brown, 1976) has been the source of DNA to construct a pMB9 derivative carrying the three genes. The resulting multicopy plasmid contains a 6-kb *Eco*RI restriction fragment encoding *pheA*, *tyrA*, and *aroF* (Zurawski et al., 1978). The *aroF tyrA* operon has been independently cloned from chromosomal DNA digested with *Eco*RI; similarly, a plasmid carrying *aroF* and part of *tyrA* has been obtained from a BamHI chromosomal digest (H. Camakaris, unpublished results).

Strains harboring the multicopy plasmid with the 6-kb *Eco*RI insert produce chorismate mutase-prephenate dehydrogenase at high levels, thereby facilitating enzyme purification (Bhosale et al., 1982). The enzyme yield is considerably greater than that expected on the basis of increase in *tyrA* gene dosage alone, a finding suggesting escape of transcription from repression control. When *tyrR* mutants are transformed with the plasmid, considerably less chorismate mutase T-prephenate dehydrogenase is produced. Restriction analysis indicates that in the *tyrR* strain the plasmid undergoes modification involving the deletion of all or part of the *aroF tyrA* operon or the insertion of transposon DNA into either the vector or the *aroF tyrA* operon (Rood et al., 1980).

A lambda specialized transducing phage carrying *tyrR* (Cobbett and Pittard, 1980) has been used as a source of DNA for the in vitro construction of a multicopy plasmid carrying *tyrR*. Further subcloning has localized the gene within a 2.1 kb region. Analysis of plasmid-encoded proteins indicates that the *tyrR* gene codes for a 63,000 dalton polypeptide (Cornish et al., 1982).

V. CONCLUSION

Tyrosine biosynthesis in different organisms varies not only with respect to the intermediates in the pathway but also with respect to the complexity of its regulation. The pretyrosine pathway is virtually unregulated. Control of tyrosine biosynthesis in organisms limited to this pathway appears to be rudimentary and indirect through modulation of the prephenate dehydratase activity, an enzyme that competes for available chorismate. The pretyrosine pathway is considered primitive in the evolutionary sense (Jensen and Pierson, 1975). Simple regulatory patterns evidently survive in organisms where increased complexity would offer no great selective advantage. The photoautotrophic blue-green algae, for example, exist in an environment that is relatively poor in nutrients, so most tyrosine would have to be produced endogenously. On the other hand, *E. coli* inhabits an environment subject to enormous fluctuations in the levels of exogenous amino acids and other nutrients and therefore has evolved sensitive control mechanisms to optimize the use of such resources while they last and to allow a rapid switchover to endogenous synthesis upon demand. Such complexity is well exemplified in the regulation of tyrosine biosynthesis.

Expression of the *aroF tyrA* operon is controlled by repression. Specific activities of the enzymes vary over a 50-fold range in response to the concentration of active repressor, that is, tyrosine complexed with the *tyrR* gene product. In addition, both DAHP synthase (Tyr) and chorismate mutase-prephenate dehydrogenase are sensitive to feedback inhibition by tyrosine. Additional control of the *aroF tyrA* operon may allow a further twofold to fourfold increase in DAHP synthase (Tyr) activity. DAHP synthase (Tyr) is also particularly sensitive to proteolysis and possibly to other regulatory controls attuned to the physiological state of the cell (Tribe et al., 1976; H. Camakaris et al., 1980).

Regulation of the capacity for tyrosine uptake by repression of both the common and the tyrosine-specific transport systems accentuates the ability of *E. coli* to respond rapidly and effectively to changes in external resources.

REFERENCES

Baker, T. I. (1968) *Genetics 58,* 351–359.

Bhosale, S., Sneddon, M., and Morrison, J. (1982) *Proc. Aust. Biochem. Soc. 11,* 24.

Brown, K. D. (1968) *Genetics 60,* 31–48.

Brown, K. D., and Somerville, R. L. (1971) *J. Bacteriol. 108,* 386–399.

Calhoun, D. H., Pierson, D. L., and Jensen, R. A. (1973) *J. Bacteriol. 113,* 241–251.

Camakaris, H., and Pittard, J. (1973) *J. Bacteriol. 115,* 1135–1144.

Camakaris, H., Camakaris, J., and Pittard, J. (1980) *J. Bacteriol. 143,* 613–620.

Camakaris, J., and Pittard, J. (1971) *J. Bacteriol. 107,* 406–414.

Camakaris, H., and Pittard, J. (1982) *J. Bacteriol. 150,* 70–75.

Casadaban, M. J., and Cohen, S. N. (1979) *Proc. Nat. Acad. Sci. U.S.A. 76,* 4530–4533.

Catcheside, D. E. A. (1969) *Biochem. Biophys. Res. Comm. 36,* 651–656.

Champney, W. S., and Jensen, R. A. (1970) *J. Biol. Chem. 245,* 3763–3770.

Chandra, P., and Vining, L. C. (1968) *Can. J. Microbiol. 14,* 573–578.

Cobbett, C. S., and Pittard, J. (1980) *J. Bacteriol. 144,* 877–883.

Cornish, E. C., Davidson, B. E., and Pittard, J. (1982) *J. Bacteriol. 152,* 1276–1279.

Cotton, R. G. H., and Gibson, F. (1965) *Biochim. Biophys. Acta 100,* 76–88.

Cotton, R. G. H., and Gibson, F. (1967) *Biochim. Biophys. Acta 147,* 222–237.

Dayan, J., and Sprinson, D. B. (1971) *J. Bacteriol. 108,* 1174–1180.

Dopheide, T. A. A., Crewther, P., and Davidson, B. E. (1972) *J. Biol. Chem. 247,* 4447–4452.

Egan, A. F., and Gibson, F. (1966) *Biochim. Biophys. Acta 130,* 276–277.

Ely, B., and Pittard, J. (1979) *J. Bacteriol. 138,* 933–943.

Fazel, A. M., and Jensen, R. A. (1979) *J. Bacteriol. 138,* 805–815.

Fazel, A. M., Bowen, J. R., and Jensen, R. A. (1980) *Proc. Nat. Acad. Sci. U.S.A. 77,* 1270–1273.

Gelfand, D. H., and Rudo, N. (1977) *J. Bacteriol. 130,* 441–444.

Gollub, E., and Sprinson, D. B. (1969) *Biochem. Biophys. Res. Comm. 35,* 389–395.

Gollub, E. G., Liu, K. P., and Sprinson, D. B. (1973) *J. Bacteriol. 115,* 1094–1102.

Heinonen, J., Artz, S. W., and Zalkin, H. (1972) *J. Bacteriol. 112,* 1254–1263.

Huang, L., Montoya, A. L., and Nester, E. W. (1975) *J. Biol. Chem. 250,* 7675–7681.

Im, S. W. K., and Pittard, J. (1973) *J. Bacteriol. 115,* 1145–1150.

Im, S. W. K., Davidson, H., and Pittard, J. (1971) *J. Bacteriol. 108,* 400–409.

Jensen, R. A., and Pierson, D. L. (1975) *Nature 254,* 667–671.

Jensen, R. A., and Stenmark, S. L. (1975) *J. Mol. Evol. 4,* 249–259.

Koch, G. L. E., Shaw, D. C., and Gibson, F. (1971) *Biochim. Biophys. Acta 229,* 795–804.

Lingens, F., Goebel, W., and Uesseler, H. (1967) *Eur. J. Biochem. 2,* 442–447.
Lorence, J. H., and Nester, E. W. (1967) *Biochemistry 6,* 1541–1553.
Mattern, I. E., and Pittard, J. (1971) *J. Bacteriol. 107,* 8–15.
Morse, D. E., and Morse, A. N. C. (1976) *J. Mol. Biol. 103,* 209–226.
Nester, E. W., and Jensen, R. A. (1966) *J. Bacteriol. 91,* 1594–1598.
Nester, E. W., Jensen, R. A., and Nasser, D. S. (1969) *J. Bacteriol. 97,* 83–90.
Patel, N., Pierson, D. L., and Jensen, R. A. (1977) *J. Biol. Chem. 252,* 5839–5846.
Pittard, J., and Wallace, B. J. (1966) *J. Bacteriol. 91,* 1494–1508.
Powell, J. T., and Morrison, J. F. (1978) *J. Bacteriol. 136,* 1–4.
Rebello, J. L., and Jensen, R. A. (1970) *J. Biol. Chem. 245,* 3738–3744.
Rood, J. I., Sneddon, M. K., and Morrison, J. F. (1980) *J. Bacteriol. 144,* 552–559.
Schmit, J. C., Artz, S. W., and Zalkin, H. (1970) *J. Biol. Chem. 245,* 4019–4027.
Shiio, I., and Sugimoto, S. (1976) *J. Biochem. 79,* 173–183.
Sprinson, D. B., Gollub, E. G., Hu, R. C., and Liu, K. P. (1976) *Acta Microbiol. Acad. Sci. Hung. 23,* 167–170.
Stemark, S. L., Pierson, D. L., Jensen, R. A., and Glover, G. I. (1974) *Nature 247,* 290–292.
Stenmark-Cox, S., and Jensen, R. A. (1975) *Arch. Biochim. Biophys. 167,* 540–546.
Sugimoto, S., Nakagawa, M., Tsuchida, T., and Shiio, I. (1973) *Agr. Biol. Chem. 37,* 2327–2336.
Tribe, D. E., Camakaris, H., and Pittard, J. (1976) *J. Bacteriol. 127,* 1085–1097.
Turnbough, C. L., Jr., Neill, R. J., Landsberg, R., and Ames, B. N. (1979) *J. Biol. Chem. 254,* 5111–5119.
Wallace, B. J., and Pittard, J. (1969) *J. Bacteriol. 97,* 1234–1241.
Whipp, M. J., and Pittard, A. J. (1977) *J. Bacteriol. 132,* 453–461.
Yanofsky, C., and Soll, L. (1977) *J. Mol. Biol. 113,* 663–677.
Zalkin, H. (1967) *Biochim. Biophys. Acta 148,* 609–621.
Zurawski, G., and Brown, K. D. (1976) *J. Mol. Biol. 102,* 311–324.
Zurawski, G., Brown, K., Killingly, D., and Yanofsky, C. (1978) *Proc. Nat. Acad. Sci. U.S.A. 75,* 4271–4275.

CHAPTER

20

Tryptophan: Biosynthesis, Regulation, and Large-Scale Production

Ronald L. Somerville

I. INTRODUCTION

Genetic analysis of tryptophan biosynthesis in prokaryotes was once viewed as a moderately useful adjunct to standard biochemical methods for dissecting the pathways along which the low molecular weight intermediates were shuttled. Largely through the efforts of Yanofsky and his co-workers, the tryptophan field has taken on a life of its own that reaches far beyond the initial limited goals of the early workers in the field of biochemical genetics; current efforts almost invariably involve the application of chemical procedures to provide answers to problems defined in genetic and physiological terms, rather than the other way around. To name but a few, the tryptophan field has contributed in vital ways to problems of protein structure and synthesis, coding, suppression, transcriptional and translational control, protein–protein and protein–nucleic acid interaction, catalysis, mutagenesis, genetic recombination, viral genetics, molecular cloning, and transmembrane phenomena. It is reasonable to expect that the field will continue to enlarge and diversify and that the conceptual insights produced in the course of such efforts will strongly influence the future of biotechnology.

Tryptophan is the second least abundant of the common amino acids, generally constituting 1% or less of the average protein mass. Some proteins contain no tryptophan residues. Two of the tryptophan biosynthetic enzymes in *Escherichia coli,* anthranilate synthase (520 amino acids) and the α subunit of tryptophan synthase (268 amino acids), are devoid of tryptophan (Yanofsky et al., 1967; Nichols et al., 1981b). Proteins with unusually high tryptophan content are also known; 38 of the 1021 residues (3.7%) in β-galactosidase are tryptophan (Fowler and Zabin, 1977). The significance of such variations in the incidence of tryptophan residues in protein has actually been established in certain cases. Tryptophan residues are located at or near the catalytic centers of a variety of enzymes (Vallee and Riordan, 1969). Tryptophan residues at certain locations might also be incompatible with the biological activity of a protein. This issue is being addressed in the *trpA* gene–tryptophan synthase system through the use of specific suppressor mutations (Murgola, 1981).

Tryptophan biosynthesis has been extensively reviewed during the past decade (Margolin, 1971; Platt, 1978); much of the emphasis in recent years has been upon the regulatory mechanisms that control *trp* gene expression in prokaryotes (Imamoto, 1973; Crawford and Stauffer, 1980; Yanofsky, 1981; Kolter and Yanofsky, 1982). Earlier reviews dwell upon comparative aspects (Crawford, 1975, 1980) and on various aspects of tryptophan pathway enzymology (Yanofsky, 1960; Yanofsky and Crawford, 1972; Zalkin, 1973; Miles, 1979).

II. THE ENZYMES OF TRYPTOPHAN BIOSYNTHESIS

The biosynthesis of tryptophan from chorismate proceeds in all organisms according to the scheme shown in Fig. 20.1. Each of the five proteins involved in this conversion has been obtained in highly purified form from at least one organism (Table 20.1). Tryptophan is the most costly of the amino acids; 78 moles of ATP are required to synthesize one mole of tryptophan (Atkinson, 1977). This energy cost is 20% higher than that for the biosynthesis of the second most costly amino acid, phenylalanine. Tryptophan biosynthesis requires approximately double the biosynthetic energy of the average amino acid.

Between different genera there is striking diversity with respect to the molecular nature of the individual *trp* biosynthetic enzymes and their modes of regulation (Crawford, 1975, 1980). For example, certain multifunctional enzymes catalyze chemically distinct steps in tryptophan biosynthesis. The steps are sometimes consecutive and sometimes not. A particular step in tryptophan biosynthesis may be catalyzed in one organism by a protein product of a single gene, while in a different species the same reaction may be catalyzed by a multimeric protein encoded by two genes. There is considerable scope for the exploitation of this range of diversity in the design of production strains. If one could, by genetic engineering, bypass a strictly regulated step by introducing from another organism a segment of DNA encoding a functionally identical but unregulated catalyst, the net flux of intermediates along the tryptophan pathway might be enhanced. In the following sections, each step in the conversion of chorismate to tryptophan is briefly discussed.

A. Anthranilate Synthase

Anthranilate synthase catalyzes the conversion of chorismate and ammonia to anthranilate and pyruvate. Physiologically, ammonia is generated *in situ* by cleavage of the amide bond of glutamine with the intermediate formation of a γ-glutamyl enzyme. The glutamine amidotransferase that funnels ammonia into tryptophan synthesis is now designated *trpG*, even though *trpG* may not be a distinct gene. Amidotransferase is either a protein specialized with respect to glutamine cleavage or part of a bifunctional enzyme that catalyzes a subsequent reaction of tryptophan biosynthesis. In *E. coli* and *S. typhimurium,* glutaminase activity is part of a bifunctional, two-domain protein that also catalyzes the conversion of anthranilate to phosphoribosyl anthranilate. In *Bacillus subtilis, Serratia marcescens,* and *Pseudomonas putida* the glutaminase of anthranilate synthesis resides within a separate protein with no other apparent function. In *Neurospora crassa,* glutaminase activity is part of a protein that converts phosphoribosyl anthranilate to indoleglycerol phosphate.

Anthranilate synthase is the major feedback-sensitive enzyme in tryptophan biosynthesis. By virtue of a ligand-binding site specific for tryptophan, the physiological feedback effector, organisms effectively regulate how much chorismate is channeled toward tryptophan biosynthesis. In *E. coli* the affinity toward tryptophan of anthranilate synthase is subject to genetic modification. Missense mutations that alter in both directions the parameters of tryptophan binding have been characterized (Pabst et al., 1973). Although none of the other enzymes specific for tryptophan biosynthesis have feedback-sensitive sites, this does not mean that other reactions cannot be modulated by tryptophan. In effect, any protein that is tightly aggregated to anthranilate synthase is potentially subject to tryptophan inhibition. In *E. coli* and *S. typhimurium,* glutaminase activity and phosphoribosyl transferase activity can both be inhibited by tryptophan, albeit to differing degrees, because of conformational changes communicated via tightly complexed anthranilate synthase to the other catalytic centers. In *N. crassa,* however, the phosphoribosyl anthranilate isomerase–indoleglycerol phosphate synthase activity of the complex that forms in association with anthranilate synthase is not subject to tryptophan inhibition, nor is the state of aggregation of the complex altered in the presence of tryptophan (J. DeMoss, personal communication).

The inhibition of anthranilate synthase by tryptophan is competitive

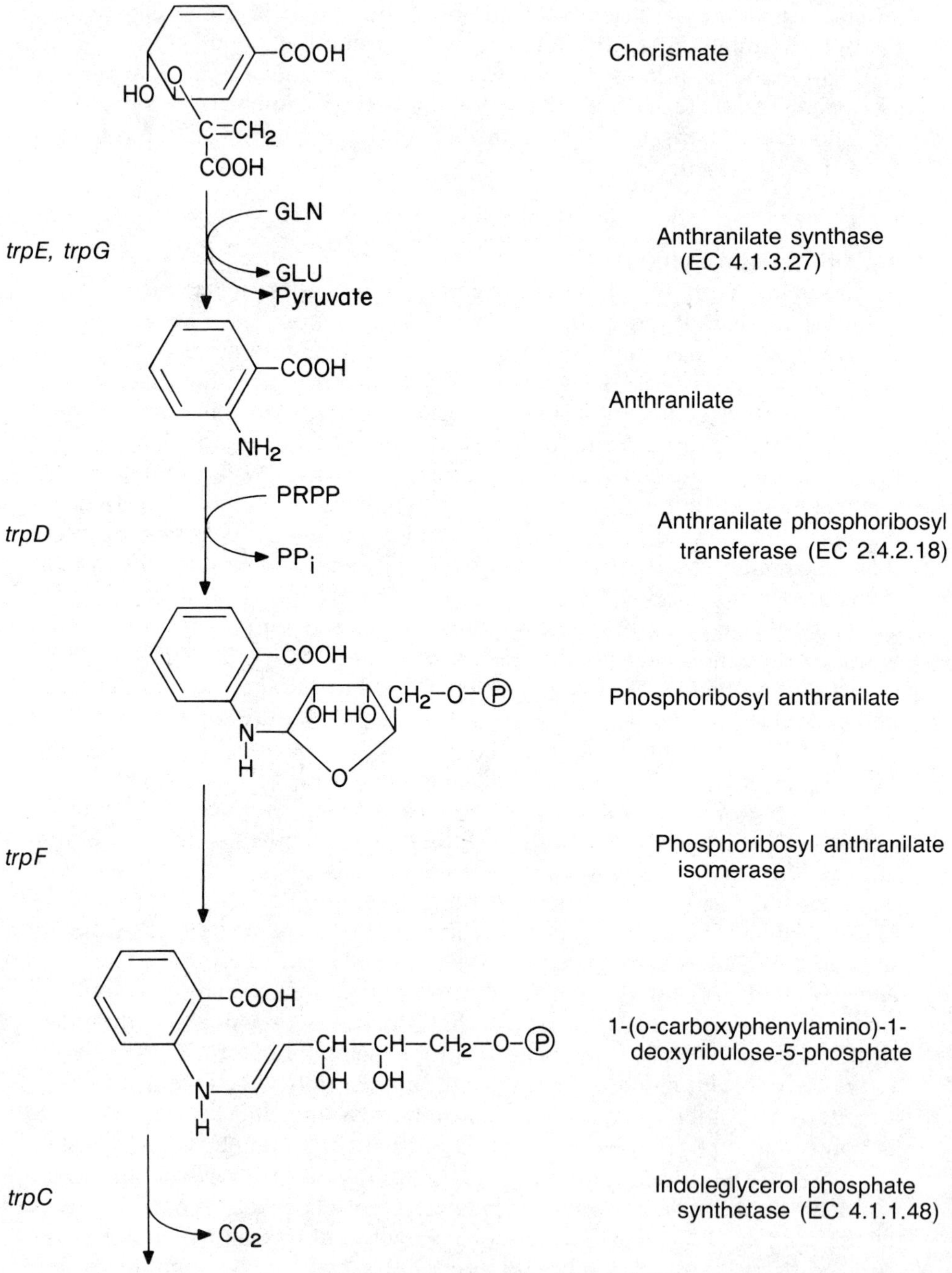
Chorismate
trpE, trpG
GLN
GLU
Pyruvate
Anthranilate synthase (EC 4.1.3.27)
Anthranilate
trpD
PRPP
PP$_i$
Anthranilate phosphoribosyl transferase (EC 2.4.2.18)
Phosphoribosyl anthranilate
trpF
Phosphoribosyl anthranilate isomerase
1-(o-carboxyphenylamino)-1-deoxyribulose-5-phosphate
trpC
CO_2
Indoleglycerol phosphate synthetase (EC 4.1.1.48)

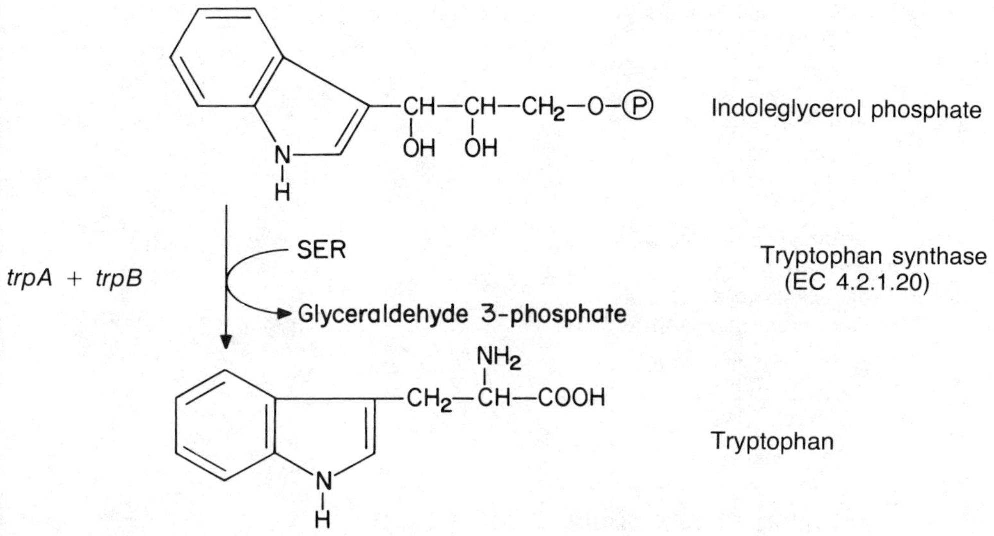

FIGURE 20.1 The terminal reactions of tryptophan biosynthesis. Chorismate, the end product of general aromatic biosynthesis (Chapter 17), is converted in five discrete steps to L-tryptophan. In addition to chorismate, L-serine, 5-phosphoribosyl-1-pyrophosphate, and glutamine are required for efficient tryptophan biosynthesis. Under certain circumstances, NH_3 can serve as a source of the ring nitrogen of the indole nucleus (see text). For a description of the biosynthetic pathways leading to serine and glutamine, see Chapters 6 and 1.

with respect to chorismate both in wild-type cells and in missense mutants where the ligand-binding parameters for tryptophan are altered. This fact in no way implies that tryptophan and chorismate are alternative occupants of the same site. The most logical explanation for the existing data is to assume that there are alternative conformational states associated with chorismate or tryptophan binding and that enzyme molecules in one conformational state are ineffective at binding one or the other ligand. Mutations that desensitize anthranilate synthase with respect to tryptophan inhibition are likely to enhance chorismate binding; mutations that enhance the binding of tryptophan are likely to allosterically diminish chorismate binding.

The complete nucleotide and amino acid sequences of anthranilate synthase of *E. coli* and *S. typhimurium* have been determined (Yanofsky and van Cleemput, 1982). Each protein contains 520 amino acids, none of which is tryptophan. The two proteins differ in 65 amino acid residues. Most of the differences are clustered in the amino-terminal half of the molecule. No amino acid switches associated with alterations in catalytic activity, feedback sensitivity, or ability to aggregate with functionally related proteins are established.

TABLE 20.1 Purification and Properties of the *trp* Enzymes from *E. coli*

Protein	*Reference*
Tryptophan synthase α	Gschwind et al. (1979)
Tryptophan synthase β_2	Miles and Moriguchi (1977)
Tryptophan synthase $\alpha_2\beta_2$	Tschopp and Kirschner (1980)
Phosphoribosyl anthranilate isomerase–indoleglycerol phosphate synthase	Bisswanger et al. (1969)
Anthranilate synthase–anthranilate phosphoribosyl transferase	Henderson et al. (1970)
Anthranilate synthase	Ito et al. (1969)

B. Anthranilate Phosphoribosyl Transferase

This protein is frequently referred to as anthranilate synthase component II because it forms with anthranilate synthase the stable tetrameric functional complex I_2II_2. In *E. coli,* phosphoribosyl transferase is bifunctional. Two structural domains, one having amidotransferase activity and the other catalyzing the condensation of anthranilate with the ribose-5-phosphate moiety of phosphoribosyl pyrophosphate, are encoded within a single polypeptide chain of 531 amino acids. Approximately 192 amino acids at the N-terminus form the amidotransferase, the remainder the phosphoribosyl transferase (Nichols et al., 1980). The equimolar synthesis of anthranilate synthase components I and II is governed by translational coupling (Oppenheim and Yanofsky, 1980), a mechanism wherein the translation of *trpD* mRNA requires the efficient translation of the end of *trpE* mRNA. Translational coupling is thought to be related to the fact that the *trpE–trpD* intercistronic region consists of overlapping start and stop codons. Although *trpE* mutants will grow on exogenously added anthranilate, a fact implying efficient access by anthranilate to the catalytic center of phosphoribosyl transferase, it is not known whether free anthranilate dissociates from the surface of the I_2II_2 complex in the course of normal tryptophan biosynthesis in wild-type cells. It is conceivable that most, if not all, of the anthranilate formed by component I is channeled to the catalytic center of component II. This might enhance the overall efficiency of tryptophan biosynthesis via catalytic facilitation, a process first postulated for the *arom* cluster of *N. crassa* (Case and Giles, 1974) (Chapter 17). Both catalytic centers of component II are feedback-inhibited by tryptophan, but only in the tetrameric aggregate. This is attributable to conformational changes communicated to all subunits of the aggregate following the binding of tryptophan to component I, the protein product of the *trpE* gene. The *trpD* gene product has no intrinsic tryptophan-binding capacity of its own (Pabst et al., 1973), despite early indications to the contrary (Ito and Yanofsky, 1969).

C. Phosphoribosyl Anthranilate Isomerase–indoleglycerol Phosphate Synthase

The *trpC* gene encodes a bifunctional two-domain polypeptide of 452 amino acid residues that catalyzes two sequential reactions, the Amadori rearrangement of phosphoribosyl anthranilate to 1-(o-carboxyphenylamino) 1-deoxyribulose 5-phosphate and the decarboxylation and ring closure of this intermediate to indoleglycerol phosphate, the immediate precursor of tryptophan. The two active centers are distinct and nonoverlapping (Creighton, 1970).

The amino terminal half of the polypeptide forms the indoleglycerol phosphate synthase domain (sometimes encountered as a monofunctional protein in nonenteric microorganisms), while the carboxy-terminal half of the peptide forms the isomerase domain (also encountered as a monofunctional enzyme in other organisms; the structural gene of this enzyme is then designated *trpF*) (Kirschner et al., 1980). In nongrowing cultures of *E. coli* the *trpC* product is inactivated, presumably by intracellular proteolysis (Mosteller et al., 1977). The *trpC* polypeptide of *E. coli* has been crystallized (White et al., 1982).

D. Tryptophan Synthase

In prokaryotes, tryptophan synthase is a tetramer containing two α subunits of 268 amino acids each and one β_2 subunit of two identical chains with 396 residues each. The $\alpha_2\beta_2$ tetramer catalyzes the pyridoxal phosphate–dependent conversion of indoleglycerol phosphate and serine to tryptophan and glyceraldehyde-3-phosphate. Each subunit alone catalyzes a partial reaction: α the reversible conversion of indoleglycerol phosphate to indole and glyceraldehyde-3-phosphate; β_2 the condensation of indole and serine to tryptophan (Miles, 1979).

Tryptophan synthase of lower eukaryotes such as *Saccharomyces cerevisiae* catalyzes the same overall and partial reactions, but the catalytic centers are organized differently (Zalkin and Yanofsky, 1982). Two identical subunits of 707 amino acid residues each contain fused α and β domains. The NH_2-terminal segment of 239 amino acids shows substantial homology to the α subunit of *E. coli* and related enterics, while the carboxy-terminal segment of 389 residues is homologous to the β chain of *E. coli*. A connector of 28 amino acid residues is homologous to neither subunit from the enteric organisms.

In prokaryotes, both α and β subunits of tryptophan synthase are composed of two independently folding domains (Miles, 1980), which can be separated following limited tryptic digestion. The tryptic fragments fold independently and reassociate to give an active enzyme, although the trypsin-cleaved β_2 subunit is unable to form a complex with the α subunit (Högberg-Raibaud and Goldberg, 1977).

III. REGULATION OF GENE EXPRESSION

A. Repression

The *trp* operon of *E. coli* and related enterics has been the most intensively studied of any biosynthetic pathway. Because the *E. coli* system is best known, I will emphasize its structural and functional properties. Comparative information from other organisms has been reviewed (Crawford 1975, 1980).

The *trp* operon encompasses roughly 7000 nucleotides (Fig. 20.2). The five *trp* genes are transcribed as a polycistronic mRNA, especially under conditions of tryptophan starvation. The complete nucleotide sequence of the *E. coli trp* operon is known (Yanofsky et al., 1981), as is the complete or partial sequence of several other genes of tryptophan metabolism, namely, *trpR* (Singleton et al., 1980; Gunsalus and Yanofsky, 1980), *ttrC* (Young, 1979), *trpS* (Hall et al., 1982), *tna* (Deeley and Yanofsky, 1981), and *aroH* (Zurawski et al., 1981).

The primary promoter of the *trp* operon is about 200 nucleotides upstream of *trpE,* the first structural gene. Transcription initiation at the *trp* promoter is regulated by repressor–operator interactions. If the *trp* operator, a 22-base-pair target within the *trp* promoter, is occluded by a repressor, RNA polymerase cannot form a productive complex with the promoter (Zalkin et al., 1974). Trp repressor binding to the *trp* operator is controlled by tryptophan: when a sufficient level of this effector is present, repressor–operator interaction is favored; when tryptophan is absent, repressor–operator interaction does not occur.

Because of its strength and the ease with which it can be regulated, the *trp* promoter of *E. coli* has gained considerable favor in the design of systems for the expression of foreign (mainly eukaryotic) genes in bacteria. These include the genes for fowl plague virus hemagglutinin (Emtage et al., 1980), capsid protein VP_3 of foot-and-mouth disease virus (Kleid et al., 1981), vesicular stomatitis virus glycoprotein (Rose and Shafferman, 1981), Rous sarcoma virus *src* gene product (McGrath and Levinson, 1982), human interferons (Sheppard et al., 1981; Windass et al., 1982), hepatitis virus core antigen (Edman et al., 1981), and human tissue-type plasminogen activator (Pennica et al., 1983). Specialized plasmid vectors wherein the *trp* promoter–operator region is either optimally juxtaposed to ribosome-binding sequences (Windass et al., 1982) or available to facilitate gene expression in each of three possible reading frames have been described (Tacon et al., 1980).

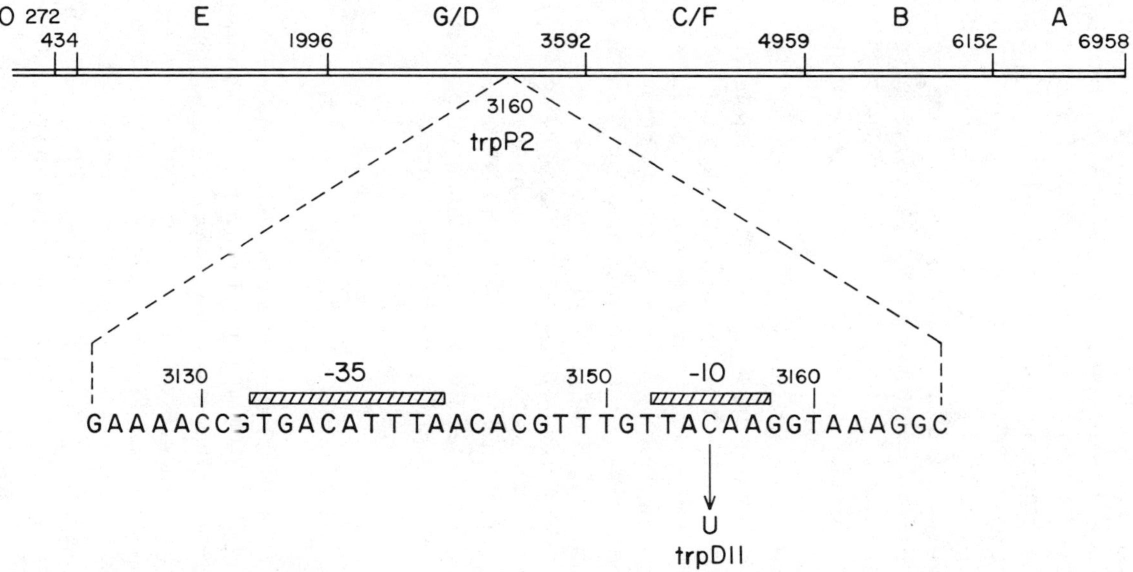

FIGURE 20.2 Low-resolution diagram of *trp* operon DNA. The numbers above the double line refer to the coordinates of the promoter–operator region (272), the leader region (434), etc. The letters refer to the protein coding regions discussed in the text. In expanded form in the lower part of the figure are shown the base sequence of the messenger-equivalent strand in and around the internal promoter (*trpP*2), the proposed −10 and −35 regions of *trpP*2, and one of several base pair changes that enhances the efficiency of the *trpP*2 promoter (adapted from Yanofsky et al. (1981); the *trpD*11 change is unpublished work of K. Brechling and B. Hausler from the author's laboratory).

B. Attenuation

Although repressor–operator interaction is quantitatively the most important factor governing the expression of the *trp* operon, a second control process, attenuation, also strongly influences the rate of *trp* gene expression. RNA polymerase molecules that initiate transcription at the *trp* promoter must traverse a 162-nucleotide leader region before encountering the coding sequence of the first structural gene. The leader region contains a potential transcription termination site. Whether or not transcription terminates in the *trp* leader region depends upon the availability of aminoacylated $tRNA^{Trp}$. When $tRNA^{Trp}$ is uncharged, transcription proceeds into the *trpE* coding region, producing a full-length polycistronic mRNA. When cells have adequate supplies of tryptophanyl-$tRNA^{Trp}$, transcription terminates in the leader region, thereby leading to short transcripts with no *trp* enzyme coding properties.

A plausible model for attenuation (Fig. 20.3) (Yanofsky, 1981; Kolter and Yanofsky, 1982) is based in part on the fact that Trp leader mRNA contains regions of dyad symmetry that potentially accommodate the formation of two major alternative stem and loop structures: one such secondary structure, the terminator, affects RNA polymerase by acting as a transcription termination signal; an alternative structure (preemptor) favors readthrough. Whichever secondary structure forms is coupled to events within a coding region for a 14-amino-acid leader peptide having two tandem UGG (Trp) codons. If enough tryptophanyl-tRNA is available for translation of the leader peptide, the terminator forms, and attenuation occurs. When tryptophanyl-$tRNA^{Trp}$ is not available for leader peptide synthesis, ribosome stalling at the site of the tandem Trp codons disfavors termination, and readthrough occurs.

C. Transcription Termination at the Distal End of the *trp* Operon

Following the last structural gene of the *trp* operon is a DNA region of approximately 300 base pairs that encodes a complex set of transcription termination signals. The sequences of this region from *E. coli* (Wu and Platt, 1978) and the related enterics *Klebsiella aerogenes* and *S. typhimurium* (Nichols et al., 1981a) are known. In vivo termination of *trp* mRNA occurs efficiently at a site called *trpt,* lying 36 nucleotides downsteam of *trpA*. Conserved sequences in the DNA of the three organisms include a G + C–rich region with high potential for formation of a stem–loop RNA secondary structure followed by an A + T–rich sequence. In *E. coli* the latter encodes the 3′ terminal sequence ($CAUUUU_{OH}$) of *trp* mRNA. Several deletions with endpoints outside of the 3′ end of *trp* mRNA effectively relieve termination (Guarente et al., 1979). Further in vitro studies led to the recognition of a second site, *trpt′*, located 250 nucleotides downsteam from *trpt* (Wu et al., 1981). Deletion of *trpt′* reduces or eliminates in vivo termination at

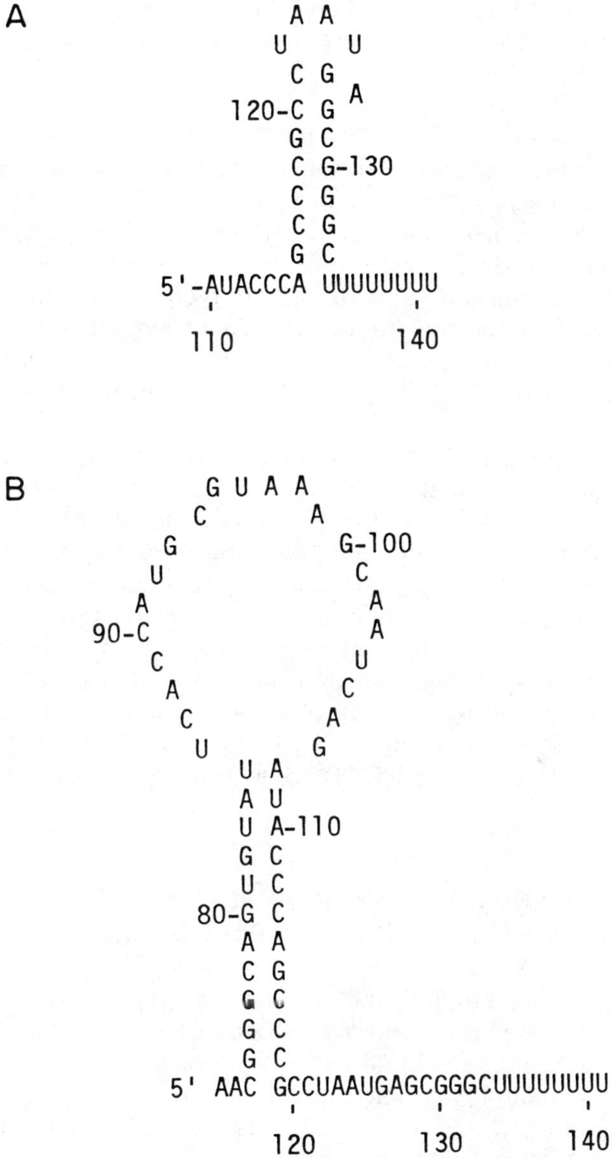

FIGURE 20.3 Alternative stem and loop structures in *trp* messenger RNA. Nucleotide 141 is the 3′-hydroxyl terminus of the leader transcript. Structure A is the terminator stem and loop associated with the cessation of transcription by RNA polymerase under conditions of tryptophan excess. Structure B forms under conditions of tryptophan starvation. This prevents the terminator loop from forming, thereby allowing RNA polymerase to proceed into the structural genes. The two tandem Trp codons referred to in the text (not shown above) lie between coordinates 54 and 59. (Adapted from Yanofsky, 1981).

trpt. The functional relationship between these sites, RNA polymerase, and two accessory factors, *rho* and *nusA* protein, remains to be elucidated.

D. Internal Promoters Within the *trp* Operon

Cells grown under conditions of maximum repression for tryptophan lose coordinacy in the production of the polypeptides of the *trp* operon. The protein products of *trpE* and *trpD* are reduced to barely detectable levels, but the remaining three enzymes are present in significant amounts. First noted in *S. typhimurium* (Bauerle and Margolin, 1966) and confirmed for *E. coli* and related enterics (Morse and Yanofsky, 1968), this phenomenon has been traced to the presence of a low-level internal promoter, *trpP2*. The sequence of the *E. coli trpP2* has been deduced (Horowitz and Platt, 1982).

There are at least 12 sites within the *trp* operon of *E. coli* where single-base-pair changes or other minor structural alterations generate promoters (McPartland, 1974). At one of these sites, designated *trpD*11, a C → T change (in mRNA terms a CAA (gln) codon is converted to a UAA (ochre) codon) results in premature polypeptide chain termination. At the DNA level the *trpD*11 mutation generates a promoter equal in strength to that of the primary *trp* promoter. A total of 20 mutants with single-base-pair changes at the *trpD*11 site have been isolated; 14 retain full promoter activity, while in the remaining 6, downstream protein production is at or below wild-type level. The *trpD*11 site is likely to be functionally related to *trpP2* (K. Brechling, B. Hausler, and R. L. Somerville, unpublished results) (Fig. 20.4).

IV. TRYPTOPHAN ANALOGS FOR ANALYSIS OF BIOSYNTHETIC AND REGULATORY PHENOMENA

A wide range of tryptophan analogs have been synthesized (Wilcox, 1974) or isolated from nature. Almost no phase of tryptophan metabolism could have been adequately explored without the use of analogs. Their use in the selection of regulatory mutants, in the construction of production strains, and in the production process itself is expected to remain high.

Analogs of metabolites are usually employed in a way that causes a wild-type organism to behave as if it had a certain type of mutational lesion. The experimentalist aims to design media whose analog content imposes some sort of loss in functional capability, that is, a "phenocopy" of a potential hereditary change. In many cases the target(s) of a given analog within the cell's metabolic machinery is unknown; frequently, several processes become inhibited at once. This can lead to severe difficulties in the interpretation of results obtained with whole cells. Analog studies that include rigorous parallel studies in cell-free systems deserve the most consideration, and these are emphasized in this section.

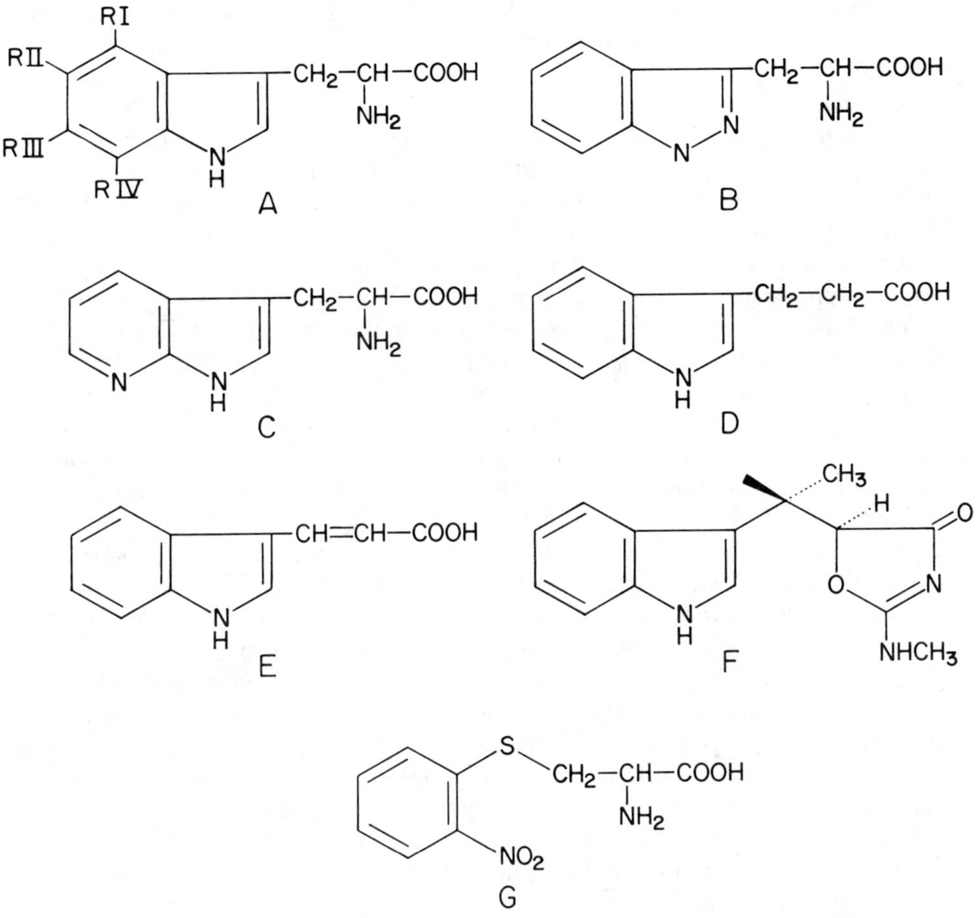

FIGURE 20.4 The analogs of tryptophan most widely employed. The methyl- and halogen-substituted tryptophans are related to structure A; in 4-methyltryptophan, RI = CH_3, RII = RIII = RIV = H; in 5-methyltryptophan, RII = CH_3, RI = RIII = RIV = H; the fluorotryptophans and chlorotryptophans follow the same design and nomenclature system. B is tryptazan; C is 7-azatryptophan; D is indolepropionate; E is indoleacrylate; F is indolmycin; G is S-o-nitrophenyl cysteine.

Proteins with tryptophan-binding sites are the logical candidates for cellular elements involved in tryptophan analog sensitivity and resistance. Permeability is the first such area of interest; many analogs of tryptophan are transported into cells by general or specific carriers whose normal function is to facilitate the passage of tryptophan into or out of the cytoplasm. Furthermore, if an analog bears a close enough resemblance to tryptophan, it may interact with tryptophanyl-tRNA synthetase and ultimately become incorporated into protein. One must be careful not to generalize from studies on a single organism about the ability of a given tryptophan analog to end up in protein. Different genera (or even different species within the same genera) may differ in their ability to make proteins containing analogs of tryptophan.

The principal point of short-term control over metabolite flow leading to tryptophan is anthranilate synthase, which is feedback-modulated by tryptophan and some tryptophan analogs. Many organisms also synthesize 3-deoxy-D-*arabino*-heptulosonate 7-phosphate (DAHP) synthases that are feedback-inhibited by tryptophan. False feedback inhibition by tryptophan analogs and its genetic reversal or potentiation has received much attention. A more slowly acting mechanism of control is repression of Trp enzyme formation, mediated by Trp repressor in response to tryptophan or various tryptophan analogs. Alterations in Trp repressor protein or any of its operator targets can drastically modify the cytoplasmic content of tryptophan biosynthetic enzymes. Another tryptophan-responsive system is tryptophanase, present in many enteric bacteria (Snell, 1975). This degradative enzyme is inducible by tryptophan and some tryptophan analogs. The enzyme is not expressed unless cyclic AMP and the catabolite gene activator protein are present. Tryptophanase degrades tryptophan to indole, pyruvate, and ammonia. Many tryptophan analogs are substrates for tryptophanase.

In the following sections I selectively discuss several tryptophan analogs that have been used over the past three decades to elucidate various facets of tryptophan metabolism. In most cases, workers have employed the commercially available, chemically synthesized racemic mixtures. Is it probably safe to assume that the D-isomers are generally inert, but in certain cases this may not be the case. Rigorous studies with optically pure D- and L-isomers of tryptophan analogs could force us to revise currently fashionable models for tryptophan regulatory phenomena. The structures of the analogs are given in Fig. 20.4.

A. 5-Methyltryptophan

This tryptophan analog received considerable early attention because it is cleaved by tryptophanase (Beerstecher and Edmonds, 1951). 5-Methyltryptophan-resistant mutants, selected on the basis of previous findings that this analog represses the *trp* enzymes, defined *trpR*, the structural gene for the Trp repressor (Cohen and Jacob, 1959). In *trpR* mutants, *trp* enzyme levels are 10-fold to 20-fold higher than in wild-type, even when cells are cultivated in media with excess tryptophan; *trpR* mutants are readily scored because they excrete tryptophan. The elevated intracellular levels of the *trp* enzymes, particularly anthranilate synthase, enable cells to continue funneling chorismate into tryptophan synthesis. At first sight, one might guess that feedback inhibition would prevent tryptophan biosynthesis even at high enzyme levels, but this is clearly not the case. Further isolates of 5-methyltryptophan-resistant *trpR* mutants were described in *E. coli* (Ito and Crawford, 1965) and in *S. typhimurium* (Bauerle and Margolin, 1966; Balbinder et al., 1970).

The principal target site for Trp repressor is *trpO,* situated within the *trp* promoter. Thus resistance to 5-methyltryptophan is a phenotype also displayed by *trpO* mutants that reduce or eliminate the ability of this DNA region to bind Trp repressor (Hiraga et al., 1967; Hiraga, 1969; Bennett and Yanofsky, 1978).

5-Methyltryptophan functions not only as a false corepressor of the *trp* operon but also as a false feedback inhibitor of anthranilate synthase (Moyed, 1960). Alterations within *trpE* that reduce the susceptibility of anthranilate synthase to feedback inhibition have been described (Somerville and Yanofsky, 1965). Enhanced suceptibility to inhibition by 5-methyltryptophan can also arise by mutations within *trpE* (Kuhn et al., 1972). Enhanced and reduced susceptibility to feedback regulation of anthranilate synthase has been attributed to alterations in the conformational equilibrium or state of aggregation of the tetrameric protein catalyzing the first two reactions of tryptophan biosynthesis (Pabst et al., 1973). This is confirmed through the isolation of mutants with lesions within *trpE* that affect amino acids that are relevant to the assembly of the *trpE* and *trpD* encoded polypeptides (Bauerle and Patterson, 1982). From *trpE* mutants hypersensitive to 5-methyltryptophan, secondary mutants have been isolated that display absolutely no sensitivity to tryptophan or its analogs (Pabst et al., 1973). Mutational desensitization of the tryptophan-inhibitable DAHP synthase has also been observed (Pittard et al., 1969).

Three permeases transport tryptophan from the medium into the cytoplasm: a high capacity, broad specificity general aromatic transport system (K_m, 0.4 μM) (Brown, 1970; Kuhn and Somerville, 1974); a system specific for tryptophan and certain tyrptophan analogs (K_m, 3 μM); and a catabolic permease (K_m, 70 μM), inducible in parallel with tryptophanase (Boezi and DeMoss, 1961). 5-Methyltryptophan resistance in *E. coli* and *S. typhimurium* can result from mutational lesions in *aroP,* the structural gene for the general aromatic permease (Ames, 1964, Brown, 1970). Since the general aromatic permease is under catabolite repression control (Alper and Ames, 1978), *cya* and *crp* mutants are relatively insensitive to tryptophan analogs such as 5-methyltryptophan.

Resistance to 5-methyltryptophan can also arise through lesions in loci whose gene products are unknown, such as *mtr* (Hiraga et al., 1968). The *mtr* gene of *E. coli* is not linked to any of the known *trp* genes; in *mtr* mutants, *trp* operon repression and tryptophan uptake are virtually unaltered. Another locus, *trpT,* was identified in *S. typhimurium* (Balbinder et al., 1970). The map location of *trpT* is unknown. This gene does not correspond to any previously described regulatory gene, although mutations in *trpT* lead to high level constitutive synthesis of the *trp* enzymes.

5-Methyltryptophan incorporation into protein appears to be species-specific. Tryptophanyl-tRNA synthetase from *E. coli* fails to charge $tRNA^{Trp}$

with 5-methyltryptophan, although the analog inhibits the enzyme (Doolittle and Yanofsky, 1968) with a K_i for 5-methyltryptophan of 0.7 mM (Squires et al., 1973). Since at the time of these studies the role of attenuation was not appreciated, the negative results were used to rule out a role for $tRNA^{Trp}$ in the control of *trp* operon expression, particularly since 5-methyltryptophan functions with great efficiency as a corepressor (Rose et al., 1973). In whole cell systems of *E. coli* 15 (Lark, 1969) and *E. coli* B (Ezekiel et al., 1973), 5-methyltryptophan can be incorporated into protein. The incorporation of 5-methyltryptophan into protein by *B. subtilis* and *Staphylococcus aureus* has also been reported (Barlatti and Ciferri, 1970; Imsande, 1973). To my knowledge, no systematic attempt has been made to resolve the apparent contradiction implicit in these results. It seems important to recognize differences between genera and even within individual strains of the same organism.

The use of 5-methyltryptophan-containing media has helped to advance our understanding of the molecular genetics of the *trp* system. Whereas wild-type *E. coli* is unaffected by 5-methyltryptophan at 0.1 μg/ml or less, bradytrophs and suppressed nonsense mutants fail to form colonies at this concentration (Allen and Yanofsky, 1963). One is thus able to treat such leaky mutants in selection procedures as if they were completely blocked in one of the steps of tryptophan biosynthesis. The presumed basis of hypersensitivity to 5-methyltryptophan is a combination of false feedback inhibition and false repression, resulting in diminished catalytic capability for one or more of the *trp* enzymes. The classical study of second-site mutation within *trpA* (Yanofsky et al., 1964) stands out as an example of the degree of sophistication attainable in the *trp* system.

Another example of the use of 5-methyltryptophan in selective media is the selection of second-site mutations in *trpA* that restore a full length, albeit enzymatically inactive, α-chain (Brammar et al., 1967; Murgola and Yanofsky, 1974). Mutants that do not produce a functional tryptophan synthase α-chain cannot activate the uncomplexed β_2 subunit in the conversion of indole to tryptophan. Consequently, such strains cannot grow on minimal agar containing indole (2 μg/ml) plus 5-methyltryptophan (100 μg/ml) because the latter compound effectively represses tryptophan synthase β_2 levels below those needed for utilization of indole for tryptophan biosynthesis. Mutational events that enable *trpA* nonsense or frameshift mutants to produce a tryptophan synthase α subunit allow the double mutants to grow on this selective media, even if the α subunit cannot convert indoleglycerol phosphate to indole.

5-Methyltryptophan has also been used to probe the expression of *tna* (Yudkin, 1976). *E. coli* strains utilize indole as a source of tryptophan, provided that they are capable of synthesizing the β subunit of tryptophan synthase. In *trpB* mutants, indole is still converted to tryptophan if the tryptophanase level is sufficiently high, since this catabolic enzyme also catalyzes the condensation of indole and serine to tryptophan (Snell, 1975). 5-Methyltryptophan (50 μg/ml) induces tryptophanase, especially under conditions where catabolite repression is avoided, thereby enabling *trpB* tna^+ strains to grow on indole. Thus 5-methyltryptophan–indole media are used to map and clone *tna* (Taylor and Yudkin, 1978; Deeley and Yanofsky, 1981).

B. Other Ring-substituted Methyltryptophans

Various applications and properties of 4-, 6-, and 7-methyltryptophan have served to gain new insight into some facets of tryptophan biosynthesis or its control.

The ring-methylated tryptophans can function as false corepressors and false feedback inhibitors. Some are incorporated into protein as analogs of tryptophan and some are substrates, inhibitors, or inducers of tryptophanase. In some cases the available data are obtained indirectly and/or are incomplete. A summary of the limited information available is found in Table 20.2. Usually, the organism studied has been *E. coli*.

C. Halogen-substituted Tryptophan Analogs

Fluorinated tryptophan analogs have been utilized to elucidate structure–function relationships in proteins. The van der Waals radius of ^{19}F is only 0.15 Å larger than that of ^{1}H, but ^{19}F is much more electronegative. Fluorotryptophans are readily incorporated into protein in place of tryptophan; some of the resulting proteins have been analyzed by ^{19}F-nmr spectroscopy (Browne et al., 1970; Pratt and Ho, 1975). 4-Fluorotryptophan is superior to the 5- and 6-isomers in supporting the growth of *trp* mutants. Enzymes substituted with this analog more closely resemble native enzymes than those in which tryptophan is replaced with either the 5- or 6-isomer. Extensive substitution with 4-fluorotryptophan in D-lactate dehydrogenase increases the activity of this enzyme (Pratt and Ho, 1975).

The fluorinated analogs of tryptophan are effective corepressors (Pauley et al., 1978; Doolittle and Yanofsky, 1968). They also inhibit both anthranilate synthase and the tryptophan-specific DAHP synthase (Kuhn et al., 1972; Görisch and Lingens, 1971). Analogs of tryptophan substituted with Cl in positions 4, 5, 6, and 7 are effective substrates for tryptophanase (Snell, 1975).

D. Tryptazan and 7-Azatryptophan

When carbon 2 of the indole nucleus is replaced by nitrogen, the resulting indazole-containing tryptophan analog is known as tryptazan. This compound and the 7-N homolog, 7-azatryptophan, are both efficiently incorporated in place of tryptophan into protein (Schlesinger, 1968), presumably with loss of viability (Rabinovitz et al., 1969). 7-Azatryptophan competes with tryptophan in the charging of $tRNA^{Trp}$ and for incorporation into protein (Mosteller and Yanofsky, 1971; Squires et al., 1973), but the compound is inactive as

TABLE 20.2 Properties of Methyltryptophans

	4-Methyl-tryptophan	*6-Methyl-tryptophan*	*7-Methyl-tryptophan*	*Reference*
Substrate of tryptophanase	+	+	+	Snell (1975)
Corepressor of *trp* operon	+	+	−	Lester and Yanofsky (1961), Doolittle and Yanofsky (1968), Held and Smith (1970), Mosteller and Yanofsky (1971), Rose et al. (1973), Pauley et al. (1978)
Inhibitor of anthranilate synthase	+	+	−	Held and Smith (1970), Kuhn et al. (1972)
Inhibitor of DAHP synthase	+	−		Görisch and Lingens (1971)
Inhibitor of aromatic amino acid uptake	+	+	(+)	Brown (1970), Gorisch and Lingens (1971)
Charged to $tRNA^{Trp}$	+	−		Mosteller and Yanofsky (1971), Morse and Morse (1976)
Incorporated into protein	+	−		Mosteller and Yanofsky (1971)

a corepressor. 7-Azatryptophan is also a weak allosteric inhibitor of anthranilate synthase (Kuhn et al., 1972). The charging of 7-azatryptophan to $tRNA^{Trp}$ enhances termination at the *trp* attenuator (Morse and Morse, 1976). By including 7-azatryptophan in selective media in combination with indole it is possible to enrich for regulatory mutants defective in transcription termination at the attenuator (Stauffer et al., 1978).

E. Indolepropionate and Indoleacrylate

These two analogs both cause rapid and coordinate derepression of the *trp* operon (Baker and Yanofsky, 1968; Doolittle and Yanofsky, 1968), through direct binding to Trp repressor (Squires et al., 1975). Indolepropionate has no effect on anthranilate synthase (Pauley et al., 1978). The ability of these analogs to act as gratuitous inducers of the *trp* operon has been exploited for *trp* mRNA production (Morse et al., 1969) and as a practical means of derepressing systems in which some desired mammalian structural gene is under the control of the *trp* promoter in genetically engineered production systems (Emtage et al., 1980). Indoleacrylate and indolepropionate are two of the best competitive inhibitors of tryptophanase (Snell, 1975).

F. Indolmycin

This tryptophan analog, a natural substance produced by *Streptomyces griseus* (Schach von Wittenau and Els, 1961), lacks a functional amino or carboxyl group, yet is a powerful competitive inhibitor of tryptophanyl-tRNA synthetase (Werner et al., 1976). Cells with increased levels of this enzyme are indolmycin-resistant (Bogosian et al., 1983). Indolmycin serves as a corepressor for transcription initiation at the *trp* promoter, although the analog does not interact with highly purified Trp repressor. This result hints at the existence of a hitherto undefined control mechanism for *trp* operon expression that may function in parallel with the classical repression and attenuation mechanisms.

G. S-o-Nitrophenyl cysteine

This tryptophan analog facilitates the direct spectrophotometric assay of tryptophanase (Suelter et al., 1976). This synthetic compound also allows a simple and sensitive chromogenic assay for pyridoxal phosphate (Suelter and Snell, 1979).

V. CLONING OF *trp* GENES

Small homogeneous DNA molecules enriched for the *trp* genes have been key ingredients in the analysis of how *trp* gene expression is regulated. The lambdoid phage Ø80 (Matsushiro, 1961a) undergoes reciprocal recombination with a site in the host DNA near the *trp* operon to form lysogens from which specialized *trp* transducing phages may be isolated (Matsushiro, 1961b). One of the most widely used *trp*-transducing phages is Ø80pt190, which contains the entire *trp* operon (Deeb et al., 1967).

Deletion mutants isolated from strains lysogenic for Ø80-lambda hybrid phages were important in substantiating Campbell's model for prophage insertion into the genome of host cells (Franklin et al., 1965). The construction of *trp–lac* operon fusions by an extension of this technique (Miller et al., 1970) facilitated the analysis of *trp* regulatory elements. The use of Ø80 transducing phages containing short blocks of *trp* operon DNA as probes for *trp* mRNA led to information on transcription in vivo (Imamoto et al., 1966).

Other types of specialized transducing phages, enriched for either the proximal or distal halves of the *trp* operon, were isolated from an abnormal lambda lysogen whose prophage was inserted within *trpC* (Shimada et al., 1973). These specialized transducing phages have been used to analyze *trp* operon expression (Rose et al., 1973). The nucleotide sequence of the secondary attachment site for phage lambda within *trpC* has been determined (Christie and Platt, 1979).

The direct cloning from chromosomal DNA of three separate *Hind*III fragments of the *E. coli trp* operon into specially designed lambda vehicles (Hopkins et al., 1976) led to amplified production of anthranilate synthase to the point at which this enzyme constituted 25% of the soluble cell protein. Attempts to construct phage with chromosomal *EcoR*I fragments containing the *trp* operon were unsuccessful (Borck et al., 1976), presumably because the distance between adjoining *EcoR*I sites flanking the *trp* operon (37.9 kb) is too large to be compatible with the packaging of phage lambda (Bouché, 1982).

Plasmids carrying *trp* operon DNA have been constructed in several laboratories. One of the first *trp* plasmids was pVH151 (Hershfield et al., 1974), constructed by ligation of *EcoR*I fragments from colicin E1 DNA and Ø80pt190 DNA. Another widely used *trp* plasmid is pVH153, isolated following the ligation of *EcoR*I fragments of DNA from a spontaneous deletion derivative of colE1 and a lambdoid phage, ED 10 (Helinski et al., 1977). A further series of plasmids with *trp* DNA are those made by connecting DNA from the *trp* transducing phages lambda EA60-3 or EA51 to plasmids pSC101, RSF1010, RSF2124, RP4, pBR313, and pBR345 (Nagahari et al., 1977b; Imanaka et al., 1980; Enger-Valk et al., 1980). The latter plasmids possess selectable antibiotic resistance markers in addition to their ability to confer a Trp^+ phenotype. The majority of the *trp* plasmids currently in use have been constructed through restriction endonuclease cleavage of specialized transducing phage DNA and have not been cloned directly from genomic DNA. The work of Bouché (1982) gives a moderately complete report of the distribution of cleavage sites in and around the *trp* operon of

E. coli. Transfer of *trp* DNA between *E. coli* and *Pseudomonas aeruginosa* has been accomplished through the in vitro construction of recombinant plasmids based on the wide-ranging P-group R factor RP4 (Nagahari et al., 1977a; Hedges et al., 1977). Repression of *E. coli trp* genes seems not to occur in the cytoplasm of *P. aeruginosa,* and the expression of the *P. aeruginosa trpAB* genes in *E. coli* is similar to the inducibility pattern of *Pseudomonas*. It is probable that *Pseudomonas* promoters are poorly recognized in *E. coli,* whereas the expression of *E. coli* genes in *Pseudomonas* is very efficient. This fact plus the absence of *E. coli* control systems in *Pseudomonas* implies that significant hyperproduction of proteins whose synthesis is driven by an *E. coli trp* promoter should occur in *Pseudomonas*.

The structural gene for the tryptophan-repressible DAHP synthase, *aroH,* lies within a 24-kb *EcoR*I fragment. By a variety of in vitro and in vivo deletion events the *aroH* gene was localized within a 1.93-kb *EcoR*I-*BamH*I fragment. Structural studies of this gene and its control regions are well advanced (Zurawski et al., 1981).

The structural gene for *E. coli* tryptophanyl-tRNA synthetase (*trpS*) was identified within the plasmid bank of Clarke and Carbon (1976) by virtue of its ability to complement a *trpS* mutation (Omnaas et al., 1979). Subcloning experiments localized the *trpS* gene to a 2.6-kb *EcoR*I-*Sal*I fragment, from which the complete sequence of the gene was determined (Hall et al., 1982).

The structural gene for $tRNA^{Trp}$, designated *ttrC*, is located within the distal part of the *rrnC* operon, whose promoter also drives the production of $tRNA_2^{Glu}$ and $tRNA_1^{Asp}$ as well as 16s, 23s, and 5s rRNA (Morgan et al., 1978). Originally identified within the Clarke-Carbon plasmid pLC22-36, the structural gene for $tRNA^{Trp}$ has also been cloned directly into RSF2124 (Sekiya et al., 1980). The sequence of $tRNA^{Trp}$ DNA has been reported (Young, 1979) and agrees with earlier RNA sequencing studies (Hirsh, 1970, 1971).

E. coli produces only one species of $tRNA^{Trp}$. Mutational alterations in *ttrC* distant from the anticodon region can lead to a $tRNA^{Trp}$ able to read both the UGG and UGA codons (Hirsh, 1970), while a recessive lethal mutation in the anticodon region leads to a species of $tRNA^{Trp}$ (su^+7 tRNA) that can decode both UAA and UGA by charging the suppressor tRNA with glutamine. This amino acid is subsequently inserted into polypeptide linkage (Soll, 1974; Yaniv et al., 1974).

The structural gene for tryptophanase was cloned by an in vivo genetic engineering procedure involving phage Mu–mediated transposition of the chromosomal *tnaA* gene onto the transmissible plasmid RP4 (Deeley and Yanofsky, 1981). The structural gene was localized within a 3.2-kb *Hind*III-*BamH*I fragment and sequenced. Tryptophanase is a polypeptide of 471 amino acids whose synthesis is subject to induction by tryptophan. Transcription initiated at the tryptophanase promoter involves the cAMP-dependent, catabolite gene activator protein-facilitated binding of RNA polymerase to a specific target site within the promoter (Deeley and Yanofsky, 1982). A low-affinity tryptophan-specific permease is likely to be part of the same operon as *tna* (Edwards and Yudkin, 1982).

The cloning of the gene encoding the Trp repressor was accomplished following the isolation of *trpR* specialized transducing phage from secondary site lysogens (Roeder and Somerville 1979a, b). DNA sequencing studies established that the Trp repressor is 108 amino acids in length (Gunsalus and Yanofsky, 1980). The discovery of a sequence upstream of the *trpR* coding region with strong homology to the primary *trp* operator (Singleton et al., 1980) suggested that the *trpR* gene was autogenously regulated at the transcriptional level. This has been experimentally confirmed (Bogosian et al., 1981; Haydock, 1983). The existence within the normal promoter for *trpR* of a target sequence for the Trp repressor obviously imposes limitations on the methods that can be employed to achieve hyperproduction of this protein. In vitro constructions, wherein alternative strong promoters unresponsive to Trp repressor are substituted for the natural *trpR* promoter, have led to production strains from which homogeneous Trp repressor protein has been prepared (Somerville et al., 1982; Haydock, 1983; Joachimiak et al., 1983).

VI. OUTLOOK

Structural studies have long dominated the tryptophan field. This kind of work has become vastly more efficient with the advent of techniques for DNA sequencing. The next few years will undoubtedly yield complete information on the primary structure of all of the proteins from prokaryotes and lower eukaryotes that are relevant to tryptophan biosynthesis, degradation, and the regulation of these processes. The ability to manipulate DNA molecules in vitro will lead to production strains for any desired protein or nucleic acid molecule relevant to tryptophan metabolism. This in turn will lead to even more sophisticated physicochemical studies at the molecular or even atomic level.

The bulk production of tryptophan is mainly of current importance to the pharmaceutical industry. As the costs of production are lowered, the use of this amino acid to supplement animal feeds is expected to grow. Tryptophan now sells at about US $43 per pound; the current annual volume is 225,000 lb, representing a market value of $9.7 million. Lysine and methionine, now used around the world in swine and poultry rations, sell at $1.40 and $2.10 per pound, respectively; their combined market value is in excess of $550 million. The economic incentive for the development of cheaper, more efficient production schemes for tryptophan is therefore considerable. Given the relatively complete nature of our understanding of tryptophan biosynthesis in microorganisms, it seems safe to predict continued effort on fermentative methods involving the use of genetically optimized microbial strains for the conversion of inexpensive feedstocks to tryptophan.

In recent efforts along these lines, Tribe and Pittard (1979) and Aiba et al., (1980; 1982) combined in production strains of *E. coli* a variety of mutations designed to increase the levels of the tryptophan enzymes and/or overcome known limitations on the flow of intermediates along the tryptophan pathway. Because *E. coli* is, by definition, tryptophanase-positive, it was considered necessary to include *tna* mutations in the genetic backgrounds of their strains. Not surprisingly, maximum specific productivity for tryptophan was attained in strains that were mutationally desensitized with respect to feedback inhibition and where the diversion of metabolites along competing pathways had been minimized through mutation. An important aspect of such studies is their potential for identifying unknown factors or steps that might restrict the cellular conversion of glucose and ammonia to the desired end product, for example, the decay of DAHP synthase (Tyr) as a rate-limiting factor in tryptophan production (Tribe and Pittard, 1979).

Within the next few years the technology for the production of tryptophan by fermentation should have improved to the point at which its market value will be low enough to make it economically feasible to include tryptophan as a routine supplement in rations formulated for domestic animals. Future research in the commercial aspects of tryptophan production should now begin to focus on secondary and tertiary uses of this amino acid in contexts that are now impractical because of its high cost. Of particular note is the field of pharmaceutical intermediates. There are many indole alkaloids with direct or indirect therapeutic uses whose production could be improved by a more complete knowledge of their pathways of biosynthesis from tryptophan.

Finally, one should not assume that the sophisticated models that have been put forward to explain the regulation of tryptophan biosynthesis in prokaryotes can account in full for all of the genetic and physiological properties of tryptophan biosynthesis in microorganisms. On the basis of recent experience in the field we can expect that an expansion of knowledge about regulatory mechanisms such as repression and attenuation will enable researchers to pose increasingly incisive experimental questions in the area of metabolic control. Mechanisms of positive control, for example, are just beginning to be explored (Bogosian et al., 1983). One may confidently predict that future work in the tryptophan field will produce unexpected and interesting new results with applications in biotechnology.

ACKNOWLEDGMENTS

Research in the author's laboratory has been supported by grants from the United States Public Health Service. This is Journal Paper No. 9388 of the Purdue University Agriculture Experiment Station.

REFERENCES

Aiba, S., Imanaka, T., and Tsunekawa, H. (1980) *Biotechnology Lett. 2,* 525–530.
Aiba, S., Tsunekawa, H., and Imanaka, T. (1982) *Appl. Environ. Microbiol. 43,* 289–297.
Allen, M. K., and Yanofsky, C. (1963) *Genetics 48,* 1056–1083.
Alper, M. D., and Ames, B. N. (1978) *J. Bacteriol. 133,* 149–157.
Ames, G. F. (1964) *Arch. Biochem. Biophys. 104,* 1–18.
Atkinson, D. E. (1977) *Cellular Energy Metabolism and Its Regulation.* Academic Press, New York.
Baker, R. F., and Yanofsky, C. (1968) *Proc. Nat. Acad. Sci. U.S.A. 60,* 313–320.
Balbinder, E., Callahan, R., McCann, P. P., Cordaro, J. C., Weber, A. R., Smith, A. M., and Angelosanto, F. (1970) *Genetics 66,* 31–53.
Barlatti, S., and Ciferri, O. (1970) *J. Bacteriol. 101,* 166–172.
Bauerle, R. H., and Margolin, P. (1966) *Proc. Nat. Acad. Sci. U.S.A. 56,* 111–118.
Bauerle, R., and Patterson, T. (1982) *Genetics 100,* s3.
Beerstecher, E., Jr., and Edmonds, E. J. (1951) *J. Biol. Chem. 192,* 497–504.
Bennett, G. N., and Yanofsky, C. (1978) *J. Mol. Biol. 121,* 179–192.
Bisswanger, H., Kirschner, K., Cohn, W., Hager, V., and Hansson, E. (1979) *Biochemistry 18,* 5946–5953.
Boezi, J. A., and DeMoss, R. D. (1961) *Biochim. Biophys. Acta 49,* 471–484.
Bogosian, G., Bertrand, K., and Somerville, R. L. (1981) *J. Mol. Biol. 149,* 821–825.
Bogosian, G., Haydock, P. V., and Somerville, R. L. (1983) *J. Bacteriol. 153,* 1120–1123.
Borck, K., Beggs, J. D., Brammar, W. J., Hopkins, A. S., and Murray, N. E. (1976) *Mol. Gen. Genet. 146,* 199–207.
Bouché, J. P. (1982) *J. Mol. Biol. 154,* 1–20.
Brammar, W. J., Berger, H., and Yanofsky, C. (1967) *Proc. Nat. Acad. Sci. U.S.A. 58,* 1499–1506.
Brown, K. D. (1970) *J. Bacteriol. 104,* 177–188.
Browne, D. T., Kenyon, G. L., and Hegeman, G. D. (1970) *Biochem. Biophys. Res. Comm. 39,* 13–19.
Case, M. E., and Giles, N. H. (1974) *Genetics 77,* 613–626.
Christie, G. E., and Platt, T. (1979) *Cell 16,* 407–413.
Clarke, L., and Carbon, J. (1976) *Cell 9,* 91–99.
Cohen, G. N., and Jacob, F. (1959) *Compt. Rend. 248,* 3490–3492.
Crawford, I. P. (1975) *Bacteriol. Rev. 39,* 87–120.
Crawford, I. P. (1980) *CRC Crit. Rev. Biochem. 8,* 175–189.
Crawford, I. P., and Stauffer, G. V. (1980) *Ann. Rev. Biochem. 49,* 163–195.
Creighton, T. E. (1970) *Biochem. J. 120,* 699–707.
Deeb, S. S., Okamoto, K., and Hall, B. D. (1967) *Virology 31,* 289–295.
Deeley, M. C., and Yanofsky, C. (1981) *J. Bacteriol. 147,* 787–796.
Deeley, M. C., and Yanofsky, C. (1982) *J. Bacteriol. 151,* 942–951.
Doolittle, W. F., and Yanofsky, C. (1968) *J. Bacteriol. 95,* 1283–1294.
Edman, J. C., Hallewell, R. A., Valenzuela, P., Goodman, H. M., and Rutter, W. J. (1981) *Nature 291,* 503–506.
Edwards, R. M., and Yudkin, M. D. (1982) *Biochem. J. 204,* 617–619.

Emtage, J. S., Tacon, W. C. A., Catlin, G. H., Jenkins, B., Porter, A. G., and Carey, N. H. (1980) *Nature 283,* 171–174.

Enger-Valk, B. E., Heyneker, H. L., Oosterbaan, R. A., and Pouwels, P. H. (1980) *Gene 9,* 69–85.

Ezekiel, D. H., Carlson, N., and Jenkins, T. (1973) *Biochim. Biophys. Acta 312,* 751–764.

Fowler, A. V., and Zabin, I. (1977) *Proc. Nat. Acad. Sci. U.S.A. 74,* 1507–1510.

Franklin, N. C., Dove, W. F., and Yanofsky, C. (1965) *Biochem. Biophys. Res. Comm. 18,* 910–923.

Görisch, H., and Lingens, F. (1971) *Biochim. Biophys. Acta 242,* 630–636.

Gschwind, H.-P., Gschwind, U., Paul, C. H., and Kirschner, K. (1979) *Eur. J. Biochem. 96,* 403–416.

Guarente, L. P., Beckwith, J., Wu, A. M., and Platt, T. (1979) *J. Mol. Biol. 133,* 189–197.

Gunsalus, R. P., and Yanofsky, C. (1980) *Proc. Nat. Acad. Sci. U.S.A. 77,* 7117–7121.

Hall, C. V., van Cleemput, M., Muench, K. H., and Yanofsky, C. (1982) *J. Biol. Chem. 257,* 6132–6136.

Haydock, P. V. (1983) "Purification of Trp Repressor Protein of *Escherichia coli*: Physical and Immunochemical Characterization," Ph.D. thesis, Purdue University, West Lafayette, IN.

Hedges, R. W., Jacob, A. E., and Crawford, I. P. (1977) *Nature 267,* 283–284.

Held, W. A., and Smith, O. H. (1970) *J. Bacteriol. 101,* 209–217.

Helinski, D. R., Hershfield, V., Figurski, D., and Meyer, R. J. (1977) in *Recombinant Molecules: Impact on Science and Society,* (Beers, R. F., and Basset, E. G., eds.) pp. 151–165. Raven Press, New York.

Henderson, E. J., Nagano, H., Zalkin, H., and Hwang, L. H. (1970) *J. Biol. Chem. 245,* 1416–1423.

Hershfield, V., Boyer, H. W., Yanofsky, C., Lovett, M. A., and Helinski, D. R. (1974) *Proc. Nat. Acad. Sci. U.S.A. 71,* 3455–3459.

Hiraga, S. (1969) *J. Mol. Biol. 39,* 159–179.

Hiraga, S., Ito, K., Hamada, K., and Yura, T. (1967) *Biochem. Biophys. Res. Comm. 26,* 522–527.

Hiraga, S., Ito, K., Matsuyama, T., Ozaki, H., and Yura, T. (1968) *J. Bacteriol. 96,* 1880–1881.

Hirsh, D. (1970) *Nature 228,* 57.

Hirsh, D. (1971) *J. Mol. Biol. 58,* 439–458.

Högberg-Raibaud, A., and Goldberg, M. E. (1977) *Proc. Nat. Acad. Sci. U.S.A. 74,* 442–446.

Hopkins, A. S., Murray, N. E., and Brammar, W. J. (1976) *J. Mol. Biol. 107,* 549–569.

Horowitz, H., and Platt, T. (1982) *J. Mol. Biol. 156,* 257–267.

Imamoto, F. (1973) *Progr. Nucl. Acid Res. Mol. Biol. 13,* 339–407.

Imamoto, F., Ito, J., and Yanofsky, C. (1966) *Cold Spring Harbor Symp. Quant. Biol. 31,* 235–249.

Imanaka, T., Tsunekawa, H., and Aiba, S. (1980) *J. Gen. Microbiol. 118,* 253–261.

Imsande, J. (1973) *Genetics 75,* 1–17.

Ito, J., and Crawford, I. P. (1965) *Genetics 52,* 1303–1316.

Ito, J., and Yanofsky, C. (1969) *J. Bacteriol. 97,* 734–742.

Ito, J., Cox, E. C., and Yanofsky, C. (1969) *J. Bacteriol. 97,* 725–733.
Joachimiak, A., Kelley, R. L., Gunsalus, R. P., Yanofsky, C., and Sigler, P. B. (1983) *Proc. Nat. Acad. Sci. 80,* 668–672.
Kirschner, K., Szadkowski, H., Henschen, A., and Lottspeich, F. (1980) *J. Mol. Biol. 143,* 395–409.
Kleid, D. G., Yansura, D., Small, B., Dowbenko, D., Moore, D. M., Grubman, M. J., McKercher, P. D., Morgan, D. O., Robertson, B. H., and Bachrach, H. L. (1981) *Science 214,* 1125–1129.
Kolter, R., and Yanofsky, C. (1982) *Ann. Rev. Genet. 16,* 113–114.
Kuhn, J., and Somerville, R. L. (1974) *Biochim. Biophys. Acta 332,* 298–312.
Kuhn, J. C., Pabst, M. J., and Somerville, R. L. (1972) *J. Bacteriol. 112,* 93–101.
Lark, K. G. (1969) *J. Bacteriol. 97,* 980–982.
Lester, G., and Yanofsky, C. (1961) *J. Bacteriol. 81,* 81–90.
Margolin, P. (1971) in *Metabolic Pathways* Vol. V (Vogel, H. J., ed.) pp. 389–446, Academic Press, New York.
Matsushiro, A. (1961a) *Biken's J. 4,* 133–135.
Matsushiro, A. (1961b) *Biken's J. 4,* 141–144.
McGrath, J. P., and Levinson, A. D. (1982) *Nature 295,* 423–425.
McPartland, A. A. (1974) "Isolation and Characterization of Mutants Harboring High Efficiency Transcription Initiation Signals Within Structural Genes of the *trp* Operon of *Escherichia coli,*" Ph.D. thesis, Purdue University, West Lafayette, IN.
Miles, E. W. (1979) in *Advances in Enzymology* Vol. 49 (Meister, A., ed.) pp. 127–186, John Wiley and Sons, New York.
Miles, E. W. (1980) in *Biochemical and Medical Aspects of Tryptophan Metabolism* (Hayaishi, O., Ishimura, Y., and Kido, R., eds.) pp. 137–147, Elsevier-North Holland, Amsterdam.
Miles, E. W., and Moriguchi, M. (1977) *J. Biol. Chem. 252,* 6594–6599.
Miller, J. H., Reznikoff, W. S., Silverstone, A. E., Ippen, K., Singer, E. R., and Beckwith, J. R. (1970) *J. Bacteriol. 104,* 1273–1279.
Morgan, A., Ikemura, T., Lindahl, L., Fallon, A. M., and Nomura, M. (1978) *Cell 13,* 335–344.
Morse, D. E., and Morse, A. N. C. (1976) *J. Mol. Biol. 103,* 209–226.
Morse, D. E., and Yanofsky, C. (1968) *J. Mol. Biol. 38,* 447–451.
Morse, D. E., Mosteller, R. D., and Yanofsky, C. (1969) *Cold Spring Harbor Symp. Quant. Biol. 34,* 725–739.
Mosteller, R. D., and Yanofsky, C. (1971) *J. Bacteriol. 105,* 268–275.
Mosteller, R. D., Nishimoto, K. R., and Goldstein, R. V. (1977) *J. Bacteriol. 131,* 153–162.
Moyed, H. S. (1960) *J. Biol. Chem. 235,* 1098–1102.
Murgola, E. J. (1981) *J. Mol. Biol. 149,* 1–13.
Murgola, E. J., and Yanofsky, C. (1974) *J. Mol. Biol. 86,* 775–784.
Nagahari, K., Sano, Y., and Sakaguchi, K. (1977a) *Nature 266,* 745–746.
Nagahari, K., Tanaka, T., Hishinuma, F., Kuroda, M., and Sakaguchi, K. (1977b) *Gene 1,* 141–152.
Nichols, B. P., Miozzari, G. F., van Cleemput, M., Bennett, G. N., and Yanofsky, C. (1980) *J. Mol. Biol. 142,* 503–517.
Nichols, B. P., Blumenberg, M., and Yanofsky, C. (1981a) *Nucl. Acids Res. 9,* 1743–1755.

Nichols, B. P., van Cleemput, M., and Yanofsky, C. (1981b) *J. Mol. Biol. 146,* 45–54.

Omnaas, J., Bonura, M., Chang, P., Safille, P. A., and Muench, K. H. (1979) *Fed. Proc. 38,* 821.

Oppenheim, D. S., and Yanofsky, C. (1980) *Genetics 95,* 785–795.

Pabst, M. J., Kuhn, J. C., and Somerville, R. L. (1973) *J. Biol. Chem. 248,* 901–914.

Pauley, R. J., Fredricks, W. W., and Smith, O. H. (1978) *J. Bacteriol. 136,* 219–226.

Pennica, D., Holmes, W. E., Kohr, W. J., Harkins, R. N., Vehar, G. A., Ward, C. A., Bennett, W. F., Yelverton, E., Seeburg, P. H., Heynecker, H. L., Goeddel, D. V., and Collen, D. (1983) *Nature 301,* 214–221.

Pittard, J., Camakaris, J., and Wallace, B. J. (1969) *J. Bacteriol. 97,* 1242–1247.

Platt, T. (1978) in *The Operon* (Miller, J. H., and Reznikoff, W. S., eds.) pp. 263–302, Cold Spring Harbor Laboratory, Cold Spring Harbor, N.Y.

Pratt, E. A., and Ho, C. (1975) *Biochemistry 14,* 3035–3040.

Rabinovitz, M., Finkleman, A., Reagan, R. L., and Breitman, T. R. (1969) *J. Bacteriol. 99,* 336–338.

Roeder, W. D., and Somerville, R. L. (1979a) *Fed. Proc. 38,* 396.

Roeder, W. D., and Somerville, R. L. (1979b) *Mol. Gen. Genet. 176,* 361–368.

Rose, J. K., and Shafferman, A. (1981) *Proc. Nat. Acad. Sci. U.S.A. 78,* 6670–6674.

Rose, J. K., Squires, C. L., Yanofsky, C., Yang, H.-L., and Zubay, G. (1973) *Nature New Biol. 245,* 133–137.

Schach von Wittenau, M., and Els, H. (1961) *J. Amer. Chem. Soc. 83,* 4678–4680.

Schlesinger, S. (1968) *J. Biol. Chem. 243,* 3877–3883.

Sekiya, T., Mori, M., Takahashi, N., and Nishimura, S. (1980) *Nucl. Acids Res. 8,* 3809–3827.

Shepard, H. M., Leung, D., Stebbing, N., and Goeddel, D. V. (1981) *Nature 294,* 563–565.

Shimada, K., Weisberg, R. A., and Gottesman, M. E. (1973) *J. Mol. Biol. 80,* 297–314.

Singleton, C. K., Roeder, W. D., Bogosian, G., Somerville, R. L., and Weith, H. L. (1980) *Nucl. Acids Res. 8,* 1551–1560.

Snell, E. E. (1975) *Adv. Enzymol. 42,* 287–333

Soll, L. (1974) *J. Mol. Biol. 86,* 233–243.

Somerville, R. L., and Yanofsky, C. (1965) *J. Mol. Biol. 11,* 747–759.

Somerville, R. L., Bogosian, G., Haydock, P., and Johnson, D. (1982) in *Abstracts of the XII International Congress of Biochemistry,* Perth, Australia, p. 239.

Squires, C. L., Rose, J. K., Yanofsky, C., Yang, H.-L., and Zubay, G. (1973) *Nature New Biol. 245,* 131–133.

Squires, C. L., Lee, F. D., and Yanofsky, C. (1975) *J. Mol. Biol. 92,* 93–111.

Stauffer, G. V., Zurawski, G., and Yanofsky, C. (1978) *Proc. Nat. Acad. Sci. U.S.A. 75,* 4833–4837.

Suelter, C. H., and Snell, E. E. (1979) *Meth. Enzymol. 62,* 561–568.

Suelter, C. H., Wang, J., and Snell, E. E. (1976) *FEBS Lett. 66,* 230–232.

Tacon, W., Carey, N., and Emtage, S. (1980) *Mol. Gen. Genet. 177,* 427–438.

Taylor, H. V., and Yudkin, M. D. (1978) *Mol. Gen. Genet. 165,* 195–102.

Tribe, D. E., and Pittard, J. (1979) *Appl. Environ. Microbiol. 38,* 181–190.

Tschopp, J., and Kirschner, K. (1980) *Biochemistry 19,* 4514–4521.
Vallee, B. L., and Riordan, J. F. (1969) *Ann. Rev. Biochem. 38,* 733–794.
Werner, R. G., Thorpe, L. F., Reuter, W., and Nierhaus, K. H. (1976) *Eur. J. Biochem. 68,* 1–3.
White, J. L., Grutter, M. G., Wilson, E., Thaller, C., Ford, G. C., Smit, J. D. G., Jansonius, J. N., and Kirschner, K. (1982) *FEBS Lett. 148,* 87–90.
Wilcox, M. (1974) *Anal. Biochem. 59,* 436–440.
Windass, J. D., Newton, C. R., de Maeyer-Guignard, J., Moore, V. E., Markham, A. F., and Edge, M. D. (1982) *Nucl. Acids Res. 10,* 6639–6657.
Wu, A. M., and Platt, T. (1978) *Proc. Nat. Acad. Sci. U.S.A. 75,* 5442–5446.
Wu, A. M., Christie, G. E., and Platt, T. (1981) *Proc. Nat. Acad. Sci. U.S.A. 78,* 2913–2917.
Yaniv, M., Folk, W. R., Berg, P., and Soll, L. (1974) *J. Mol. Biol. 86,* 245–260.
Yanofsky, C. (1960) *Bacteriol. Rev. 24,* 221–245.
Yanofsky, C. (1981) *Nature 289,* 751–758.
Yanofsky, C., and Crawford, I. P. (1972) in *The Enzymes* Vol. VII (Boyer, P. D., ed.) pp. 1–31, Academic Press, New York.
Yanofsky, C., and van Cleemput, M. (1982) *J. Mol. Biol. 155,* 235–246.
Yanofsky, C., Horn, V., and Thorpe, D. (1964) *Science 146,* 1593–1594.
Yanofsky, C., Drapeau, G. R., Guest, J. R., and Carlton, B. C. (1967) *Proc. Nat. Acad. Sci. U.S.A. 57,* 296–298.
Yanofsky, C., Platt, T., Crawford, I. P., Nichols, B. P., Christie, G. E., Horowitz, H., van Cleemput, M., and Wu, A. M. (1981) *Nucl. Acids Res. 9,* 6647–6668.
Young, R. A. (1979) *J. Biol. Chem. 254,* 12725–12731.
Yudkin, M. D. (1976) *J. Gen. Microbiol. 92,* 125–132.
Zalkin, H. (1973) in *Advances in Enzymology,* Vol. 38 (Meister, A., ed.) pp. 1–39, John Wiley, New York.
Zalkin, H., and Yanofsky, C. (1982) *J. Biol. Chem. 257,* 1491–1500.
Zalkin, H., Yanofsky, C., and Squires, C. L. (1974) *J. Biol. Chem. 249,* 465–475.
Zurawski, G., Gunsalus, R. P., Brown, K. D., and Yanofsky, C. (1981) *J. Mol. Biol. 145,* 47–73.

CHAPTER
21

Histidine Biosynthesis and its Regulation

Stanley W. Artz
Donald Holzschu

I. INTRODUCTION

The enzymes catalyzing the biosynthesis of histidine (Fig. 21.1) in *Salmonella typhimurium* and *Escherichia coli* are encoded by the nine contiguous genes of the histidine operon, located at min 44 on the chromosome. The first enzyme, ATP phosphoribosyl transferase, catalyzes the condensation of ATP and phosphoribosyl pyrophosphate, forming N-1-(5′-phosphoribosyl)-ATP. Pyrophosphate is cleaved from this product, which then undergoes a ring opening followed by an Amadori rearrangement. An amidotransfer from glutamine follows with closure of the imidazole ring to form imidazole glycerol phosphate. This compound is dehydrated, and a transamination reaction yields histidinol phosphate. Dephosphorylation yields histidinol, which is oxidized to histidine.

The expenditure of metabolic resources by the cell to synthesize histidine from precursor molecules is substantial and unnecessary under growth conditions of exogenous histidine supply (Brenner and Ames, 1971). Accordingly, the enteric bacteria have developed mechanisms for regulating histidine biosynthesis. These include: operon repression mediated at an attenuator

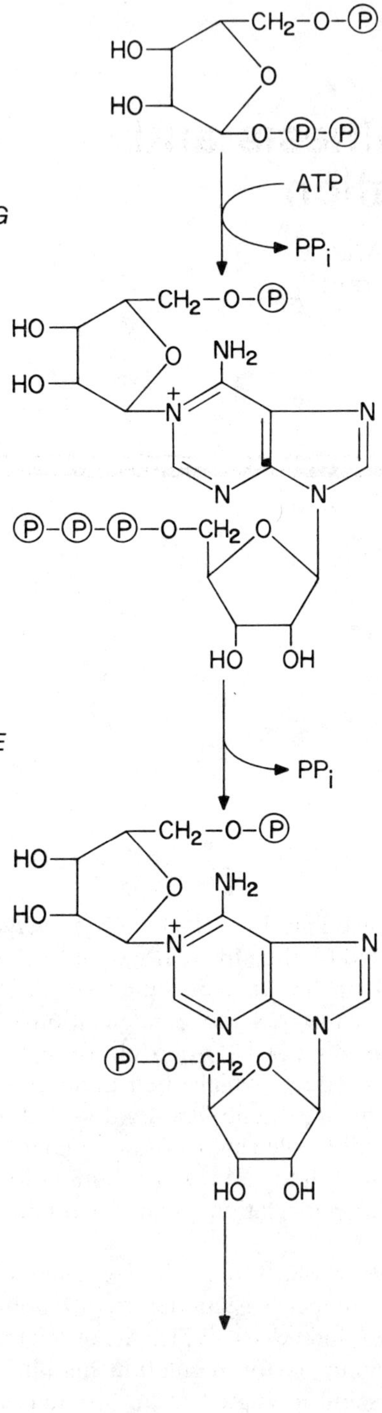

Phosphoribosyl pyrophosphate

hisG ATP phosphoribosyl transferase (EC 2.4.2.17)

N[1]-(5′-Phosphoribosyl)-ATP

hisE Phosphoribosyl-ATP pyrophosphohydrolase (EC 3.5.4.19)

N[1]-(5′-Phosphoribosyl)-AMP

hisI Phosphoribosyl-AMP cyclohydrolase (EC 3.5.4.19)

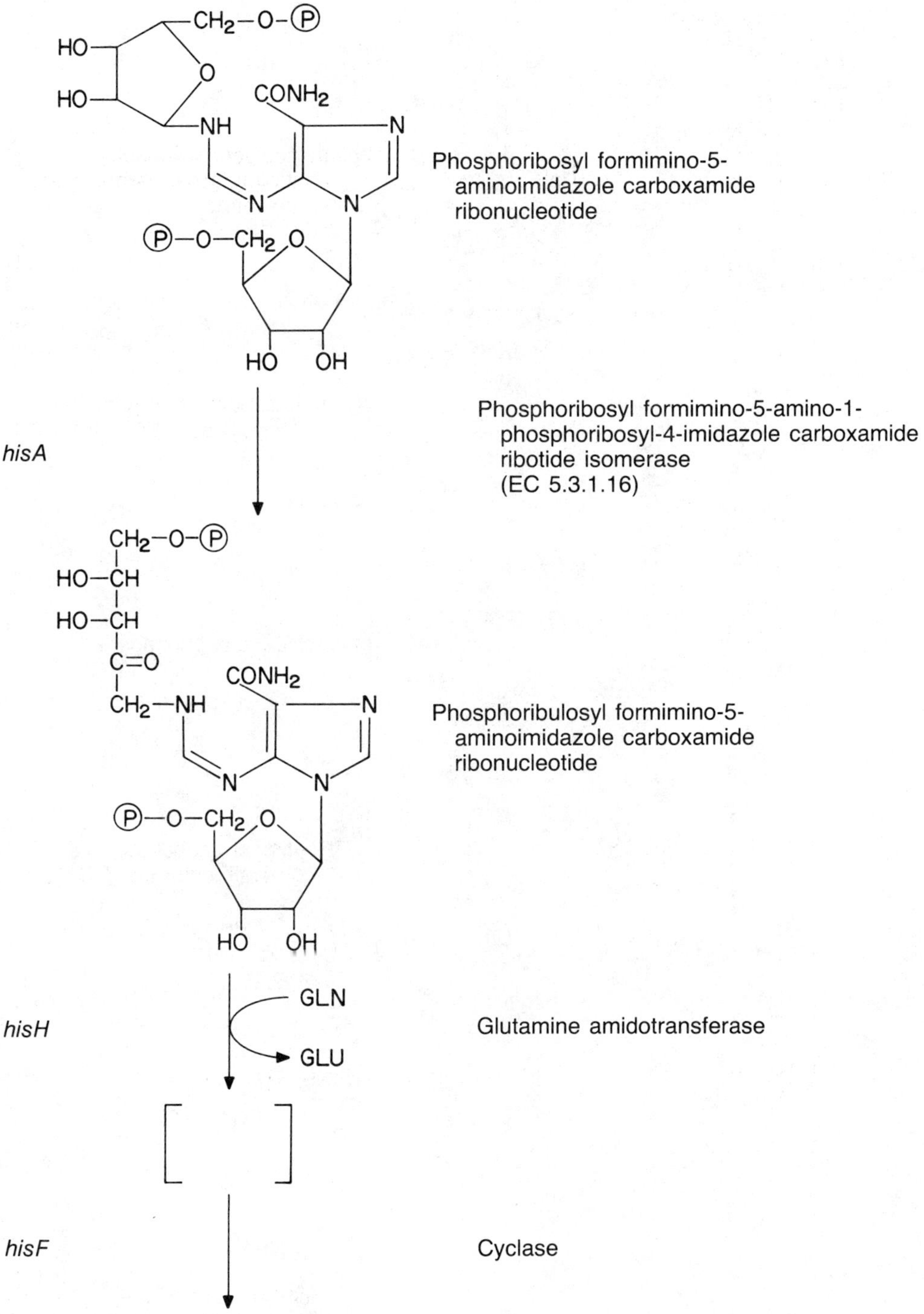
Phosphoribosyl formimino-5-aminoimidazole carboxamide ribonucleotide
hisA
Phosphoribosyl formimino-5-amino-1-phosphoribosyl-4-imidazole carboxamide ribotide isomerase (EC 5.3.1.16)
Phosphoribulosyl formimino-5-aminoimidazole carboxamide ribonucleotide
hisH
GLN
GLU
Glutamine amidotransferase
hisF
Cyclase

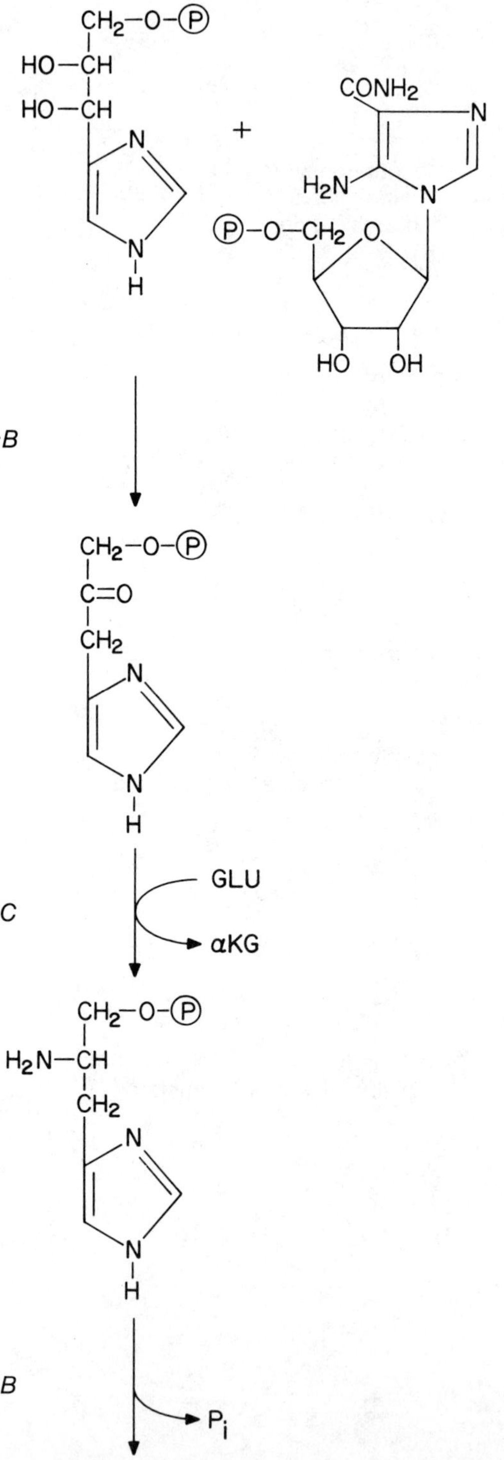

Imidazoleglycerol phosphate + 5-aminoimidazole carboxamide ribonucleotide

hisB Imidazoleglycerol phosphate dehydratase (EC 4.2.1.19)

Imidazoleacetol phosphate

hisC Histidinol phosphate aminotransferase (EC 2.6.1.9)

Histidinol phosphate

hisB Histidinol phosphate phosphatase (EC 3.1.3.15)

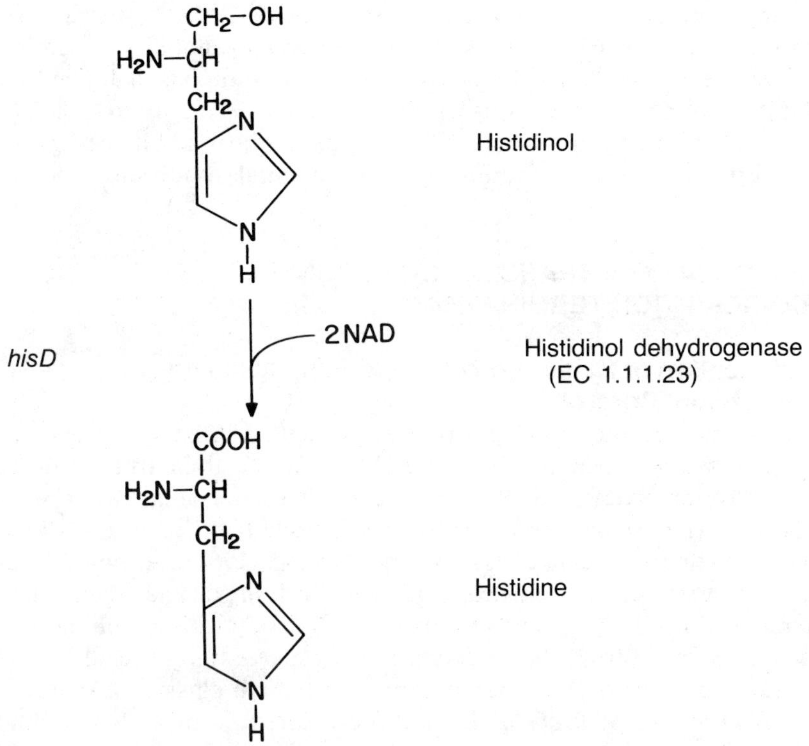

FIGURE 21.1 Pathway of histidine biosynthesis. The brackets indicate an intermediate of unknown structure (Martin et al., 1971). The order of genes in the *his* operon is *hisOGDCBHAFIE*. Transcription is initiated within the *hisO* regulatory region and proceeds to the right.

site in response to the availability of histidyl-tRNA^{His}, a transcriptional control mechanism responsive to the general availability of amino acids involving the general signal molecule guanosine tetraphosphate, and feedback inhibition of the first enzyme of the pathway. These control mechanisms serve to maintain appropriate histidine levels during balanced growth and to adjust these levels to changing environmental conditions.

II. ATTENUATION: TRANSLATIONAL CONTROL OF TRANSCRIPTION TERMINATION

A. Implication of Histidyl-tRNA^{His} and Early Ideas on Translational Control

The hallmark in the study of *his* operon regulation has been evidence that the repression signal is histidyl-tRNA^{His} rather than free histidine. The concept of aminoacylated tRNA as a corepressor for amino acid biosynthetic operons (Ames and Garry, 1959) was included in the formulation of the repressor–operator model (Jacob and Monod, 1961). Although there was earlier suggestive, physiological evidence (Schlesinger and Magasanik, 1964), the most compelling evidence for the role of histidyl-tRNA^{His} in the regulatory mechanism came from the biochemical characterization of regulatory mutants resulting in elevated *his* operon expression. Five classes of mutants—designated *hisR, hisS, hisT, hisU,* and *hisW* (the fate of *hisV* is unknown)—were isolated with the mutations located at distinct chromosomal sites separate from the *his* operon. Each of these classes alters either the synthesis, aminoacylation, modification, or maturation of tRNA^{His} (Brenner and Ames, 1971; Lewis and Ames, 1972; Singer et al., 1972; DeLorenzo et al., 1972; Brenchley and Ingraham, 1973; Bossi et al., 1978). The *hisR* gene is thought to encode the precursor of mature tRNA^{His}; *hisR* mutations reduce specifically the amount of tRNA^{His} (and therefore histidyl-tRNA^{His}) without altering the primary structure of mature tRNA^{His}. The *hisS* gene encodes histidyl-tRNA synthetase; *hisS* mutations likewise reduce the amount of histidyl-tRNA^{His}. The *hisU* and *hisW* genes apparently encode tRNA maturation (processing) enzymes; *hisU* and *hisW* mutations reduce the amount of functional tRNA^{His} (and therefore histidyl-tRNA^{His}) and alter the levels of other tRNAs. The classic study of Lewis and Ames (1972) demonstrated a quantitative, inverse correlation between the intracellular amount of histidyl-tRNA^{His} and the level of *his* operon expression in *hisR, hisS, hisU,* and *hisW* mutants as well as in a physiologically derepressed strain. This analysis confirmed the idea that histidyl-tRNA^{His} is the repression signal. A key exception to the correlation was the effect of *hisT* mutations. The *hisT* gene encodes pseudouridylate synthase I, an enzyme that modifies uridine residues to pseudouridine in the anticodon regions of about half the tRNAs in the cell, including tRNA^{His}. Although constitutively derepressed, *hisT* mutants contain wild-type levels of histidyl-tRNA^{His}. The role of the *hisT*-specified pseudouridine modifications will be considered in Section IIC(iii).

Following the ascendancy of the repressor–operator model of gene regulation, and in particular the biochemical evidence convincingly demonstrating the *lac* repressor–operator interaction, much attention was directed toward conforming observations made with other systems to this model. In the case of *his* regulation, so-called operator-constitutive mutations were isolated and located adjacent to the first structural gene of the operon (Roth et al., 1966a; Ely, 1974). Possibilities for the hypothetical repressor with which histidyl-tRNA^{His} would interact included histidyl-tRNA synthetase (DeLorenzo and Ames, 1970) and ATP phosphoribosyltransferase (Goldberger, 1974). We now know that the regulatory site mutations alter the *his* attenuator; attenuator-constitutive mutations are indistinguishable from operator-constitutive mutations in a diploid analysis. The term "operator" should now be reserved to define a repressor binding site, when one has been rigorously identified, rather than as a general term to describe a site identified only by mutations causing elevated, constitutive expression of a gene.

Despite the impact of the repressor–operator model, several of the histidine workers considered gene regulatory mechanisms involving histidyl-tRNA^{His} in its normal role in protein synthesis. For example (Roth et al., 1966b),

> One can also think of a number of models of regulation in which histidine tRNA is involved directly in a control at the translation level. Thus the operator region could be in RNA. It could be a control region near (or at) the beginning of the large message corresponding to the histidine operon. . . . One type of model that seems attractive is to have the initiation of reading controlled by having a histidine codon in some critical spot of the RNA operator so that charged tRNA(his) causes a block while the uncharged tRNA(his) allows initiation to occur.

These somewhat vague ideas were refined and a more detailed model devised by Robert Martin and Bruce Ames. The model, although never published, was presented at an international meeting (Martin et al., 1967) and was known to workers in the area of amino acid regulation (Umbarger, 1980; Yanofsky, 1981).

Expression of the *his* structural genes was proposed to be controlled by ribosomes translating a His-codon-containing RNA operator sequence overlapping with the start point of translation of the first structural gene (Fig. 21.2). With unlimiting histidyl-tRNA^{His} present, a ribosome initiating at the AUG (1′) of the RNA operator would proceed to the UAA (4′) termination codon and block initiation by ribosomes at the AUG (1) start of the structural genes. Repression would result, and the peptide FMet-His-His would be released into the cytoplasm. With limiting histidyl-tRNA^{His}

RNA Operator

```
              1'    2'    3'    4'
mRNA        A U G C A U C A U U A A U G X Y Z . . . . . . etc
Peptide   F M e t-H i s-H i s       1    2    3    4
```

First structural gene

FIGURE 21.2 The hypothetical RNA operator model.

present, the ribosome would stall at the His (2′, 3′) codons of the RNA operator, allowing ribosomes access to the initiation codon for the structural genes. This situation would result in derepression.

In retrospect, these ideas on translational control were remarkable for their prescience. The model predicted a mechanism of regulation by histidyl-tRNA^{His} involving translation of a histidine-rich leader peptide. It was also pointed out that, by varying the sequence of the RNA operator, the mechanism could be made to work for any amino acid biosynthetic gene or operon including the *ilv* operon, which was known to be multivalently repressed by the three branched-chain amino acids. This was an especially attractive simplification for *ilv* gene expression, since a repressor–operator mechanism for this operon would require either multiple repressors, each interacting with a different amino acid and a complex operator site, or a single repressor with distinct sites for binding the different amino acids. In addition, the model presaged the existence of overlapping translation stop–start codons (Sanger et al., 1977). Although we now know that ribosomes physically protect more extended stretches of mRNA, the overlapping codons were at the time thought necessary for the ribosome competition mechanism to work. It was pointed out that any of the three stop codons could be utilized: UAAUG, UGAUG, or UAGUG. Finally, the model was potentially flawed. Overlapping stop–start codons, rather than causing competition, may actually signal ribosomes to reinitiate translation without dissociating at the end of a peptide-coding sequence (Oppenheim and Yanofsky, 1980). On the other hand, free ribosomal subunits apparently can initiate translation de novo at internal cistrons in polycistronic mRNAs, so it is not entirely obvious how regulating translation initiation of the first gene would necessarily control expression of subsequent genes in an operon. Similar translational control mechanisms, however, could work for prokaryotic, monocistronic mRNAs and operons with specialized intercistronic sequences or, conceivably, for eukaryotic transcription units where internal translation starts apparently do not exist. A remarkably analogous mechanism of translational control has recently been proposed for the *Staphylococcus aureus* (and *Bacillus subtilis*) erythromycin-resistance gene (Horinouchi and Weisblum, 1980).

In 1974, Kasai reported that the *hisO1242* deletion mutation removes a transcriptional barrier, or attenuator site, located within the wild-type *his* operon control region. In vivo the *hisO1242* mutation elicits maximum, constitutive *his* operon expression that is totally independent of histidyl-$tRNA^{His}$ control. Kasai's results focused attention on a transcription termination mechanism of *his* operon regulation, as opposed to a transcription initiation repressor–operator mechanism. The term "attenuator" has been generally adopted as the designation for a regulated, bacterial transcription termination site.

Following the discovery that attenuation is coupled to translation of the *his* control region (Artz and Broach, 1975; see also Section IIC(ii)), the idea of regulation by histidyl-$tRNA^{His}$ through translation of a histidine-rich leader peptide received renewed interest in the laboratory of Bruce Ames. When we analyzed the first complete sequence of the leader RNA of the *E. coli trp* operon (Bertrand et al., 1975), we found a potential 14-amino-acid peptide-coding region containing two adjacent Trp codons. Although a revision of the *trp* leader RNA sequence made the assignment of the peptide sequence ambiguous (Squires et al., 1976), it became obvious that knowledge of the sequence of the *his* operon control region was essential. Appropriate DNA templates (Voll, 1972; Smith and Tong, 1974) for locating the predicted His-codon-containing leader region were forwarded to Wayne Barnes, who determined the *S. typhimurium* sequence (Barnes, 1978, 1981). The *E. coli* sequence was independently reported at the same time (DiNocera et al., 1978). Each sequence contains a potential 16-amino-acid peptide-coding region with seven contiguous His codons.

B. The Regulatory Model

The DNA sequence of the sense strand of the region preceding the first structural gene (*hisG*) of the *S. typhimurium his* operon is shown in Fig. 21.3. Important features noted include the start point of transcription, the histidine-rich leader peptide, the attenuator sequence, and the attenuator-deletion mutation *hisO1242*. Also noted are differences between the *S. typhimurium* and *E. coli* sequences. The two sequences are highly conserved in the region around the presumed Pribnow box and from the beginning of the leader peptide gene through the attenuator.

The regulatory model for *his* attenuation (Johnston et al., 1980) is patterned after the model proposed for the *trp* operon (Lee and Yanofsky, 1977). The regulatory mechanism is based upon formation of alternative RNA secondary structures that either include or exclude formation of an RNA attenuator stem structure (Fig. 21.4). RNA secondary structures are

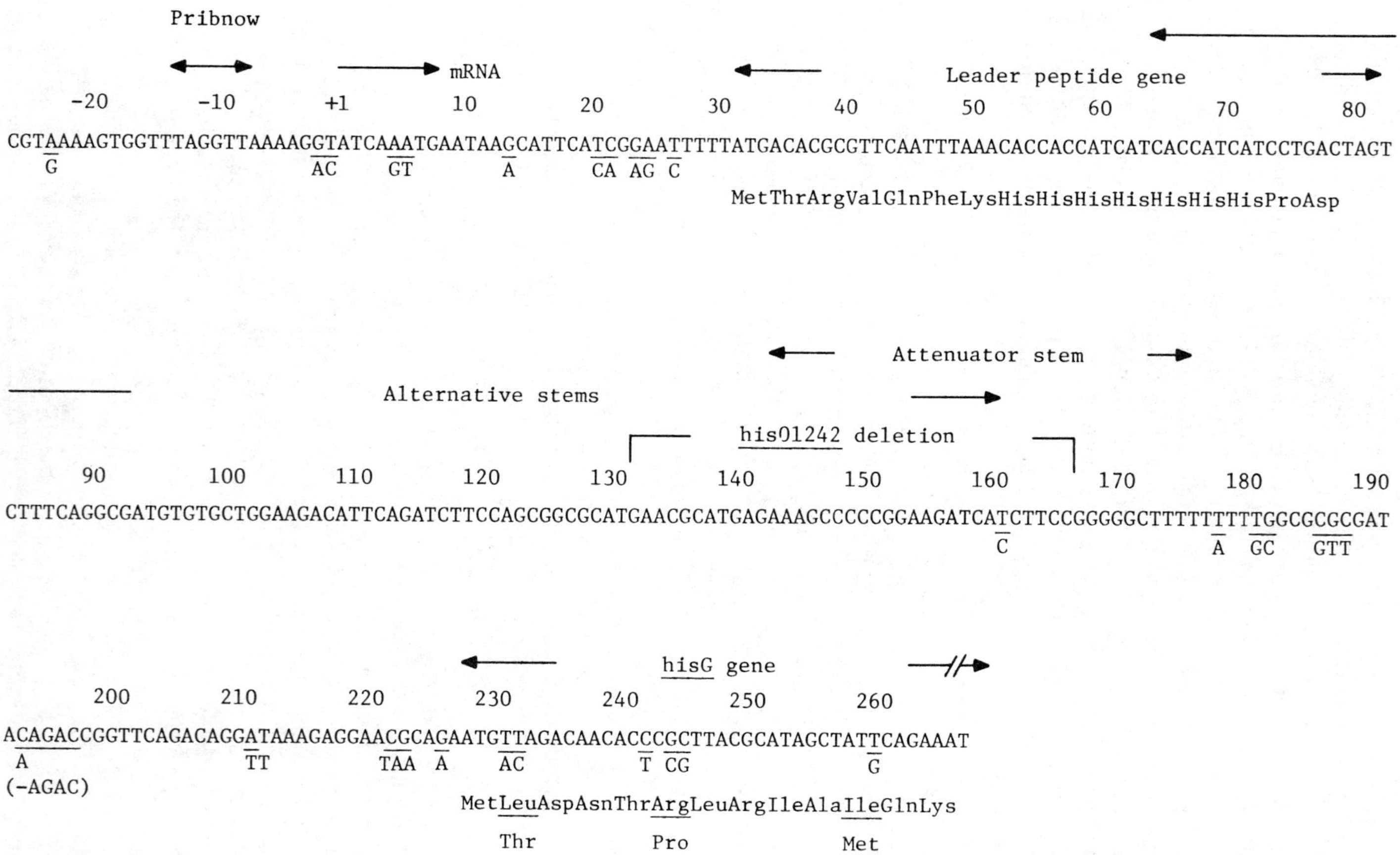

FIGURE 21.3 DNA sequence of the *his* leader region. Underlined bases and amino acids indicate differences in the *S. typhimurium* sequence shown from the *E. coli* sequence. (Adapted from Barnes, 1978; DiNocera et al., 1978; Johnston and Roth, 1981a; Frunzio et al., 1981; Freedman and Schimmel, 1981; Verde et al., 1981).

assumed to form in the order in which they are synthesized by RNA polymerase. Position of the ribosome translating the leader peptide sequence determines which set of alternative structures forms. The leading edge of the ribosome is assumed to prevent or disrupt secondary structure formation about 16 bases ahead of the codon occupying the amino-acyl (A) site.

If the concentration of histidyl-tRNA^{His} in the cell is high, because of an abundance of histidine, then the seven His codons are translated rapidly, and the ribosome terminates translation at the end of the peptide. In this situation (Fig. 20.4a), a set of RNA secondary structures including the attenuator stem structure forms, RNA polymerase terminates transcription, and the operon is repressed. However, if the concentration of histidyl-tRNA^{His} is low, then the ribosome lags behind as it translates the His codons. Impeded progress of the ribosome allows formation of a set of RNA secondary structures that excludes the attenuator stem (Fig. 21.4b), transcription termination fails to occur, and the operon is derepressed. If the ribosome stalls before the His codons (or the leader peptide sequence is not translated at all), formation of RNA secondary structures is dictated by transcription alone, the attenuator stem structure again is included (Fig. 20.4c), and the operon is repressed. This last feature allows derepression to be responsive primarily to histidine starvation as opposed to starvation for amino acids preceding the run of histidines in the peptide sequence. Similarly, if the ribosome stalls at codons following the His codons, repression presumably would ensue, owing to formation of the attenuator stem structure (Fig. 20.4a).

C. Support for the Model

On the basis of available information the translational-control-of-transcription-attenuation model appears to be substantially correct. The all-or-nothing nature of the mechanism described in Fig. 21.4 is, however, an oversimplification. The mechanism is necessarily a kinetic one, depending on the relative rates of transcription and translation of the leader RNA. Slightly different RNA secondary structures might form, and the extent of ribosome protection cannot be precisely defined. Evidence described below supports the proposed model and provides further insight into its function.

i. Leader RNA. The startpoint of transcription of the *his* message (Fig. 21.3) has been localized both in vitro and in vivo, and a 175 nucleotide leader RNA transcript terminated with a poly-U sequence at its 3′ end has been identified (Frunzio et al., 1981; Freedman and Schimmel, 1981; Verde et al., 1981). The *hisO1242* mutation deletes part of the G-C rich region of dyad symmetry in the DNA immediately preceding the point of transcription termination (Johnston et al., 1980) (Fig. 21.3). Although the mechanism of transcription termination is not completely understood, the G-C-rich RNA

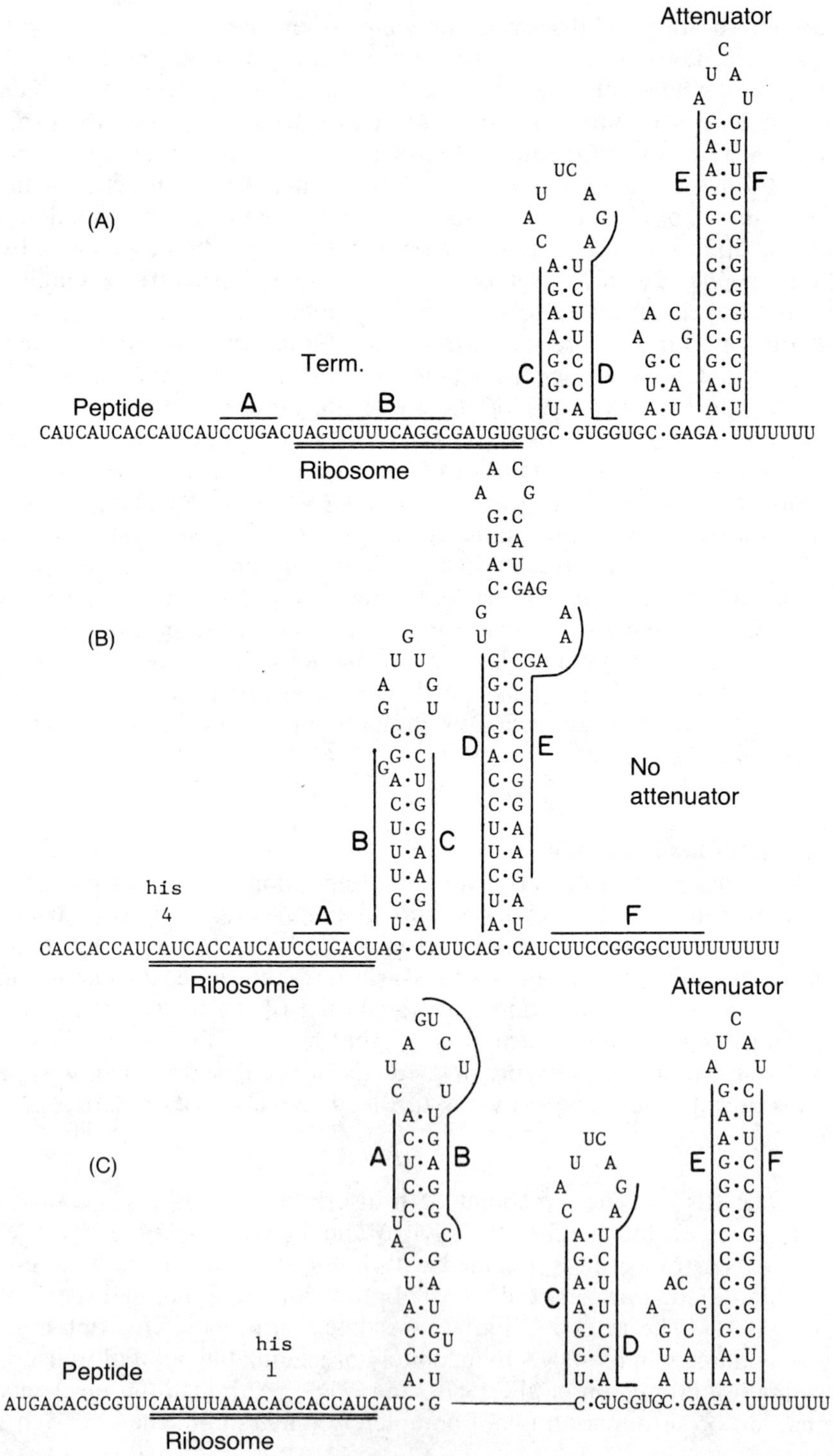
(A)
Attenuator
Term.
Peptide
A
B
C
D
E
F
CAUCAUCACCAUCAUCCUGACUAGUCUUUCAGGCGAUGUGUGC·GUGGUGC·GAGA·UUUUUUU
Ribosome
(B)
his
4
A
B
C
D
E
F
No attenuator
CACCACCAUCAUCACCAUCAUCCUGACUAG·CAUUCAG·CAUCUUCCGGGCUUUUUUUU
Ribosome
(C)
Attenuator
Peptide
his
1
A
B
C
D
E
F
AUGACACGCGUUCAAUUUAAACACCACCAUCAUC·G
C·GUGGUGC·GAGA·UUUUUUU
Ribosome

stem structure that can form and the adjacent poly-U sequence appear to be essential for strong (rho-independent) transcription termination. RNA stem structures enhance pausing by RNA polymerase, and the poly-U sequence may facilitate dissociation of the RNA from the RNA polymerase·DNA·RNA ternary termination complex (Platt, 1981).

ii. Leader Peptide. In a DNA-directed, tRNA-dependent, in vitro protein-synthesizing system, translation is required to obtain transcription of the *his* structural genes when the DNA template contains the wild-type (*hisO*$^+$) regulatory region (Artz and Broach, 1975). When the DNA template contains the *hisO1242* attenuator–deletion mutation (Fig. 21.3), the requirement for translation is obviated. These results provided initial evidence for translational-control-of-transcription-attenuation and predicted the situation shown in Fig. 21.4c before the sequence of the *hisO* region was determined. Knowledge of the DNA sequence of the *his* regulatory region (Fig. 21.3) and, in particular, the potential leader peptide gene, has allowed these studies to be extended (S. Artz, unpublished results). A summary of the results is shown in Table 21.1 and Fig. 21.5).

Table 21.1 illustrates the effects of adding different combinations of amino acids to the in vitro synthesizing system. Either the ten different amino acids predicted to be coded by the leader peptide gene (Condition A), the ten missing from the sequence (Condition B), or no amino acids (Condition C) are added. With the *hisO*$^+$ DNA template, the presence of leader peptide amino acids stimulates transcription of *his* structural gene mRNA about tenfold above that observed when these amino acids are absent (line 1). In vitro derepression caused by pseudouridine-deficient tRNA (isolated from a *hisT* mutant) requires the presence of the leader peptide amino acids (line 2). The twofold to threefold derepression observed is similar to that reported previously during normal coupled in vitro protein synthesis (Artz and Broach, 1975). When the DNA template contains the *hisO1242* attenuator deletion mutation, *his* expression in vitro is independent of leader peptide translation and pseudouridine-deficient tRNA (lines 3 and 4). This experimental control is crucial to the interpretation that in vitro effects on *hisO*$^+$ expression of leader peptide amino acids, and pseudouridine-deficient tRNA are exerted specifically at the attenuator.

◀ **FIGURE 21.4 Alternative secondary structures of the *his* leader RNA. The letters (A to F) denote the stretches of sequence involved in the alternative structures. The length of RNA assumed to be excluded from secondary structure formation by the translating ribosome is that above the double lines. The ribosome is proposed to exclude participation in secondary structure formation of the 16 ribonucleotides 3′ to the codon occupying the aminoacyl site. Ribosome aminoacyl site positioned at (A) termination codon (UAG); (B) fourth His codon (CAU); (C) Gln codon (CAA) (from Johnston and Roth, 1981).**

TABLE 21.1 Effects of Leader Peptide Amino Acids and Pseudouridine-Deficient (*hisT*) tRNA on *his* mRNA Synthesis in vitro.

		Relative his *mRNA*		
			Condition	
Template	*tRNA*	*A*	*B*	*C*
hisO$^+$	*hisT*$^+$	≡ 1.0	0.10	0.07
hisO	*hisT*	2.4	0.13	0.11
hisO1242	*hisT*$^+$	4.2	4.3	3.8
hisO1242	*hisT*	4.6	3.9	3.6

Condition refers to which amino acids are present in the cell-free protein synthesizing system during transcription: A, ten amino acids contained in the leader peptide sequence; B, ten amino acids not contained in leader peptide sequence; C, no amino acids. After transcription is completed, *his* structural gene mRNA is quantitated by its ability to be translated into active histidinol dehydrogenase in the presence of all 20 amino acids (Artz and Broach, 1975).

These results illustrate two important features central to the regulatory mechanism. First, continuation of transcription beyond the attenuator is stimulated by the presence of those amino acids predicted from the DNA sequence to be contained in the leader peptide. Second, the data indicate that derepression caused by pseudouridine-deficient tRNA depends on translation of the leader peptide gene. While it is apparent how levels of histidyl-tRNA^{His} can influence the mechanism, it has not been so obvious how pseudouridine-deficient tRNA would cause derepression, since tRNA^{His} is charged normally in *hisT* mutants (Lewis and Ames, 1972). Direct evidence that *hisT* tRNA results in a reduced rate of translation will be described in the following section.

The position of the ribosome translating the leader peptide gene influences attenuation (Fig. 21.5). In this experiment, amino acids are added to the in vitro protein-synthesizing system in groups based on their order of occurrence in the leader peptide sequence—that is: Met, or Met·Thr, or Met·Thr·Arg, or Met·Thr·Arg·Val, etc. The results indicate that stimulation of transcription past the attenuator begins to occur when the ribosomal aminoacyl (A) site is positioned at the sixth codon, or that specifying Phe, in the peptide sequence. These findings provide functional evidence that extent of translation of the leader peptide gene regulates attenuation. Analogous genetic evidence has also been obtained (Johnston and Roth, 1981b). Mutations generating chain-terminating, nonsense codons prior to the His codons in the leader peptide sequence result in increased attenuation in proportion to their distance from the AB stem structure in Fig. 21.4. Mutations apparently resulting in defective translation initiation of the leader peptide also cause increased attenuation. The effects of these mutations correspond to the excessive attenuation observed when leader peptide amino acids are withheld from the in vitro protein-synthesizing system (Table 21.1, Fig. 21.5) (Artz and Broach, 1975). A rationale for this excessive attenuation will be discussed in Section IIC(iv).

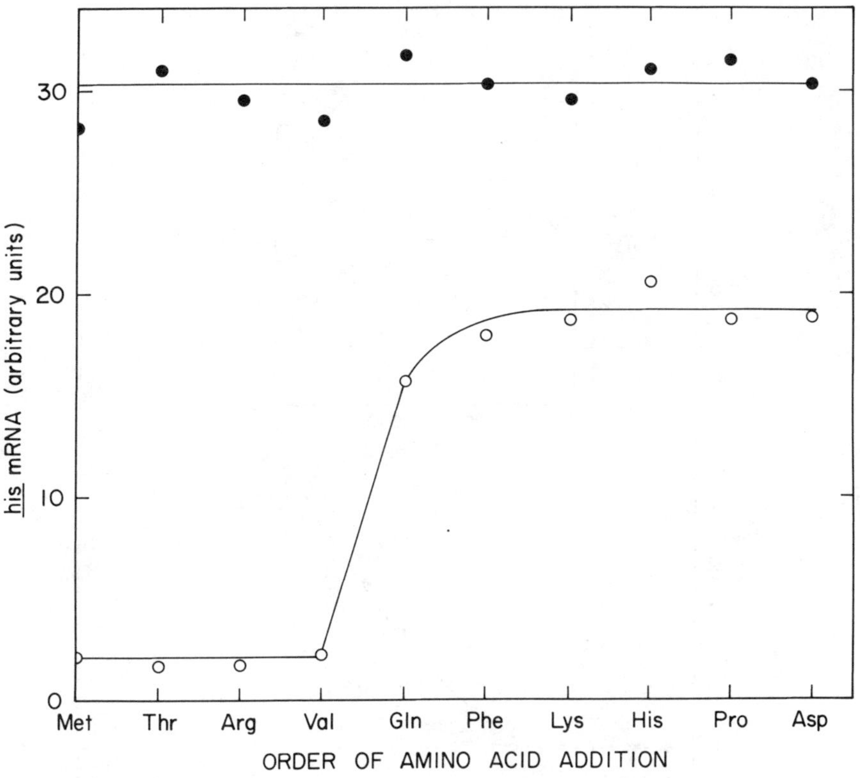

FIGURE 21.5 Effects of position of the ribosome translating the leader peptide gene on *his* mRNA synthesis in vitro. Amino acids were added initially to the cell-free protein-synthesizing system as described in the text. The *his* structural gene mRNA was quantitated as indicated in Table 21.1. Reaction mixtures contained pseudouridine-deficient (*hisT*) tRNA. Open symbols: *hisO*$^+$ DNA template. Closed symbols: *hisO1242* DNA template.

iii. Effects of the *hisT* Mutation on Translation Rate and Translational Fidelity. The *hisT* mutation results in loss of pseudouridine base modifications in the anticodon regions of about half of the tRNA species in *S. typhimurium,* including tRNA^{His} (Turnbough et al., 1979). The mutation leads to derepressed *his* operon expression as well as other pleiotropic effects on cellular regulation. Effects of pseudouridine-deficient tRNA on translation rate in vivo have been investigated by using measurements in the lag of β-galactosidase appearance following *lac* operon induction in isogenic *hisT*$^+$ and *hisT* strains (Palmer et al., 1983). Results of a typical experiment are shown in Fig. 21.6. Early regions of the induction curves are shown, and the data are replotted by the method of Schleif et al. (1973). The replot converts the

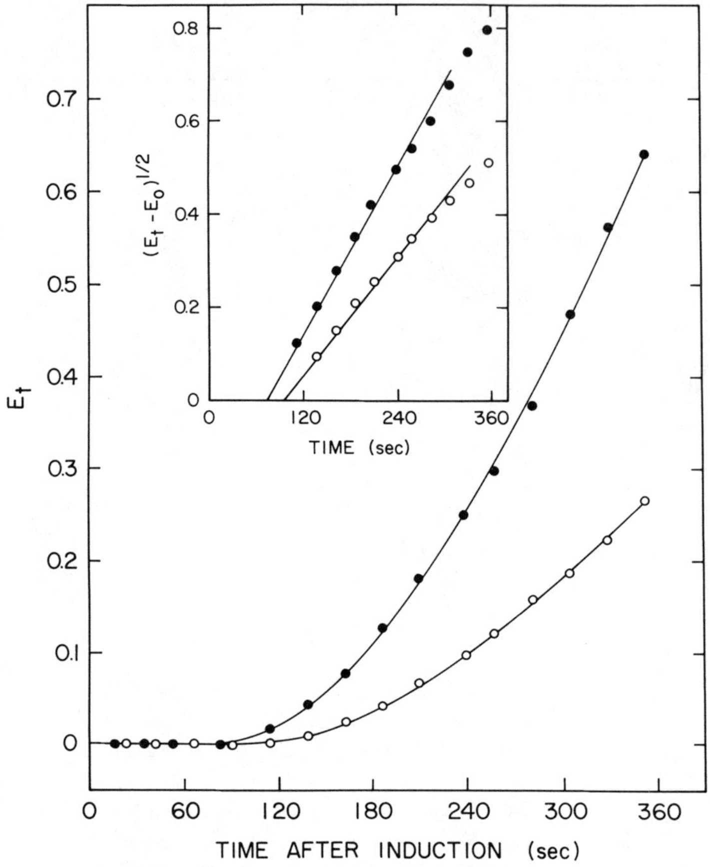

FIGURE 21.6 β-Galactosidase induction kinetics in *hisT*$^+$ (closed symbols) and *hisT1504* (open symbols). After induction with 1.0 mM isopropylthiogalactoside, culture samples were added to chloramphenicol at the indicated times. E_t and E_0 indicate β-galactosidase specific activity at different times after induction and at zero time, respectively (Palmer et al., 1983).

curves to straight lines, which extrapolate to the time of completion of the first β-galactosidase subunit chain. It is apparent that this time is longer for the *hisT* mutant than for the *hisT*$^+$ strain. By using 1021 amino acids as the chain length of β-galactosidase, translation rates of about 13 and 10 amino acids per second are calculated for the *hisT*$^+$ and *hisT* strains, respectively. Experiments with transcription inhibitors confirm that the reduced rate of protein synthesis occurs primarily at the translational level. Analogous in vitro experiments with the DNA-directed, tRNA-dependent, protein-synthesizing system using a *lac* DNA template and purified, bulk tRNA isolated from *hisT* or *hisT*$^+$ strains also confirm a reduced rate of translation similar to that observed in vivo. Since the only variable in such experiments is the type of tRNA added, it can be concluded that lack of the *hisT*-specified pseudouridine modifications leads directly to slowed translation rate.

Thus pseudouridine-deficient tRNA in *hisT* mutants leads to *his* operon derepression apparently by slowing translation of the leader peptide gene. Since pseudouridine-deficient $tRNA^{His}$ is charged normally by histidyl-tRNA synthetase, it seems likely that slowed translation results from defective interaction of the undermodified tRNA with the ribosome or defective codon–anticodon interaction.

Slowed translation in *hisT* mutants is accompanied by an increase in translational fidelity (P. Blum and S. Artz, in preparation). The *hisT* mutation reduces aminoglycoside antibiotic suppression of a number of *his* and *lac* nonsense and missense mutations. An example is shown in Fig. 21.7. The *hisT* mutation also results in impaired function of a glutamine-inserting amber suppressor tRNA (Johnston et al., 1980). These findings are consistent with kinetic proofreading models of the mechanism of translational fidelity (Hopfield, 1974; Ninio, 1974) and suggest that at least one function of the pseudouridine modification is to aid in determining an appropriate compromise between maximum efficiency (rate) and accuracy of translation.

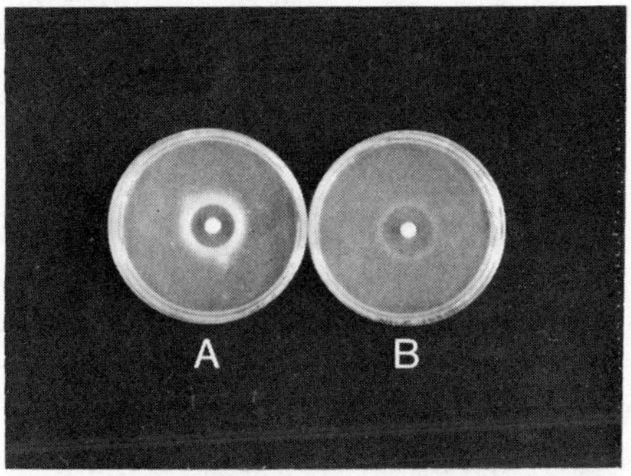

FIGURE 21.7 Phenotypic suppression of a *lacZ* ochre, nonsense mutation by kanamycin in *S. typhimurium* $hisT^+$ and *hisT* strains. Minimal salts medium plates, supplemented with lactose, are overlaid with (A) $lacZ_{oc}$ $hisT^+$ and (B) $lacZ_{oc}$ *hisT1504* and incubated for two days. The filter paper disc in the center of each plate contains kanamycin.

iv. Alternative RNA Secondary Structures and Kinetic Considerations. The importance of alternative RNA secondary structures in the mechanism is supported by genetic evidence (Johnston and Roth, 1981b). Mutations that reduce stability of the BC and/or DE RNA stem structures (Fig. 21.4) result in excessive attenuation. Presumably, the BC and DE stems form less frequently as a consequence of such mutations making formation of the alternative attenuator (EF) stem more probable. The finding that the *hisO1242* attenuator deletion mutation suppresses one of these mutations is consistent with the idea that BC and DE stem formation plays a role in preventing attenuator stem formation.

In the in vitro experiments described in Section IIC(ii) (Table 21.1) the presence or absence of leader peptide amino acids corresponds to conditions that specify the leader RNA structures shown in Figs. 21.4a and 21.4c, respectively. Although both structures contain the attenuator stem, excessive attenuation occurs when the leader peptide gene fails to be translated at all. This result is consistent with the relatively high basal level of *his* operon expression observed under repressing conditions in vivo (Brenner and Ames, 1971). Repression (Fig. 21.4a) therefore corresponds to an intermediate level of attenuation suggesting that even when the leader peptide is translated in an unimpeded way, RNA polymerase often advances sufficiently ahead to get past the attenuator while the ribosome is still translating the His codons. That is, the repressed level of expression is an average one apparently depending upon random variations in the kinetics of transcription and translation of the *his* leader region.

For derepression to occur by the alternative RNA secondary structures mechanism requires a remarkably precise coordination of movement between the RNA polymerase transcribing the leader region and the ribosome translating the leader peptide gene. Derepression can occur only if the ribosome is positioned to block *potential* formation of the attenuator stem before RNA polymerase has transcribed the attenuator sequence. A plausible order of events required for derepression is as follows: RNA polymerase initiates transcription, synthesizes the translation initiation sequence, and proceeds ahead while a ribosome initiates translation of the leader peptide gene. Although translation initiation requires some finite (unknown) time, the ribosome must initiate sufficiently rapidly and follow closely enough behind RNA polymerase to disrupt AB stem formation (Fig. 21.4c), allowing BC (Fig. 21.4b) to form, before the polymerase has transcribed the D sequence. If CD is allowed to form, EF is then determined, and repression would ensue (Fig. 21.4c). Subsequent progress of the ribosome through the His codons must be delayed by limiting histidyl-$tRNA^{His}$ a sufficient amount of time to enable RNA polymerase to transcribe the E sequence (forming DE) before the ribosome can disrupt the BC stem (Fig. 21.4b). If the ribosome progresses too far, disrupting BC before E is transcribed, CD will form; EF is then determined, and repression would again ensue (Fig. 21.4a).

As an initial attempt to assess the validity of this scenario, we have measured in vivo the time delay in translating His codons associated with derepression of the *his* operon (D. Palmer and S. Artz, in preparation). Cultures of F′*lac*-containing, His^- cells are subjected to varying conditions of histidine starvation, and the level of *his* operon expression, as well as the increased delay in appearance of β-galactosidase, is determined. The β-galactosidase induction data are treated as shown in Fig. 21.5, and a summary of results is shown in Fig. 21.8. The data indicate that maximum

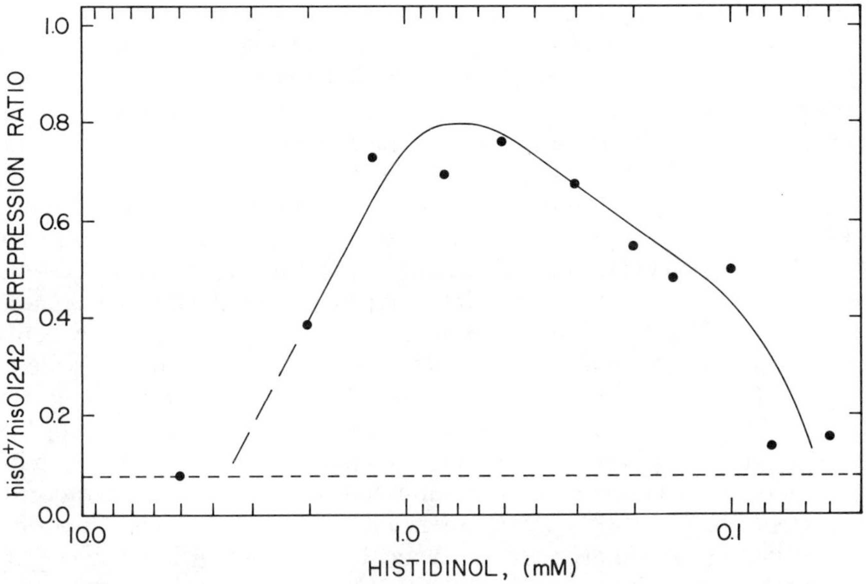

FIGURE 21.8 Correlation of *his* operon derepression with β-galactosidase induction lags during histidine starvation in vivo. A *his* auxotroph is grown at 37° in a minimal salts medium containing different concentrations of histidinol to elicit histidine starvation. Histidine starvation increases from left to right (corresponding to histidinol concentrations of 5.0 mM to 0.1 mM). Relative *his* operon derepression is determined from measurements of the differential rate of synthesis of histidinol phosphate phosphatase. β-Galactosidase induction lags are determined as described in Figure 21.6. The net change is determined by subtracting the induction lag measured at each histidinol concentration from the induction lag measured in an unstarved, control culture containing excess histidine. The dashed line indicates the repressed level of *his* operon expression (≡ 1.0).

his operon derepression during physiological histidine starvation corresponds to an increased induction lag in β-galactosidase synthesis of about 30 seconds. Since the β-galactosidase subunit chain contains 34 His residues, we calculate that maximum derepression occurs when ribosomes are delayed for about 0.9 s per His codon. During derepression, RNA polymerase must transcribe a maximum distance of about 80 bases (from the beginning of the B sequence through the E sequence; see Fig. 21.4b), requiring, at an average rate of 45 bases per s, a maximum time of about 1.8 s. Derepression is proposed to be most probable when the ribosome reaches the fourth His codon (Fig. 21.4b) while RNA polymerase completes transcription of the E sequence. At a delay time of 0.9 s per His codon, the 3.6 s available provides adequate time for the alternative RNA secondary structure mechanism to work as proposed.

Although the measured delay is sufficient for the mechanism to work, the 3.6 s per four His codons actually provides at least twice the time required for RNA polymerase to transcribe the maximum distance of 80 bases. Derepression could, in fact, be most probable when the ribosome reaches the second His codon. Alternatively, RNA polymerase might "pause" in its progress toward the attenuator. "Pausing" by RNA polymerase at a specific site before the *trp* attenuator sequence has been observed in vitro (Platt, 1981; Winkler and Yanofsky, 1981). The existence of such "pause" sites could generally be required to allow the lagging ribosome time to catch up and be in a position to exert its regulatory effect. The longer-than-expected translation delay interval we observed in vivo could thus be necessary to account for one or more transcriptional pauses plus the 80-base distance traversed by RNA polymerase. A requirement for pausing by RNA polymerase during transcription of the *his* leader RNA is further suggested by evidence indicating that the ribosome-binding site of the leader peptide gene is a relatively weak one (as compared to ribosome-binding sites of the *his* operon structural genes). Expression of the *his* operon in the coupled in vitro transcription–translation system is saturated at a lower concentration of *hisO*$^{+}$ DNA than *hisO1242* DNA (Artz and Broach, 1975). This result can now most easily be interpreted to mean that translation initiation of the leader peptide gene becomes the limiting factor at high concentrations of *hisO*$^{+}$ DNA in vitro. Expression of the *his* operon increases and then declines as histidine starvation becomes more severe (corresponding to increasing β-galactosidase induction lags) (Fig. 21.8). The decline results largely from increased attenuation because the *hisO1242* attenuator deletion mutation mitigates the effect. The *ratio* of *his* operon expression in *hisO*$^{+}$ and *hisO1242* strains is dependent upon the degree of histidine starvation (corresponding to decreasing histidinol in the medium) (Fig. 21.9). Relative to that in the *hisO1242* strain, *his* expression in the *hisO*$^{+}$ strain first increases, reaching about 80% of the fully derepressed *hisO1242* level, and then decreases approaching the repressed ratio. The indication is that stalling of a ribosome at the earliest His codons of the leader peptide gene results in less than maximal derepression because of a tendency to form the set of RNA secondary structures including the attenuator stem (Fig. 21.4c). The overall mechanism therefore appears to be both positive and negative with respect to position of the ribosome translating the run of His codons. As a physiological rationale, we propose that following the onset of a severe histidine starvation it may be more economical for the cells to obtain histidine initially through turnover of proteins than by derepressing the *his* operon.

III. REGULATION BY GUANOSINE 5′-DIPHOSPHATE 3′-DIPHOSPHATE (ppGpp)

In *rel*$^{+}$ bacteria, ppGpp appears to be a general signal molecule that redirects cellular metabolism in response to amino acid deprivation (Cashel and Gallant, 1974; Stephens et al., 1975; Cozzone, 1980; Atherly, 1981). Expression

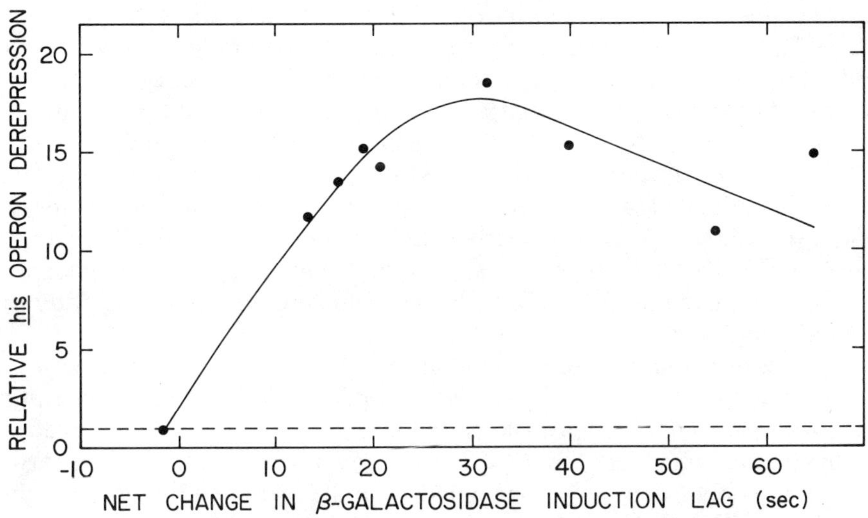

FIGURE 21.9 Derepression of the *his* operon in a *hisO*$^{+}$ strain relative to constitutive expression in a *hisO1242* strain as a function of increasing histidine starvation. A pair of isogenic *hisO*$^{+}$/*hisO1242 his–lac* fusion strains with β-galactosidase synthesis under control of the *his* regulatory region (P. Blum and S. Artz, in preparation) is grown at 37° in a minimal medium containing different concentrations of histidinol. Histidine starvation increases from left to right. The differential rate of β-galactosidase synthesis is determined in both strains at each histidinol concentration. The ratio *hisO*$^{+}$/*hisO1242* expression at each histidinol concentration is plotted. The horizontal dashed line indicates the repressed ratio of expression measured in cultures grown with excess histidine.

of the *S. typhimurium his* operon is strongly dependent on the presence of ppGpp in a DNA-dependent in vitro protein-synthesizing system (Stephens et al., 1975). With S-30 extracts of a *relA* mutant (defective in ppGpp synthesis) of *S. typhimurium,* addition of 0.2 mM ppGpp elicits about a tenfold stimulation of *his* operon expression. With S-30 extracts of *relA*$^{+}$ strains, the stimulation by ppGpp is only about twofold, consistent with the endogenous level of synthesis of ppGpp observed in these extracts. The stimulatory effect of ppGpp occurs at the level of transcription of *his* mRNA and before the action of rifamycin, a fact suggesting that ppGpp activates transcription initiation by RNA polymerase. The guanosine tetraphosphate is equally effective when the DNA template contains either the wild-type *his* regulatory region or the *hisO1242* attenuator deletion mutation, indicating that the attenuation mechanism is independent of ppGpp.

The effects on *his* operon expression of conditions that alter ppGpp levels in vivo have also been studied. Both the rate and extent of *his* operon derepression are diminished in a *relA* mutant, compared to a *relA*$^{+}$ strain, following histidine starvation in a medium containing the other 19 amino acids (Stephens et al., 1975). Expression of the *his* operon increases following limitation for amino acids other than histidine (Winkler et al., 1978). The increase correlates with ppGpp levels and is attenuator-independent. In a *spoT* mutant (defective in ppGpp degradation) the ppGpp pool is elevated about threefold during balanced growth in an amino acid–rich medium, attenuator-independent expression of the *his* operon is similarly increased (Winkler et al., 1979). Taken together, these results indicate that the positive effect of ppGpp on *his* operon expression is physiologically significant in mediating adjustment of histidine biosynthesis to environmental conditions of amino acid supply.

A mutation altering the sequence of the Pribnow box of the *E. coli lac* promoter diminishes the dependence on ppGpp for maximal transcription of the *lac* operon in vitro (Primakoff and Artz, 1979). Although the molecular mechanism by which ppGpp interacts to activate transcription from certain promoters is not known, the mechanism may be mediated by direct binding of ppGpp to RNA polymerase, or in conjunction with an accessory regulatory factor that binds to RNA polymerase or DNA.

To define the target sequence for ppGpp regulation in the *S. typhimurium his* promoter region, we have extended the sequence of the wild-type promoter and have isolated a collection of spontaneous mutants with decreased promoter expression (Fig. 21.10) (D. Holzschu, R. Shand, P. Blum, and S. W. Artz,

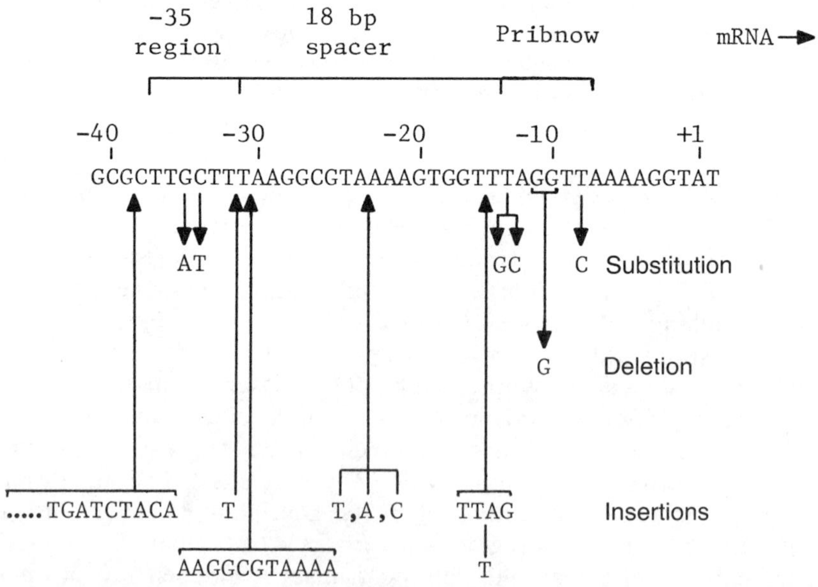

FIGURE 21.10 Mutations that decrease expression of the *S. typhimurium his* promoter.

in preparation). All of the base pair substitution mutations and a one-base pair deletion mutation alter the sequence in the region located either at around −10 or −35 with respect to the startpoint of transcription, consistent with the known importance of these two regions for RNA polymerase interaction. Seven different insertion mutations increase the space between the −10 and −35 regions, and one large insertion mutation (greater than 500 base pairs) is located just upstream from the −35 region. The mutations decrease transcription of the *his* promoter from about 5 fold to 400 fold. Which of the mutations may alter ppGpp regulation of the *his* promoter is currently under study.

IV. CONTROL OF ATP PHOSPHORIBOSYL TRANSFERASE ACTIVITY

The activity of the first enzyme of the pathway, ATP phosphoribosyl transferase, encoded in *hisG,* is inhibited by histidine (Martin, 1963). Mutants defective in feedback inhibition overproduce and excrete histidine, while wild-type cells do not, a finding indicating that the wild-type enzyme is inhibited under physiological conditions (15 μM histidine) (Brenner and Ames, 1971).

The enzyme has been purified and is a hexamer under catalytic conditions. Wild-type and feedback-resistant enzymes bind six molecules of histidine, with only the wild-type enzyme undergoing a conformational change. Enzyme inhibition by histidine is noncompetitive with respect to both substrates with a K_i of approximately 60 μM at 25°, pH 8.5, and 380 μM at 37°, pH 8.5 (Martin, 1963; Sterboul et al., 1977). High concentrations of histidine totally inhibit enzyme activity in a positively cooperative manner.

The greater than physiological concentrations of histidine required to inhibit ATP phosphoribosyl transferase activity in vitro suggest that other factors may be involved in modulating enzyme activity in vivo. Inhibition by histidine in vitro is enhanced by the presence of AMP, ADP, and ppGpp (Morton and Parsons, 1977a,b). Physiological concentrations of histidine lower the apparent AMP dissociation constant over 20-fold, a finding suggesting charge regulation of histidine biosynthesis. Guanosine tetraphosphate acts synergistically with histidine to inhibit enzyme activity but is not inhibitory in the absence of histidine. Since ppGpp is a positive effector of *his* operon expression, inhibition of the transferase by ppGpp at first seems contradictory but can be rationalized. If histidine is not the limiting amino acid, the positive effect of ppGpp on *his* operon expression is counterproductive. The increased inhibition of ATP phosphoribosyltransferase by histidine in the presence of ppGpp would counteract the effect on operon expression. On the other hand, if histidine is the limiting amino acid, then feedback inhibition would not occur, and maximal synthesis of histidine would ensue.

V. AUTOGENOUS REGULATION

ATP phosphoribosyltransferase has been implicated in autogenous regulation of *his* operon expression by both in vivo and in vitro experiments (Goldberger, 1974). Purified ATP phosphoribosyltransferase inhibits *his* mRNA synthesis in vitro (Blasi et al., 1973; DiNocera et al., 1975) and binds specifically to the attenuator sequence of the *his* control region (Meyers et al., 1975). These results seemingly point to a role for the enzyme as a transcription termination factor. On the other hand, mutations that delete the *hisG* gene have no apparent effect on regulation of the *his* operon in vivo (Scott et al., 1975; Goitein and Parsons, 1980). Although it may be possible to accommodate an accessory, regulatory role for ATP phosphoribosyltransferase (Artz and Broach, 1975), these discrepancies are at present very difficult to rationalize, and further evidence is required to conclude that the enzyme is involved in regulation at the genetic level.

VI. SUMMARY

This chapter has reviewed current information and ideas on the regulation of histidine biosynthesis in *S. typhimurium* and *E. coli*. Although the translational-control-of-transcription-attenuation model is a relatively recent development in bacterial gene regulation, the mechanism had its seeds in the early work on *his* operon regulation. Discovery of this mechanism provided the first alternative means of regulating gene expression different from mechanism involving DNA-binding proteins that modulate transcription initiation by RNA polymerase. Since the translational-control-of-transcription-attenuation mechanism requires only the basic components of the protein synthetic apparatus, it is tempting to speculate that this type of mechanism may have arisen early in the evolution of genetic regulatory mechanisms. It also seems plausible that similar mechanisms might regulate expression of genes involved in areas of metabolism other than amino acid biosynthesis. Guanosine tetraphosphate is a positive effector of *his* operon expression. Although the physiological implications of this regulation seem clear, understanding the mechanism by which ppGpp acts to stimulate transcription at the *his* promoter awaits the characterization of mutations altering the mechanism as well as further biochemical analysis. The flow of metabolites through the histidine biosynthetic pathway is controlled through inhibition of ATP phosphoribosyl transferase activity. The mechanism is apparently a complex one involving ppGpp and the energy-charge state of the cell in addition to feedback inhibition by histidine.

ACKNOWLEDGMENT

We thank Bruce Ames for providing information on the original Martin and Ames translational control model and for reading this manuscript. We also thank John

Ingraham for making helpful comments. Research from this laboratory was supported by USPHS Grant GM27307.

REFERENCES

Ames, B. N., and Garry, B. (1959) *Proc. Nat. Acad. Sci. U.S.A. 45,* 1453–1461.

Artz, S. W., and Broach, J. R. (1975) *Proc. Nat. Acad. Sci. U.S.A. 72,* 3453–3457.

Atherly, A. G. (1981) in *Microbiology* (Schlessinger, D., ed.) pp. 7–10, American Society for Microbiology, Washington, D.C.

Barnes, W. M. (1978) *Proc. Nat. Acad. Sci. U.S.A. 75,* 4281–4285.

Barnes, W. M. (1981) *J. Bacteriol. 147,* 124–134.

Bertrand, K., Korn, L., Lee, F., Platt, T., Squires, C. L., Squires, C., and Yanofsky, C. (1975) *Science 189,* 22–26.

Blasi, F., Bruni, C. B., Avitabile, A., Deeley, R. G., Goldberger, R. F., and Meyers, M. M. (1973) *Proc. Nat. Acad. Sci. U.S.A. 70,* 2692–2696.

Bossi, L., Ciampi, M. S., and Cortese, R. (1978) *J. Bacteriol. 134,* 612–620.

Brenchley, J. E., and Ingraham, J. L. (1973) *J. Bacteriol. 114,* 528–536.

Brenner, M., and Ames, B. N. (1971) in *Metabolic Pathways* (Vogel, H. J., ed.) Vol. V, pp. 349–387, Academic Press, New York.

Cashel, M., and Gallant, J. (1974) in *Ribosomes,* (Nomura, M., Tissieres, A., and Lengyel, P., eds.) pp. 733–745, Cold Spring Harbor Laboratory, Cold Spring Harbor, N.Y.

Cozzone, A. J. (1980) *Biochimie 62,* 647–664.

DeLorenzo, F., and Ames, B. N. (1970) *J. Biol. Chem. 245,* 1710–1716.

DeLorenzo, F., Straus, D. S., and Ames, B. N. (1972) *J. Biol. Chem. 247,* 2302–2307.

DiNocera, P. P., Avitabile, A., and Blasi, F. (1975) *J. Biol. Chem. 250,* 8376–8381.

DiNocera, P. P., Blasi, F., DiLauro, R., Frunzio, R., and Bruni, C. B. (1978) *Proc. Nat. Acad. Sci. U.S.A. 75,* 4276–4280.

Ely, B. (1974) *Genetics 78,* 593–606.

Freedman, R., and Schimmel, P. (1981) *J. Biol. Chem. 256,* 10747–10750.

Frunzio, R., Bruni, C. B., and Blasi, F. (1981) *Proc. Nat. Acad. Sci. U.S.A. 78,* 2767–2771.

Goitein, R. K., and Parsons, S. M. (1980) *J. Bacteriol. 144,* 337–345.

Goldberger, R. F. (1974) *Science 183,* 810–816.

Hopfield, J. J. (1974) *Proc. Nat. Acad. Sci. U.S.A. 71,* 4135–4139.

Horinouchi, S., and Weisblum, B. (1980) *Proc. Nat. Acad. Sci. U.S.A. 77,* 7079–7083.

Jacob, F., and Monod, J. (1961) *J. Mol. Biol. 3,* 318–356.

Johnston, H. M., and Roth, J. R. (1981a) *J. Mol. Biol. 145,* 713–734.

Johnston, H. M., and Roth, J. R. (1981b) *J. Mol. Biol. 145,* 735–756.

Johnston, H. M., Barnes, W. M., Chumley, F. G., Bossi, L., and Roth, J. R. (1980) *Proc. Nat. Acad. Sci. U.S.A. 77,* 508–512.

Kasai, T. (1974) *Nature* (London) *249,* 523–527.

Lee, F., and Yanofsky, C. (1977) *Proc. Nat. Acad. Sci. U.S.A. 74,* 4365–4369.

Lewis, J. A., and Ames, B. N. (1972) *J. Mol. Biol. 66,* 131–142.

Martin, R. G. (1963) *J. Biol. Chem. 238,* 257–268.

Martin, R. G., Ames, B. N., and Hartman, P. E. (1967) *Int. Congr. Biochem. Abstr. 2,* 261.

Martin, R. G., Berberich, M. A., Ames, B. N., Davis, W. W., Goldberger, R. F., and Yourno, J. D. (1971) *Meth. Enzymol. 17,* 3–44.

Meyers, M. M., Blasi, F., Bruni, C. B., Deeley, R. G., Kovach, J. S., Levinthal, M., Mullinix, K. P., Vogel, T., and Goldberger, R. F. (1975) *Nucl. Acid Res. 2,* 2021–2036.

Morton, D. P., and Parsons, S. M. (1977a) *Arch. Biochem. Biophys. 181,* 643–648.

Morton, D. P., and Parsons, S. M. (1977b) *Biochem. Biophys. Res. Comm. 74,* 172–177.

Nino, J. (1974) *J. Mol. Biol. 84,* 297–313.

Oppenheim, D. S., and Yanofsky, C. (1980) *Genetics 95,* 785–795.

Palmer, D. T., Blum, P. H., and Artz, S. W. (1983) *J. Bacteriol. 153,* 357–363.

Platt, T. (1981) *Cell 24,* 10–23.

Primakoff, P., and Artz, S. W. (1979) *Proc. Nat. Acad. Sci. U.S.A. 76,* 1726–1730.

Roth, J. R., Anton, D. N., and Hartman, P. E. (1966a) *J. Mol. Biol. 22,* 305–323.

Roth, J. R., Silbert, D. F., Fink, G. R., Voll, M. J., Anton, D., Hartman, P. E., and Ames, B. N. (1966b) *Cold Spring Harbor Symp. Quant. Biol. 31,* 383–392.

Sanger, F., Air, G. M., Barrell, B. G., Brown, N. L., Coulson, A. R., Fiddes, J. C., Hutchinson, C. A., III, Slocombe, P. M., and Smith, M. (1977) *Nature* (London) *265,* 687–695.

Schleif, R., Hess, W., Finkelstein, S., and Ellis, D. (1973) *J. Bacteriol. 115,* 9–14.

Schlesinger, S., and Magasanik, B. (1964) *J. Mol. Biol. 9,* 670–682.

Scott, J. F., Roth, J. R., and Artz, S. W. (1975) *Proc. Nat. Acad. Sci. U.S.A. 72,* 5021–5025.

Singer, C. E., Smith, G. R., Cortese, R., and Ames, B. N. (1972) *Nature New Biol. 238,* 72–74.

Smith, G. R., and Tong, B. (1974) *J. Bacteriol. 120,* 1223–1226.

Squires, C., Lee, F., Bertrand, K., Squires, C. L., Bronson, M. J., and Yanofsky, C. (1976) *J. Mol. Biol. 103,* 351–381.

Stephens, J. C., Artz, S. W., and Ames, B. N. (1975) *Proc. Nat. Acad. Sci. U.S.A. 72,* 4389–4393.

Sterboul, C. C., Kleeman, J. E., and Parsons, S. M. (1977) *Arch. Biochem. Biophys. 181,* 632–642.

Turnbough, C. L., Jr., Neill, R. J., Landsberg, R., and Ames, B. N. (1979) *J. Biol. Chem. 254,* 5111–5119.

Umbarger, H. E. (1980) in *Transfer RNA: Biological Aspects* (Soll, D., Abelson, J. N., and Schimmel, P. R., eds.) pp. 453–467, Cold Spring Harbor Laboratory, Cold Spring Harbor, New York.

Verde, P., Frunzio, R., DiNocera, P. P., Blasi, F., and Bruni, C. B. (1981) *Nucl. Acids Res. 9,* 2075–2086.

Voll, M. J. (1972) *J. Bacteriol. 109,* 741–750.

Winkler, M. E., and Yanofsky, C. (1981) *Biochemistry 20,* 3738–3744.

Winkler, M. E., Roth, D. J., and Hartman, P. E. (1978) *J. Bacteriol. 133,* 830–843.

Winkler, M. E., Zawodny, R. V., and Hartman, P. E. (1979) *J. Bacteriol. 139,* 993–1000.

Yanofsky, C. (1981) *Nature* (London) *289,* 751–758.

CHAPTER
22

Production of Amino Acids

Fumihiro Yoshinaga
Shigeru Nakamori

I. INTRODUCTION

The first report of the production of an amino acid (K. Ikeda, 1908) describes the preparation from sea tangles of the flavoring agent sodium glutamate. This work led to the commercial production of monosodium glutamate as a seasoning agent from acid hydrolysates of wheat gluten or soybean protein. For the next 50 years, amino acid production was synonymous with glutamate production.

When the nutritional importance of amino acids other than glutamate was discovered, these amino acids were obtained as by-products of glutamate production. As the demand for these amino acids increased, new processes for their production were developed involving chemical or biotechnological syntheses. The latter term covers fermentative and enzymatic production processes. Today, all kinds of amino acids are manufactured on an industrial scale either by these two methods or by classical hydrolysis and refinement methods from natural protein sources. This chapter provides a broad outline of the uses and production methods currently employed in the field of bulk amino acid preparation.

II. USES OF AMINO ACIDS

Before discussing production methods it is useful to touch upon the uses of amino acids. This provides a better understanding of the rationale which underlies their production.

A. Food

Amino acids are used as food additives. The worldwide output of glutamate, used as a seasoning, is at an annual level of about 270,000 tons. Glycine, cysteine, and D,L-alanine are also used as food additives, and mixtures of amino acids serve as flavor enhancers in the food industry.

B. Feedstuffs

The amino acid balance of soy or corn protein for animal feed is markedly improved by the addition of the nutritionally limiting amino acids, methionine and lysine. For use as animal feed supplements, D,L-methionine and lysine monohydrochloride are produced globally at the rate of 180,000 and 25,000 tons per year, respectively. Tryptophan and threonine would also come into widespread use for the same purpose if their production costs were to be reduced.

C. Medical Applications

Amino acids are used therapeutically for nutritional and pharmaceutical purposes. A typical example is the intravenous infusion of amino acids to supply these nutrients before and after surgical treatments. Regimes involving single amino acids as therapeutic agents include the use of glutamine and histidine to treat peptic ulcers; dihydroxyphenylalanine for Parkinson's disease; and arginine, citrulline, and ornithine to treat hepatic diseases.

D. Raw Materials for Chemical Processes

Certain acyl derivations of amino acids, especially of glutamate, are used as surface active agents in mild soaps and shampoos. Another use of glutamate is for the production of synthetic leather, a polymer of methyl glutamate.

D-Phenylglycine and D-p-hydroxyphenylglycine are intermediates in the semichemical synthesis of β-lactam antibiotics.

Amino acids are also important raw materials for peptide synthesis. Aspartame, aspartyl phenylalanine methyl ester, is a recently developed sweetener prepared from the individual component amino acids. In the future, large quantities of amino acids will undoubtedly be in demand as raw materials for biologically active and other useful peptides.

III. PRODUCTION OF AMINO ACIDS

The estimated world output of amino acids in 1979 and the methods by which they are produced are shown in Table 22.1. We distinguish four basic methods: extraction, chemical synthesis, fermentation, and enzymatic methods. Fermentation procedures are subclassified in methods using direct accumulation of amino acids, fermentation with wild-type or mutant organisms, among which there are auxotrophic and regulatory mutants, and fermentation procedures that involve the conversion of precursors to amino acids.

Extraction methods require the purification of amino acids from acid hydrolysates of proteins such as soybean meal or human hair. Tyrosine, leucine, cystine, and hydroxyproline are now commercially produced by this method.

Almost every amino acid can be prepared by chemical synthesis, even though that is not always commercially practical, because long and complicated

TABLE 22.1 The World Supply of Amino Acids in 1979

Amino Acid	*Method of Production**	*Amount (tons/year)*
Alanine	E	130
Arginine	F	500
Asparagine	Ex, C	50
Aspartate	E	250
Citrulline	F	50
Cystine (Cysteine)	Ex	700
Dihydroxyphenylalanine	C	200
Glutamate	F	270,000
Glutamine	F	500
Glycine	C	6000
Histidine	F	200
Hydroxyproline	Ex	50
Isoleucine	F	150
Leucine	Ex, F	150
Lysine	F	25,000
Methionine	C, E	150
D,L-Methionine	C	180,000
Ornithine	F	50
Phenylalanine	C, F	150
Proline	F	100
Serine	F, C	50
Threonine	F	160
Tryptophan	C	200
Tyrosine	Ex	100
Valine	F, C	150

* E = enzymatic method, F = fermentation, Ex = extraction, C = chemical synthesis.

routes are frequently necessary to obtain the L-isomers of the amino acids. Racemic mixtures and diasteroisomers are usually formed in the course of chemical synthesis. Those amino acids that have no enantiomers, like glycine, or that can be converted from the D- to the L-forms by animals, for example, methionine or alanine, or that are difficult to produce by fermentation, for example, tryptophan or phenylalanine, are produced by chemical synthesis (Kaneko et al., 1974).

Fermentation and enzymatic methods involve biotechnology, and these methods constitute the main emphasis of this chapter. Amino acids produced by fermentation using wild-type organisms are glutamate and glutamine. Recently, however, wild-type strains have been replaced by superior mutants. The first successful industrial fermentation using an auxotroph was ornithine production by citrulline (arginine) auxotrophs of *Micrococcus glutamicus* (Kinoshita et al., 1957b). It followed the production of lysine by homoserine (methionine plus threonine) auxotrophs, homoserine by threonine auxotrophs, citrulline by arginine auxotrophs, proline by isoleucine auxotrophs and so on. Amino acid overproduction by auxotrophs usually involves the release of feedback inhibition or repression by restricting the availability of those amino acids that act as effectors.

The overproduction of amino acids by fermentation using regulatory mutants is generally attributable to a change in the amino acid sequence of a repressor protein or of a regulatory site on a key enzyme. The first successful use of regulatory mutants was in the production of threonine by *Escherichia coli* and *Brevibacterium flavum* mutants resistant to α-amino-β-hydroxyvalerate, an analog of threonine (Shiio and Nakamori, 1969, 1970). Lysine, arginine, histidine, isoleucine, and valine producers were also obtained as amino acid analog-resistant mutants and were successfully exploited for practical production. Lately, multiple mutants with auxotrophic and altered regulatory properties have proven to be superior production strains in many cases.

Precursor conversion fermentations received much attention in early studies of amino acid production. The regulated step in the biosynthesis of each amino acid is generally the first enzyme. The feeding of a precursor to a culture in which the biosynthetic enzyme levels are high enough to effectively desensitize regulated enzymes is an effective strategy for production of specific amino acids. Such production methods are reported for threonine from D,L-homoserine, isoleucine from D,L-threonine, histidine from histidinol, tryptophan from anthranilate or indole, and serine from glycine. However, only serine production from glycine is practical because none of the other precursors are cheaply available.

The enzymatic methods are essentially the same as those of precursor conversion fermentation, except that whole cells are used in lieu of isolated enzymes to catalyze one or two reactions. Recently developed immobilization techniques for enzymes or whole cells are practically applicable. Some examples are the production of aspartate, alanine, cysteine, lysine, tyrosine, dihydroxyphenylalanine, tryptophan, D-phenylglycine and D-p-hydroxy-phenylglycine. By this method, optical isomers of the amino acids can be obtained at high concentrations.

In the following sections, production methods for each amino acid are described. Although a large number of reports on amino acid production have been published, only a few of those, especially those which are relevant to practical production, are referred to in this chapter. The conversion of a commercial carbon source into an amino acid is generally, but not always, accomplished through the biosynthetic pathways outlined in the preceding chapters. Therefore we will detail the production pathways only when they deviate from the standard biosynthetic pathways. For more details and fundamental aspects of amino acid production, see Yamada et al. (1972a), Kinoshita (1976), Nakayama (1976), Hirose and Okada (1979), and Hirose et al. (1979).

A. Alanine

Fermentative production of alanine was investigated in the early studies on amino acid production. While many kinds of microorganisms produce alanine by direct fermentation (Kinoshita et al., 1957a; Yamada and Hirose, 1960), in many cases the product is the D,L-form. Today, alanine is produced enzymatically from aspartate (Watanabe et al., 1965; Okumura et al., 1969). *Pseudomonas dacunhae* is a potent producer that stoichiometrically converts aspartate to alanine (Chibata et al., 1965). Whole cells of the organism are used as a source of aspartate β-decarboxylase, which catalyzes the reaction shown in Figure 22.1. The continuous production of alanine is achieved by immobilized cells of *Pseudomonas dacunhae* (Yamamoto et al., 1979).

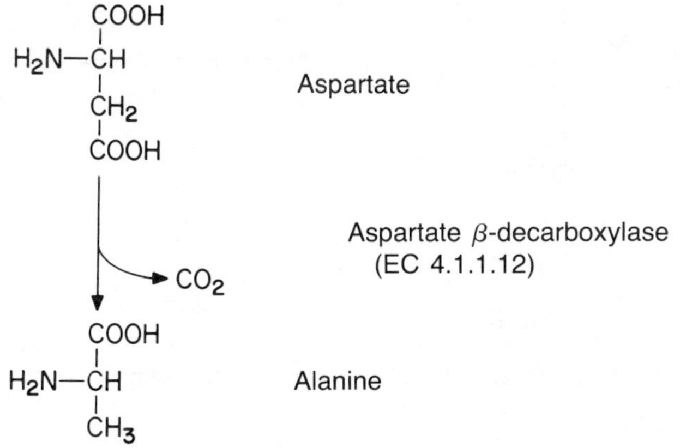

FIGURE 22.1 Production of alanine from aspartate by immobilized cells of *Pseudomonas dacunhae*.

B. Arginine

Arginine is produced by direct fermentation of cane molasses or glucose with regulatory mutants of *Corynebacterium glutamicum* (Nakayama and Yoshida, 1972), *Brevibacterium flavum* (Kubota et al., 1973), *Bacillus subtilis* (Kato et al., 1977), or *Serratia marcescens* (Kisumi et al., 1978). The yields vary from 10 to 35 g of arginine per liter of medium. The particular mutant of *Brevibacterium flavum* overproduces arginine owing to increased levels and K_m values of the first biosynthetic enzyme, N-acetyl glutamate kinase (Morioka et al., 1977). The particular mutant of *Corynebacterium glutamicum* overproduces arginine owing to a change in the permeability for this amino acid.

C. Aspartate

Synthesis of aspartate from fumarate and ammonia (Fig. 22.2) was among the first established amino acid production methods. The enzyme aspartate ammonia lyase of *Bacillus megaterium, Escherichia coli,* and *Pseudomonas trifolii* converts fumarate and ammonia stoichiometrically to aspartate (Kinoshita et al., 1958a; Kitahara et al., 1959; Kisumi et al., 1960; Takahashi et al., 1963). The process is now applied for practical production of aspartate. A new method for continuous commercial-scale production of this amino acid uses the immobilized enzyme or immobilized *Escherichia coli* cells (Tosa et al., 1973; Chibata et al., 1974).

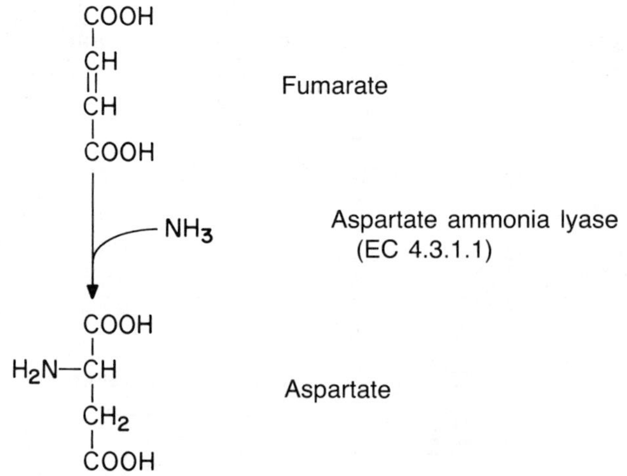

FIGURE 22.2 Production of aspartate from fumarate and ammonia.

D. Citrulline

Citrulline is an intermediate in arginine biosynthesis. In combination with arginine and ornithine, citrulline is used to treat hepatic diseases. Citrulline is produced by direct fermentation of glucose using arginine auxotrophs of *Bacillus subtilis* (Okumura et al., 1964), *Micrococcus glutamicus* (Nakayama and Hagino, 1966), or *Corynebacterium* (Tokoro et al., 1969). An enzymatic method for citrulline production (Kakimoto et al., 1971) involves a *Pseudomonas putida* strain rich in arginine deiminase (EC 3.5.3.6). This enzyme catalyzes the hydrolysis of arginine to citrulline and ammonia. A continuous citrulline production method uses immobilized cells of the *Pseudomonas* strain (Yamamoto et al., 1974).

E. Cysteine and Cystine

Cysteine and cystine are now produced by extraction from acid hydrolysates of human hair. No fermentative process has yet been reported.

Cysteine can also be produced from β-chloroalanine, ammonia, and sodium sulfide by an enzymatic method using cystathionine γ-lyase (EC 4.4.1.1) of *Klebsiella aerogenes* (Kumagai et al., 1974). The maximum yield of cysteine (48.5 g/*l*) is 80% based on the starting material, β-chloroalanine.

Another intriguing method for cysteine production is the enzymatic conversion of D,L-2-amino-thiazoline-4-carboxylate to cysteine by enzymes from *Pseudomonas thiazolinophilum* (Fig. 22.3) (Sano et al., 1977a; Sano and Mitsugi, 1978). The maximum yield of cysteine is 31.4 g/*l* from 40 g/*l* of starting material. Cystine can also be obtained by this method. 2-Amino-thiazoline-4-carboxylate is an intermediate of the chemical synthesis of D,L-cysteine.

F. Dihydroxyphenylalanine

Dihydroxyphenylalanine (L-DOPA), a drug used in the treatment of Parkinson's disease, is now chemically synthesized. Dihydroxyphenylalanine can also be produced by an enzymatic method using tyrosine phenol lyase (Fig. 22.4) from *Erwinia herbicola* (Enei et al., 1971, 1972, 1973). By using pyrocatechol and D,L-serine or pyruvate plus ammonia as substrates, 58.5 g of dihydroxyphenylalanine per liter of medium are obtained. Tyrosine phenol lyase from *Escherichia intermedia* has been isolated and biochemically characterized (Kumagai et al., 1969).

G. Glutamate

The demand for glutamate as a seasoning grew in 1950–1960. Many studies on glutamate production by fermentative processes have been conducted. Early attempts focused on α-ketoglutarate as a substrate for amination with bacterial enzymes (Otsuka et al., 1957) or for transamination with alanine or aspartate, because α-ketoglutarate could be obtained cheaply by fermentative processes (Masuo and Wakisaka, 1955). However, glutamate production by these processes was not practical because of the high costs of such two-step fermentations.

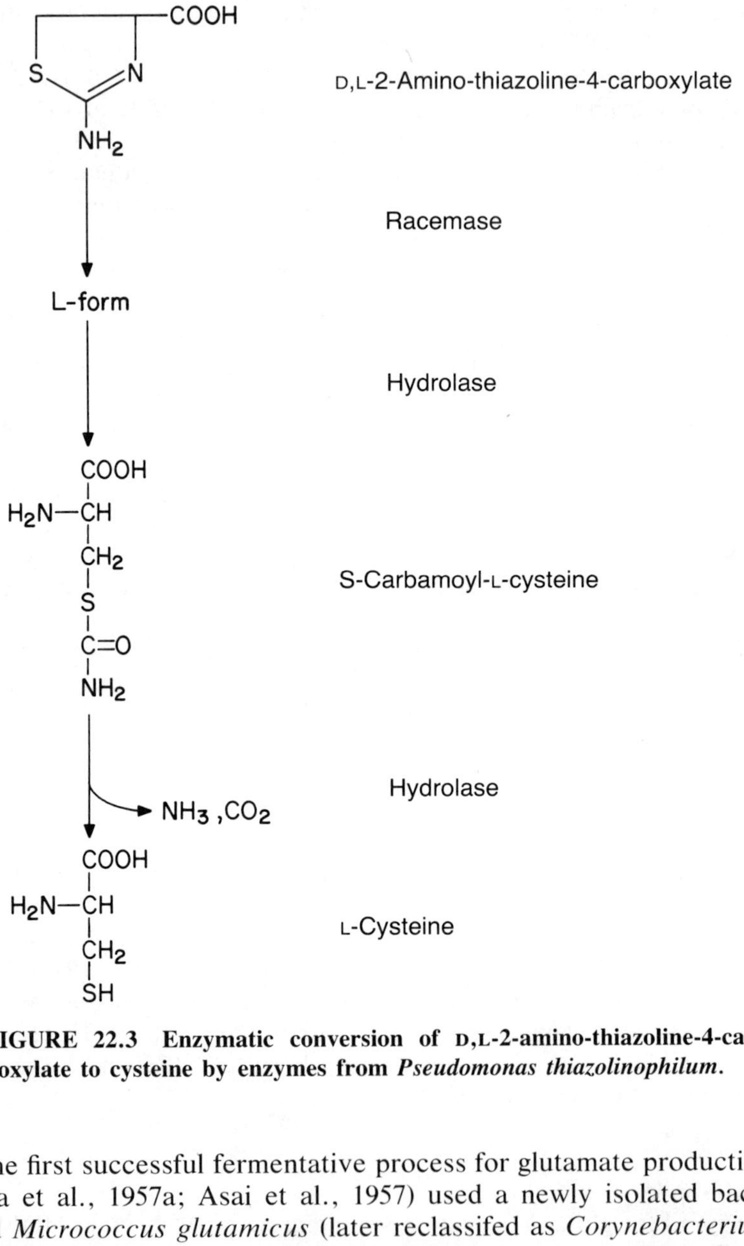

FIGURE 22.3 Enzymatic conversion of D,L-2-amino-thiazoline-4-carboxylate to cysteine by enzymes from *Pseudomonas thiazolinophilum*.

The first successful fermentative process for glutamate production (Kinoshita et al., 1957a; Asai et al., 1957) used a newly isolated bacterium named *Micrococcus glutamicus* (later reclassifed as *Corynebacterium glutamicum*), which produces about 30 g of glutamate per liter of medium from 10% glucose (Kinoshita et al., 1957a). This process was used immediately for commercial production. This advance was of great value not only for glutamate production, but also because it triggered many studies on fermentative production of other amino acids.

A large number of glutamate-producing microorganisms have subsequently

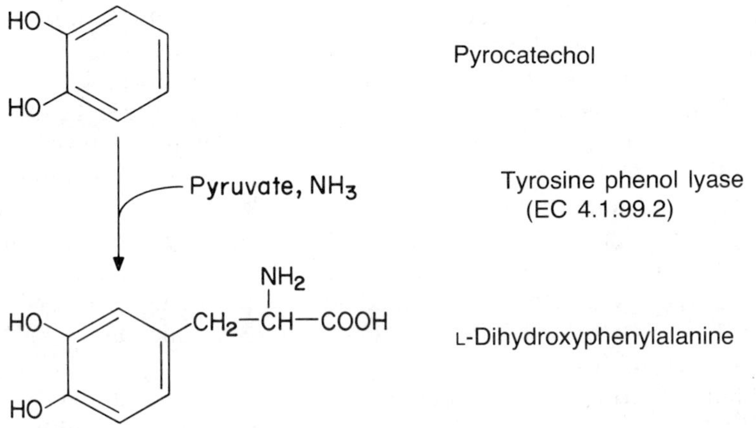

FIGURE 22.4 Production of dihydroxyphenylalanine from pyrocatechol.

been found, and some of them have been used in commercial-scale production, for example, *Brevibacterium divaricatum* (Su and Yamada, 1960), *Brevibacterium flavum* and *lactofermentum* (Okumura et al., 1962), *Microbacterium ammoniaphilum* (Miyai et al., 1963a; 1963b), and *Brevibacterium thiogenitalis* (Okazaki et al., 1967). Carbon sources are usually glucose or cane molasses, and yields vary from 25 to 63 g/*l*.

All bacteria used for glutamate production require biotin for growth. High yields of glutamate are observed only upon cultivation in media containing a limited amount of biotin (2.5–5.0 μg/*l*) (Tanaka et al., 1960). Lactate and α-ketoglutarate are produced upon growth in a medium containing more than 10 μg of biotin per liter. Glutamate can also be produced by bacteria that do not require biotin; however, the yields are generally low (Perlman and O'Brien, 1958; Aida et al., 1960; Tsugawa et al., 1967; Takahashi et al., 1967).

Glutamate is produced by adenine auxotrophs of *Corynebacterium glutamicum* in a biotin-rich medium under conditions of limited availability of adenine (Tanaka et al., 1965). Oleate-requiring mutants of *Brevibacterium thiogenitalis* produce 50 g of glutamate per liter from 10% glucose in a biotin-rich medium by limiting the amount of oleate (Kanzaki et al., 1967). Glycerol auxotrophs of *Brevibacterium alkalolyticum* produce glutamate when limiting amounts of glycerol are furnished in media containing glucose or n-paraffin as carbon sources (Nakao et al., 1972). Glutamate productivity is correlated with the activities of the glutamate-synthesizing enzymes; the yields of glutamate increase in mutants of *Brevibacterium flavum* with elevated enzyme levels (Shiio et al., 1969). Temperature-sensitive (Momose and Takagi, 1978) or vitamin P–resistant (Nakazawa et al., 1981) mutants of *Brevibacterium lactofermentum* as well as lysozyme-sensitive mutants of *Corynebacterium glutamicum* (Katsumata and Takayama, 1981) produce glutamate in biotin-rich medium.

Although cane and beet molasses are good carbon sources, it is necessary for efficient glutamate production to counteract the deleterious effects of the high biotin levels of these feedstocks. This can be accomplished through the addition of penicillin or some other antibiotics to the culture medium (Somerson and Phillips, 1961). Since penicillin inhibits bacterial cell wall synthesis, this finding has been significant not only for the improvement of glutamate production but also for the elucidation of the mechanism by which glutamate is excreted. Penicillin also improves glutamate production from hydrocarbons or acetate as carbon sources. Cationic and nonionic surface active agents also stimulate glutamate production (Yamada et al., 1962; Udagawa et al., 1962; Shiio et al., 1963a; Takinami et al., 1963; Shibukawa et al., 1964).

Although various carbon sources other than carbohydrates can be fermented to glutamate, no practical production from organic acids, alcohols, or hydrocarbons has been reported. Glutamate can be produced from fumarate by various microorganisms; the best yields are obtained with *Bacillus pumilis* (Aoki et al., 1959). This organism produces glutamate also from γ-aminobutyrate (Tsunoda and Shiio, 1959). Glutamate can be produced from pyrrolidone carboxylate (Kawai and Uemura, 1966; see also Chapter 3). The best microorganism to produce glutamate from D,L-hydantoin-5-propionate, an intermediate in the chemical synthesis of D,L-glutamate, appears to be *Bacillus brevis* (Tsugawa et al., 1966). Several *Brevibacteria* and *Corynebacteria* that require biotin for growth produce glutamate from acetate (Shiio et al., 1959; Tsunoda et al., 1961b).

Glutamate can be produced from ethanol in 60% yield using a strain of *Brevibacterium* sp. (Oki et al., 1968). Glutamate can also be produced by fermentation from methanol (Oki et al., 1973).

Glutamate production from hydrocarbons has been studied actively from 1960 to 1973. However, because of recent high world oil prices, these studies have diminished. Bacteria that produce amino acids from hydrocarbons have been isolated (Shiio et al., 1963b; Iizuka and Komagata, 1964), and hydrocarbons can be fermented to glutamate with high yields (Imada and Yamada, 1969; Yamada et al., 1963; Otsuka et al., 1964; Tanaka et al., 1968a; S. Ikeda et al., 1969; Shiio and Uchio, 1969; Kobayashi et al., 1971; Nakao et al., 1972).

The mechanism of glutamate production has been extensively investigated in parallel with advances in glutamate fermentation procedures. Since high yields of glutamate are obtained only when biotin requiring strains are cultured in biotin-poor medium, the effect of biotin on the enzymes of glutamate biosynthesis has been studied. However, no remarkable effects of biotin have been observed. Increased productivity is a result of increased permeability of the cell membranes to glutamate (Shiio et al., 1962a,b). Biotin availability is an important factor for the biosynthesis of cell membranes. This finding agrees well with other studies showing that effective production of glutamate occurs only when the permeability of cell membranes has been altered by penicillin or surface active agents, or when oleate or glycerol auxotrophs with presumptive membrane assembly defects are grown in biotin-rich medium. Biosynthesis of glutamate in *Brevibacterium flavum* follows the pathways discussed in Chapter 1. Regulation of glutamate biosynthetic enzymes in this organism has been studied (Shiio et al., 1969; Shiio, 1978; Shiio and Ujigawa, 1978).

H. Glutamine

Glutamine biosynthesis from glutamate is catalyzed by glutamine synthetase (Chapter 1). Glutamine is now produced by direct fermentation. The first such fermentation (Tsunoda et al., 1961a) was with glutamate-producing bacteria like *Brevibacterium flavum* and *Brevibacterium lactofermentum* that also produce glutamine from glucose. Glutamine productivity has been improved by using sulfaguanidine-resistant mutants of *Brevibacterium flavum* (Yoshinaga et al., 1975). Glutamate production by *Corynebacterium glutamicum* has been altered to a combination of glutamine and N-acetylglutamine production (Nakanishi et al., 1975, 1977). Glutamine is also produced from glucose and ammonium fumarate by *Flavobacterium regenes* (Ujimaru et al., 1978).

I. Histidine

Histidine is synthesized in ten steps from phosphoribosylpyrophosphate and ATP (Chapter 21). The first enzyme of histidine biosynthesis, ATP phosphoribosyl transferase, is allosterically inhibited by histidine. The synthesis of all histidine biosynthetic enzymes is also controlled by histidine.

Histidine auxotrophs of *Brevibacterium flavum* produce histidinol, the intermediate precursor of histidine, and histidinol can be converted to histidine by several microorganisms (Kubota et al., 1971a). Histidine is produced by direct fermentation using regulatory mutants of *Corynebacterium glutamicum* or *Brevibacterium flavum* that are resistant to histidine analogs (Araki and Nakayama, 1971; Kamijo et al., 1973). The histidine yields are increased by using secondary mutants that are resistant to other amino acid analogs (Araki et al., 1974). The ATP phosphoribosyl transferase of histidine-producing mutants is insensitive to feedback inhibition by histidine (Araki and Nakayama, 1974). A potent histidine-producing strain of *Serratia marcescens* is a multiple mutant deficient in histidase, derepressed for the histidine biosynthetic enzymes, and has an altered ATP phosphoribosyl transferase that is insensitive to feedback inhibition by histidine (Kisumi et al., 1977a). A revertant of a histidine auxotroph of *Streptomyces coelicolor* also overproduces histidine (Derkos-Sojak et al., 1977).

J. Isoleucine

The biosynthetic pathways to isoleucine, leucine, and valine in microorganisms are discussed in detail in Chapters 14, 15, and 16. The first enzyme of isoleucine biosynthesis, threonine deaminase, is inhibited by isoleucine, and its formation is controlled by isoleucine, leucine, and valine.

The production of isoleucine has been approached in three ways: by precursor conversion fermentation, by direct fermentation, and by enzymatic methods. Typical examples of precursor conversion fermentation are the production of isoleucine from D,L-threonine with *Bacillus subtilis* (Hayashi et al., 1959), from D-threonine with *Pseudomonas aeruginosa* (Kisumi, 1962), or from D,L-α-hydroxybutyrate with *Bacillus megaterium* (Wada, 1974). However, none of these methods are commercially practical owing to the expense of the precursors.

Isoleucine is produced by direct fermentation from glucose with *Brevibacterium flavum* mutants resistant to α-amino-β-hydroxyvalerate (Shiio and Nakamori, 1970). With O-methylthreonine-resistant mutants isolated from the α-amino-β-hydroxyvalerate-resistant strains the productivity was raised to 14.5 g/*l* from 10% glucose (Shiio et al., 1973b). In these isoleucine producers, homoserine dehydrogenase (Chapter 9) is insensitive to feedback inhibition by threonine, and the levels of threonine deaminase are twofold higher than in the parent strain. Ethionine resistant mutants have been selected (S. Ikeda et al., 1976) from a threonine-producing *Brevibacterium flavum* mutant resistant to α-amino-β-hydroxyvalerate (Shiio and Nakamori, 1970). These double mutants produce isoleucine effectively from acetate as well as from glucose. Isoleucine producers are also among *Serratia marcescens* (Kisumi et al., 1977b) and *Corynebacterium glutamicum* (Kase and Nakayama, 1977) mutants resistant to isoleucine analogs.

Isoleucine can be produced continuously from D,L-threonine by an enzymatic method using immobilized cells of *Serratia marcescens* (Wada et al., 1979).

K. Leucine

In *Brevibacterium lactofermentum,* leucine biosynthesis is regulated by feedback inhibition and by leucine control on the formation of the leucine biosynthetic enzymes (Tsuchida and Momose, 1975).

Leucine is currently produced by extraction from soybean protein hydrolyzates and to a lesser degree by direct fermentation. *Salmonella typhimurium* mutants resistant to leucine analogs excrete leucine (Calvo and Umbarger, 1964; Calvo and Calvo, 1967). Leucine accumulation also occurs in isoleucine auxotrophs of *Brevibacterium* (Yoshinaga et al., 1967a;b) and *Corynebacterium* (Ishii et al., 1967a). However, the amount of leucine excreted by these mutants is comparatively small.

Leucine regulatory mutants show greatly improved leucine productivity. *Serratia marcescens* mutants resistant to α-aminobutyrate produce leucine (Kisumi et al., 1973), and 2-thiazolealanine-resistant, isoleucine plus methionine double auxotrophs of *Brevibacterium lactofermentum* (Tsuchida et al., 1975a) accumulate 28 g of leucine per liter of medium from 10%

glucose. In this strain of *Brevibacterium* the first enzyme of leucine biosynthesis, α-isopropylmalate synthase, is constitutive and feedback-insensitive.

L. Lysine

Lysine biosynthesis is discussed in detail in Chapters 9, 12, and 13. Lysine is now produced by direct fermentation with auxotrophic and regulatory mutants and by enzymatic methods. The production of lysine and ornithine represents the first industrial utilization of auxotrophs for amino acid production. The first report describes lysine production by homoserine auxotrophs of *Micrococcus glutamicus* grown at suboptimal concentrations of the required amino acids methionine and threonine (Kinoshita et al., 1958b).

Various auxotrophs of *Brevibacterium flavum* (Sano and Shiio, 1967, 1970; Shiio and Sano, 1969) overproduce lysine because aspartate kinase, the first enzyme of lysine, threonine, and methionine biosynthesis, is not inhibited owing to the limited supply of threonine (Nakayama et al., 1961a; Shiio and Miyajima, 1969). Regulatory mutants have also been utilized for lysine production. *Brevibacterium flavum* mutants resistant to the lysine analog S-2-aminoethylcysteine (Sano and Shiio, 1970) have aspartate kinase activity that is insensitive to feedback inhibition (Shiio et al., 1970a). Lysine production has been further improved with *Brevibacterium lactofermentum* leucine and alanine auxotrophs derived from S-2-aminoethylcysteine-resistant mutants (Tosaka et al., 1978a, b).

An enzymatic method for lysine production uses the yeast *Cryptococcus laurentii*, which hydrolyzes α-amino-caprolactam, a synthetic intermediate in the manufacture of nylon, to lysine (Fukumura, 1976). A strain of *Achromobacter obae* can racemize D,L-α-amino-caprolactam, and cells of both organisms are combined in an efficient enzymatic process for lysine production (Fukumura,1977).

M. Ornithine

Ornithine, an intermediate in arginine biosynthesis (Chapters 4 and 5), is produced at yields of 25 g/*l* from 10% glucose, when *Micrococcus glutamicus* citrulline (arginine) auxotrophs are cultivated with suboptimal concentrations of arginine (Kinoshita et al., 1957b). This and the aforementioned production of lysine by homoserine auxotrophs represent the first exploitations of auxotrophic mutants for industrial amino acid production. The molecular basis for ornithine overproduction in the mutants is the release, by limitation of the intracellular arginine, of feedback inhibition of N-acetylglutamate kinase. This enzyme catalyzes the regulated step of arginine biosynthesis in *Micrococcus glutamicus* (Udaka and Kinoshita, 1958). Ornithine is also produced by arginine auxotrophs of *Brevibacterium lactofermentum* (Okumura et al., 1968) and *Arthrobacter citreus* (Shibuya et al., 1967). Arginine auxotrophs of *Corynebacterium hydrocarboclastus* (Ishii et al., 1967b) and *Arthrobacter parafienus* (Tanaka et al., 1968b) produce ornithine from hydrocarbons.

N. Phenylalanine

Phenylalanine biosynthesis is discussed in detail in Chapter 18. Phenylalanine is now produced by chemical synthesis and by direct fermentation in appropriately constructed auxotrophic and regulatory mutants. Tyrosine auxotrophs of *Escherichia coli* excrete phenylalanine (Davis, 1950), and tyrosine auxotrophs of *Micrococcus glutamicus* produce 6.3 g/*l* of phenylalanine when cultivated with suboptimal amounts of tyrosine (Nakayama et al., 1961b).

Phenylalanine is excreted by *Escherichia coli* mutants resistant to the phenylalanine analog α-2-thienylalanine (Adelberg, 1958). Phenylalanine can be produced with various strains of phenylalanine analog–resistant mutants of *Brevibacteria, Corynebacteria, Arthrobacter, Bacillus,* or *Candida* (Okumura et al., 1972). Phenylalanine is accumulated in a m-fluorophenylalanine-resistant mutant of *Brevibacterium flavum* because the prephenate dehydratase of the mutant is desensitized to feedback inhibition (Shiio and Sugimoto, 1976). Large amounts of phenylalanine (23.1 g/*l*) are produced from 13% glucose by a tyrosine auxotroph, p-fluorophenylalanine plus 5-methyltryptophan resistant mutant of *Brevibacterium lactofermentum* (Tsuchida et al., 1975c).

O. Proline

Proline biosynthesis is discussed in detail in Chapter 3. Proline is now produced exclusively by direct fermentation using either wild-type, auxotrophs or analog resistant-mutants of *Brevibacterium flavum* (Yoshinaga et al., 1966; Tsuchida et al., 1974; Nakamori et al., 1979), *Brevibacterium* sp. (Yamatodani et al., 1967), *Kurthia catenaforma* (Kato et al., 1972), *Corynebacterium melassecola* (Kanemitsu, 1971), or *Corynebacterium glutamicum* (Nakanishi et al., 1973; Araki et al., 1975). The first successful use of mutants employed an isoleucine auxotroph of *Brevibacterium flavum,* which produced 12 g/*l* of proline from 10% glucose (Yoshinaga et al., 1966). Proline overproduction in this mutant is attributed to an excess supply of ATP to glutamate kinase, the first enzyme of proline biosynthesis (Yoshinaga et al., 1967b; Yoshinaga, 1969). The yield of proline rises to 30 g/*l* when secondary mutants that are resistant to sulfaguanidine are employed (Tsuchida et al., 1974). Yields of 40 g/*l* are observed when using D,L-3,4-dehydroproline-resistant strains selected from the isoleucine-requiring sulfaguanidine-resistant mutant (Nakamori et al., 1979).

P. Serine

Serine production from glycine is a typical example of precursor conversion fermentation. Direct production of serine by fermentation has not yet been reported. Among many microorganisms screened for serine production (Kubota et al., 1971b), the best thus far is *Corynebacterium glycinophilum*, which produces 10 g/*l* of serine from 20 g/*l* of glycine (Kubota et al., 1972). A single enzyme, serine hydroxymethyl transferase, converts glycine to serine (Chapter 6; see also Kubota et al., 1975). Serine can also be produced from glycine by *Nocardia butanica* (Tanaka et al., 1978), *Sarcina albida* (Ema et al., 1978) and from methanol by *Arthrobacter globiformis* (Tani et al., 1978).

Q. Threonine

Threonine biosynthesis (Chapters 9 and 10) is mainly controlled by feedback inhibition and by transcriptional control of the biosynthetic enzymes mediated by threonine and methionine. The removal of the controls should favor threonine production. Attempts to circumvent the control points involve three approaches: conversion of D,L-homoserine to threonine and direct fermentation using either methionine auxotrophs or regulatory mutants that are resistant to threonine analogs.

Homoserine conversion fermentation (Watanabe and Shimura, 1960) is not used for practical production because of the high cost of homoserine. In direct fermentation with methionine auxotrophs, methionine starvation alleviates the transcriptional control on homoserine dehydrogenase synthesis (Huang, 1961a; Kase et al., 1971; Hirakawa et al., 1973). However, the threonine yields are low. While multiple auxotrophs of *Candida guilliermondii* (Tsukada and Sugimori, 1971) and particularly an isoleucine leaky mutant of *Arthrobacter parafineus* (Kase and Nakayama, 1974) produce more threonine, the yields are still lower than the levels obtained with regulatory mutants.

High yields of threonine are obtained by direct fermentation with *Escherichia coli* (Cohen and Patte, 1963) and *Brevibacterium flavum* (Shiio and Nakamori, 1969, 1970) regulatory mutants that are resistant to the threonine analog α-amino-β-hydroxyvalerate. The work of Shiio and Nakamori represents the first successful application of regulatory mutants to the practical production of amino acids. *Brevibacterium flavum* mutants that are resistant to α-amino-β-hydroxyvalerate produce 15–18 g/*l* of threonine from 10% glucose (Nakamori and Shiio, 1972, 1973). The homoserine dehydrogenase of these mutants is 1000-fold less sensitive to feedback inhibition by threonine than the enzyme from the parental strain (Shiio et al., 1970b). Suitable threonine producers are also found among α-amino-β-hydroxyvalerate-resistant mutants of *Proteus rettgeri* (Nakamura and Aida, 1970), *Corynebacterium glutamicum* (Kase and Nakayama, 1972) and *Serratia marcescens* (Komatsubara et al., 1978).

Recently, genetic engineering technology has been applied to amino acid production (Miwa et al., 1981). By amplifying genes encoding threonine biosynthetic enzymes, especially aspartate kinase–homoserine dehydrogenase, which has been mutationally desensitized to feedback inhibition by threonine, productivity for threonine has been improved twofold to threefold over the level obtained with the α-amino-β-hydroxyvalerate-resistant mutant of *Escherichia coli* (Shiio and Nakamori, 1969).

R. Tryptophan

The industrial production of tryptophan is currently accomplished by chemical synthesis (Komachiya, 1961). A large number of biotechnological methods have been tested for the production of tryptophan during the last 20 years. However, a truly effective process has not yet been developed.

Tryptophan biosynthesis and its regulation are discussed in detail in Chapter 20. Production trials include precursor conversion fermentation, direct fermentation, and enzymatic methods. Tryptophan can be produced from anthranilate and indole by precursor conversion fermentation using the yeast *Hansenula* (Terui et al., 1962; Terui and Niizu, 1969). Production of tryptophan by direct fermentation is achieved with 5-methyltryptophan-resistant mutants of glutamate-producing bacteria (Shiio et al., 1972) or *Bacillus subtilis* (Shiio et al., 1973a). Tryptophan producers are also among tryptophan analog–resistant mutants selected from tyrosine and phenylalanine auxotrophs of *Corynebacterium glutamicum* (Hagino and Nakayama, 1975).

Tryptophan producers have also been constructed from *Escherichia coli,* in which several characteristics related to tryptophan overproduction, namely, *trpE, trpR, trpS, aroG, aroF, tyrR,* and *tnaA,* are altered and *trpE* is amplified by in vivo recombinant DNA technology (Tribe and Pittard, 1979; Tsunekawa et al., 1980). However, the yields of tryptophan from such multiple mutants are still too low for practical production purposes.

Many microorganisms have been screened for production of tryptophanase (Nakazawa et al., 1972). This enzyme can convert indole, pyruvate, and ammonia to tryptophan. With cells of *Proteus rettgeri,* 90 g/*l* of tryptophan are obtained from 60 g/*l* indole and 80 g/*l* pyruvate (Nakazawa et al., 1973). The bacterial tryptophanase has been purified, crystallized, and biochemically characterized (Yoshida et al., 1974). Tryptophan can be produced by using immobilized tryptophanase (Fukui et al., 1975). Also, 7 g/*l* of tryptophan are produced in 82% yield from D,L-hydantoin-5-propionate with an enzyme from *Flavobacterium aminogenes* (Sano et al., 1977b). D,L-Hydantoin-5-propionate is a precursor in the chemical synthesis of D,L-tryptophan and can easily be racemized chemically or enzymatically.

S. Tyrosine

Tyrosine is now produced mainly by extraction from acid hydrolysates of soybean proteins, although production by direct fermentation or enzymatic methods has also been reported. Tyrosine can be produced by direct fermentation using tyrosine analog–resistant mutants of glutamate-producing bacteria (Hagino and Nakayama, 1973; Sugimoto et al., 1973). An enzymatic method using intact cells of *Erwinia herbicola* produces 60 g/*l* of tyrosine from phenol and D,L-serine or pyruvate plus ammonia (Enei et al., 1971, 1972). The synthesis of tyrosine from phenol and serine is catalyzed by tyrosine phenol lyase (Yamada et al., 1972b) in a reaction similar to the one shown in Fig. 22.4.

T. Valine

Valine biosynthesis is discussed in detail in Chapter 14. Valine production is now by direct fermentation. Although some wild-type bacteria produce valine (Sugizaki, 1959; Udaka and Kinoshita, 1959), large amounts of this amino acid are accumulated by isoleucine and leucine auxotrophs of *Corynebacterium glutamicum* when cultivated with suboptimal amounts of isoleucine and leucine (Nakayama et al., 1961c). Valine is also accumulated by auxotrophs of *Escherichia coli* (Huang, 1961b), *Brevibacterium lactofermentum* (Kajiwara et al., 1968), and *Corynebacterium hydrocarboclastus* (Ishii et al., 1967a).

Valine is excreted by valine analog–resistant mutants of *Escherichia coli* (Karlström, 1965) and *Salmonella typhimurium* (Calvo et al., 1969). *Serratia marcescens* mutants resistant to α-aminobutyrate produce 8 g/*l* of valine (Kisumi et al., 1969). Potent valine producers are among 2-thiazolealanine-resistant mutants of *Brevibacterium lactofermentum* (Tsuchida et al., 1975b). The best strains produce 31 g/*l* of valine from 10% glucose. In the 2-thiazolealanine-resistant mutants the acetohydroxy acid synthase, the key enzyme of valine biosynthesis, is derepressed and desensitized to feedback inhibition by valine.

U. D-Phenylglycine and D-p-Hydroxyphenylglycine

The production methods described above are for the biologically active L-form of each amino acid. While some D-amino acids occur as cell wall components in microorganisms, D-phenylglycine and D-p-hydroxyphenylglycine are parts of semisynthetic β-lactam antibiotics, such as ampicillin and amoxycillin.

Many microorganisms can convert the hydantoin intermediates of D,L-amino acid chemical synthesis to the corresponding N-carbamoyl-D-amino acids, from which the D-amino acids can be produced by additional chemical treatment (Yamada et al., 1978). D-Amino acids can also be formed directly from the corresponding hydantoins; by precursor conversion fermentation a bacterium belonging to the genus *Pseudomonas* produces 24 g/*l* of D-p-hydroxyphenylglycine from 30 g/*l* of D,L-5-(p-hydroxyphenyl) hydantoin, a molar yield of 91% (Yokozeki et al., 1979, 1981).

IV. LARGE-SCALE FERMENTATION

Large-scale submerged cultivation techniques are necessary in the industrial processes for amino acid production. The technique, based on developments for large-scale penicillin production, has been improved with the growth of the amino acid industry. Important features of large-scale fermentation are maintenance of pure culture conditions, aeration and agitation, and control of pH and temperature of the culture medium. Details of large-scale fermentation methods are described by Yamada et al. (1972a) and Hirose et al. (1979).

V. ISOLATION AND PURIFICATION OF AMINO ACIDS

Isolation and purification of amino acids from fermentation broths are important steps in practical production. Techniques to obtain individual amino acids from hydrolysates of proteins (Dunn and Rockland, 1951) generally differ from the way by which amino acids are obtained from fermentation broth because these broths contain cells, biopolymers originating from cells, sugars, salts, pigments, byproduct amino acids, and many other impurities. Therefore processes for the isolation and purification of amino acids from fermentation broth include separation of cells, adsorption and desorption by ion exchanger, electrodialysis, condensation, decolorization, and crystallization (Yamada et al., 1972a). Since amino acids are used for food, feed, and medicines, especially severe quality controls are necessary in their production.

VI. CONCLUSION

The remarkable progress in the technology of amino acid production, based on biochemistry and genetics, has made the amino acid industry essential for human life today. Because amino acids will be required in ever-increasing amounts in the future, new technologies for more economically supplying these compounds will continue to receive a major share of research attention.

REFERENCES

Adelberg, E. A. (1958) *J. Bacteriol. 76,* 326.

Aida, K., Oishi, K., and Asai, Y. (1960) *Nippon Nogei Kagaku Kaishi* (Japan) *34,* 70–77.

Aoki, R., Kondo, Y., Tsunoda, T., and Ogawa, T. (1959) *Nippon Nogei Kagaku Kaishi* (Japan) *33,* 843–845.

Araki, K., and Nakayama, K. (1971) *Agr. Biol. Chem. 35,* 2081–2088.

Araki, K., and Nakayama, K. (1974) *Agr. Biol. Chem. 38,* 2219–2225.

Araki, K., Kato, F., Arai, Y., and Nakayama, K. (1974) *Agr. Biol. Chem. 38,* 189–194.

Araki, K., Takasawa, Y., and Nakajima, J. (1975) *Agr. Biol. Chem. 39,* 1193–1200.

Asai, T., Aida, K., and Oishi, K. (1957) *Bull. Agr. Chem. Soc.* (Japan) *21,* 134.

Calvo, R. A. and Calvo, J. M. (1967) *Science 156,* 1107–1109.

Calvo, J. M., and Umbarger, H. E. (1964) *Fed. Proc. 23,* 377.

Calvo, J. M., Freundlich, M., and Umbarger, H. E. (1969) *J. Bacteriol. 97,* 1272–1282.

Chibata, I., Kakimoto, T., and Kato, J. (1965) *Appl. Microbiol. 13,* 638–645.

Chibata, I., Tosa, T., and Sato, T. (1974) *Appl. Microbiol. 27,* 878–885.

Cohen, G. N., and Patte, J.-C. (1963) *Cold Spring Harbor Symp. Quant. Biol. 28,* 513–516.

David, B. D. (1950) *Experientia 6,* 41–50.

Derkos-Sojak, V., Korajlija, J. and Delic, V. (1977) *Appl. Environ. Microbiol. 34,* 621–625.

Dunn, M. S. and Rockland, L. B. (1951) in *Amino Acids and Proteins* (Greenberg, D. M. ed.) Charles C. Thomas Publisher, Springfield, 213–255.

Ema, M., Kakimoto, T., and Chibata, I. (1978) *Abst. Ann. Meet. Agr. Chem. Soc.* (Japan), p. 260.

Enei, H., Matsui, H., Okumura, S., and Yamada, H. (1971) *Biochem. Biophys. Res. Comm. 43,* 1345–1349.

Enei, H., Nakazawa, H., Matsui, H., and Okumura, S. (1972) *FEBS Lett. 21,* 39–41.

Enei, H., Matsui, H., Nakazawa, H., Okumura, S., and Yamada, H. (1973) *Agr. Biol. Chem. 37,* 493–499.

Fukui, S., Ikeda, S., Fujimura, M., Yamada, H., and Kumagai, H. (1975) *Eur. J. Appl. Microbiol. 1,* 25–39.

Fukumura, T. (1976) *Agr. Biol. Chem. 40,* 1687–1693.

Fukumura, T. (1977) *Agr. Biol. Chem. 41,* 1327–1330.

Hagino, H., and Nakayama, K. (1973) *Agr. Biol. Chem. 37,* 2007–2011.

Hagino, H., and Nakayama, K. (1975) *Agr. Biol. Chem. 39,* 343–349.

Hayashi, K., Watanabe, Y., Fujii, Y., and Shimura, K. (1959) *Amino Acids* (Japan) *1,* 89–93.

Hirakawa, T., Tanaka, T., and Watanabe, K. (1973) *Agr. Biol. Chem. 37,* 123–130.

Hirose, Y., and Okada, H. (1979) in *Microbial Technology* (Peppler, H. J., and Perlman, D., eds.) Vol. I, pp. 211–240, Academic Press, New York.

Hirose, Y., Sano, K., and Shibai, H. (1979) *Annual Report on Fermentation Processes* (Perlman, D., ed.) Vol. II, p. 155–189, Academic Press, New York.

Huang, H. T. (1961a) *Appl. Microbiol. 9,* 419–424.

Huang, H. T. (1961b) US Patent 2,975,105.

Iizuka, H., and Komagata, K. (1964) *J. Gen. Appl. Microbiol. 10,* 207–221.

Ikeda, K. (1908) *J. Tokyo Chem. Soc. 30,* 820–836.

Ikeda, S., Hirose, Y., Kobayashi, K., and Kinoshita, K. (1969) *Agr. Biol. Chem. 33,* 1042–1046.

Ikeda, S., Fujita, I., and Yoshinaga, F. (1976) *Agr. Biol. Chem. 40,* 511–516.

Imada, Y., and Yamada, K. (1969) *Agr. Biol. Chem. 33,* 1326–1332.

Ishii, R., Otsuka, S., and Shiio, I. (1967a) *J. Gen. Appl. Microbiol. 13,* 217–225.

Ishii, R., Otsuka, S., and Shiio, I. (1967b) *J. Gen. Appl. Microbiol. 13,* 303–312.

Kajiwara, T., Yamashita, H., Yoshinaga, F., Okumura, S., and Kinoshita, K. (1968) *Abst. Ann. Meet. Agric. Chem. Soc.* (Japan), p. 158.
Kakimoto, T., Shibatani, T., Nishimura, N., and Chibata, I. (1971) *Appl. Microbiol. 22,* 992–999.
Kamijo, H., Mihara, O., and Kubota, K. (1973) *Abst. Ann. Meet. Agr. Chem. Soc.* (Japan), p. 112.
Kaneko, T., Izumi, Y., Chibata, I., and Itoh, T. (eds.) (1974) in *Synthetic Production and Utilization of Amino Acids,* Kodansha, Tokyo, and Halsted Press, New York.
Kanemitsu, O. (1971) *Abst. Ann. Meet. Agr. Chem. Soc.* (Japan), p. 156.
Kanzaki, T., Isobe, K., Okazaki, H., Motizuki, K., and Fukuda, H. (1967) *Agr. Biol. Chem. 31,* 1307–1313.
Karlström, O. (1965) *Biotechnol. Bioeng. 7,* 245–268.
Kase, H., and Nakayama, K. (1972) *Agr. Biol. Chem. 36,* 1611–1621.
Kase, H., and Nakayama, K. (1974) *Agr. Biol. Chem. 37,* 1643–1649.
Kase, H., and Nakayama, K. (1977) *Agr. Biol. Chem. 41,* 109–116.
Kase, H., Tanaka, H., and Nakayama, K. (1971) *Agr. Biol. Chem. 35,* 2089–2096.
Kato, J., Kisumi, M., and Chibata, I. (1972) *Appl. Microbiol. 23,* 758–764.
Kato, J., Kisumi, M., Takagi, T., and Chibata, I. (1977) *Appl. Environ. Microbiol. 34,* 689–694.
Katsumata, R. and Takayama, K. (1981) *Abst. Ann. Meet. Agr. Chem. Soc.* (Japan), p. 296.
Kawai, Y., and Uemura, T. (1966) *Agr. Biol. Chem. 30,* 438–446.
Kinoshita, S., Udaka, S., and Shimono, M. (1957a) *J. Gen. Appl. Microbiol. 3,* 193–205.
Kinoshita, S., Nakayama, K., and Udaka, S. (1957b) *J. Gen. Appl. Microbiol. 3,* 276–277.
Kinoshita, S., Nakayama, K., and Kitada, S. (1958a) *Hakko Kyokaishi* (Japan) *16,* 517–520.
Kinoshita, S., Nakayama, K., and Kitada, S. (1958b) *J. Gen. Appl. Microbiol. 4,* 128–129.
Kinoshita, S. (1976) in *Microbiology-1976* (Schlessinger, D., ed.) American Society for Microbiology, Washington, D.C., 526–527.
Kisumi, M. (1962) *J. Biochem. 52,* 390–399.
Kisumi, M., Ashikaga, Y., and Chibata, I. (1960) *Bull. Agr. Chem. Soc.* (Japan) *24,* 296–305.
Kisumi, M., Komatsubara, S., and Chibata, I. (1969) *Amino Acid and Nucleic Acid* (Japan) *19,* 1–8.
Kisumi, M., Komatsubara, S., and Chibata, I. (1973) *J. Biochem. 73,* 107–115.
Kisumi, M., Nakanishi, N., Takagi, T., and Chibata, I. (1977a) *Appl. Environ. Microbiol. 34,* 465–472.
Kisumi, M., Komatsubara, S., and Chibata, I. (1977b) *Appl. Environ. Microbiol. 34,* 647–653.
Kisumi, M., Takagi, T., and Chibata, I. (1978) *J. Biochem. 84,* 881–890.
Kitahara, K., Fukui, S., and Misawa, M. (1959) *J. Gen. Appl. Microbiol. 5,* 74–77.
Kobayashi, K., Ikeda, S., Takinami, K., Hirose, Y., and Shiro, T. (1971) *Agr. Biol. Chem. 35,* 1241–1247.
Komachiya, Y. (1961) *Nippon Nogeikagaku Kaishi* (Japan) *35,* 8–10.

Komatsubara, S., Kisumi, M., Murata, K., and Chibata, I. (1978) *Appl. Environ. Microbiol. 35,* 834–840.

Kubota, K., Shiro, T., and Okumura, S. (1971a) *J. Gen. Appl. Microbiol. 17,* 1–12.

Kubota, K., Kageyama, K., Shiro, T., and Okumura, S. (1971b) *J. Gen. Appl. Microbiol. 17,* 167–168.

Kubota, K., Kageyama, K., Maeyashiki, I., Yamada, K., and Okumura, S. (1972) *J. Gen. Appl. Microbiol. 18,* 365–375.

Kubota, K., Onoda, T., Kamijo, H., Yoshinaga, F., and Okumura, S. (1973) *J. Gen. Appl. Microbiol. 19,* 339–352.

Kubota, K., Yokoi, M., and Hirose, Y. (1975) *Abst. Ann. Meet. Soc. Ferment. Tech.* (Japan), p. 80.

Kumagai, H., Matsui, H., Ohgishi, H., Ogata, K., Yamada, H., Ueno, T., and Fukami, H. (1969) *Biochem. Biophys. Res. Comm. 34,* 266–270.

Kumagai, H., Choi, Y.-T., Sejima, S., and Yamada, H. (1974) *Biochem. Biophys. Res. Comm. 59,* 789–795.

Masuo, E. and Wakisaka, Y. (1955) *Abst. Ann. Meet. Agr. Chem. Soc.* (Japan), p. 12.

Miwa, K., Tsuchida, T., Kurahashi, O., Nakamori, S., Sano, K., and Momose, H. (1981) *Abst. Ann. Meet. Agr. Chem. Soc.* (Japan), p. 8.

Miyai, K., Tsuruo, I., Kodaira, R., Hayakawa, S., Akimoto, T., and Goto, K. (1963a) *Nippon Nogei Kagaku Kaishi 37,* 32–36.

Miyai, K., Kodaira, R., Tsuruo, I., Goto, K., and Akimoto, T. (1963b) *Agr. Biol. Chem. 27,* 243–248.

Momose, H., and Takagi, T. (1978) *Agr. Biol. Chem. 42,* 1911–1917.

Morioka, H., Ishii, K., and Takinami, K. (1977) *Abst. Ann. Meet. Agr. Chem. Soc.* (Japan), p. 213.

Nakamori, S., and Shiio, I. (1972) *Agr. Biol. Chem. 36,* 1209–1216.

Nakamori, S., and Shiio, I. (1973) *Agr. Biol. Chem. 37,* 653–659.

Nakamori, S., Morioka, H., Yoshinaga, F., and Yamanaka, S. (1982) *Agr. Biol. Chem. 46,* 487–491.

Nakamura, T. and Aida, K. (1970) *Abst. Ann. Meet. Agr. Chem. Soc.* (Japan), p. 178.

Nakanishi, T., Yokote, Y., and Taketsugu, Y. (1973) *J. Ferment. Technol. 51,* 742–749.

Nakanishi, T., Nakajima, J., and Kanda, K. (1975) *J. Ferment. Technol. 53,* 543–550.

Nakanishi, T., Taketsugu, Y., and Nakajima, J. (1977) *J. Ferment. Technol. 55,* 224–232.

Nakao, Y., Kikuchi, M., Suzuki, M., and Doi, M. (1972) *Agr. Biol. Chem. 36,* 490–496.

Nakayama, K. (1976) *Proc. Biochem. 11*(2), 4–9.

Nakayama, K. and Yoshida, H. (1972) *Agr. Biol. Chem. 36,* 1675–1684.

Nakayama, K., Kitada, S., and Kinoshita, S. (1961a) *J. Gen. Appl. Microbiol. 7,* 145–154.

Nakayama, K., Sato, Z., and Kinoshita, S. (1961b) *Nippon Nogei Kagaku Kaishi* (Japan) *35,* 142–145.

Nakayama, K., Kitada, S., and Kinoshita, S. (1961c) *J. Gen. Appl. Microbiol. 7,* 52–69.

Nakayama, K. and Hagino, H. (1966) *Nippon Nogei Kagaku Kaishi* (Japan) *40,* 397–380.

Nakazawa, H., Enei, H., Okumura, S., Yoshida, H., and Yamada, H. (1972) *FEBS Lett. 25,* 43–45.

Nakazawa, H., Enei, H., Okumura, S., and Yamada, H. (1973) *Amino Acid and Nucleic Acid* (Japan) *27,* 19–24.

Nakazawa, H., Akutsu, E., and Kamiya, H. (1981) *Abst. Ann. Meet. Soc. Ferment. Tech.* (Japan), p. 46.

Okazaki, H., Kanzaki, T., Sugawara, A. and Fukuda, H. (1967) *Agr. Biol. Chem. 31,* 1410–1415.

Oki, T., Sayama, Y., Nishimura, Y. and Ozaki, A. (1968) *Agr. Biol. Chem. 32,* 119–120.

Oki, T., Kitai, A., Kowno, K., Ozaki, A. (1973) *J. Gen. Appl. Microbiol. 19,* 79–83.

Okumura, S., Tsugawa, R., Tsunoda, T., Kono, K., Matsui, T., and Miyachi, N. (1962) *Nippon Nogei Kagaku Kaishi* (Japan) *36,* 141–159.

Okumura, S., Shibuya, M., Konishi, S., Ishida, M. and Shiro, T. (1964) *Agr. Biol. Chem. 28,* 742–743.

Okumura, S., Shibuya, M., Yoshinaga, F., Maeyashiki, I., and Konishi, S. (1968) Japanese Patent 527,567.

Okumura, S., Yoshinaga, F., and Yoshinaga, Y. (1969) U.S. Patent 3,463,704.

Okumura, S., Otsuka, S., Yamanoi, A., Yoshinaga, F., Honda, T. Kubota, K., and Tsuchida, T. (1972) U.S. Patent 3,660,235.

Otsuka, S., Yazaki, H., Nagase, H., and Sakagushi, K. (1957) *J. Gen. Appl. Microbiol. 3,* 35–53.

Otsuka, S., Ishii, R., Shiio, I., and Katsuya, N. (1964) *J. Gen. Appl. Microbiol. 10,* 179–180.

Perlman, D., and O'Brien, E. (1958) *J. Bacteriol. 75,* 611.

Sano, K., and Mitsugi, K. (1978) *Agr. Biol. Chem. 42,* 2315–2321.

Sano, K., and Shiio, I. (1967) *J. Gen. Appl. Microbiol. 13,* 349–358.

Sano, K., and Shiio, I. (1970) *J. Gen. Appl. Microbiol. 16,* 373–391.

Sano, K., Yokozeki, K., Tamura, F., Yasuda, N., Noda, I., and Mitsugi, K. (1977a) *Appl. Environ. Microbiol. 34,* 806–810.

Sano, K., Yokozeki, K., Eguchi, C., Kagawa, T., Noda, I., and Mitsugi, K. (1977b) *Agr. Biol. Chem. 41,* 819–825.

Shibukawa, M., Osawa, T., Nobukuni, T., and Yamamoto, S. (1964) Nippon Nogei Kagaku Kaishi (Japan) *38,* 285–290.

Shibuya, M., Yoshinaga, F., and Okumura, S. (1967) *Abst. Ann. Meet. Agric. Chem. Soc.* (Japan), p. 178.

Shiio, I. (1978) *Seikagaku* (Japan) *50,* 1–16.

Shiio, I., and Miyajima, R. (1969) *J. Biochem. 65,* 849–859.

Shiio, I., and Nakamori, S. (1969) *Agr. Biol. Chem. 33,* 1152–1160.

Shiio, I., and Nakamori, S. (1970) *Agr. Biol. Chem. 34,* 448–456.

Shiio, I., and Sano, K. (1969) *J. Gen. Appl. Microbiol. 15,* 267–287.

Shiio, I., and Sugimoto, S. (1976) *J. Biochem. 79,* 173–183.

Shiio, I., and Uchio, R. (1969) *J. Gen. Appl. Microbiol. 15,* 65–84.

Shiio, I., and Ujigawa, K. (1978) *J. Biochem. 84,* 647–657.

Shiio, I., Mitsugi, K., and Tsunoda, T. (1959) *J. Biochem. 46,* 1665–1671.

Shiio, I., Otsuka, S., and Takahashi, M. (1962a) *J. Biochem. 51,* 56–62.

Shiio, I., Narui, K., Yahaba, N., and Takahashi, M. (1962b) *J. Biochem. 51,* 109–111.

Shiio, I., Otsuka, S., and Katsuya, N. (1963a) *J. Biochem. 53,* 333–340.

Shiio, I., Otsuka, S., Ishii, R., Katsuya, N., and Iizuka, H. (1963b) *J. Gen. Appl. Microbiol. 9,* 23–30.

Shiio, I., Momose, H., and Oyama, A. (1969) *J. Gen. Appl. Microbiol. 15,* 27–40.

Shiio, I., Miyajima, R., and Sano, K. (1970a) *J. Biochem. 68,* 701–710.

Shiio, I., Miyajima, R., and Nakamori, S. (1970b) *J. Biochem. 68,* 859–866.

Shiio, I., Sato, H., and Nakagawa, M. (1972) *Agr. Biol. Chem. 36,* 2315–2322.

Shiio, I., Ishii, K., and Yokozeki, K. (1973a) *Agr. Biol. Chem. 37,* 1991–2000.

Shiio, I., Sasaki, A., Nakamori, S., and Sano, K. (1973b) *Agr. Biol. Chem. 37,* 2053–2061.

Somerson, N. L., and Phillips, T. (1961) U.S. Patent 3,080,297.

Su, Y., and Yamada, K. (1960) *Amino Acids* (Japan) *1,* 38.

Sugimoto, S., Nakagawa, M., Tsuchida, T., and Shiio, I. (1973) *Agr. Biol. Chem. 37,* 2327–2336.

Sugizaki, Z. (1959) *J. Gen. Appl. Microbiol. 5,* 138–149.

Takahashi, M., Hino, T., Okumura, S., Miura, Y., and Ishida, M. (1963) *Amino Acids* (Japan) *7,* 23–28.

Takahashi, H., Hayakawa, S., Shibukawa, M., Ohsawa, T., and Miyaii, K. (1967) *Nippon Nogei Kagaku Kaishi* (Japan) *41,* 560–566.

Takinami, K., Okada, H., and Tsunoda, T. (1963) *Agr. Biol. Chem. 27,* 858–863.

Tanaka, K., Iwasaki, T., and Kinoshita, S. (1960) *Nippon Nogei Kagaku Kaishi* (Japan) *34,* 593–600.

Tanaka, K., Oshima, K., Kimura, K., and Kinoshita, S. (1965) *Amino Acid and Nucleic Acid* (Japan) *11,* 106–110.

Tanaka, K., Kimura, K., Yamaguchi, T., and Kinoshita, S. (1968a) *Amino Acid and Nucleic Acid* (Japan) *17,* 136–142.

Tanaka, K., Oshima, K., Tokoro, H., and Kinoshita, S. (1968b) *Amino Acid and Nucleic Acid* (Japan) *18,* 57–63.

Tanaka, K., Araki, K., and Nakayama, K. (1978) *Abst. Ann. Meet. Soc. Ferment. Tech.* (Japan), p. 73.

Tani, Y., Kanagawa, T., Hanponkittikun, A., Ogata, K., and Yamada, H. (1978) *Agr. Biol. Chem. 42,* 2275–2279.

Terui, G., and Niizu, H. (1969) *Biotech. Bioeng. Symp. 1,* 33–52.

Terui, G., Enatsu, T., and Okazaki, Y. (1962) *Hakko Kogaku Zasshi* (Japan) *40,* 120–124.

Tokoro, H., Oshima, K., Tanaka, K., and Kinoshita, S. (1969) *Amino Acid and Nucleic Acid* (Japan) *19,* 115–120.

Tosa, T., Sato, T., Mori, T., Matuo, Y., and Chibata, I. (1973) *Biotechnol. Bioeng. 15,* 69–84.

Tosaka, O., Hirakawa, H., Takinami, K., and Hirose, Y. (1978a) *Agr. Biol. Chem. 42,* 1501–1506.

Tosaka, O., Hirakawa, H., Yoshihara, Y., Takinami, K., and Hirose, Y. (1978b) *Agr. Biol. Chem. 42,* 1773–1778.

Tribe, D. E., and Pittard, J. (1979) *Appl. Environ. Microbiol. 38,* 181–190.

Tsuchida, T., and Momose, H. (1975) *Agr. Biol. Chem. 39,* 2193–2198.

Tsuchida, T., Kubota, K., Yoshinaga, F., and Hirose, Y. (1974) *Abst. 23th. Symp. Amino Acid and Nucleic Acid* (Japan), p. 4.

Tsuchida, T., Yoshinaga, F., Kubota, K., Momose, H., and Okumura, S. (1975a) *Agr. Biol. Chem. 39,* 1149–1153.

Tsuchida, T., Yoshinaga, F., Kubota, K., and Momose, H. (1975b) *Agr. Biol. Chem. 39,* 1319–1322.

Tsuchida, T., Matsui, H., Enei, H., and Yoshinaga, F. (1975c) U.S. Patent 3,909,353.

Tsugawa, R., Okumura, S., Ito, T., and Katsuya, N. (1966) *Amino Acid and Nucleic Acid* (Japan) *12,* 42–48.

Tsugawa, R., Kobayaski, T., Yamada, K., and Okumura, S. (1967) *Abst. Ann. Meet. Agr. Chem. Soc.* (Japan), p. 173.

Tsukada, Y., and Sugimori, T. (1971) *Agr. Biol. Chem. 35,* 1–7.

Tsunekawa, H., Imanaka, T., and Aiba, S. (1980) *Abst. Ann. Meet. Hakko Kogakukai* (Japan), p. 170.

Tsunoda, T., and Shiio, I. (1959) *J. Biochem. 46,* 1011–1021.

Tsunoda, T., Tsuchiya, T., Konishita, K., Okada, H., and Kawamoto, A. (1961a) *Nippon Nogei Kagaku Kaishi* (Japan) *35,* 269–275.

Tsunoda, T., Shiio, I., and Mitsugi, K. (1961b) *J. Gen. Appl. Microbiol. 7,* 18–29.

Udagawa, K., Abe, S., and Kinoshita, S. (1962) *Hakko Kogaku Zasshi 40,* 614–619.

Udaka, S., and Kinoshita, S. (1958) *J. Gen. Appl. Microbiol. 4,* 272–282.

Udaka, S., and Kinoshita, S. (1959) *J. Gen. Appl. Microbiol. 5,* 159–174.

Ujimaru, T., Nabe, K., Yamada, S., and Chibata, I. (1978) *Abst. Ann. Meet. Agric. Chem. Soc.* (Japan), p. 261.

Wada, H. (1974) *Nogeikagaku Kaishi 48,* 359–366.

Wada, M., Uchida, Y., Kato, J., and Chibata, I. (1979) *Abst. Ann. Meet. Agr. Chem. Soc.* (Japan), p. 30.

Watanabe, Y., and Shimura, K. (1960) *J. Biochem. 47,* 266–268.

Watanabe, S., Isshiki, S., Osawa, T., and Yamamoto, S. (1965) *Hakko Kogaku Zasshi 43,* 95–101.

Yamada, K., and Hirose, Y. (1960) *Bull. Agr. Chem. Soc.* (Japan) *24,* 621–627.

Yamada, K., Takahashi, J., and Nakamura, J. (1962) *Hakko Kyokaishi* (Japan) *20,* 348–350.

Yamada, K., Takahashi, J., Kobayashi, K., and Imada, Y. (1963) *Amino Acids* (Japan) *7,* 79–85.

Yamada, K., Kinoshita, S., Tsunoda, T., and Aida, K. (eds.) (1972a) *Microbial Production of Amino Acids,* Kodansha, Tokyo, and Halsted Press, New York.

Yamada, H., Kumagai, H., Kashima, N., Torii, H., Enei, H., and Okumura, S. (1972b) *Biochem. Biophys. Res. Commun.,* 370–374.

Yamada, H., Takahashi, S., Kii, Y., and Kumagai, H. (1978) *J. Ferment. Technol. 56,* 484–491.

Yamamoto, K., Sato, T., Tosa, T., and Chibata, I. (1974) *Biotechnol. Bioeng. 16,* 1589–1599.

Yamamoto, K., Tosa, T., and Chibata, I. (1979) *Abst. Ann. Meet. Agr. Chem. Soc.* (Japan), p. 30.

Yamatodani, S., Suzuki, M., and Nakao, Y. (1967) *Amino Acid and Nucleic Acid* (Japan) *16,* 126–133.

Yokozeki, K., Nakamori, S., Eguchi, C., Yoshinaga, F., and Mitsugi, K. (1979) *Abst. Ann. Meet. Agr. Chem. Soc.* (Japan), p. 178.

Yokozeki, K., Nakamori, S., Eguchi, C., Yamanaka, S., and Yoshinaga, F. (1981) *Abst. Ann. Meet. Agr. Chem. Soc.* (Japan), p. 18.

Yoshida, H., Utagawa, T., Kumagai, H., and Yamada, H. (1974) *Agr. Biol. Chem. 38,* 2065–2072.

Yoshida, H., Araki, K., and Nakayama, K. (1979) *Agri. Biol. Chem. 43,* 105–111.

Yoshinaga, F., Konishi, S., Okumura, S., and Katsuya, N. (1966) *J. Gen. Appl. Microbiol. 12,* 219–228.

Yoshinaga, F., Yoshihara, Y., Okumura, S., and Katsuya, N. (1967a) *J. Gen. Appl. Microbiol. 13,* 25–37.

Yoshinaga, F., Takeda, Y., and Okumura, S. (1967b) *Biochem. Biophys. Res. Comm. 27,* 143–149.

Yoshinaga, F. (1969) *J. Gen. Appl. Microbiol. 15,* 387–398.

Yoshinaga, F., Tsuchida, T., Kikuchi, K., and Okumura, S. (1975) U.S. Patent 3,886,039.

INDEX